Lecture Notes in Computer Science

Commenced Publication in 1973
Founding and Former Series Editors:
Gerhard Goos, Juris Hartmanis, and Jan van Leeuwen

T0230170

Claudia Bauzer Medeiros   Max Egenhofer
Elisa Bertino (Eds.)

Advances in Spatial
and Temporal Databases

9th International Symposium, SSTD 2005
Angra dos Reis, Brazil, August 22-24, 2005
Proceedings

 Springer

Volume Editors

Claudia Bauzer Medeiros
University of Campinas, Institute of Computing
CP 6176, 13084-971 Campinas, Brazil
E-mail: cmbm@ic.unicamp.br

Max Egenhofer
University of Maine
National Center for Geographic Information and Analysis
348 Boardman Hall, Orono, ME 04469-5711, USA
E-mail: max@spatial.maine.edu

Elisa Bertino
Purdue University, Department of Computer Science
West Lafayette, IN, USA
E-mail: bertino@cerias.purdue.edu

Library of Congress Control Number: 2005929997

CR Subject Classification (1998): H.2, H.3, H.4, I.2.4

ISSN   0302-9743
ISBN-10   3-540-28127-4 Springer Berlin Heidelberg New York
ISBN-13   978-3-540-28127-6 Springer Berlin Heidelberg New York

Springer is a part of Springer Science+Business Media

springeronline.com

© Springer-Verlag Berlin Heidelberg 2005
Printed in Germany

Typesetting: Camera-ready by author, data conversion by Scientific Publishing Services, Chennai, India
Printed on acid-free paper   SPIN: 11535331   06/3142   5 4 3 2 1 0

Preface

It is our great pleasure to introduce the papers of the proceedings of the 9th International Symposium on Spatial and Temporal Databases – SSTD 2005. This year's symposium continues the tradition of being the premier forum for the presentation of research results and experience reports on leading edge issues of spatial and temporal database systems, including data models, systems, applications and theory. The mission of the symposium is to share innovative solutions that fulfill the needs of novel applications and heterogeneous environments and identify new directions for future research and development. SSTD 2005 gives researchers and practitioners a unique opportunity to share their perspectives with others interested in the various aspects of database systems for managing spatial and temporal data and for supporting their applications.

A total of 77 papers were submitted this year from several countries. After a thorough review process, the program committee accepted 24 papers covering a variety of topics, including indexing techniques and query processing, mobile environments and moving objects, and spatial and temporal data streams. We are very pleased with the variety of the symposium's topics, and we are proud of the resulting strong program.

Many people contributed to the success of the SSTD 2005 program. First of all, we would like to thank the authors for providing the content of the program, and all the members of the program committee and the additional reviewers, for their detailed comments. Philippe Rigaux was of help in adding functions to his program MyReview, which was used in the reviewing process. We would also like to express our gratitude to Gilberto Câmara, the general chair of SSTD 2005, for his constant guidance and advice on many organizational aspects of the symposium and for his work on the local arrangements. Finally, we would like to thank our sponsors (notably INPE – the Brazilian National Institute for Space Research) who have enabled us to hold a successful meeting. We are also grateful for the support of the Brazilian Computer Society (SBC).

We hope that you find this program to be both beneficial and enjoyable and that the symposium provides you with the opportunity to meet other researchers and practitioners from institutions around the world. Enjoy!!

August 2005

Claudia Bauzer Medeiros,
Max Egenhofer,
Elisa Bertino

Organization

SSTD 2005 was organized by the Department of Image Processing of INPE – the National Institute for Space Research (Instituto Nacional de Pesquisas Espaciais), Brazil.

Executive Committee

General Conference Chair: Gilberto Câmara (Department of Image Processing, INPE, Brazil)

Program Chairs: Elisa Bertino (Department of Computer Sciences, Purdue University, USA)

Max Egenhofer (NCGIA, University of Maine, USA)

Claudia Bauzer Medeiros (Institute of Computing, University of Campinas, Brazil)

Program Committee

Amr El Abbadi	UC Santa Barbara, USA
Walid G. Aref	Purdue University, USA
Alessandro Artale	University of Bolzano, Italy
Alberto Belussi	University of Verona, Italy
Michela Bertolotto	University College Dublin, Ireland
Gilberto Camara	National Institute for Space Research, Brazil
Marco Casanova	Dept Informatics, PUC-Rio, Brazil
Barbara Catania	University of Genoa, Italy
Christophe Claramunt	Naval Academy Research Institute, France
Matt Duckham	University of Melbourne, Australia
Fred Fonseca	Penn State University, USA
Fosca Giannotti	CNR, Italy
Ralf Hartmut Güting	University of Hagen, Germany
Kathleen Hornsby	University of Maine, USA
Christian S. Jensen	Aalborg University, Denmark
Christopher Jones	Cardiff University, UK
Daniel Keim	University of Constance, Germany
Eamonn Keogh	University of California, Riverside, USA
George Kollios	Boston University, USA
Bart Kuijpers	University of Limburg, Belgium
Mario Nascimento	University of Alberta, Canada
Raymond Ng	University of British Columbia, Canada

Silvia Nittel   NCGIA, University of Maine, USA
Beng Chin Ooi   National University of Singapore, Singapore
Peter van Oosterom   Delft Univ. of Technology, The Netherlands
Dimitris Papadias   Hong Kong Univ. of Science and Technology,
                                       Hong Kong
Jignesh Patel   University of Michigan, USA
Sunil Prabhakar   Purdue University, USA
Philippe Rigaux   University Paris IX, France
Andrea Rodríguez-Tastets   University of Concepción, Chile
Ana Carolina Salgado   University of Pernambuco, Brazil
George Samaras   University of Cyprus, Cyprus
Peter Scheuermann   Northwestern University, USA
Markus Schneider   University of Florida, USA
Bernhard Seeger   University of Marburg, Germany
Cyrus Shahabi   University of Southern California, USA
Shashi Shekhar   University of Minnesota, Minneapolis, USA
Rick Snodgrass   University of Arizona, USA
Stefano Spaccapietra   EPFL Lausanne, Switzerland
Paolo Terenziani   Università del Piemonte Orientale, Italy
Yannis Theodoridis   University of Piraeus, Greece
Nectaria Tryfona   Computer Technology Institute, Greece
Michalis Vazirgiannis   Athens Univ. of Economics and Business, Greece
Agnes Voisard   Fraunhofer ISST and FU Berlin, Germany
Ouri Wolfson   University of Illinois at Chicago, USA

Additional Referees

Nagender Bandi   Feifei Li   Joern Schneidewind
Benjamin Bustos   Xiang Lian   Chengyu Sun
Hu Cao   Dan Lin   Valeria C. Times
Reynold Cheng   Juhong Liu   Goce Trajcevski
Alminas Civilis   Andrei Lopatenko   Yicheng Tu
Stephen Cole   Anna Maddalena   Tao Wan
Carlo Combi   Florian Mansmann   David Yang
Stephane Coulondre   Ahmed Metwally   Kiyoung Yang
Pier Luigi Dragotti   Mohamed Mokbel   Huabei Yin
Ying Feng   Kyriakos Mouratidis   Man Lung Yiu
Robson N. Fidalgo   Guillaume Noel   Yuni Xia
Elias Frentzos   Andrea Nucita   Xiaopeng Xiong
Marios Hadjieleftheriou   Nikos Pelekis   Bo Xu
Christoph Heinz   Paola Podesta   Mingwu Zhang
Xuegang Huang   Gianna Reggio   Hartmut Ziegler
Mohammad R. Kolah-   Mehdi Sharifzadeh
douzan   Sarvjeet Singh

Sponsoring Institutions

Sponsor – National Institute of Space Research – INPE – Brazil
   Support – Brazilian Computer Society (SBC)

Table of Contents

Advanced Query Processing II

Indexing Schemes and Structures

Novel Applications and Real Systems

Moving Objects and Mobile Environments

Advanced Query Processing III

Selectivity Estimation of High Dimensional Window Queries via Clustering

Christian Böhm, Hans-Peter Kriegel, Peer Kröger, and Petra Linhart

Institute for Computer Science, University of Munich, Germany
{boehm, kriegel, kroegerp, linhart}@dbs.ifi.lmu.de

Abstract. Query optimization is an important functionality of modern database systems and often based on estimating the selectivity of queries before actually executing them. Well-known techniques for estimating the result set size of a query are sampling and histogram-based solutions. Sampling-based approaches heavily depend on the size of the drawn sample which causes a trade-off between the quality of the estimation and the time in which the estimation can be executed for large data sets. Histogram-based techniques eliminate this problem but are limited to low-dimensional data sets. They either assume that all attributes are independent which is rarely true for real-world data or else get very inefficient for high-dimensional data. In this paper we present the first multivariate parametric method for estimating the selectivity of window queries for large and high-dimensional data sets. We use clustering to compress the data by generating a precise model of the data using multivariate Gaussian distributions. Additionally, we show efficient techniques to evaluate a window query against the Gaussian distributions we generated. Our experimental evaluation shows that this approach is significantly more efficient for multidimensional data than all previous approaches.

1 Introduction

The storage and management of vectors of a multidimensional feature space has become an important basic functionality of a database system. Advanced applications such as multimedia [1], CAD [2], molecular biology [3], etc. require efficient and effective methods for content based similarity search and data mining. Such methods are typically based on feature vectors of moderate or high dimensionality. Although a vast number of index structures [4,5] and access methods [6] for vector data has been proposed, database management systems do not yet support the storage and retrieval of vector data in the same way as relational data from applications such as accounting and billing. In order to give full support to advanced applications the database system needs efficient and effective techniques for query optimization. One of the most important challenges in query optimization is the estimation of the selectivity of a query predicate. While a number of techniques to model the data distribution and thus to estimate the selectivity are known for one- and low-dimensional data spaces, this is still an unsolved problem for data spaces of medium to high dimensionality.

C. Bauzer Medeiros et al. (Eds.): SSTD 2005, LNCS 3633, pp. 1–18, 2005.

Three different paradigms of data modelling for selectivity estimation in general can be distinguished: Histograms, sampling, and parametric techniques. Of those three, only sampling can be directly applied without modification in higher dimensional data spaces. Many different sampling methods have been proposed. They share the common idea to evaluate the predicate on top of a small subset of the actual database objects and to extrapolate the observed selectivity. The well-known techniques differ in the way how the sample is drawn as well as in the determination of the suitable size of the sample. The general drawback of sampling techniques is that the accuracy of the result is strictly limited by the sample rate. To get an accurate estimation of the selectivity, a large sample (>10%) of the database is required. To evaluate the query on top of the large sample is not much cheaper than to evaluate it on the original data set which limits its usefulness for query optimization.

Histogram techniques, the most prevalent paradigm to model the data distribution in the one-dimensional case, have a different problem. This concept is very difficult to be carried over to the multidimensional case, even for low or moderate dimensional data. One way to adapt one-dimensional histograms to multidimensional data is to describe the distribution of the individual attributes of the vectors independently by usual histograms. These histograms are sometimes called marginal distributions. In this case, the selectivity of multidimensional queries can be determined easily provided that the attributes are statistically independent, i.e. neither correlated nor clustered. Real-world data sets, however, rarely fulfill this condition. Another approach is to partition the data space by a multidimensional grid and to assign a histogram bin to each grid cell. This approach may be possible for two- and three-dimensional spaces. However, for higher dimensional data this method becomes inefficient and ineffective since the number of grid cells is exponential in the dimensionality. Techniques of dimensionality reduction such as Fourier transformation, wavelets, principal component analysis or space-filling curves (Z-ordering, Hilbert) may reduce this problem to some extent. The possible problem reduction, however, is limited by the intrinsic dimensionality of the data set.

The idea of parametric techniques is to describe the data distribution by curves (functions) which have been fitted into the data set. In most cases Gaussian functions (normal distributions) are used. Instead of using one single Gaussian, a set of multivariate Gaussians can be fitted into the data set which makes the technique more accurate. Each Gaussian is then described by three parameters (mean, variance and the relative weight of the Gaussian in the ensemble). This approach can be transferred into the multidimensional case by two techniques. Like described above for histograms, the marginal distribution of each attribute can be modelled independently by a set of Gaussians. The multidimensional query selectivity can be estimated by combining the marginal distributions. This approach leads to similar problems like marginal histograms.

Therefore, our solution is different. Our technique directly models the multidimensional data distribution by a set of multivariate Gaussian functions. There are two options to use the Gaussian primitives: The Gaussians can either be used with a matrix containing both variances and covariances or with a vector of the multivariate variances only. As we will discuss later, both approaches have their advantages and disadvantages. When using Gaussians with covariance matrix, the data distribution can be described more accurately by a single primitive. On the other side, more storage is needed for the covariance matrices ($O(d^2)$ for each Gaussian) compared to the variance vector approach ($O(d)$ for each Gaussian). Moreover, the processing cost for reading the parameters and for the determination of the estimated selectivity is much higher when covariance matrices are used. Let us note that, unlike the approaches using marginal distributions, our Gaussian technique with no covariance matrix does not rely on the attribute independence assumption. This technique assumes attribute independence for each individual Gaussian primitive only, but places no constraints to the overall data distribution. We will discuss this issue in more detail in Section 4, an experimental validation is given in Section 5.

To obtain a collection of Gaussians distributions we apply a clustering algorithm. Clustering is the task of grouping vectors into different subsets (the clusters) such that the intra-cluster similarity is maximized and the inter-cluster similarity is minimized. That means points belonging to the same cluster are close together whereas points of different clusters are far away from each other. Many different algorithms have been proposed such as k-means [7], single-link [8], density-based clustering [9,10] and many others. Most of these algorithms use a point as a representative of each cluster. In contrast, the EM clustering algorithm (expectation maximization) [11] uses a multivariate Gaussian function as a cluster representative. We will discuss the suitability of different variants of the EM algorithm for our problem of getting a good approximation of the actual data distribution.

To summarize our contribution, we propose in this paper a new cost model for estimating the selectivity of multidimensional queries on top of vector data of medium to high dimensionality. The data distribution is represented by a set of multivariate Gaussian functions that have been determined using the EM clustering algorithm. We develop two methods for estimating the selectivity of window queries and range queries using the multivariate Gaussians. We demonstrate experimentally the superiority of our approach over competitive cost models based on histograms and sampling. The remainder of our paper is organized as follows: In Section 2 we discuss related work on selectivity estimation and point out our contribution. Section 3 and 4 describes in detail our proposed methods to find a representation of the data distribution by an ensemble of multivariate Gaussian functions using EM clustering and to estimate the selectivity on top of this model. Section 5 contains the experimental evaluation, and section 6 concludes our paper.

2   Related Work

In this chapter, we review current approaches for selectivity estimation and discuss their potentials.

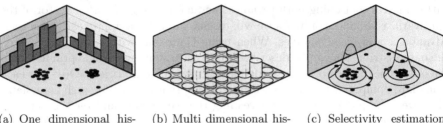

(a) One dimensional histograms

(b) Multi dimensional histograms

(c) Selectivity estimation via clustering

Fig. 1. Visualization of different concepts for selectivity estimation

2.1 Review

Recent work on selectivity estimation can be categorized into three classes, namely histogram-based methods, sampling-based methods, and parametric methods. In the following, we review and discuss the most important representatives of each class briefly.

Histogram-based Methods. The most widespread approach for selectivity estimation in practice is the use of histograms. In general, the data space is partitioned into buckets, and the frequency of points inside each bucket is computed. We can distinguish between one-dimensional and multi-dimensional histograms.

Selectivity estimation using one-dimensional histograms is based on the assumption that the attributes of the data set are independent, i.e. there is no correlation between different dimensions of the feature space. For each dimension, a histogram is built and the selectivity of a window query q is estimated in each dimension separately. The selectivity of q in the full-dimensional space is evaluated by multiplying the selectivity estimations for each attribute. Equi-width histograms [12] compute buckets of fixed size and variable point frequency, whereas equi-depth histograms [13] compute buckets of variable size and fixed point frequency.

With growing dimensionality of the feature space, the recombination of one-dimensional buckets becomes costly. Thus, in recent years, multi-dimensional histograms have been investigated. Multi-dimensional equi-depth histograms [14] partition the feature space into multi-dimensional buckets with variable size and fixed point frequency. In [14] an algorithm to construct multi-dimensional equi-depth histograms is presented that iteratively partitions the data space along each attribute into a fixed number of buckets, where the order of attributes is fixed. The selectivity of a window query q is estimated analogously to one-dimensional histograms taking the buckets into account that intersect with q. The algorithm MHIST [15] partitions the data space along the single attributes in a similar way, but decides in each step which attribute is partitioned rather than processing the attributes in a fixed order.

STHoles [16] is a recent approach that proposes hierarchically organized multi-dimensional histograms. A histogram may contain another histogram completely, or may be completely covered by part of another histogram. The is-part of hierarchy of the histograms is represented as a tree where each node represents a bucket. Using this hierarchical concept, a non-uniform distribution inside a bucket b can be adopted more accurately by several smaller buckets inside (that are part of) b. STHoles histograms are constructed using a set of sample queries as reference. Regions in the data space, that are queried more frequently, can thus be represented in more detail through a larger number of buckets. The histograms are refined after each query. However, the refinement procedure takes care that no more than a fixed upper bound of buckets is generated. If this upper bound is violated temporarily during reorganization, some buckets are merged.

In [17] the authors propose another strategy of computing multidimensional histograms using Wavelet transformation. In particular, the authors show how to apply a Wavelet transformation to one dimensional data sets. The data space is split evenly in a recursive fashion. The Wavelet coefficients are computed for each bucket. The resulting grid can be more fine grained than for traditional histograms because using the Wavelet coefficients the data is compressed more efficiently. For higher dimensional data, the authors in [17] suggest to split each attribute recursively in a given order.

Sampling-based Methods. A second approach for estimating the selectivity of queries is based on sampling. Usually, the selectivity of a query q is estimated on a small sample of the database and is then extrapolated onto the entire database. The simplest way of computing a sample is the well-known random sampling method. A more data driven variant is adaptive sampling [18,19].

A similar approach called 'double sampling' is proposed in [20]. The main difference to the adaptive sampling method is a different estimation of the sample size. In fact, the sample size is reduced using a two-way sampling procedure. However, there is no hint on how to choose the size of the first sample.

Parametric Methods. In [21], a method called Adaptive Selectivity Estimator (ASE) is proposed that tries to approximate the distribution of the data objects along one attribute using an appropriate polynomial function. This function is adopted and refined taking predefined queries into account, and minimizes the squared error between the real and the estimated selectivity. ASE is evaluated in [21] using one- and two-dimensional data sets only.

2.2 Discussion

As noticed above, current approaches for selectivity estimation have severe drawbacks. Sampling techniques suffer from the fact, that the accuracy of the result is strictly constrained by the sample rate. High sample rates on the other hand are quite inefficient and limit the usefulness of sampling techniques for query optimization. One-dimensional histograms (cf. Figure 1(a)) rely on the attribute independence assumption, i.e. on the assumption that the attributes are neither correlated nor clustered. This is quite unrealistic in real-world data sets which rarely fulfill this condition. Multi-dimensional histograms (cf. 1(b)) become inefficient and ineffective for higher dimensionalities since the number of grid cells is exponential in the dimensionality. Techniques of dimensionality reduction are (at least) limited by the intrinsic dimensionality of the data set. Similar problems are prevalent using parametric methods.

In this paper, we propose the use of clustering to get an accurate characterization of the data by means of a collection of multivariate Gaussians (cf. 1(c)). Our two methods are called SEC (Selectivity Estimation via Clustering) and SEC+ and both use different variants of the EM clustering algorithm to extract a collection of Gaussian distributions. For SEC each Gaussian is represented by the mean value, the variances and the covariances and the relative weight of the Gaussian in the ensemble. SEC+ uses the same representation but leaves out the covariances. Based on these representations, SEC and SEC+ efficiently and effectively estimate the selectivity of window queries in spatial data. We empirically show that especially SEC+ yields significantly more accurate results than comparative methods, especially when applied to higher dimensional data.

3   SEC: Selectivity Estimation via Clustering

The overall goal of representing a given data set for selectivity estimation is to find a model of the data that is as compact as possible (low amount of storage necessary) and as accurate as possible (for accurate selectivity estimations). The key idea of our new approach is to use a clustering algorithm to gain an accurate description of the data and then use this description for selectivity estimation. In this section, we describe both the clustering process and the method for selectivity estimation in detail.

3.1   Describing the Data via Clustering

Clustering has gained a lot of attention from the data mining research community over the past decades. In particular, clustering is the task of grouping objects of a data set into classes (clusters), while maximizing intra-cluster similarity and minimizing inter-cluster similarity. An overview over recent work on clustering can be found e.g. in [22]. Often clustering algorithms can also be used to obtain a compact representation of a data set. An efficient way to represent a data set for selectivity estimation, is to use a mixture of different distribution functions. The most prominent algorithm that tries to describe the data by multiple distribution functions is the EM algorithm [11]. In the following, we describe a variant of this algorithm which is used by our selectivity estimation method SEC.

Let $\mathcal{D}$ be a set of d-dimensional points, i.e. $\mathcal{D} \subseteq \mathbb{R}^d$. The general idea of the EM algorithm is to describe the data by a mixture M of k Gaussian distributions. Note that the EM algorithm can also be seen as a variant of k-means clustering. Instead of assigning each object to a cluster as is the case for k-means-based clustering algorithms, it assigns each object to a cluster according to a weight representing the probability of membership.

Each cluster $C \in M$ is a tuple $C = (\mu_C, \Sigma_C)$, where

- μ_C is the mean value of all points in C and
- Σ_C is the $d \times d$ covariance matrix of all points in C.

To compute the probability distributions, we need the following concepts. The probability with which a point $x \in \mathcal{D}$ belongs to a Gaussian distribution $C = (\mu_C, \Sigma_C)$ can be computed by:

$$P(x|C) = \frac{1}{\sqrt{(2\pi)^d |\Sigma_C|}} e^{-\frac{1}{2}(x-\mu_C)^\mathrm{T}(\Sigma_C)^{-1}(x-\mu_C)}.$$

The combined probability for k clusters can then be computed by:

$$P(x) = \sum_{i=1}^{k} w_{C_i} P(x|C_i),$$

where w_{C_i} is the fraction of points that belongs to cluster $C_i = (\mu_{C_i}, \Sigma_{C_i})$, i.e. w_{C_i} is the weight of C_i.

Then the probability that a point $x \in \mathcal{D}$ belongs to a cluster C can be computed by the rule of Bayes:

$$P(C|x) = w_C \frac{P(x|C_i)}{P(x)}.$$

The accuracy of a mixture $M = (C_1, \ldots, C_k)$ of k Gaussian distributions which describes how good the model approximates the actual data set can be computed by:

$$E(M) = \sum_{x \in \mathcal{D}} \log\left(P(x)\right).$$

The higher the value of $E(M)$, the more likely it is that the data set $\mathcal{D}$ has been generated by the mixture M of k Gaussian distributions. Thus, the aim of the EM algorithm is to optimize the parameters of M, i.e. the parameters of the k Gaussian distributions $C_1, \ldots, C_k$, such that $E(M)$ is maximized. For that purpose, the algorithm proceeds in four steps:

1. *Initialization*
   Since the clusters, i.e. Gaussian distributions $C_1, \ldots, C_k$, are unknown at the beginning, a set of k initial clusters are built randomly. For that purpose, each point $x \in \mathcal{D}$ is randomly assigned to one cluster C_i. An initial model is produced by computing μ_C and Σ_C for each cluster $C \in M$
2. *Expectation*
   Based on the current model, the parameters μ_C and Σ_C can be computed for each cluster $C \in M$ and the accuracy $E(M)$ of this mixture M is obtained.
3. *Maximization*
   In this step the accuracy of the mixture is improved via a recomputation of the parameters of each of the k clusters. Given a mixture M of k Gaussians, the parameters μ_C, Σ_C, and w_C of each cluster $C \in M$ are recomputed.

The resulting mixture M' has an equal or higher accuracy than M, i.e. $E(M) \leq E(M')$. For improving the mixture, the parameters are recomputed as follows:

$$w_C = \frac{1}{|\mathcal{D}|} \sum_{x \in \mathcal{D}} P(C|x),$$

$$\mu_C = \frac{\sum_{x \in \mathcal{D}} x \cdot P(C|x)}{\sum_{x \in \mathcal{D}} P(C|x)},$$

$$\Sigma_C = \frac{\sum_{x \in \mathcal{D}} P(C|x)(x - \mu_C)(x - \mu_C)^{\mathbf{T}}}{\sum_{x \in \mathcal{D}} P(C|x)}.$$

4. *Iteration*

Step 2 and 3 are iterated until the accuracy of the improved mixture M' differs from the accuracy of the previous mixture M by a smaller value than a user specified threshold ε, i.e. until $|E(M) - E(M')| < \varepsilon$.

The result of the EM algorithm is a set of k d-dimensional Gaussian distributions, each represented by the mean value μ and the covariance matrix Σ. The assignment of a point $x \in \mathcal{D}$ to a cluster C is given by the probability $P(C|x)$. We thus can compute how likely a point is assigned to each of the k clusters.

The accuracy of the result of the EM algorithm, i.e. the accuracy of the resulting mixture, depends on the initial mixture, i.e. on step 1 of the algorithm, and on the choice of k. In [23] a method for producing a good initial mixture is presented which is based on multiple sampling. It is empirically shown that using this method the EM algorithm achieves accurate clustering results. The authors further propose a method for determining a suitable number of clusters, i.e. a suitable value for k.

3.2   Selectivity Estimation of Window Queries

As discussed in the previous subsection, we describe the data distribution using k Gaussian distributions each represented by a mean value and a covariance matrix. Let us note that this representation does not rely on the unrealistic attribute independence assumption nor has it problems in higher dimensions such as exponential storage cost that must be compensated by less accuracy.

In the following, we assume that M is a mixture of k Gaussian distributions computed by the EM algorithm applied on the database $\mathcal{D}$ as described above. We will also call M a model that describes the distribution of the objects in $\mathcal{D}$ and we will use the two notions *Gaussian distribution* and *cluster* interchangeably for a given $C \in M$. A window query Q is a list of d pairs (l^i, u^i), where l^i and u^i are the lower and upper bounds, respectively, of Q in the i-th dimension, where $1 \leq i \leq d$. The center of Q is denoted by c_Q.

Intuitively, a good estimation of the selectivity of a query Q is the integral $\mathcal{I}(Q, C)$ of the intersection of Q and each $C \in M$. A straightforward idea to estimate the selectivity of a query Q using the model M is the following. For each cluster $C \in M$ we can compute the probability that the center c_Q is in

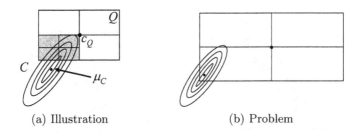

(a) Illustration   (b) Problem

Fig. 2. The naive approach to selectivity estimation

C, i.e. $P(c_Q|C)$. This probability can then be multiplied with the volume of the query. The resulting integral of the intersection of Q and cluster C is a first approximation of the selectivity of Q. If the integral is above a threshold ε, it may be interesting to further improve the estimation. We can achieve this by decomposing Q into 2^d rectangles Q_i of equal size and computing the integrals of the intersection of C with each resulting rectangle Q_i. This can be iteratively continued until all decomposed Q_i have an integral above ε. Then the selectivity of Q w.r.t. C can be computed by the sum of the integrals of the decomposed windows Q_i having an integral above ε, multiplied by the weight of the cluster w_C. The overall selectivity of Q is then simply the sum over all $C \in M$. This approach is illustrated in Figure 2(a). The query Q is decomposed into four smaller windows. One of them (marked in gray) is further decomposed. The selectivity of Q w.r.t. C is the sum of the integrals of the intersection of each gray window with C.

We called this approach SEC (Selectivity Estimation via Clustering). The next chapter will present an approach called SEC+ that proposes certain improvements over the basic version SEC.

4   SEC+: Improved Selectivity Estimation via Clustering

Unfortunately, the simple idea of decomposing the query window rises several problems. First of all, the iterative decomposition of Q into 2^d rectangles is quite inefficient and requires high storage cost. For an accurate estimation, however, we probably need multiple decompositions, i.e. several iterations of the decomposition process. Secondly, representing a window query only by its center has drawbacks, too. Especially in higher dimensional spaces, the center of Q may be far away from any $C \in M$ even if Q contains C. This is illustrated in Figure 2(b). Although query Q contains a large part of C, the center of Q is too far away from μ_C and the probability $P(c_Q|C)$ is far too small. Thus, multiplication of $P(c_Q|C)$ with the volume of Q will yield a very small value, most likely below a reasonable threshold ε. A third problem is that the storage cost in SEC for a single cluster is relatively high ($d^2 + d$ values per cluster). Therefore, we modify in SEC+ our data model, the clustering algorithm, as well as our method of selectivity estimation.

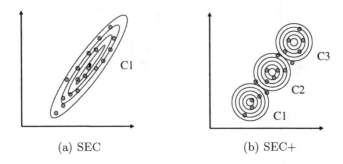

Fig. 3. Visualization of the model created for SEC (a) and SEC+ (b)

To reduce the storage cost per cluster our idea is to store only the diagonal elements of the covariance matrix, i.e. we store only the variance values of single attributes but no covariances between different attributes. This means that the Gaussian functions are oriented in an axis-parallel way rather than arbitrarily. Note that this does not mean we assume attribute independence for the complete data space (which would be unacceptable as discussed before). We assume only that the points which are associated to a common cluster observe the attribute independence assumption. This is much easier to motivate than demanding global independence for the complete data set because (1) the individual clusters contain data which is locally selected and (2) we can modify the EM algorithm to determine preferably clusters in which the assumption is fulfilled. (3) Due to saved storage cost we can maintain considerably more individual clusters in our model which generally allows a better adaptation to the real data distribution. In our SEC+-model a cluster is represented by its d-dimensional mean vector μ_C and a d-dimensional variance vector $var_C = (var_C^1, ..., var_C^d)$.

To guarantee that the EM-algorithm determines a good approximation of the real data distribution, we adapt the probability density function $P(x|C)$ for the clusters in order to use diagonal matrices only:

$$P(x|C) = \frac{1}{\sqrt{(2\pi)^d \prod_{1 \leq j \leq d} var_C^j}} e^{-\frac{1}{2} \sum_{1 \leq j \leq d}(x_j - \mu_{C,j})^2 var_C^j}.$$

This additionally makes the clustering algorithm more efficient as the variance vector is easier to determine and to invert (for computing the determinant) than a quadratic covariance matrix. Moreover, since our new model with axis-parallel Gaussians is now reflected in the algorithm and in the accuracy measure $E(M)$ the EM algorithm searches for an optimal model according to the new demands.

Figure 3 shows that our new model is not constraining the accuracy we reach for selectivity estimation. In case of strong correlations in the data set, the algorithm simply assigns more Gaussian functions to the data set. Due to the dramatically reduced storage cost for an individual Gaussian function, the

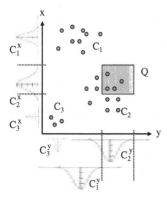

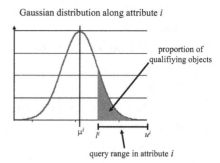

(a) One-dimensional Gaussian distributions for three clusters.

(b) Intersection of a query and a Gaussian distribution in attribute i.

Fig. 4. Illustration of selectivity estimation using SEC+

overall storage requirement for the complete model is still much lower. Note that the algorithm may assign up to d times more clusters without any extra storage cost. We will evaluate this issue experimentally in Section 5 and show that a higher number of axis-parallel Gaussians can even represent data distributions exhibiting non axis-parallel clusters more accurate than a lower number of non axis-parallel Gaussians.

Due to our new model, the method of selectivity estimation given a window query can also be improved with respect to efficiency and effectiveness in SEC+. Still, the integrals of the intersections of the query window and the Gaussian distributions are computed. However, each multi-dimensional Gaussian distribution C is split into all one-dimensional distributions and the selectivity is determined using these one-dimensional distributions. This methods avoids the problems of the first approach. We will highlight the procedure of SEC+ in the following.

Instead of measuring the selectivity of a query Q by the probability $P(c_Q|C)$ of a cluster $C \in M$, we split the d-dimensional Gaussian distribution into d corresponding one-dimensional Gaussian distributions C^i. This is visualized in Figure 4(a). The integral $\mathcal{I}(Q, C)$ is the product of the integrals of all one-dimensional distributions, i.e. $\mathcal{I}(Q, C) = \prod_{i=1}^{d} \mathcal{I}(Q, C^i)$. Let us note, that this requires the assumption that the attributes are independent for the points belonging to that cluster. However, as discussed above this is no serious constraint. Figure 4(b) illustrates the integral of the intersection of a one-dimensional Gaussian and the query Q for an attribute i. The query in that attribute is given by the interval $[l^i, u^i]$. The integral $\mathcal{I}(Q, C^i)$ measures the proportion of qualifying points, i.e. points of $\mathcal{D}$ that match the query Q.

Given the d-dimensional Gaussian distribution $C = (\mu_C, var_C)$, the d corresponding one-dimensional Gaussian distributions can easily be obtained. These one-dimensional Gaussian distributions are represented by the mean μ_C^i and by

---

SEC+ (SetOfObjects $\mathcal{D}$, Query Q)

   Compute model M of $\mathcal{D}$ by EM($\mathcal{D}$);

   for each cluster $C_i \in M$ **do**

     for each dimension j of $\mathcal{D}$ **do**

       Compute $\mathcal{I}(Q, C_i^j)$

     end for

       Compute $\mathcal{I}(Q, C_j) = \prod_{i=1}^{d} \mathcal{I}(Q, C_j^i)$

   end for

   Compute $\text{SEC+}(Q, M) = \sum_{i=1}^{k} w_{C_i} \cdot \mathcal{I}(Q, C_i)$.

Fig. 5. Pseudocode of SEC+

the standard deviation s_C^i, i.e. $C^i = (\mu_C^i, s_C^i)$. The mean value μ_C^i is simply the i-th component of μ_C. The standard deviation s_C^i can be computed as follows:

$$s_C^i = \sqrt{var_C^i},$$

where $var_C^i \in var_C$ is the variance of attribute i. Obviously, at this point, we do not need the covariance matrix Σ_C, but only the variances var_C^i which has the above discussed advantages. The integral $\mathcal{I}(Q, C^i)$ can then be computed rather straightforward. We simply materialize the standard Gaussian distribution Φ with $\mu = 0$ and $\sigma = 1$ in a table. The integral can then be computed as follows:

$$\mathcal{I}(Q, C^i) = \left| \frac{\Phi(u_i) - \mu_C^i}{\sigma_C^i} - \frac{\Phi(l_i) - \mu_C^i}{\sigma_C^i} \right|.$$

The selectivity of Q w.r.t. a cluster C is then the product of all attribute-wise integrals, i.e.

$$\mathcal{I}(Q, C) = \prod_{i=1}^{d} \mathcal{I}(Q, C^i).$$

The overall selectivity of a query Q is estimated as the weighted sum of selectivities of Q w.r.t. all $C_i \in M$, formally

$$\text{SEC+}(Q, M) = \sum_{c \in M} w_C \cdot \mathcal{I}(Q, C).$$

The pseudo code of our method SEC+ is given in Figure 5. In the next section we will show experimentally that SEC+ is superior to SEC and to other comparative methods.

5 Experimental Evaluation

In this section, we present a broad experimental evaluation of SEC, SEC+ and comparative methods on synthetic and real-world data sets. We used randomly

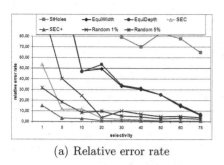

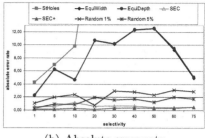

(a) Relative error rate   (b) Absolute error rate

Fig. 6. Results on synthetic data set with Gaussian non-axis parallel ellipsoid clusters

generated window queries throughout all our experiments. To judge the accuracy of each selectivity estimation method we used two measurements to compute the error rate of each method, the relative error rate and the absolute error rate. Let S_Q be the true selectivity and S'_Q the estimated selectivity of a query Q. Let n be the number of tuples in the considered data set. The relative error rate $E_r(q)$ measures the error of the estimation w.r.t. the true selectivity, formally

$$E_r(Q) = \frac{|S_Q - S'_Q|}{S_Q}$$

The absolute error rate $E_a(Q)$ measures the error of the estimation w.r.t. the size of the database, formally

$$E_a(Q) = \frac{|S_Q - S'_Q|}{n}$$

We compared SEC+ to several competitive methods, including random sampling using 1% of the database as sampling rate (indicated in the diagrams by "Random 1%"), random sampling using 5% of the database as sampling rate (indicated in the diagrams by "Random 5%"), one-dimensional equi-width histograms using 30 intervals per dimension (indicated in the diagrams by "Equi-Width"), one-dimensional equi-depth histograms using an interval capacity of 5% of the data set (indicated in the diagrams by "EquiDepth"), and multidimensional histograms (STHoles) using 1,000 randomly generated sample queries to establish the histogram as proposed in [16]. In [23] a method to choose the parameter k and the intial clustering for any variant of the EM algorithm is described. We used this method to determine k and the intial clustering for SEC and SEC+.

5.1   Comparison of SEC and SEC+

In Figure 6, we compared SEC and SEC+ with its competitors. SEC is the variant that uses covariances throughout the EM-clustering process, whereas SEC+

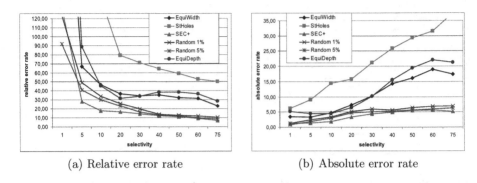

(a) Relative error rate   (b) Absolute error rate

Fig. 7. Results on the "Abalone" data set

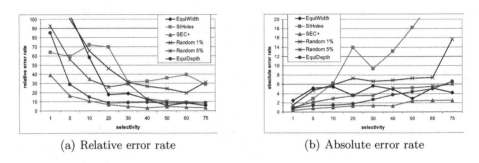

(a) Relative error rate   (b) Absolute error rate

Fig. 8. Results on the gene expression data set

only uses variances. All our experiments show that SEC+ performs as good or even better than SEC. For a comparison of SEC, SEC+ and other techniques, we applied all methods on a data set of 10,000 5-dimensional tuples containing several non-axis parallel Gaussian clusters. This data set was chosen because it seems to favor SEC which uses covariances over SEC+. SEC outperforms all competitive methods besides SEC+. However, as illustrated in Figure 6, the use of the covariances during clustering does not achieve a gain in accuracy compared to the improved SEC+ algorithm. In case of queries with lower selectivity, SEC+ even outperforms SEC which uses covariances during clustering. Let us note that we needed less storage for SEC+ than for SEC in our experiments but achieved better results when using SEC+ rather than SEC. This result was repeated in all other experiments, justifying the use of SEC+. Thus, throughout the rest of our evaluation, we will only show the results of SEC+.

5.2 Accuracy of SEC+

Comparison with other methods. We applied SEC+ and the comparative methods on several real-world data sets. Due to space limitations, we focus on two data sets. The first one is the "Abalone" benchmark data set from the UCI

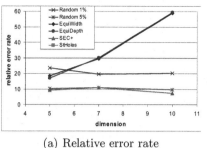

(a) Relative error rate

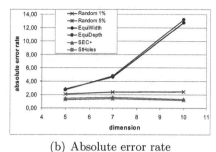

(b) Absolute error rate

Fig. 9. Comparison of the accuracy w.r.t. dimensionality of the data set

Machine Learning Database Repository[1]. It contains approximately 4,200 objects in a 8-dimensional feature space. The second data set is a gene expression data set from our project partners[2] and contains approximately 1500 objects in a 5-dimensional feature space. We evaluated the error rates of SEC+ and the comparative methods w.r.t. the selectivity (in %) of the queries. The results on the "Abalone" data set are depicted in Figure 7. SEC+ outperforms all other methods regarding relative and absolute error rates. Only for very selective queries (<5%), random sampling with 5% sampling rate is slightly better. However, a sampling rate of 5% is rather high for large databases. A similar observation can be made from Figure 8 illustrating the results on the gene expression data set. Again, SEC+ outperforms all other comparative methods w.r.t. both the relative error rate and the absolute error rate, even for very selective queries. The histogram based approaches perform slightly better than on the "Abalone" data set, especially compared to the sampling based approaches. We guess that this can be explained with the lower dimensionality of the gene expression data set. Both experiments show, that SEC+ outperforms competitive approaches in terms of accuracy especially in high dimensional spatial data.

Accuracy w.r.t. data dimensionality. We evaluated the accuracy of SEC+ and the five comparative methods w.r.t. the dimensionality of the data set using synthetic data of 10,000 tuples and a sample query q having a selectivity $S_q = 10\%$. The results are visualized in Figure 9. As expected, we can observe that the accuracy of random sampling methods are independent of the dimensionality of the data set, whereas the accuracy of histogram-based methods detoriates with increasing dimensionality. Let us note that StHoles is not shown in the charts because its error rates are far above the interval shown here. It can also be seen that the accuracy of SEC+ is independent of the data dimensionality. In a 5-, 7- and 10-dimensional feature spaces, SEC+ performs better than all other techniques besides 5% random sampling. But even 5% random sampling which is already very inefficient for large data sets, is only as good as SEC+.

---

[1] http://www.ics.uci.edu/~mlearn/MLRepository.html
[2] Genomatix SW GmbH: http://www.genomatix.de/

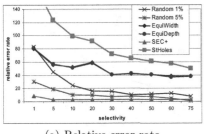
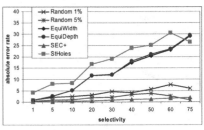

(a) Relative error rate   (b) Absolute error rate

Fig. 10. Results on the synthetic data set with 20% noise

Influence of noisy data. Next, we tested the influence of noisy data on the accuracy of SEC+ and the competitive methods. Since SEC+ relies on clustering, noisy data may cause problems in generating an accurate compression of the data distribution and thus may influence the selectivity estimation. Figure 10 illustrates the error rates of SEC+ and the comparative methods w.r.t. the selectivity of the query on a synthetic data set of 10,000 tuples of 5 dimensions with 80% of the data belonging to clusters and 20% noise objects. As it can be seen, SEC+ is quite robust against noisy data. For a broad range of query selectivity, SEC+ outperforms its competitors w.r.t. the relative and absolute error rates. Again, the sampling based methods are ranked second followed by one-dimensional histograms. Equi-depth and equi-width histograms produce nearly the same results in that experiment.

6   Conclusions

Advanced database applications rely on accurate and efficient query optimization. One key step for query optimization is the estimation of the selectivity of a given query. Recent approaches for selectivity estimation have problems with medium to high dimensional data spaces and/or usually require a high sampling rate to achieve accurate results.

   In this paper, we proposed two new methods for selectivity estimation of spatial window queries called SEC (Selectivity Estimation via Clustering) and SEC+. Our solutions are based on modelling the data through a set of multivariate Gaussian functions which are computed using different variants of the EM clustering algorithm. Two techniques to derive an accurate estimation of the query size using the generated models are discussed in detail. A broad experimental evaluation illustrates that SEC+ outperforms existing approaches in terms of accuracy. In particular, SEC+ is robust against the dimensionality of the data space and can handle noisy data effectively.

References

1. Faloutsos, C., Barber, R., Flickner, M., Hafner, J., Niblack, W., Petkovic, D., Equitz, W.: "Efficient and Effective Querying by Image Content". Journal of Intelligent Information Systems **3** (1994) 231–262
2. Mehrotra, R., Gary, J.: "Feature-Based Retrieval of Similar Shapes". In: Proc. 9th Int. Conf. on Data Engineering, Vienna, Austria,. (1993) 108–115
3. Shoichet, B.K., Bodian, D.L., Kuntz, I.D.: "Molecular Docking Using Shape Descriptors". Journal of Computational Chemistry **13** (1992) 380–397
4. Berchtold, S., Keim, D.A., Kriegel, H.P.: "The X-Tree: An Index Structure for High-Dimensional Data". In: Proc. 22nd Int. Conf. on Very Large Databases (VLDB'96). (1996)
5. Lin, K.I., Jagadish, H.V., Faloutsos, C.: "The TV-tree an index structure for high-dimensional data". VLDB Journal: Very Large Data Bases **3** (1994) 517–542
6. Weber, R., Schek, H.J., Blott, S.: "A Quantitative Analysis and Performance Study for Similarity-Search Methods in High-Dimensional Spaces". In: Proc. 24th Int. Conf. on Very Large Databases (VLDB'98). (1998) 194–205
7. McQueen, J.: "Some Methods for Classification and Analysis of Multivariate Observations". In: 5th Berkeley Symp. Math. Statist. Prob. Volume 1. (1967) 281–297
8. Sibson, R.: "SLINK: An Optimally Efficient Algorithm for the Single-Link Cluster Method". The Computer Journal **16** (1973) 30–34
9. Ester, M., Kriegel, H.P., Sander, J., Xu, X.: "A Density-Based Algorithm for Discovering Clusters in Large Spatial Databases with Noise". In: Proc. 2nd Int. Conf. on Knowledge Discovery and Data Mining (KDD'96), Portland, OR. (1996) 291–316
10. Ankerst, M., Breunig, M.M., Kriegel, H.P., Sander, J.: "OPTICS: Ordering Points to Identify the Clustering Structure". In: Proc. ACM Int. Conf. on Management of Data (SIGMOD'99). (1999)
11. Dempster, A.P., Laird, N.M., Rubin, D.B.: "Maximum Likelihood from Incomplete Data via the EM Algorithm". Journal of the Royal Statistical Society, Series B **39** (1977) 1–31
12. Selinger, P.G., Astrahan, M.M., Chamberlin, D.D., Lorie, R.A., Price, T.G.: "Access Path Selection in a Relational Database Management System". In: Proc. ACM Int. Conf. on Management of Data (SIGMOD'79). (1979)
13. Piatetsky-Shapiro, G., Connell, C.: "Accurate estimation of the number of tuples satisfying a condition". In: Proc. ACM Int. Conf. on Management of Data (SIGMOD'84). (1984)
14. Muralikrishna, M., De Witt, D.J.: "Equi-Depth Histograms For Estimating Selectivity Factors For Muli-Dimensional Queries". In: Proc. ACM Int. Conf. on Management of Data (SIGMOD'88). (1988)
15. Poosala, V., Ioannidis, Y.E.: "Selectivity Estimation without the Attribute Value Independence Assumption". In: Proc. 23rd Int. Conf. on Very Large Databases (VLDB'97). (1997)
16. Bruno, N., Chaudhuri, S., Gravan, L.: "STHoles: a Multidimensional Workload-aware Histogram". In: Proc. ACM Int. Conf. on Management of Data (SIGMOD'01). (2001)
17. Matias, Y., Vitter, J.S., Wang, M.: "Wavelet-Based Histograms for Selectivity Estimation". In: Proc. ACM Int. Conf. on Management of Data (SIGMOD'98). (1998) 448–459

18. Lipton, R., Naughton, J.: "Query size estimation by adaptive sampling". In: Proc. ACM Symp. on Principles of Database Systems (PODS'90). (1990)
19. Lipton, R., Naughton, J., Schneider, D.: "Practical selectivity estimation through adaptive sampling". In: Proc. ACM Int. Conf. on Management of Data (SIGMOD'90). (1990)
20. Hou, W.C., Ozsoyoglu, G., Dodgu, E.: "Error-constrained Count Query: Evaluation in Relational Databases". In: Proc. ACM Int. Conf. on Management of Data (SIGMOD'91). (1991)
21. Chen, C.M., Roussopoulos, N.: "Adaptive Selectivity Estimation Using Query Feedback". In: Proc. ACM Int. Conf. on Management of Data (SIGMOD'94). (1994)
22. Han, J., Kamber, M.: Data Mining: Concepts and Techniques. Academic Press (2001)
23. Fayyad, U., Reina, C., Bradley, P.: "Initialization of Iterative Refinement Clustering Algorithms". In: Proc. 4th Int. Conf. on Knowledge Discovery and Data Mining (KDD'98). (1998)

Spatio-temporal Histograms*

Hicham G. Elmongui, Mohamed F. Mokbel, and Walid G. Aref

Department of Computer Science, Purdue University, West Lafayette, IN, USA
{elmongui, mokbel, aref}@cs.purdue.edu

Abstract. This paper presents a framework for building and continuously maintaining spatio-temporal histograms (ST-Histograms, for short). ST-Histograms are used for selectivity estimation of continuous pipelined query operators. Unlike traditional histograms that examine and/or sample all incoming data tuples, ST-Histograms are built by monitoring the *actual* selectivities of the outstanding continuous queries. ST-Histograms have three main features: (1) The ST-Histograms are built with (almost) no overhead to the system. We use only feedback (i.e., the actual selectivity) from the existing continuous queries. (2) Rather than wasting system resources in maintaining accurate histograms for the whole spatial space, we only maintain accurate histograms for that part of the space that is relevant to the current existing queries. The rest of the space has less accurate histograms. (3) The ST-Histograms are equipped with a periodicity detection procedure that predicts the future execution of the continuous queries. Hence, the query processing engine can continuously adapt the continuous query pipeline to reflect this prediction. Experimental results based on a real implementation inside a data stream management system show a superior performance of ST-Histograms in terms of providing accurate operator selectivity estimations with no extra overhead.

1 Introduction

The rapid increase in spatio-temporal applications calls for new query processing and query optimization techniques to deal with both the spatial and temporal domains. Examples of these applications include location-aware services [34], traffic monitoring [37], and enhanced 911 service [1]. These applications have two main characteristics: (1) A highly dynamic environment where data from mobile objects (e.g., moving vehicles in road networks) are received continuously. (2) Queries in these spatio-temporal applications are mostly *continuous* (e.g., monitoring queries). Continuous queries require continuous evaluation as the query area and/or the data are continuously moving.

Most of the previous work on continuous spatio-temporal queries (e.g., see [19,22,25,26,27,42,44,47,48]) focus on developing out-of-the-box algorithms (i.e., algorithms built on top of database management systems (DBMSs)). Having out-of-the-box algorithms bypass completely the role of the query optimizer

---

* This work was supported in part by the National Science Foundation under Grants IIS-0093116, IIS-0209120, and 0010044-CCR.

C. Bauzer Medeiros et al. (Eds.): SSTD 2005, LNCS 3633, pp. 19–36, 2005.

in DBMSs, thus severely limiting the query performance. Recently, within the PLACE server (*Pervasive Location-Aware Computing Environments*) [35], more attention is given to embed the functionality of continuous query processing into existing database and data stream query engines. The main idea is to furnish existing query processors with a set of *spatio-temporal* operators that can be combined with traditional operators to support efficient execution of a wide variety of continuous spatio-temporal queries. Our previous work in PLACE widens the scope of research in continuous spatio-temporal queries to include system-oriented support for continuous spatio-temporal queries (e.g., query optimization, adaptive query processing, and query scalability [35]).

In this paper, we focus on query optimization for continuous spatio-temporal queries. In particular, we are concerned with two main functionalities for optimizing the execution of continuous spatio-temporal queries: (1) Building a new *optimal* query plan for each newly submitted continuous query, and (2) Continuously monitoring the performance of continuous queries to make sure that the original *optimal* query plan maintains its optimality. Once the query optimizer discovers that the original query plan become suboptimal, the query optimizer tunes the suboptimal plan to another optimal one. To support these functionalities, we propose to build and continuously maintain spatio-temporal histograms (ST-Histograms, for short). Instead of monitoring the whole spatial space as in traditional histograms, ST-Histograms are *query-driven* where they monitor only the spatial space that is covered by at least one outstanding continuous spatio-temporal query. ST-histograms continuously maintain *spatio-temporal selectivity* estimations that are used by the query optimizer to decide on the optimality of various candidate query execution plans.

The proposed ST-Histograms start by an initial estimate of the *spatio-temporal selectivity* of the underlying spatial space. The accuracy of the initial estimation is continuously enhanced based on monitoring the execution of the continuous outstanding spatio-temporal queries. One of the attractive features of an ST-Histogram is that its accuracy (and hence the efficiency of the query execution) increases with the increase in the number of outstanding continuous queries. Moreover, the ST-Histogram consults some data mining techniques for periodicity detection (e.g., [13]) to provide better *spatio-temporal selectivity* with less overhead. All the algorithms and ideas in this paper are implemented as part of the PLACE project [4,35] currently being developed at Purdue University.

1.1   Motivation

Spatio-temporal databases provide the ideal infrastructure for keeping track of and answering continuous queries on moving objects. To find an execution plan for a continuous query, the query optimizer needs to know (estimates of) the selectivity of any range that a query covers.

The distribution of the moving objects change with time. For instance, many cars go to downtown from 9am to 5pm leaving the suburbs with fewer cars. At night, most of the cars park, and hence deregister from the database. Consequently the number of the cars in the database is less. Obviously, building a

```
SELECT M.ID
FROM MovingObjects M
WHERE M.Type = "Truck"
INSIDE Area A
```

Fig. 1. Query Q

histogram once and using it for a long period is not enough. We need to maintain a spatio-temporal histogram and to reflect the change in the objects distribution on the histogram.

Not only the density of the moving objects change with time, but also there is some kind of periodicity in their behavior (as time repeats itself). For example, many people travel on weekends, yielding more traffic on the highways on Friday and Sunday evenings than on other week-evenings. Also, lots of traffic and congestion occur during the rush hour everyday. By detecting such patterns in the distribution of the moving objects, we believe that we can enhance the selectivity estimation.

We illustrate the importance of having an ST-Histogram by the following example. Consider the query Q in Figure 1 that returns the ID of any truck whose location is inside an area A. The INSIDE query operator is proposed in [33] to check whether or not a moving object is in a certain range. Initially, at time t_1, the query optimizer finds that the selectivity of the INSIDE operator is less than the selectivity of the WHERE clause (Figure 2(a)). Thus the query optimizer picks up the query execution plan in Figure 2(b) to be used to answer the query. At time t_2, many vehicles enter the area A and this increases the selectivity of the INSIDE operator, and meanwhile the number of trucks decreases (Figure 2(c)). Using an ST-Histogram, the query optimizer is able to recognize that the current plan is suboptimal. The query optimizer calls for changing the current execution plan to the plan in Figure 2(d). Notice that query re-optimization is a non-trivial process, especially that the space of the possible execution plans is large.

1.2 Challenges and Paper Outline

The main challenges for ST-Histograms are the following:

- The large number of the moving objects, which is a computing challenge and a scalability challenge.
- Keeping the overhead of maintaining the ST-Histograms low and not to hurt the execution of the continuous queries.
- Having an accurate selectivity estimation with the frequent change in the data distribution over the time.

The rest of this paper is organized as follows. Section 2 highlights some of the related work in the areas of maintaining histograms and continuous queries. The architecture of our spatio-temporal histogram is given in Section 3. The role

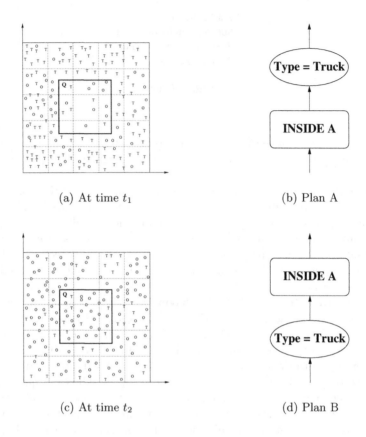

(a) At time t_1

(b) Plan A

(c) At time t_2

(d) Plan B

Fig. 2. Moving vehicles (T = truck, O = other) on the spatial space and the corresponding query execution plan

of the query executor is shown in Section 4. Section 5 introduces the histogram manager and the theory behind constructing and refining ST-Histograms. We discuss some query optimization issues in Section 6. In Section 7, we demonstrate by experiments the accuracy of ST-Histograms. Finally, Section 8 concludes the paper.

2   Related Work

Traditional histograms have been used extensively as a means for selectivity estimation in relational databases (e.g., [10,11,15,17,18,24,28,29,31,32,36,38,39,40,41]). Currently, state-of-the-art histograms are *query-driven* (e.g. [2,21,23]). The main idea is to use a feedback from the query execution engine to estimate the data distribution. Thus, the cost of building the histograms is reduced where histograms are built during the regular process of query execution.

With the emergence of spatial applications, several approaches are proposed for the selectivity estimation of spatial operations (e.g., [3,5,6,7,14,30,43,46]). These approaches deal with the selectivity estimation in various data structures, e.g., selectivity estimation for quad trees [5], selectivity estimation for R-tree [6], and selectivity estimation for point data [8].

Recently, many research efforts focus on developing spatio-temporal histograms for various spatio-temporal operations. Spatio-temporal histograms are first proposed to provide selectivity estimation for predictive spatio-temporal queries [12]. The main focus is on one-dimensional moving objects. The selectivity estimation of multi-dimensional objects is computed by multiplying the selectivity estimations of each single dimension. Similar idea in the context of predictive spatio-temporal queries is introduced in [48]. However, the main focus is on multi-dimensional moving *rectangles*. Other work on selectivity estimation of spatio-temporal queries relies on duality transformation [20], the existence of a secondary index structure [20], or clustering approaches [50]. The state-of-the-art approach for spatio-temporal selectivity estimation is Venn sampling [49]. Venn sampling is a sampling technique that is not based on histograms where it aims to reduce the number of samples needed for *perfect* estimation. The main idea is allow each moving object to be aware of a set of some *pivot queries*. Moving objects update their locations and speed only when they start/cease to satisfy pivot queries.

3 Architecture

Figure 3 gives the big picture. Spatio-temporal queries are submitted to the query optimizer to generate adequate query execution plans. The query optimizer (e.g., System R [45] and Volcano [16]) picks the best execution plan based on the *total cost*. ST-Histograms provide the query optimizer with the selectivity

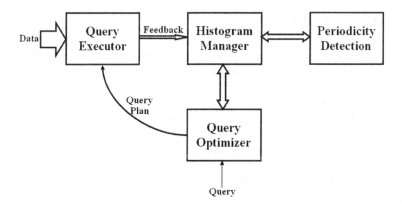

Fig. 3. The big picture

estimates used in calculating the total cost. During the execution of a spatio-temporal query, feedbacks are sent to the histogram manager. These feedbacks are typically the actual query selectivity; i.e., the fraction of the input data that is part of the query answer.

The histogram manager uses these feedbacks to refine the selectivity estimates online. The online refinement serves both the new incoming queries and the outstanding continuous queries. New incoming queries find a more accurate histogram, whereas the execution plan of outstanding continuous queries may be changed adaptively when the environment changes.

The ST-Histograms proposed in this paper consult an online periodicity mining technique (Section 6.1) to see if a periodic pattern appears in the selectivity of any region. Whenever such periodicity is detected, ST-Histograms take it into account to get more accurate selectivity estimates.

4   Query Executor

The spatial space is mapped with an $N \times N$ grid. When a moving object registers with a grid cell, the moving object sends periodic updates about its location. Hence, each grid cell is only aware of the objects inside it. Also, when a query registers with the system, the grid cells that overlap with the query are notified. Only when a change in the moving objects happens in those grid cells, the thread that executes this query is notified. In other words, each grid cell is aware of only the objects that are inside it and the queries that overlaps with it.

The query executor uses the plan provided by the optimizer to execute the query on the input data (Figure 3). Each INSIDE operator keeps track of the ratio of the number of its output tuples to the number of its input tuples. In PLACE [35], this ratio is part of the logic of the INSIDE operator. Thus this does not invoke substantial overhead on the executor. Periodically, such ratio is reported to the histogram manager as the selectivity of the INSIDE operator.

5   Histogram Manager

For streaming applications, we cannot afford storing the incoming data. The continuous query model does not allow for scanning the whole data in order to build the histogram. In fact, an ST-Histogram is built and refined progressively. We use feedback from the query result to update the spatio-temporal histogram online. Periodically, each operator reports the actual selectivity of its monitored range. These statistics are inherently computed in the operators. They do not impose additional overhead on the query executor.

Definition 1. *Dark cell:* *is a grid cell corresponding to a region with which no query overlaps.*

Definition 2. *Lit cell:* *is a grid cell corresponding to a region with which one or more queries overlap.*

Initially, the whole space is assumed to be dark, where darkness represents the unawareness of the selectivity. Queries act as spots of light. They light up a region with their feedback about the region's selectivity. We distribute this selectivity uniformly over the lit region. For the remaining dark regions, we consult the online periodicity mining technique (Section 6.1). If a region exhibits some kind of periodicity, its current selectivity can be estimated according to such periodicity. In other words, the periodic behavior of a region shades this region with little light when the corresponding grid cell is currently dark. The regions that neither fall inside the query regions nor exhibit any periodicity will stay dark. The selectivities of the remaining dark regions are estimated such that they complement the selectivities of those lit and shaded regions. This estimate is uniformly distributed over the ST-Histogram buckets that correspond to the dark regions.

The ST-Histogram is represented by a two-dimensional array. Each element of the array holds the selectivity of the corresponding histogram cell. We assume that a variable holds the total number of moving objects in the database.

5.1 Constructing the Histogram

The ST-Histogram is grid-based of size NxN grid cells. The grid divides the universe uniformly into a number of disjoint cells. We denote the *current* view of the ST-Histogram with $\mathcal{H}$, where $\mathcal{H}[r, c]$ is the selectivity estimate of the grid cell $\mathcal{G}[r, c]$. Starting with all grid cells being dark, the selectivity estimate of each bucket is initialized uniformly according to Equation 1. With the successive feedbacks from the operators, better selectivity estimates are obtained due to a clearer view of the coverage area.

$$\mathcal{H}[r, c] = \frac{1}{N^2} \qquad \text{for all } r, c \in \{1, 2, \dots, N\} \tag{1}$$

A query q is represented by a rectilinear rectangular region $\mathcal{R}_q$. Let $\mathcal{F}_q(r, c)$ be a scalar function that returns the fraction of the grid cell $\mathcal{G}[r, c]$ that is covered by $\mathcal{R}_q$. Hence, the selectivity estimate of a query q is calculated as in Equation 2. Similarly, let $\mathcal{F}_{\mathbf{dark}}(i, j)$ be the dark fraction of $\mathcal{G}[r, c]$. Thus the selectivity estimate of the whole dark area is calculated as in Equation 3.

$$SelEst(q) = \sum_{i=1}^{N} \sum_{j=1}^{N} \mathcal{H}[i, j] \mathcal{F}_q(i, j) \tag{2}$$

$$SelEst(\mathbf{dark}) = \sum_{i=1}^{N} \sum_{j=1}^{N} \mathcal{H}[i, j] \mathcal{F}_{\mathbf{dark}}(i, j) \tag{3}$$

5.2 Refining the Histogram

When the histogram manager receives feedback from the query engine, it updates the histogram to reflect the newly reported statistics. Queries act as light

RefineHistogram($\mathcal{H}, q, \mathcal{S}$)
    $Diff = \mathcal{S} - SelEst(q)$;
    if $Diff > 0$
        $Diff = min(Diff, SelEst(\textbf{dark}))$;
    AddDiff($\mathcal{H}, q, Diff$);
    AddDiff($\mathcal{H}, \textbf{dark}, -Diff$);

AddDiff($\mathcal{H}, q, Diff$)
    for $i = 1 : N$
        for $j = 1 : N$
            $\mathcal{H}(i,j) = \mathcal{H}[i,j] + \mathcal{R}_q(i,j) * Diff$;

Fig. 4. Procedure for refining the ST-Histogram

spots; they eliminate the darkness from a histogram region. The intensity of the light spot a query offers to a histogram region is proportional to the fraction of the histogram region illuminated by the query. When queries overlap, many light spots are directed on the overlapped histogram region. The more the light intensity of a histogram region, the better accuracy of the refinement of the selectivity estimate of this histogram region.

Definition 3. *The normalized rate of a query q for a grid cell $\mathcal{G}[r,c]$ is defined as the ratio between the selectivity estimation of the part of q that overlaps $\mathcal{G}[r,c]$ and the selectivity estimation of q. We denote this normalized rate by $\mathcal{R}_q(r,c)$.*

$$\mathcal{R}_q(r,c) = \frac{\mathcal{F}_q(r,c)\mathcal{H}[r,c]}{\sum_{i=1}^{N}\sum_{j=1}^{N}\mathcal{F}_q(i,j)\mathcal{H}[i,j]}$$

$$= \frac{\mathcal{F}_q(r,c)\mathcal{H}[r,c]}{SelEst(q)} \tag{4}$$

$$\mathcal{R}_q(r,c) = \frac{\mathcal{F}_q(r,c)}{\sum_{i=1}^{N}\sum_{j=1}^{N}\mathcal{F}_q(i,j)} \qquad \text{When } SelEst(q) = 0 \tag{5}$$

The actual selectivity that a query reports is assumed to be distributed uniformly on the query range. Figure 4 gives the procedure to refine the histogram when a feedback from the query engine reports the actual selectivity $\mathcal{S}$ of a query q.

First, the grid cells overlapped by q will have their values changed according to the difference of $\mathcal{S}$ and the current selectivity estimate of q. Typically, this difference is distributed according to the normalized rate of q for each of these grid cells. Next, all the grid cells that have dark portions will be modified similarly to conform with the unity invariance of $\mathcal{H}$ ($\sum_{i=1}^{N}\sum_{j=1}^{N}\mathcal{H}[i,j] = 1$). Hence, the selectivity estimation of the dark portions is the upper bound for the difference when the difference is positive.

Example. We illustrate the histogram refinement by the example given in Figure 5(a). In this example, we have six continuous queries mapped to a 5x5 grid

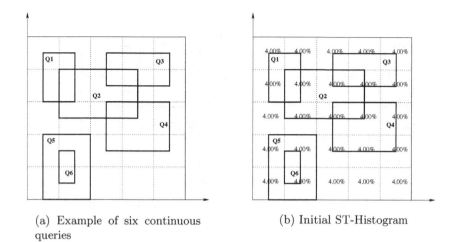

(a) Example of six continuous queries

(b) Initial ST-Histogram

Fig. 5. Example of six continuous queries with an initial histogram

buckets. Each of these queries returns the vehicles of type "Truck" in its covering region (Figure 1). $Q2$ is overlapped by $Q1, Q3$, and $Q4$. $Q6$ is contained in $Q5$. Each grid bucket starts with a selectivity estimate of 4% (Figure 5(b)). Consider when the histogram manager receives feedback from the query executor that Q_1 reports its selectivity as 10%.

The selectivity estimate of Q_1 is 6% according to Equation 2. The difference 10-6 = 4% is distributed among those grid cells overlapped by Q_1. Consider the upper left grid cell C_{ul}. The normalized rate of Q_1 for C_{ul}'s is 0.25*0.04/0.06 = 0.1667. Thus $\mathcal{H}[1, 1] = 0.04 + 0.1667*0.04 = 0.0467 = 4.67\%$. We still need to modify the histogram in order to reach the unity invariance. The increase (or decrease) of the selectivity estimate in a lit region should be accompanied with the decrease (or increase) of the selectivity estimate in the dark region. So, we decrease the selectivity estimate of the dark regions uniformly as much as the increase in the lit regions (4%). For instance, the lower-right bucket, C_{lr}, consists of 1/13.25 of the dark area. The current normalized rate of the **dark** region for C_{lr} is 0.0755. The new value for $\mathcal{H}[5, 5]$ will be 0.04 - 0.0755*0.04 = 0.0370 = 3.70\%. Figure 6(a) gives the histogram after refinement. The upper-left bucket has also a dark portion that results in decreasing $\mathcal{H}[1, 1]$.

Figures 6(b) and 6(f) give the successive updates to the histogram due to the subsequent feedbacks that the histogram manager receives as follows: Q_2 reports 20%, Q_3 reports 15%, Q_4 reports 10%, Q_5 reports 10%, and Q_6 reports 3%.

As a validity check, note that after refining the histogram, we just get a better selectivity estimate for the query. The new estimate is not the same as $\mathcal{S}$. Also, better selectivity estimates are obtained for the dark regions. With the succession of the feedbacks that report (almost) the same selectivity, the estimate for the query converges to the feedback.

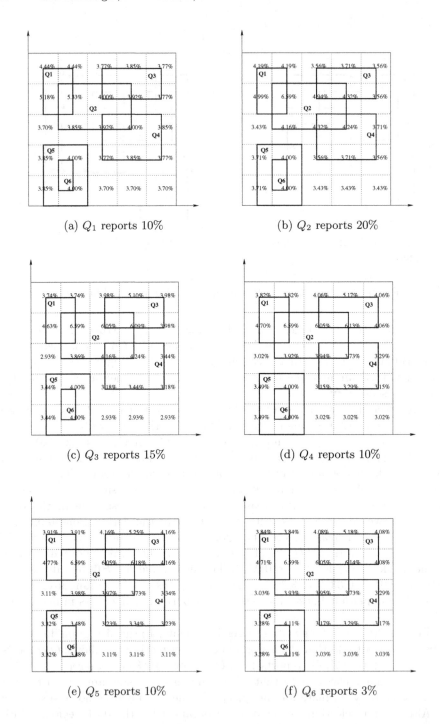

Fig. 6. Successive refinements of the ST-Histogram after receiving the feedback

6   Query Optimization Issues

Inspired by the fact that time repeats itself, we use online periodicity detection to detect periodic patterns in the distribution of the moving objects over the time. This helps in enhancing the selectivity estimation in ST-Histograms. In fact, this is an instance of where data mining can fit evenly in query optimization.

6.1   Online Periodicity Mining

Periodicity mining is defined as the process of discovering frequent periodic patterns in an attempt towards predicting the future behavior in time series data [13]. In our context, a time series for each region in the space is formed by collecting the selectivity values over time, whether they are exact values collected from the queries, or estimates computed by the previous technique. If the selectivity values are quantized into nominal levels, and each level (e.g., high, medium, low) is denoted by a symbol (e.g., a, b, c), then the time series can be considered a sequence over a finite alphabet $\Sigma = \{a, b, c, \cdots \}$.

Periodically, the histogram manager tries to see if any periodic pattern exists in any cell of the histogram. We use the periodicity mining technique in [13]. When we detect a periodic pattern in a dark cell of the grid, we no longer compute its selectivity estimate as being uniformly distributed on all the dark regions. Indeed, we call such cell "a shaded cell". Shaded cells are treated as if there is a query that covers the whole cell. Such query reports the selectivity of the cell according to the periodic pattern of its selectivity.

Note that the periodicity behavior of a region is considered only if this region does not fall inside any query. In other words, the intense of the periodicity light is too little to affect the total light intensity of an already lighted region, yet is enough to shade a dark region.

6.2   Dynamic Query Optimization

The equipment of the DBMS with ST-Histograms enables the existence of an adaptive query processor. With the online update of the ST-Histograms, we are able to detect when the currently executed query plan is suboptimal. In this case, a *need to re-optimize* flag is raised and the optimizer is reinvoked to compute a new optimal query execution plan to continue with. Hence, the already existing queries tune their pipeline for the current workload. Moreover, the new queries get benefit of the enhanced selectivity estimations (versus having one static histogram).

7   Experimental Results

We perform experiments to illustrate the efficiency of predicting the selectivity estimation using ST-Histograms. We use the Network-based Generator of Moving Objects [9] to generate a set of moving objects. The input to the generator is

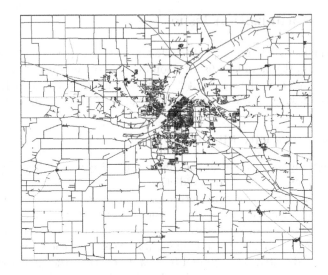

Fig. 7. Road network map of the Greater Lafayette Area City

the road map of the Greater Lafayette Area (Home of Purdue University) given in Figure 7. The output of the generator is a set of moving points that move on the road network of the given city. Moving objects can be cars, cyclists, pedestrians, etc. We generate 5K moving objects and up to 80 continuous queries over 10x10 grid. Each moving object or query reports its new information (if changed) every 10 seconds.

7.1 Effect of Query Size on the Prediction

We measure the accuracy of the selectivity estimation of the existing queries by monitoring the relative error in estimating their selectivities. Let S_i and E_i be the actual and estimated selectivity of the ith query ($1 \leq i \leq M$), respectively, where M is the number of the queries. Equation 6 gives the relative error of estimating the selectivities of the existing queries.

$$\alpha = \sqrt{\frac{1}{M} \sum_{i=1}^{M} \left(\frac{S_i - E_i}{S_i}\right)^2} \tag{6}$$

Figures 8(a) and 8(b) give the performance of estimating the selectivity of the existing queries. α measures how accurate the prediction of the selectivity of the existing queries. From this experiment, we notice that smaller queries suffer from less accuracy than moderate to larger queries. When the queries are too small, many moving objects enter and exit the query range with high frequency. The high frequency of moving in and out the query range results in larger relative error. For a query of size 0.25% of the whole area, the average relative error is

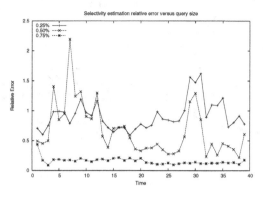

(a) Small sized queries

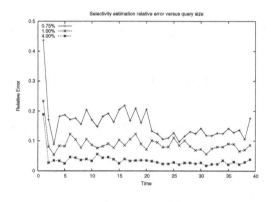

(b) Moderate sized queries

Fig. 8. Performance of estimating the selectivity of the existing queries

91.1% whereas for a query size of 0.5% the average relative error is 67.4%. This error is due to the existence of big dark portions in the grid cells corresponding to the queries. The smaller the dark portion, the less the relative error.

Moderate sized queries do not suffer from the fast moving objects. For moderate sized queries (1% of the whole range), the histogram manager is able to estimate the selectivity of the existing queries with an average relative error of 8.5%. The larger the query size the more accurate the estimation. Queries of size 4% give a relative error of 3.1%.

7.2   Coverage and the Prediction

A new query may come to the system in any area, whether lit or dark. We measure the accuracy of the prediction process for the selectivity of the new

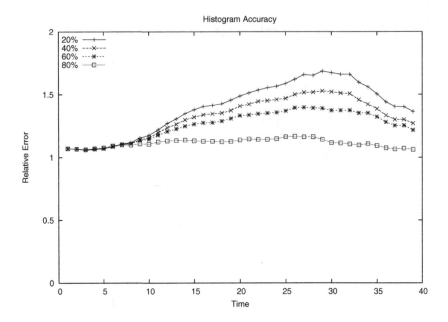

Fig. 9. Performance of estimating the selectivity of the new arriving queries

queries using the accuracy of the whole ST-Histogram. Equation 7 gives the average of the ratio between the estimated and the actual selectivities of all the histogram buckets, where S_{ij} is the actual selectivity of the grid cell $\mathcal{G}[i,j]$.

$$\beta = \frac{1}{N^2} \sum_{i=1}^{N} \sum_{j=1}^{N} \frac{\mathcal{H}[i,j]}{S_{ij}} \tag{7}$$

The selectivity estimate of existing queries is calculated by consulting lit buckets in the ST-Histogram. However, to get a selectivity estimate for a new arriving queries, dark buckets in the ST-Histogram may be consulted, yielding to higher relative error. Figure 9 gives the performance of estimating the selectivity of a bucket in an ST-Histogram (averaged over all buckets). β measures how accurate the selectivity estimates of any new arriving query. The histogram manager is able to estimate the current selectivity for any new query with an average error of 39% when the coverage are is 20%. In this experiment, the value of the error is due to a large number of grid cells totally dark (about 80% of the grid cells). The selectivity estimate of the dark region is uniformly distributed among those dark grid cells. The more spread the queries are on the space, the less number of complete dark cells, the less the relative error. When the coverage is 40%, the error is 31%. For coverage of 60%, the error is 25%, whereas the error is 11% for 80% coverage.

8   Conclusion

In this paper, we explored the usage of spatio-temporal histograms for selectivity estimation of spatio-temporal operators. We presented a general framework for building and continuously maintaining spatio-temporal histograms. The main idea of our proposed spatio-temporal histograms is to use a *continuous feedback* from the outstanding continuous queries to maintain a spatio-temporal histogram for only those parts of the spatial space that are of interest to at least one outstanding continuous query. Parts of the spatial space that are not of interest to any of the outstanding queries do not participate in maintaining the spatio-temporal histograms, thus the overhead of continuously maintaining our spatio-temporal histograms is reduced. Our proposed spatio-temporal histograms utilize periodicity detection techniques to discover temporal periodic patterns. Discovering temporal patterns provides pre-computation of the optimal query plan over the course of execution of continuous queries. Experimental results show that our spatio-temporal histograms provide only 8.5% error for the existing queries of size 1%. An average error of 25% for new queries when the existing query coverage is 60%.

References

1. FCC: Enhanced 911 - Wireless Services. http://www.fcc.gov/911/enhanced/.
2. A. Aboulnaga and S. Chaudhuri. Self-tuning Histograms: Building Histograms without Looking at Data. In *SIGMOD '99: Proceedings of the 1999 ACM SIGMOD international conference on Management of data*, pages 181–192. ACM Press, 1999.
3. N. An, Z.-Y. Yang, and A. Sivasubramaniam. Selectivity Estimation for Spatial Joins. In *ICDE '01: Proceedings of the 17th International Conference on Data Engineering*, pages 368–375. IEEE Computer Society, 2001.
4. W. G. Aref, S. E. Hambrusch, and S. Prabhakar. Pervasive Location Aware Computing Environments (PLACE). http://www.cs.purdue.edu/place/, 2003.
5. W. G. Aref and H. Samet. Estimating Selectivity Factors of Spatial Operations. In *Proceedings, Optimization in Databases - 5th Int'l Workshop on Foundations of Models and Languages for Data and Objects*, pages 31–43, Sep 1993.
6. W. G. Aref and H. Samet. A Cost Model for Query Optimization using R-trees. In *ACM-GIS '94: Proceedings of the ACM Workshop on Advances in Geographic Information Systems*, Dec 1994.
7. A. Belussi, E. Bertino, and A. Nucita. Grid based methods for estimating spatial join selectivity. In *GIS '04: Proceedings of the 12th annual ACM international workshop on Geographic information systems*, pages 92–100. ACM Press, 2004.
8. A. Belussi and C. Faloutsos. Estimating the Selectivity of Spatial Queries Using the 'Correlation' Fractal Dimension. In *VLDB '95: Proceedings of the 21th International Conference on Very Large Data Bases*, pages 299–310. Morgan Kaufmann Publishers Inc., 1995.
9. T. Brinkhoff. A Framework for Generating Network-Based Moving Objects. *Geoinformatica*, 6(2):153–180, 2002.
10. F. Buccafurri and G. Lax. Fast range query estimation by N-level tree histograms. *Data Knowl. Eng.*, 51(2):257–275, 2004.

11. C. M. Chen and N. Roussopoulos. Adaptive Selectivity Estimation using Query Feedback. In *SIGMOD '94: Proceedings of the 1994 ACM SIGMOD international conference on Management of data*, pages 161–172. ACM Press, 1994.

12. Y.-J. Choi and C.-W. Chung. Selectivity Estimation for Spatio-temporal Queries to Moving Objects. In *SIGMOD '02: Proceedings of the 2002 ACM SIGMOD international conference on Management of data*, pages 440–451. ACM Press, 2002.

13. M. G. Elfeky, W. G. Aref, and A. K. Elmagarmid. Incremental, Online, and Merge Mining of Partial Periodic Patterns in Time-Series Databases. *IEEE Transactions on Knowledge and Data Engineering (TKDE)*, 16(3):332–342, 2004.

14. C. Faloutsos, B. Seeger, A. Traina, and J. Caetano Traina. Spatial Join Selectivity using Power Laws. In *SIGMOD '00: Proceedings of the 2000 ACM SIGMOD international conference on Management of data*, pages 177–188. ACM Press, 2000.

15. P. B. Gibbons, Y. Matias, and V. Poosala. Fast Incremental Maintenance of Approximate Histograms. In *VLDB '97: Proceedings of the 23rd International Conference on Very Large Data Bases*, pages 466–475. Morgan Kaufmann Publishers Inc., 1997.

16. G. Graefe and W. J. McKenna. The Volcano Optimizer Generator: Extensibility and Efficient Search. In *ICDE '93: Proceedings of the Ninth International Conference on Data Engineering*, pages 209–218. IEEE Computer Society, 1993.

17. P. J. Haas and A. N. Swami. Sequential Sampling Procedures for Query Size Estimation. In *SIGMOD '92: Proceedings of the 1992 ACM SIGMOD international conference on Management of data*, pages 341–350. ACM Press, 1992.

18. P. J. Haas and A. N. Swami. Sampling-Based Selectivity Estimation for Joins Using Augmented Frequent Value Statistics. In *ICDE '95: Proceedings of the Eleventh International Conference on Data Engineering*, pages 522–531. IEEE Computer Society, 1995.

19. M. Hadjieleftheriou, G. Kollios, D. Gunopulos, and V. J. Tsotras. On-Line Discovery of Dense Areas in Spatio-temporal Databases. In *SSTD '03: Proceedings of the 8th International Symposium on Spatial and Temporal Databases*, pages 306–324, Jul 2003.

20. M. Hadjieleftheriou, G. Kollios, and V. J. Tsotras. Performance Evaluation of Spatio-temporal Selectivity Estimation Techniques. In *SSDBM '03: Proceedings of the 15th International Conference on Scientific and Statistical Database Management*, pages 202–211, 2003.

21. I. F. Ilyas, H. G. Elmongui, and W. G. Aref. Adaptive Processing of Ranking Queries. Technical Report CSD-TR-05-002, Purdue University, 2005.

22. G. S. Iwerks, H. Samet, and K. Smith. Continuous K-Nearest Neighbor Queries for Continuously Moving Points with Updates. In *VLDB '03: Proceedings of the 29th International Conference on Very Large Data Bases*, Sep 2003.

23. N. Kabra and D. J. DeWitt. Efficient Mid-query Re-optimization of Sub-optimal Query execution Plans. In *SIGMOD '98: Proceedings of the 1998 ACM SIGMOD international conference on Management of data*, pages 106–117. ACM Press, 1998.

24. R. P. Kooi. *The Optimization of Queries in Relational Databases*. PhD thesis, Case Western Reserve University, 1980.

25. D. Kwon, S. Lee, and S. Lee. Indexing the Current Positions of Moving Objects Using the Lazy Update R-tree. In *MDM '02: Proceedings of the Third International Workshop on Multimedia Data Mining*, pages 113–120, Jan 2002.

26. I. Lazaridis, K. Porkaew, and S. Mehrotra. Dynamic Queries over Mobile Objects. In *EDBT '02: Proceedings of the 8th International Conference on Extending Database Technology*, pages 269–286, Mar 2002.

27. M.-L. Lee, W. Hsu, C. S. Jensen, and K. L. Teo. Supporting Frequent Updates in R-Trees: A Bottom-Up Approach. In *VLDB '03: Proceedings of the 29th International Conference on Very Large Data Bases*, Sep 2003.

28. R. J. Lipton and J. F. Naughton. Query Size Estimation by Adaptive Sampling. *Journal of Computer and System Sciences*, 51(1):18–25, 1995.

29. R. J. Lipton, J. F. Naughton, and D. A. Schneider. Practical Selectivity Estimation through Adaptive Sampling. In *SIGMOD '90: Proceedings of the 1990 ACM SIGMOD international conference on Management of data*, pages 1–11. ACM Press, 1990.

30. N. Mamoulis and D. Papadias. Selectivity Estimation of Complex Spatial Queries. In *SSTD '01: Proceedings of the 7th International Symposium on Advances in Spatial and Temporal Databases*, pages 155–174. Springer-Verlag, 2001.

31. G. S. Manku, S. Rajagopalan, and B. G. Lindsay. Approximate Medians and other Quantiles in One Pass and with Limited Memory. In *SIGMOD '98: Proceedings of the 1998 ACM SIGMOD international conference on Management of data*, pages 426–435. ACM Press, 1998.

32. Y. Matias, J. S. Vitter, and M. Wang. Wavelet-based histograms for selectivity estimation. In *SIGMOD '98: Proceedings of the 1998 ACM SIGMOD international conference on Management of data*, pages 448–459. ACM Press, 1998.

33. M. F. Mokbel and W. G. Aref. GPAC: Generic and Progressive Processing of Mobile Queries over Mobile Data. In *MDM '05: Proceedings of the Third International Workshop on Multimedia Data Mining*, May 2005.

34. M. F. Mokbel, W. G. Aref, S. E. Hambrusch, and S. Prabhakar. Towards Scalable Location-aware Services: Requirements and Research Issues. In *ACM-GIS '03: Proceedings of the 11th ACM International Symposium on Advances in Geographic Information Systems*, pages 110–117, Nov 2003.

35. M. F. Mokbel, X. Xiong, W. G. Aref, S. Hambrusch, S. Prabhakar, and M. Hammad. PLACE: A Query Processor for Handling Real-time Spatio-temporal Data Streams (Demo). In *VLDB '04: Proceedings of the Thirtieth International Conference on Very Large Data Bases*, Aug 2004.

36. M. Muralikrishna and D. J. DeWitt. Equi-depth Multidimensional Histograms. In *SIGMOD '88: Proceedings of the 1988 ACM SIGMOD international conference on Management of data*, pages 28–36. ACM Press, 1988.

37. T. Nadeem, S. Dashtinezhad, C. Liao, and L. Iftode. TrafficView: A Scalable Traffic Monitoring System. In *MDM '04: Proceedings of the 5th IEEE International Conference on Mobile Data Management*, Jan 2004.

38. G. Piatetsky-Shapiro and C. Connell. Accurate Estimation of the Number of Tuples Satisfying a Condition. In *SIGMOD '84: Proceedings of the 1984 ACM SIGMOD international conference on Management of data*, pages 256–276. ACM Press, 1984.

39. V. Poosala. *Histogram-based Estimation Techniques in Database Systems*. PhD thesis, University of Wisconsin at Madison, 1997.

40. V. Poosala, P. J. Haas, Y. E. Ioannidis, and E. J. Shekita. Improved Histograms for Selectivity Estimation of Range Predicates. In *SIGMOD '96: Proceedings of the 1996 ACM SIGMOD international conference on Management of data*, pages 294–305. ACM Press, 1996.

41. V. Poosala and Y. E. Ioannidis. Selectivity Estimation Without the Attribute Value Independence Assumption. In *VLDB '97: Proceedings of the 23rd International Conference on Very Large Data Bases*, pages 486–495. Morgan Kaufmann Publishers Inc., 1997.

42. S. Prabhakar, Y. Xia, D. V. Kalashnikov, W. G. Aref, and S. E. Hambrusch. Query Indexing and Velocity Constrained Indexing: Scalable Techniques for Continuous Queries on Moving Objects. *IEEE Transactions on Computers*, 51(10):1124–1140, Oct 2002.

43. G. Proietti and C. Faloutsos. Selectivity Estimation of Window Queries for Line Segment Datasets. In *CIKM '98: Proceedings of the seventh international conference on Information and knowledge management*, pages 340–347. ACM Press, 1998.

44. S. Saltenis, C. S. Jensen, S. T. Leutenegger, and M. A. Lopez. Indexing the Positions of Continuously Moving Objects. In *SIGMOD '00: Proceedings of the 2000 ACM SIGMOD international conference on Management of data*, pages 331–342, May 2000.

45. P. G. Selinger, M. M. Astrahan, D. D. Chamberlin, R. A. Lorie, and T. G. Price. Access Path Selection in a Relational Database Management System. In *SIGMOD '79: Proceedings of the 1979 ACM SIGMOD international conference on Management of data*, pages 23–34. ACM Press, 1979.

46. C. Sun, D. Agrawal, and A. E. Abbadi. Selectivity Estimation for Spatial Joins with Geometric Selections. In *EDBT '02: Proceedings of the 8th International Conference on Extending Database Technology*, pages 609–626. Springer-Verlag, 2002.

47. Y. Tao, D. Papadias, and Q. Shen. Continuous Nearest Neighbor Search. In *VLDB '02: Proceedings of the 28th International Conference on Very Large Data Bases*, pages 287–298, Aug 2002.

48. Y. Tao, D. Papadias, and J. Sun. The TPR*-Tree: An Optimized Spatio-temporal Access Method for Predictive Queries. In *VLDB '03: Proceedings of the 29th International Conference on Very Large Data Bases*, Sep 2003.

49. Y. Tao, D. Papadias, J. Zhai, and Q. Li. Venn Sampling: A Novel Prediction Technique for Moving Objects. In *ICDE '05: Proceedings of the 21st International Conference on Data Engineering*, April 2005.

50. Q. Zhang and X. Lin. Clustering Moving Objects for Spatio-temporal Selectivity Estimation. In *CRPIT '27: Proceedings of the fifteenth conference on Australasian database*, pages 123–130. Australian Computer Society, Inc., 2004.

GAMMA: A Framework for Moving Object Simulation*

Haibo Hu and Dik-Lun Lee

Department of Computer Science,
The Hong Kong University of Science and Technology
{haibo, dlee}@cs.ust.hk

Abstract. Simulating user mobility is crucial for mobile computing and spatial database research. However, all existing moving object generators assume a fixed and often unrealistic mobility model. In this paper, we represent the moving behavior as a trajectory in the location-temporal space and propose two generic metrics to evaluate a trajectory dataset. In this context, trajectory generation is treated as an optimization problem and a framework, GAMMA, is proposed to solve it by the genetic algorithm. We demonstrate GAMMA's practicability and flexibility by configuring it for two specific simulations, namely, cellular network trajectory and symbolic location tracking. The experimental results show that GAMMA can efficiently and robustly produce high quality moving object datasets for various simulation objectives.

1 Introduction

Simulation is a common practice for performance evaluation due to its low cost and ease of realization. Simulating user mobility has been widely accepted for the evaluation of moving object databases (MOD). However, existing spatio-temporal dataset generators [1,2,3] always consider the moving behavior either as a purely random walk [4] or a predefined mobility model [5], although some sophisticated generators further concern other factors such as obstacles or road networks.

In this paper, each moving object i is represented by a **trajectory** $T(i)$ in the location-temporal space. The space can be considered two-dimensional: the x-axis denotes the timeline whereas the y-axis denotes the location, i.e., i's spatial property. In this context, simulating moving object behavior is equivalent to generating satisfactory trajectories $T(i)$ in the location-temporal space according to the simulation objective. The degree of satisfaction, denoted as **fitness** of the trajectories, is two-fold: **individual** fitness and **global** fitness. The former designates how close each individual trajectory is to the expected moving behavior. And the latter designates how close the set of trajectories follow a certain distribution, i.e., the mobility model. All previous generators only considered

* This work is supported by the Research Grants Council, Hong Kong SAR under grants HKUST6277/04E and CITYU1204/03E.

global fitness and required the mobility model to be in a closed mathematical form. However, in many real-life applications, no closed form mobility model can be applied. In this paper, we consider the trajectory generation problem from an optimization perspective. Specifically, all possible trajectories comprise a search space, and the generator's objective is to find a set of trajectories which have the optimal values for both individual and global fitness.

We then apply genetic algorithm (GA) to solve the optimization problem and hence obtain the trajectories. The proposed generator framework is called **GAMMA**, which stands for "**G**enerating **A**rtificial **M**odeless **M**ovement by genetic-**A**lgorithm". The genetic algorithm (GA) is known as a robust global optimization technique for both numerical and non-numerical problems [6]. The distinguishing characteristic of GA from other optimization techniques is that an entire population of solutions, rather than a single solution, evolves, which makes GA an ideal optimizer for the moving object generation problem.

The rest of the paper is organized as follows. Section 2 introduces existing work and some preliminaries of genetic algorithms. Section 3 overviews the GAMMA framework and addresses the configuration of its components. In Sections 4 and 5, we demonstrate GAMMA by devising the generators respectively for two simulations with specific purposes. The empirical performance analysis of the two generators is conducted in Section 6.

2   Related Work and Preliminaries

2.1   Generating Moving Objects with the Predefined Mobility Model

In order to analyze and evaluate wireless (PCS or ad-hoc) networks, a variety of user mobility models have been proposed. They can be categorized into *geometric* and *symbolic* models, which are suitable for Euclidean space and cellular network space, respectively. The most fundamental geometric mobility model is the **random period** model [7]. In this model, each object selects a direction θ from the range $[0, 2\pi]$ and a speed from a user-defined distribution. Then it moves in the chosen direction and speed for a random period of time. The object then halts, selects a new direction and speed, and resumes its movement.

Many variations such as the **random direction** model originate from it. The most widely used model is the **random waypoint** model [8]. In this model, each object selects a random point in the simulation area as its destination, and a speed v from an input range $[vmin, vmax]$. The object then moves to the destination at the chosen speed. When it reaches the destination, it rests for a random period of time and selects a new destination and speed to resume its movement. However, the *random waypoint* model suffers from three defects: (1) the objects tend to congregate in the center of the simulation area, resulting in non-uniform network density; (2) the average speed of the objects decreases until converging to some long-term average; and (3) the objects are free to move in the space, which is not realistic. In order to address issues (1) and (2), Yoon et al. proposed a framework which stabilizes the mobility model by choosing the initial speeds from a steady-state distribution (which can be derived analytically) and subsequent speeds from the original speed distribution [9]. Jardosh et al. addressed issue (3) by allowing the existence of obstacles and restricting the objects' movement to paths which are constructed by the Voronoi diagram of obstacle vertices [7].

Symbolic mobility models exist in cellular networks. In the basic model, for each mobile user, a particular cell in the network is chosen as the destination. Whenever the user leaves the current cell, he/she moves to the neighboring cell which is closest to the destination. If the mobile user is already in the destination cell, after a certain period of time it will move to one of the neighboring cells. This continues until the next destination is chosen. Another particular model for cellular networks is the **hexagonal random walk** model [5]. In this model, the mobile user resides in a cell for a period of time with a known distribution and then moves to the next cell, which is selected from the neighboring cells with equal probabilities. Though simple and easy to implement, these models are far from real-life user moving behavior.

From a spatio-temporal database perspective, the moving object trajectories form a spatio-temporal dataset. The first and the most significant spatio-temporal dataset generator, *GSTD*, was proposed by Theodoridis et al. [1]. It defines a set of parameters that control the trajectory of the moving objects: (1) the **duration** of an object instance, i.e., the change of timestamps between consecutive instances; (2) the **shift** of an object, i.e., the change of an object's spatial location; and (3) the **resizing** of an object, i.e., the change of an object's size. These parameters at the new timestamp t' is calculated by summing up the respective current values and the respective δ values of these parameters. The initial and δ values of these parameters, as well as other parameters, can be defined by the user to generate desirable datasets. The authors extended this work to generate more realistic moving objects by introducing the notion of clustered movement and a new parameter *dinterval* (i.e., directed movement interval, an interval in which the object's $\delta(shift)$ keeps constant) [3]. By this means, the authors showed some more realistic moving behavior, such as *preferred movement*, *group movement*, and *obstructed movement*. Brinkhoff further proposed a framework for generating road network-based moving objects [2]. The moving behavior of the objects is influenced by such characteristics as the maximum speed and the maximum capacity of roads, the influence of nearby moving objects , and the route of an object.

As a conclusion, all these moving object generators require the object's behavior to follow a simple and close-form mobility model. This restriction makes the generated datasets significantly different from realistic ones and heavily confines the generators' scope of usage.

2.2   Genetic Algorithm

The genetic algorithm (GA) is a computational model inspired by evolution [6]. The algorithm encodes a solution to a specified problem on a chromosome-like data structure. GA makes a set of solutions, called the **population**, evolve into better solutions by **selection** to preserve good solutions, and by **recombination** to generate new solutions in the solution space. In this section, we show some preliminaries of a genetic algorithm; a more thorough tutorial can be found in [10].

To operate on the solutions, GA encodes every solution into a fixed-length string, called a **chromosome**. After *initialization*, the population of the solution goes through a series of evolutions, called **generations**. The solutions in the next generation are obtained by **selection**, **crossover**, and **mutation** on the solutions in this generation:

1. **selection**: each chromosome is evaluated by a **fitness** function, which measures how good the corresponding solution is to the problem. The probability that a chromosome is selected for the next generation is inversely proportional to its fitness value ranking in the whole population.
2. **crossover**: the binary operator breaks down two chromosomes at the same positions and recombines the segments to form two new chromosomes. The number of breakdown positions can be one or more.
3. **mutation**: the unary operator randomly selects a bit in the chromosome string and changes its value.

The crossover and mutation are performed with certain probabilities.

Though genetic algorithms are often used as function optimizers, the range of problems to which they can be applied is quite broad. GA has already been used to enhance the simulation effect in such areas as chip design and network protocol analysis. In [11], GA guided the random input sequences generation to achieve the same chip design verification coverage with fewer input sequences. In [12], Baldi et al. evaluated the TCP protocol by identifying network traffic patterns that lead to network sensitiveness and bottleneck. GA was used to generate the background traffic, which turned out to be a more effective network underminer than the traffic generated by statistical simulation.

3   GAMMA Framework Overview

Figure 1 illustrates the GAMMA architecture. The six gray boxes, namely, apriori knowledge, selection, crossover, mutation, evaluation and refinement, and decoding, are the **configurable components** in GAMMA. They exhibit generality and flexibility for this framework.

The modeling operation from the trajectory T to a chromosome is called **encoding**, which involves choosing appropriate *events* to represent T's behavior.[1] An event is an activity which the object performs to change its moving behavior. The inverse operation is called **decoding**, which involves interpolating the locations at a time when no events occur.

---

[1] In the sequel, unless otherwise stated, *chromosome* and *trajectory* are used interchangeably.

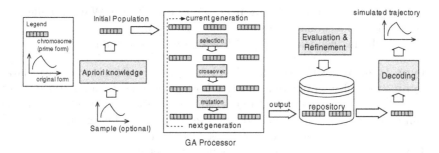

Fig. 1. The GAMMA Architecture

The core component of GAMMA is the **GA processor**, which performs the selection, crossover and mutation operators. The chromosome population evolve according to *individual fitness* for a certain number of generations before being outputted to a **chromosome repository** where they are further evaluated and refined according to *global fitness*. The initial population is obtained from some sample trajectories.

The *chromosome repository* has two purposes: (1) to buffer all the chromosomes generated so far (since the population in GA normally cannot be too large, the GA processor has to run many **iterations** to generate a required number of trajectories); (2) to store the set of intermediate chromosomes for further evaluation and refinement according to its global fitness.

3.1   Configuring GAMMA Components

The six configurable components control the output trajectories. In this subsection, we propose their configuration guidelines for various simulation objectives

Encoding and Decoding. Encoding should select events, i.e., the "keyframes", as much distinguishing as possible so that the chromosome reflects the mobility characteristics. Thus a good event is where/when the mobility behavior of the trajectory is significantly altered (hereby interpolation is imprecise). The definition of altering is dependent on the simulation objective. There is an obvious trade-off between the preciseness and conciseness for the encoding. On the other hand, decoding need to interpolate the rest of the points on the trajectory. The simplest method is to linearly interpolate the locations between two consecutive events with a constant velocity.

Fitness Function and Selection. The essential part of GAMMA is the fitness function $\mathcal{F}$, which is a mapping from the chromosome domain M to the non-negative real domain, i.e., $\mathcal{F} : M \rightarrow R^+$. The fitness function is always associated with the distance measure between two trajectories, denoted as $Dist$. To compute $Dist(T_1, T_2)$ based on time series, we: (1) assign each *location* a real value using a *space filling curve*; (2) translate the T_1 and T_2 into two conventional real-valued time-series; and (3) use the **Segmented Dynamic Time Warping** method [13] to calculate the similarity distance between T_1 and T_2.

We then distinguish three types of fitness functions as follows: [2]

- **Type 1**: the fitness function only involves trajectory T and some static data.
- **Type 2**: the fitness function involves both trajectory T and a static set of trajectories I. For example, to generate trajectories which resemble those sample trajectories I, we may define the fitness function as the inverse of the k-th nearest similarity distance between T and I.
- **Type 3**: the fitness function involves both trajectory T and the entire set of trajectories in the chromosome repository. For example, to generate a set of trajectories that have similar mobility behavior, the fitness should be defined as the inverse of the k-th nearest similarity distance between T and those in the chromosome repository.

For the Type 3 fitness function, the naive way of finding the kth nearest distance is through a sequential scan. However, as the repository size gets larger, sequential scan becomes less efficient. In this regard, we first embed all the chromosomes in an n-dimensional Euclidean space using **FastMap** [14]. Then we perform a K-nearest-neighbor search ($K \geq k$) in the embedding space. The resulting set of chromosomes is the cutdown candidate set for the sequential search.

Crossover and Mutation. In GAMMA, the chromosome is not in a canonical form (i.e., each chromosome bit is either bit 1 or 0), it doesn't even have a fixed length. The crossover and mutation operators are based on *events* rather than bits.

Mutation selects a fixed-length time period, removes the events within the period, and then generates new events. During the removal, some events may overlap with, but not completely reside in this range. If this occurs, we stochastically decide whether they are removed (i.e., the mutation time period is prolonged) or retained (i.e., the mutation time period shrinks).

The crossover operation selects a time point t and based on this pivot, switches the events of two chromosomes c_1 and c_2. There are four possible cases: (1) t is within the time period of event $i \in c_1$ and event $j \in c_2$ while $i.location = j.location$; (2) t is not within the time period of any event in c_1 or c_2; (3) same as (1), except that $i.location \neq j.location$; (4) t is within the time period of an event in c_1 (or c_2), but not within that of any event in c_2 (or c_1). The crossovers in (1)(2) are valid whereas those in the last two are not, since *events* are considered as atomic and should not be further divided for swapping. Obviously, the domain of valid t is $intersection(\hat{c}_1, \hat{c}_2)$ where $\hat{c}_1$ is an augmented trajectory of c_1 which adds a dummy event $\langle dummy\_location, i.endtime, j.begintime \rangle$

---

[2] To prevent some "super-best" chromosomes from dominating the population, the probability of a chromosome being selected for the next generation is inversely proportional to its fitness value ranking among the population, rather than to the fitness value itself.

between each two consecutive events i and j. The crossover has a potential problem if $intersection(\hat{c}_1, \hat{c}_2)$ is null. If this occurs, the crossover operation is not performed. Given the fact that the dummy event is ubiquitous, such a scenario is quite rare.

Handling Constraints and Other Apriori Knowledge. In practice, the user's moving behavior should conform to certain rules and heuristics specific to each application. In [6], the techniques of handling various types of constraints in genetic algorithms were studied. The authors listed four alternatives, namely, eliminating infeasible chromosomes, repairing infeasible chromosomes, preserving feasibility by special mutation/crossover operators, and transforming the search space into a regular one. In GAMMA, we incorporate the constraints and other heuristics as follows:

1. The mutation and crossover operation is augmented to be aware of these rules and heuristics. For example, if a piece of apriori knowledge is "the user goes to the library at most once a day," the mutation operator should respect this rule and check existing library event before creating a new one.
2. After crossover and mutation, the trajectories might invalidate certain apriori rules or heuristics, so a **chromosome repairing** scheme is devised to correct them.
3. The fitness function should be aware of apriori rules and heuristics. For example, a trajectory that violates certain constraints is assigned the lowest possible fitness value as a penalty.

Evaluation and Refinement. The weakness of most existing spatio-temporal dataset generators is the lack of a uniform metric to evaluate the generated datasets. In GAMMA, we evaluate the datasets in terms of *individual fitness* and *global fitness*:

1. **Evaluation on Individual Fitness**: by the mean and standard deviation of all the individual fitness values in the dataset.
2. **Evaluation on Global Fitness**: by the distribution of the set of trajectories.

The *global fitness* metric measures how the dataset conforms to a specified distribution, i.e., mobility model. Either the trajectories themselves or their projections on the spatial/temporal/arbitrary axis are the subjects of distribution. We distinguish two special types of distributions in GAMMA. For *uniform* (random) distribution, the global fitness is derived in the next subsection, which proposes a **hash test** as a general technique to determine whether the trajectories are significantly randomly distributed. For *normal* distributed trajectories, the global fitness is derived from a statistical T-test, which tests if a dataset conforms to a given normal distribution. In this case, the global fitness is defined as the statistical variable $T = \frac{\bar{X} - \mu_0}{S/\sqrt{n}}$.

The evaluation is followed by the refinement process, in which "bad" chromosomes that lead to low individual or global fitness values are removed or modified.

Hash Test. The *hash test*, also known as the *collision test*, is a general technique to test the randomness of a sequence of *keys* [15]. It creates a hash table containing k memory addresses. Imagine a perfectly uniform hashing function which hashes n keys ($k \gg n$) into the table. The expected number of collisions C follows the *Poisson* distribution with mean value $\lambda = n^2/2k$ [16]. If the n keys are perfectly random, the number of hashing collisions x should approximate this mean value.

In GAMMA, a *signature* function first maps the trajectory space to an integer domain, which serves as the domain of the memory address of the hash table. The hash test then collects the number of the actual collisions x and denotes $\frac{1}{x-\lambda}$ as the *global fitness* value. If this value is significantly low, the chromosome with a lower individual fitness value in each conflicted chromosome pair is modified or removed with probability $\frac{\lambda}{x}$ to resolve the collision. After the refinement, x approximates λ and the set of chromosomes is guaranteed to be random.

4   Generating Trajectories in a Cellular Space

In this and the next section, we present two concrete examples of how to configure GAMMA to generate satisfactory moving trajectories for different simulation objectives.

In spatial database research, the result of a nearest neighbor query may contain not only the nearest object, but also a vicinity region where the result is still valid [17]. It saves communication costs for subsequent queries which fall in this region. As a similar approach, a "time-parameterized nearest neighbor" (TPNN) query returns the query result's valid time period [18]. These schemes essentially deal with a cellular space, i.e., the *Voronoi Diagram*. The performance depends on how frequently the user passes through the borders of Voronoi cells[3]. As such, to measure the worse-case performance, we need to generate moving trajectories that frequently cross cell borders.

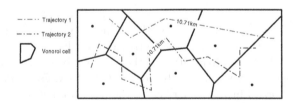

Fig. 2. Moving Trajectories in a Cellular Network

---

[3] The higher the frequency, the less effective these schemes are, as it becomes more probable that the next query goes beyond the valid region or the valid time period for this query. Figure 2 illustrates this proposition. Although trajectories 1 and 2 have the same length, " 2" passes through many more Voronoi cells (11 vs. 4). Therefore, many more queries can be answered at the client side in "1" than in "2", thanks to the semantic cache or the TPNN scheme.

The cellular space is the only input for GAMMA. It acts as the static data to compute the individual fitness. To simplify the problem, we assume that: (1) users are points; (2) their speed is constant; (3) they are free to move in the space.

4.1 Trajectory Encoding/Decoding and Fitness Function

As the speed is fixed, the user can change his/her moving direction only. We let the *events* be the time points when the user takes a turn. These positions are called **turning points**. Since the speed is fixed, the begintime and endtime of each event, can be derived from the turning points. Therefore, the projection on the spatial axis suffices to represent the trajectory. In other words, the trajectory is represented by a polyline segment interconnected by turning points.

Let $\mathcal{C}$ denote the cellular space and $E(\mathcal{C})$ denote the edges that form $\mathcal{C}$. Likewise, let $E(T) = \{T.e_1, T.e_2, ...T.e_L\}$ denote the set of line segments in trajectory T. Let $cross(a, b)$ denote the number of intersection points between line (or lines set) a and b. The fitness function $\mathcal{F}(T)$, which denotes the number of times T goes across the border of the cells in a unit length, is derived as:

$$\mathcal{F}(T) = \frac{cross(E(\mathcal{C}), E(T))}{T.length} = \frac{\sum_{i=1}^{L} cross(E(\mathcal{C}), T.e_i)}{T.length} \tag{1}$$

4.2 Mutation and Crossover

We simplify the mutation by allowing it to modify one event at a time. Three alternatives are possible: to add, remove, or to update a turning point. To crossover two trajectories, the sequences of the turning points are swapped at a certain time point $t \in intersection(\hat{T}_1, \hat{T}_2)$, which is simply the set of geometric points where T_1 and T_2 intersect. Formally, let T denote the chromosome of the trajectory. $T = \{p_1, p_2, ..., p_L\}$ where p_i are the turning points of T.

$$mutation(T) : T' = \{p_1, p_2, ..., p_i, p^*, p_{i+1}, ..., p_L\}$$
$$or \quad T' = \{p_1, p_2, ..., p_{i-1}, p_{i+1}, ..., p_L\}$$
$$or \quad T' = \{p_1, p_2, ..., p_{i-1}, p'_i, p_{i+1}..., p_L\}$$
$$crossover(T1, T2) : T1' = \{p1_1, ..., p1_i, p2_{i+1}, ..., p2_{L_1}\}$$
$$T2' = \{p2_1, ..., p2_j, p1_{j+1}, ..., p1_{L_2}\},$$

where line segment $p1_i p1_{i+1}$ intersects $p2_i p2_{i+1}$.

4.3 Evaluation and Refinement

To remove "bad" trajectories, we set the individual fitness threshold to the mean fitness value $\overline{\mathcal{F}}$. There are two ways to obtain $\overline{\mathcal{F}}$: empirically or by analysis. The former adopts the Monte Carlo method which randomly generates N trajectories T_i and let $\overline{\mathcal{F}} = \sum_{i=1}^{N} \mathcal{F}(T_i)/N$. The latter develops a mathematical model as follows. Let $Dia(i)$ [4] denote the mean distance between two points on the border of a cell i and $area(i)$ denote the area of i. We have:

$$\overline{\mathcal{F}} \approx \frac{1}{\overline{Dia(C)}} = \frac{\sum_{i=1}^{C} area(i)}{\sum_{i=1}^{C} area(i) * Dia(i)}, \qquad (2)$$

where C is the cellular space. To derive $Dia(i)$ and $area(i)$, we approximate the cell i by a circle $\mathcal{C}$ with radius r. Thus, $Dia(i)$ is "the average distance between two points on the circumference of Cir."

$$Dia(i) \approx \frac{r \cdot \int_0^{\pi} \sqrt{2 - 2\cos\theta}\, d\theta}{\int_0^{\pi} d\theta} = \frac{4r}{\pi}$$

And $area(i)$ is the area of $\mathcal{C}$, i.e., $area(i) = \pi r^2$. There are two approximations of $\mathcal{C}$, namely "optimistic" and "pessimistic," whose radii are denoted as r_{opt} and r_{pess}, respectively. The optimistic circle, which is the inscribed circle of i, achieves an upper bound estimation of $\overline{\mathcal{F}}$, whereas the pessimistic one, i.e., the circumcircle of i, achieves a lower bound estimation of $\overline{\mathcal{F}}$. Whichever approximation we apply, the mean individual fitness is:

$$\overline{\mathcal{F}} \approx \frac{\sum_{i=1}^{C} \pi r_i^2}{\sum_{i=1}^{C} 4 r_i^3} \qquad (3)$$

Regarding the global fitness, since simulation coverage is the main concern, the trajectories should distribute randomly throughout the space. Thus, the hash-test-based refinement is performed. The signature function is devised as follows. First, the space is partitioned into m-by-m even-sized rectangles. Then each chromosome is represented by an m^2-bit string, where each bit corresponds to one rectangle. A "1" on this bit means "the trajectory passes through this rectangle" while "0" means the opposite. Finally the signature is the decimal value of this binary string.

5   Simulating Real-Life Symbolic Moving Behavior

In this section, we aim to generate a set of real-life symbolic trajectories for a single user. The trajectories are symbolic in that all the locations are symbolic, such as campus, office, cafeteria, home, etc. Symbolic locations preserve the semantics of the user's moving behavior better than geometric coordinates. The generated trajectories are used for the simulation of applications such as "location prediction" and "context-aware mobile phone." The trajectories are "real-life" in that: (1) they conform to a set of mobility patterns which hide in the sample trajectories obtained from real-life location tracking; (2) they conform to the real-life constraints and heuristics.

---

[4] The abbreviation for "Diameter".

The difficulty of designing such a dataset generator is that a user's real-life moving behavior is governed by: (1) a variety of heuristics and rules from different disciplines; (2) his/her underlying mobility model, which has no perfect mathematical form. As too many factors concern the behavior, we assume that:

1. The set of locations $\mathcal{G}$ in which the user might stay is known.
2. The user is out of the scope of all *leaf locations* for the time periods not covered by any event.
3. The distances between the leaf locations in terms of transportation time are known and stored as edges in a complete graph, called a **distance graph**. Figure 3(a) illustrates set of locations and the distance graph. The digits on each line denote the time cost in minutes.[5]

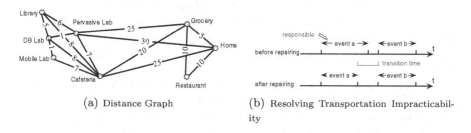

(a) Distance Graph   (b) Resolving Transportation Impracticability

Fig. 3. Real-Life Symbolic Trajectory Generation

5.1   Trajectory Encoding/Decoding and Fitness Function

Each event e designates that the user stays at $e.location$ from $e.begintime$ to $e.endtime$. Furthermore, the trajectory conforms to the **transportation practicability** constraint, i.e., the time interval between two consecutive events should be large enough to allow the user to move between the two locations:

As the objective is to generate trajectories with the same mobility patterns as the samples, we define the individual fitness as the *similarity* between the trajectory and the samples, i.e.,

$$\mathcal{F}(T) = \{KMIN_{T_i \in \mathcal{S}} Dist(T, T_i)\}^{-1},$$

where $\mathcal{S}$ is the set of sample trajectories and $KMIN$ is the operator to retrieve the kth minimal. A large fitness value means that T is close to some samples, whereas a small value means that it is probably an outlier.

5.2   Mutation, Crossover and Refinement

The mutation operator may violate the *transportation practicability* constraint. We repair the chromosome by randomly assigning the responsibility of maintaining the practicability to one of the events. As illustrated in Figure 3(b), event

---

[5] To improve legibility, we do not depict all distances in the figure. If the edge between a and b is missing, $distance(a, b)$ is equivalent to the shortest path from a to b.

a is the responsible event; it shrinks its duration to leave the intermission time between a and b equivalent to the a, b transportation time defined in the distance graph.

As for refinement, the fitness threshold is set to the mean individual fitness value $\overline{\mathcal{F}}$ of all possible chromosomes. As it is infeasible to derive $\overline{\mathcal{F}}$ analytically in the original chromosome space, we use *FastMap* to embed all sample chromosomes into an n-dimensional Euclidean space $\mathcal{R}^n$. Given a uniform distribution of the N embedding points (i.e., the images of the N sample chromosomes), the average k-nearest distance for any point is: [6]

$$\overline{Dist_k} = \sqrt[n]{\frac{kV(n/2)!}{N\pi^{n/2}}},$$
(4)

where $V = \prod_{i=1}^{n}(\mathcal{R}_i.max - \mathcal{R}_i.min)$ is the volume of the subspace $\mathcal{R}^n$.

To refine a trajectory T, it is first embedded into a point in this $\mathcal{R}^n$ space, where its k-nearest distance to the sample trajectories is obtained. If the value is greater than $\overline{Dist_k}$ obtained in Equation 4, the corresponding chromosome is dropped.

As of refinement on global fitness, we expect the set of trajectories to exhibit randomness when his/her behavior is less regulated and predictable, e.g., during the leisure hours. As such, a hash test is necessary to guarantee the randomness. We devise the signature of a trajectory as the locations at the "sampling moments":

$$signature(c) = (a_1a_2...a_n)_L = \sum_{i=1}^{n} a_i L^{n-i},$$
(5)

where L is the number of *leaf locations* and n is the number of *sampling moments*; a_i is the location id (from 0 to $L-1$) at sample moment t_i. To determine the set of *sampling moments*, we randomly generate a set of *candidate sampling moments* and choose those good moments where the location distribution of all sample chromosomes is random. The randomness is measured in terms of **entropy**:

$$entropy(t) = -\sum_{i=1}^{L} p_t^i \cdot log\, p_t^i$$
(6)

where p_t^i is the probability that the user is at location i at moment t in the samples. The higher the entropy value, the more random the distribution is at time t.

6   Experiments

In this section, we analyze the datasets generated according to the previous two sections to further study the characteristics of GAMMA.

---

[6] For an even number n, the volume of a hypersphere with radius r in n-dimensional space is $\frac{\pi^{n/2}r^n}{(n/2)!}$

6.1   Moving Object Trajectories in a Cellular Network

We generate the trajectories by GAMMA and the traditional random waypoint model (RANWAY). In RANWAY, the turning points are randomly selected. We compare their performance under the same number of CPU cycles. The two-dimensional cellular partition of the space is derived by generating 30 random service points and computing the Voronoi diagram. Regarding the signature function, the space is evenly partitioned into a three-by-three grid. Thus, the hash table size is $2^{3 \times 3} = 512$. The experimental parameters are summarized in Table 1.

Table 1. Experimental Parameters for Trajectories in a Cellular Network

Parameter	Value	Parameter	Value
population	100	generation	50
mutation probability	0.05	default iterations	20
crossover probability	0.8		

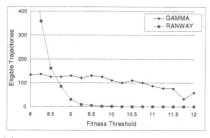

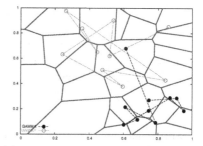

(a) Experiment 1: Eligible Trajectories vs. Fitness Threshold

(b) Sample Trajectories from GAMMA and RANWAY

Fig. 4. Moving Object Trajectories

Experiment 1: Fixed CPU Time. In this experiment, GAMMA is shown to generate trajectories with high individual fitness values. We fix the CPU time, vary the individual fitness threshold, and measure the number of eligible trajectories generated from GAMMA and RANWAY. The performance comparison is plotted in Figure 4(a). GAMMA generates $100 \times 20 = 2,000$ trajectories, while RANWAY generates 90,000 in the same period of CPU time. When the threshold increases from 8 to 12, the number of eligible trajectories generated by RANWAY decreases sharply (from 937 to 0), whereas GAMMA steadily drops by 60%. This is because: when the threshold gets higher, RANWAY is less likely to generate eligible trajectories; nevertheless GAMMA searches the trajectory space more efficiently and still obtains a satisfactory number of trajectories even when the threshold is harsh. As a concrete evident, Figure 4(b) visualizes the cellular network and two sample trajectories which are randomly picked from the two generated datasets; GAMMA is more clever in choosing areas with dense cells so that the trajectory cross the boundaries frequently. As a conclusion, GAMMA is much more efficient and robust than RANWAY for generating moving trajectories with desired characteristics.

Experiment 2: Fixed Fitness Threshold. In this experiment, we fix the fitness threshold and vary the CPU time. For each CPU time setting, we obtain m trajectories output from GAMMA and choose the best m trajectories from RANWAY for comparison. Their mean individual fitness and global fitness values are compared in Figures 5(a) and 5(b). In terms of individual fitness, the trajectories from GAMMA is always better than those from RANWAY. More importantly, as the population grows GAMMA is able to retain a high mean fitness value. Furthermore, Figure 5(b) shows that GAMMA yields a higher and more steady global fitness value than that of RANWAY as the population grows. As a final note, the population grows steadily in GAMMA as more CPU time is consumed. To conclude, GAMMA is a robust, efficient, and high-quality generator for moving trajectories.

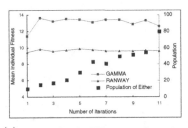 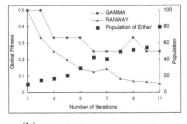

(a) Mean Individual Fitness vs. Iterations   (b) Global Fitness vs. Iterations

Fig. 5. Experiment 2: Mean Individual Fitness and Global Fitness vs. Iterations

6.2 Real-Life Symbolic Trajectories

In this experiment, we simulate the moving behavior of a graduate student from 8am to 6pm everyday as shown in Section 5. The experimental parameters are summarized in Table 2.

Table 2. Experimental Parameters for Real-life Symbolic Trajectories

Parameter	Value	Parameter	Value
population	100	generation	50
mutation probability	0.05	k	5
crossover probability	0.9	mutation granularity	1 hr
fitness threshold	150	iterations	500

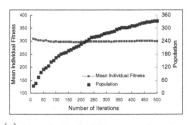
(a) Mean Individual Fitness vs. Iterations

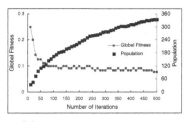
(b) Global Fitness vs. Iterations

Fig. 6. Performance of Symbolic Trajectory Generation

Twenty real trajectories are tracked and recorded as the samples, some of which are illustrated in Figure 9(a). As we analyzed, these trajectories exhibit the following mobility patterns of the user:

- He/She normally goes to campus at about 8:30~9:30 and leaves for home at about 16:00~18:00.
- In campus, he/she is mainly at the mobile lab, but occasionally goes to the database lab (for meeting or discussion) and to the pervasive lab (for some experiments) in the afternoon.
- He/She goes to the library for one or two hours at times, but without a fixed time schedule.
- He/She usually has lunch at cafeteria at noon.
- Occasionally, he/she goes to the grocery on the way home or to the campus; once in a while, he/she goes to the restaurant for dinner.

As there is no comparable generator reported in the literature, we only measured the performance of GAMMA. It ran for 500 iterations and we measured the metrics, namely, the mean individual fitness, the global fitness and the whole population in the repository. Figures 6(a) and 6(b) plot the performance curves. It is observed that when the population increases, the mean fitness value stabilizes at about 300 and the global fitness at 0.1. This implies that the quality of the trajectories produced in each iteration is almost equally good. Nevertheless, the population grows more slowly as the iteration increases, due to the increasing possibility of a hash test collision.

To demonstrate the effectiveness of GAMMA, we show all samples and generated trajectories in the embedded three-dimensional space in Figure 7. The generated trajectories, designated by the crosses, are always around some samples. To visualize from another perspective, we also project them on the spatial axis and derive the first-order Markov models of the location transitions for the samples and the generated trajectories, respectively. The location transition diagrams are partially plotted in Figures 8(a) and 8(b).[7] The state transition from each locations are similar. For example, in Figure 8(a), the user moves

---

[7] To make the figures legible, only the transitions from "library", "cafeteria", "grocery" and "home" are depicted.

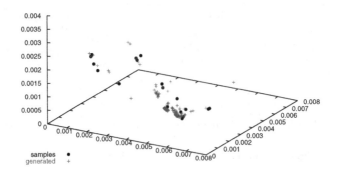

Fig. 7. Visualization of Symbolic Trajectories in a 3D space

from "home" to "grocery", "mobile lab" and "library" with probability 0.1, 0.75 and 0.15; in Figure 8(b), the destinations are "grocery", "mobile lab", "library" and "restaurant" with probability 0.1, 0.66, 0.14 and 0.1. This implies that a similar underlying mobility patterns. Therefore, we conclude that the generated trajectories resemble the samples and reserve the user's mobility patterns.

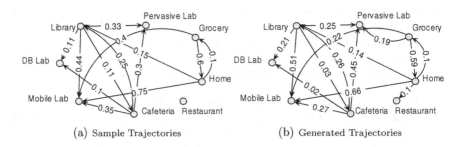

(a) Sample Trajectories   (b) Generated Trajectories

Fig. 8. First-Order Markov Model of the Trajectories

As a direct evidence of this conclusion, we also choose five generated trajectories randomly and plot them in Figure 9(b). The semantics of these trajectories conforms to common sense and they exhibit almost the same mobility patterns as the samples.

7   Conclusion and Future Work

Simulating a user's moving behavior is crucial for mobile computing and spatial database research. However, realistic object (user) moving behavior does not have a fixed and close form mobility model, which is required by all existing moving object generators. In this paper, we represent moving behavior as trajectories and present a framework, GAMMA, to generate high-quality trajectories

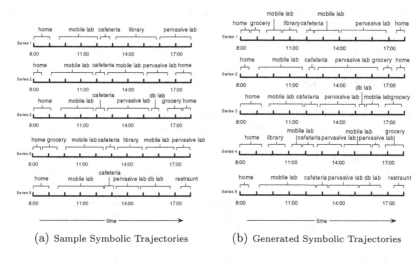

(a) Sample Symbolic Trajectories   (b) Generated Symbolic Trajectories

Fig. 9. Symbolic Trajectories Generated by GAMMA

according to individual and global fitness. We regard the data generation as an optimization problem and adopt GA to search the trajectory space. The framework is highly flexible in that the components of GAMMA can be configured to meet various simulation objectives. To demonstrate its feasibility and usage, we configure it for two specific simulations.

As for future work, we will investigate how the GAMMA parameters, such as population size and mutation probability, affect its performance. We also plan to apply the generated datasets to some simulations such as those in [17,18] and report its effect on the simulation result.

References

1. Theodoridis, Y., Silva, J., Nascimento, M.: On the generation of spatiotemporal datasets. In: 6th Intl. Symposium on Spatial Databases (SSD'99), Hong Kong, China. (1999)
2. Brinkhoff, T.: A framework for generating network-based moving objects. Geoinformatica **6** (2002) 153–180
3. Pfoser, D., Theodoridis, Y.: Generating semantics-based trajectories of moving objects. In: International Workshop on Emerging Technologies for Geo-Based Applications. (2000) 59–76
4. Wong, W.S.V., Leung, V.C.M.: An adaptive distance-based location update algorithm for next-generation pcs networks. IEEE Journal on Selected Areas in Communications **19** (2001) 1942–1952
5. Akyildiz, I., Ho, J., Lin, Y.: Movement based location update and selective paging for pcs networks. IEEE/ACM Transaction on Networking **4** (1996) 629–638
6. Chambers, L., ed.: The Practical Handbook of Genetic Algorithms Applications. Chapman&Hall/CRC (2001)

7. Jardosh, A., Belding-Royer, E., Almeroth, K., Suri, S.: Towards realistic mobility models for mobile ad hoc networks. In: Proceedings of the ACM/IEEE MobiCom Conference. (2003) 217–229
8. Broch, J., Maltz, D.A., Johnson, D., Hu, Y.C., Jetcheva, J.: A performance comparison of multi-hop wireless ad hoc network routing protocols. In: Proceedings of the ACM/IEEE MobiCom Conference. (1998) 85–97
9. Yoon, J., Liu, M., Noble, B.: Sound mobility models. In: Proceedings of the ACM/IEEE MobiCom Conference. (2003) 205–216
10. Whitley, D.: A genetic algorithm tutorial. Statistics and Computing **4** (1994) 65–85
11. Cerny, E., Faye, P., Pownall, P.: Improved design verification by random simulation guided by genetic algorithms. In: Proceedings of the International Conference on Chip Design Automation (ICDA). (2000)
12. Baldi, M., Corno, F., Rebaudengo, M., Squillero, G.: Ga-based performance analysis of network protocols. In: Proceedings of 9th International Conference on Tools with Artificial Intelligence (ICTAI). (1997) 118–124
13. Keogh, E.J., Pazzani, M.J.: Scaling up dynamic time warping to massive datasets. In: Principles of Data Mining and Knowledge Discovery, Third European Conference. (1999) 1–11
14. Faloutsos, C., Lin, K.I.: Fastmap: A fast algorithm for indexing, data-mining and visualization of traditional and multimedia datasets. In: Proceedings of the ACM SIGMOD Conference. (1995) 163–174
15. Klimasauskas, C.C.: Not knowing your random number generator could be costly: Random generators - why are they important. PC Artificial Intelligence Magazine **16** (2002) 52–58
16. L'Ecuyer, P.: Software for uniform random number generation: Distinguishing the good and the bad. In: Proceedings of the Winter Simulation Conference, IEEE Press (2001) 95–105
17. Zheng, B., Lee, D.L.: Semantic caching in location-dependent query processing. In: Proceedings of the 7th International Symposium on Spatial and Temporal Databases (SSTD). (2001) 97–116
18. Tao, Y., Papadias, D.: Time-parameterized queries in spatio-temporal databases. In: Proceedings of the ACM SIGMOD Conference. (2002) 334–345

Medoid Queries in Large Spatial Databases[*]

Kyriakos Mouratidis[1], Dimitris Papadias[1], and Spiros Papadimitriou[2]

[1] Department of Computer Science, Hong Kong University of Science and Technology,
Clear Water Bay, Hong Kong
{kyriakos, dimitris}@cs.ust.hk

[2] Department of Computer Science, Carnegie Mellon University, Pittsburgh, PA, USA
spapadim@cs.cmu.edu

Abstract. Assume that a franchise plans to open k branches in a city, so that the average distance from each residential block to the closest branch is minimized. This is an instance of the *k-medoids* problem, where residential blocks constitute the input dataset and the k branch locations correspond to the medoids. Since the problem is NP-hard, research has focused on approximate solutions. Despite an avalanche of methods for small and moderate size datasets, currently there exists no technique applicable to very large databases. In this paper, we provide efficient algorithms that utilize an existing data-partition index to achieve low CPU and I/O cost. In particular, we exploit the intrinsic grouping properties of the index in order to avoid reading the entire dataset. Furthermore, we apply our framework to solve *medoid-aggregate* queries, where k is not known in advance; instead, we are asked to compute a medoid set that leads to an average distance close to a user-specified parameter T. Compared to previous approaches, we achieve results of comparable or better quality at a small fraction of the CPU and I/O costs (seconds as opposed to hours, and tens of node accesses instead of thousands).

1 Introduction

Given a set P of points, we wish to find a set of medoids $R \subseteq P$ with cardinality k that minimizes the average Euclidean distance $||p - r(p)||$ between each point $p \in P$ and its closest medoid $r(p) \in R$. Formally, our aim is to minimize the function

$$C(R) = \frac{1}{|P|} \sum_{p \in P} ||p - r(p)||$$

under the constraint that $R \subseteq P$ and $|R| = k$. Figure 1 shows an example, where $|P| = 23$, $k = 3$, and $R = \{h, o, t\}$. Assuming that the points of P constitute residential blocks, the three medoids h, o, t constitute candidate locations for service facilities (e.g., franchise branches), so that the average distance $C(R)$ from each block to its closest facility is minimized. A related problem is the *medoid-aggregate* (MA) query, where k is not known in advance. The goal is to select a minimal set R of medoids, such that $C(R)$ best approximates an input value T. Considering again the franchise example, instead of specifying the number of facilities, we seek the minimum set of branches that leads to an average distance (between each residential block and the closest branch) of about $T = 500$ meters.

[*] Supported by grant HKUST 6180/03E from Hong Kong RGC.

C. Bauzer Medeiros et al. (Eds.): SSTD 2005, LNCS 3633, pp. 55–72, 2005.

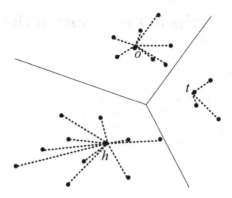

Fig. 1. Example of a k-medoid query

Efficient solutions to medoid queries are essential in several applications related to resource allocation and spatial decision making. In this paper, we propose TPAQ (*Tree-based PArtition Querying*), a strategy that avoids reading the entire dataset by exploiting the grouping properties of a data partition method on P. TPAQ initially traverses the index top-down, stopping at an appropriate level and placing the corresponding entries into groups according to proximity. Finally, it returns the most centrally located point within each group as the corresponding medoid. Compared to previous approaches, TPAQ achieves solutions of comparable or better quality, at a small fraction of the cost (seconds as opposed to hours).

The rest of the paper is organized as follows. Section 2 reviews related work. Section 3 introduces key concepts and describes the intuition and the general framework for our techniques. Section 4 considers k-medoid queries and Section 5 focuses on MA queries. Section 6 presents experimental results on both real and synthetic datasets. Finally, Section 7 concludes the paper.

2   Background

Although our techniques can be used with any data-partition method, here we assume R*-trees [1] due to their popularity. Section 2.1 overviews R*-trees and their application to nearest neighbor queries. Section 2.2 presents existing algorithms for k-medoids and related problems.

2.1   R-Trees and Nearest Neighbor Search

We illustrate our examples with the R-tree of Figure 2 assuming a capacity of four entries per node. Points that are nearby in space (e.g., a, b, c, d) are inserted into the same leaf node (N_3). Leaf nodes are recursively grouped in a bottom-up manner according to their vicinity, up to the top-most level that consists of a single root. Each node is represented as a minimum bounding rectangle (MBR) enclosing all the points in its sub-tree. The nodes of an R*-tree are meant to be compact, have small margin and

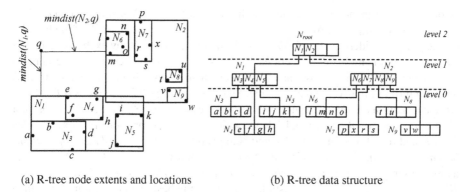

(a) R-tree node extents and locations   (b) R-tree data structure

Fig. 2. R-tree example

achieve minimal overlap (among nodes of the same level) [2]. Additionally, in practice, nodes at the same level contain a similar number of data points, due to a minimum utilization constraint (typically, 40%). These properties imply that the R-tree (or any other data-partition method based on similar concepts) provides a natural way to partition P according to object proximity and group cardinality criteria. However, with few exceptions (discussed in the next subsection) R-trees have been used exclusively for processing spatial queries such as range search, nearest neighbors and spatial joins.

A nearest neighbor (NN) query retrieves the data object that is closest to an input point q. R-tree algorithms for processing NN queries utilize some metrics to prune the search space. The most common such metric is $mindist(N, q)$, which is defined as the minimum possible distance between q and any point in the sub-tree rooted at node N. Figure 2 shows the $mindist$ between q and nodes N_1 and N_2. The algorithm of [3] traverses the tree in a depth-first (DF) manner: starting from the root, it first visits the node with the minimum $mindist$ (i.e., N_1 in our example). The process is repeated recursively until a leaf node (N_4) is reached, where the first potential nearest neighbor (point e) is found. Subsequently, the algorithm only visits entries whose minimum distance is less than $||e - q||$. In the example, N_3 and N_5 are pruned since their $mindist$ from q is greater than $||e - q||$. Similarly, when backtracking to the upper level, node N_2 is also excluded and the process terminates with e as the result. The extension to k (> 1) NNs is straightforward. Hjaltason and Samet [4] propose a best-first variation which is I/O optimal (i.e., it only visits nodes that may contain NNs) and incremental (the number of NNs does need to be known in advance).

2.2 k-Medoids and Related Problems

A number of approximation schemes for k-medoids[1] and related problems appear in the literature [5]. Most of these findings, however, are largely theoretical in nature. Kaufmann and Rousseeuw [6] propose *partitioning around medoids* (PAM), a practical

---

[1] If the selected points (R) do not necessarily belong to the dataset P (i.e., they are arbitrary points in the Euclidean space), the problem is known as *Euclidean k-medians* [5].

algorithm based on the hill climbing paradigm. In particular, PAM starts with a random set of k medoids $R_0 \subseteq P$. At each iteration i, it updates the current set R_i of medoids by exhaustively considering all *neighbor sets* R_i' that result from R_i by exchanging one of its elements with another object. For each of these $k \cdot (|P| - k)$ alternatives, it computes the function $C(R_i')$ and chooses as R_{i+1} the one that achieves the lowest value. It stops when no further improvement is possible. Since computing $C(R_i')$ requires $(O|P|)$ distance calculations, PAM is prohibitively expensive for large $|P|$. Thus, [6] also present *clustering large applications* (CLARA), which draws one or more random samples from P and runs PAM on those. Ng and Han [7] propose *clustering large applications based on randomized search* (CLARANS) as an extension to PAM. CLARANS draws a random sample of size *maxneighbors* from all the $k \cdot (|P| - k)$ possible neighbor sets R_i' of R_i. It performs *numlocal* restarts and selects the best local minimum as the final answer.

Although CLARANS is more scalable than PAM, it is inefficient for disk-resident datasets because each computation of $C(R_i')$ requires a scan of the entire database. Assuming that P is indexed with an R-tree, Ester et al. [8,9] develop *focusing on representatives* (FOR) for large datasets indexed by R-trees. FOR takes the most centrally located point of each leaf node and forms a sample set, which is considered as representative of the entire P. Then, it applies CLARANS on this sample to find the k medoids. Although FOR is more efficient than CLARANS, it still has to read the entire dataset in order to extract the representatives. Furthermore, in very large databases, the leaf level population may still be too high for the efficient application of CLARANS (the experiments of [8] use R-trees with only 50,559 points and 1,027 leaf nodes).

The k-medoid problem is related to clustering. Clustering methods designed for large databases include DBSCAN [10], BIRCH [11], CURE [12] and OPTICS [13]. However, the objective of clustering is to partition data objects in groups (clusters) such that objects within the same group are more similar to each other than points in other groups. Figure 3(a) depicts a 2-way clustering for a dataset, while Figure 3(b) shows the two medoids. Clearly, assigning a facility per cluster would not achieve the purpose of minimizing the average distance between points and facilities. Furthermore, although the number of clusters depends on the data characteristics, the number of medoids is an input parameter determined by the application requirements.

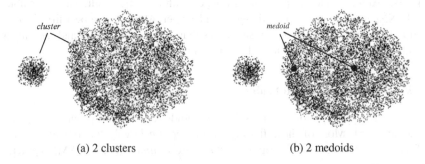

(a) 2 clusters   (b) 2 medoids

Fig. 3. Clustering versus medoids problem

Extensive work on medoids and clustering has been carried out in the areas of statistics [14,6,15], machine learning [16,17,18] and data mining [10,19]. However, the focus there is on assessing the statistical quality of a given clustering, usually based on assumptions about the data distribution [15,6,17,18]. Only few approaches aim at discovering the number of clusters dynamically [17,18]. Besides tackling a problem of different nature, existing algorithms are computationally intensive and unsuitable for disk-resident datasets. In summary, there is need for methods that fully exploit spatial access methods and can answer alternative types of medoid queries.

3   General Framework and Definitions

The TPAQ framework traverses the R-tree in a top-down manner, stopping at the topmost level that provides enough information for answering the given query. In the case of k-medoids, this decision depends on the number of entries at the level. On the other hand, for MA queries, the selection of the partitioning level is also based on the spatial extents and the expected cardinality of its entries. Next, TPAQ groups the entries of the partitioning level into *slots*. For given k, this procedure is performed by a fast algorithm that applies a single pass over the initial entries. For MA, multiple passes over the entries might be required. The last step returns the NN of each slot center as the medoid of the corresponding partition. We first provide some basic definitions, which are used throughout the paper.

Definition 1 (Extended entry). *An extended entry e consists of an R-tree entry N, augmented with information about the underlying data points, i.e.,*

$$e = \langle c, w, N \rangle$$

where the weight w is the expected number of points in the subtree rooted at N. The center c is a vector of co-ordinates that corresponds to the geometric centroid of N, assuming that the points in the sub-tree of N are uniformly distributed.

Definition 2 (Slot). *A slot s consists of a set E of extended entries, along with aggregate information about them. Formally, a slot s is defined as $s = \langle c, w, E \rangle$, where w is the expected number of points represented by s,*

$$w = \sum_{e \in E} e.w$$

and c is the weighted center of s,

$$c = \frac{1}{w} \sum_{e \in E} e.w \cdot e.c$$

A fundamental operation is the insertion of an extended entry e into a slot s. The pseudo-code for this function is shown in Figure 4. The insertion computes the new center taking into account the relative positions and weights of the slot s and the entry e, e.g., if s and e have the same weights, the new center is at the midpoint of the line segment connecting $s.c$ and $e.c$. Table 1 summarizes the frequently used symbols, along with short descriptions. In the subsequent sections, we describe the algorithmic details for each query type.

Function **InsertEntry**(extended entry e, slot s)
1. $s.c = (e.w \cdot e.c + s.w \cdot s.c)/(e.w + s.w)$
2. $s.w = e.w + s.w$
3. $s.E = s.E \cup \{e\}$

Fig. 4. The *InsertEntry* function

Table 1. Frequently used symbols

Symbol	Description
P	Set of data points
$\|p_1 - p_2\|$	Euclidean distance between points p_1 and p_2
R	Set of medoids
k	Number of medoids $k = |R|$
$r(p)$	Closest medoid of $p \in P$
$C(R)$	Average distance achieved by R
T	Target distance (for MA queries)
N	R-tree node
E	Set of entries $e_i = \langle c_i, w_i, N_i \rangle$
S	Set of slots $s_i = \langle c_i, w_i, E_i \rangle$

4 k-Medoid Queries

Given a k-medoid query, TPAQ finds the top-most level with $k' \geq k$ entries. For example, if $k = 3$ in the tree of Figure 2, TPAQ descends to level 1, which contains $k' = 7$ entries, N_3 through N_9. The weights of these entries are computed as follows. Since $|P| = 23$, the weight of the root node N_{root} is $w_{root} = 23$. Assuming that the entries of N_{root} are equally distributed between the two children N_1 and N_2, $w_1 = w_2 = N/2 = 11.5$ (whereas, the true cardinalities are 11 and 12, respectively). The process is repeated for the children of N_1 ($w_3 = w_4 = w_5 = w_1/3 = 3.83$) and N_2($w_6 = w_7 = w_8 = w_9 = w_2/4 = 2.87$). Figure 5 illustrates the algorithm for computing the initial set of entries. Note that *InitEntries* considers that k does not exceed the number of leaf nodes. This assumption is not restrictive because the lowest level typically contains several thousand nodes (e.g., in our datasets, between 3,000–60,000), which is sufficient for all ranges of k that are of practical interest. Nevertheless, if needed, larger values of k can be accommodated by conceptually splitting leaf level nodes.

The next step merges the k' initial entries in order to obtain exactly k groups. Initially, k out of the k' entries are selected as slot *seeds*, i.e., each of the chosen entries forms an initial slot. Clearly, the seed locations play an important role in the quality of the final answer. The seeds should capture the distribution of points in P, i.e., dense areas should contain many seeds. Our approach for seed selection is based on *space-filling curves*, which map a multi-dimensional space into a linear order. Among several alternatives, Hilbert curves best preserve the locality of points [20,21]. Therefore, we first Hilbert-sort the k' entries and select every other m-th entry as a seed, where $m = k'/k$.

Function **InitEntries**(P, k)
1. Load the root of the R-tree of P
2. Initialize $list = \{e\}$, where $e = \langle N_{root}.c, |P|, N_{root} \rangle$
3. While $list$ contains fewer than k extended entries
4.   Initialize an empty list $next\_level\_entries$
5.   For each $e = \langle c, w, N \rangle$ in $list$ do
6.   Let num be the number of child entries in node N
7.   For each entry N_i in node N do
8. $w_i = w/num$ // the expected cardinality of N_i
9.   Insert extended entry $\langle N_i.c, w_i, N_i \rangle$ to $next\_level\_entries$
10.   Set $list = next\_level\_entries$
11. Return $list$

Fig. 5. The *InitEntries* function

This procedure is fast and produces well-spaced seeds that follow the data distribution. Returning to our example, Figure 6 shows the level 1 MBRs (for the R-tree of Figure 2) and the output seeds $s_1 = N_4, s_2 = N_9$ and $s_3 = N_7$ according to their Hilbert order. Recall that each slot is represented by its weight (e.g., $s_1.w = w_4 = 3.83$), its center (e.g., $s_1.c$ is the centroid of N_4) and its MBR.

Then, each of the remaining $(k' - k)$ entries is inserted into the k seed slots, based on proximity criteria. More specifically, for each entry e, we choose the slot s whose weighted center $s.c$ is closest to the entry's center $e.c$. In the running example, assuming that N_3 is considered first, it is inserted into the slot s_1 using the *InsertEntry* function of Figure 4. The center of s_1 is updated to the midpoint of N_3 and N_4's centers, as shown in Figure 7(a). TPAQ proceeds in this manner, until the final slots and weighted centers are computed as shown in Figure 7(b).

After grouping all entries into exactly k slots, we find one medoid per slot by performing a nearest-neighbor query. The query point is the slot's weighted center $s.c$, and the search space is the set of entries $s.E$. Since all the levels of the R-tree down to the partition level have already been loaded in memory, the NN queries incur very few

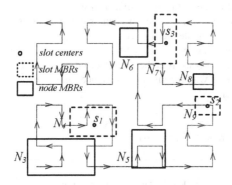

Fig. 6. Hilbert seeds on example dataset

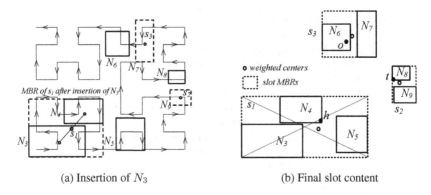

(a) Insertion of N_3 (b) Final slot content

Fig. 7. Insertion of entries into slots

node accesses and negligible CPU cost. Observe that an actual medoid (i.e., a point in P that minimizes the sum of distances) is more likely to be closer to $s.c$ than simply to the center of the MBR of s. The intuition is that $s.c$ captures information about the point distribution within s. The NN queries on these points return the final medoids $R = \{h, o, t\}$. Figure 8 shows the complete TPAQ k-medoid computation algorithm.

Note that the problem of seeding the slot table is similar to that encountered in spatial hash joins, where the number of buckets is bounded by the available main memory [22,23,24]. However, our ultimate goals are different. First, in the case of hash joins, the table capacity is an upper bound. Reaching it is desirable in order to exploit available memory as much as possible, but falling slightly short is not a problem. In contrast, we want *exactly* k slots. Second, in our case slots should minimize the average distance $C(R)$ on one dataset, whereas slot selection in spatial joins attempts to minimize the number of intersection tests that must be performed between objects that belong to different datasets.

Algorithm **TPAQ**(P, k)
1.   Initialize a set $S = \emptyset$, and empty $list$
2.   Set $E =$ the set of entries returned by *InitEntries*(P, k)
3.   Hilbert-sort the centers of the entries in E and store them in a sorted list $sorted\_list$
4.   For $i = 1$ to k do // compute the slot seeds
5.   Form a slot containing the $(i \cdot |E|/k)$-th entry of $sorted\_list$ and insert it into S
6.   For each entry e in E (apart from the ones selected as seeds) do
7.   Find the slot s in S with the minimum distance $\|e.c - s.c\|$
8.   *InsertEntry*(e, s)
9.   For each $s \in S$ do
10.   Perform a NN search at $s.c$ on the points under $s.E$
11.   Append the retrieved point to $list$
12. Return $list$

Fig. 8. The *TPAQ* algorithm

5   Medoid-Aggregate Queries

A medoid-aggregate (MA) query specifies the desired average distance T (between points and medoids), and asks for the minimal medoid set R that achieves $C(R) = T$. Consider the example of Figure 9, and assume that we know a priori all the optimal i-medoid sets R^i and the corresponding $C(R^i)$, for $i = 1, \ldots, 23$. If $C(R^4)$ is the average distance that best approximates T (compared to $C(R^i)\ \forall i \neq 4$), set R^4 is returned as the result of the query. The proposed algorithm, TPAQ-MA, is based on the fact that as the number of medoids $|R|$ increases, the corresponding $C(R)$ decreases. In a nutshell, it first descends the R-tree of P down to an appropriate (partitioning) level. Next, it estimates the value of $|R|$ that achieves the closest average distance $C(R)$ to T and returns the corresponding medoid set R.

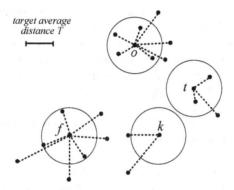

Fig. 9. A medoid-aggregate query example

The first step of TPAQ-MA is to determine the partitioning level. The algorithm selects for partitioning the top-most level whose *minimum possible average distance* (MPD) is less than or equal to T. The MPD of a level is the smallest $C(R)$ that can be achieved if partitioning takes place in this level. According to the methodology presented in Section 4, MPD equals to the $C(R)$ resulting if we extract one medoid from each entry in the level. Since computing the exact $C(R)$ requires scanning the entire dataset P, we use an estimate of $C(R)$ as the MPD. In particular, for each entry e of the level, we assume that the underlying points are distributed uniformly in its MBR[2], and that the corresponding medoid is at $e.c$. The average distance $\bar{C}(e)$ between the points in e and the $e.c$ is given by the following lemma.

Lemma 1. *If the points in e are uniformly distributed in its MBR, then their average distance from e.c is*

$$\bar{C}(e) = \frac{1}{3}\left(\frac{D}{2} + \frac{B^2}{8A}\ln\left(\frac{D+A}{D-A}\right) + \frac{A^2}{8B}\ln\left(\frac{D+B}{D-B}\right)\right),$$

where A and B are the side lengths of the MBR of e and D is its diagonal length.

---

[2] This is a reasonable assumption for low-dimensional R-trees [2].

Proof. If we translate the MBR of e so that its center $e.c$ falls at the origin $(0,0)$, $\bar{C}(e)$ is the average distance of points $(x, y) \in [-A/2, A/2] \times [-B/2, B/2]$ from $(0,0)$. Hence,

$$\bar{C}(e) = \frac{1}{AB} \int_{-A/2}^{A/2} \int_{-B/2}^{B/2} \sqrt{x^2 + y^2} \, dx dy,$$

which evaluates to the quantity of Lemma 1.

The MPD of the considered level is estimated by averaging $\bar{C}(e)$ over all $e \in E$, where E is the set of entries at the level, i.e.,

$$\mathrm{MPD} = \frac{1}{|P|} \sum_{e \in E} e.w \cdot \bar{C}(e)$$

TPAQ-MA applies the *InitEntries* function to select the top-most level that has MPD $\leq$ T. The pseudo-code of *InitEntries* is the same as shown in Figure 5, after replacing the while-condition of line 3 with the expression: "the estimated MPD is more than T". Returning to our running example, the root node N_{root} of the R-tree of P has MPD $=$ $\bar{C}(N_{root})$ higher than T. Therefore, *InitEntries* proceeds with level 2 (containing entries N_1 and N_2), whose MPD is also higher than T. Next, it loads the level 1 nodes and computes the MPD over the entries from N_3 to N_9. The MPD is less than T, and level 1 is selected for partitioning. The *InitEntries* procedure returns a list containing 7 extended entries corresponding to N_3 up to N_9.

The next step of TPAQ-MA is to determine the number of medoids that better approximate the value T. If E is the set of entries in the partitioning level, then the candidate values for $|R|$ range between 1 and $|E|$. Assuming that $C(R)$ is decreasing with respect to $|R|$, TPAQ-MA performs binary search in order to select the value of $|R|$ that yields the closest average distance to T. This procedure considers $O(\log |E|)$ different values for $|R|$, and creates slots for each of them as discussed in Section 4. Since the exact evaluation of $C(R)$ for every examined $|R|$ would be very expensive, we produce an estimate $\bar{C}(S)$ of $C(R)$ for the corresponding set of slots S. Particularly, we assume that the medoid of each slot s is located at $s.c$, and that the average distance from the points in every entry $e \in s$ equals the distance $||e.c - s.c||$. Hence, the estimated value for $C(R)$ is given by the formula

$$\bar{C}(S) = \frac{1}{|P|} \sum_{s \in S} \sum_{e \in s} e.w \cdot ||e.c - s.c||$$

where S is the set of slots produced by partitioning the entries in E into $|R|$ groups. Note that we could use a more accurate estimator assuming uniformity within each entry $e \in s$, similar to Lemma 1. However, the derived expression would be more complex and more expensive to evaluate, because now we need the average distance from $s.c$ (as opposed to the center $e.c$ of the entry's MBR). The overall TPAQ-MA algorithm is shown in Figure 10.

In the example of Figure 9, the partitioning level contains entries $E = \{N_3, N_4, N_5, N_6, N_7, N_8, N_9\}$. The binary search considers values of $|R|$ between 1 and 7. Starting with $|R| = (1 + 7)/2 = 4$, the algorithm creates S with 4 slots, as shown in

Algorithm **TPAQ-MA**(P, T)
1.   Initialize an empty *list*
2.   Set E = set of the entries at the topmost level with MPD $\leq T$
3. $low = 1; high = |E|$
4.   While $low \leq high$ do
5. $mid = (low + high)/2$
6.   Group the entries in E into mid slots
7. S = the set of created slots
8.   If $\bar{C}(S) < T$, set $high = mid$
9.   Else, set $low = mid$
10.   For each $s \in S$ do
11.   Perform a NN search at $s.c$ on the points under $s.E$
12.   Append the retrieved point to *list*
13.   Return *list*

Fig. 10. The *TPAQ-MA* algorithm

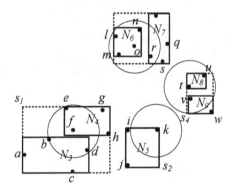

Fig. 11. Entries and final slots

Figure 11. It computes $\bar{C}(S)$, which is lower than T. It recursively continues the search for $|R| \in [1, 4]$ in the same way, and finally decides that $|R| = 4$ yields a value of $\bar{C}(S)$ that best approximates T. Next, TPAQ-MA performs a NN search at the center $s.c$ of the slots corresponding to $|R| = 4$, and returns the retrieved points (f, i, o, and k) as the result.

6   Experimental Evaluation

In this section we evaluate the performance of the proposed methods for k-medoid and medoid-aggregate queries. We use both synthetic and real datasets. The synthetic ones (SKW) follow a zipf distribution with parameter $\alpha = 0.8$, and have cardinality 256K, 512K, 1M, 2M and 4M points. The real datasets are (i) NA, with 569,120 points (available at *www.maproom.psu.edu/dcw*), and (ii) LA, with 1,314,620 points (available at *www.rtreeportal.org*). All datasets are normalized to cover the same space with extent $10^4 \times 10^4$ and indexed by an R*-tree [1] (the block size ranges between 1 and 4Kbytes). For the experiments we use a Pentium 3GHz CPU.

6.1   *k*-Medoid Queries

First, we focus on *k*-medoid queries and compare TPAQ against FOR (which, as discussed in Section 2.2, is the only other method that utilizes R-trees for computing *k*-medoids). For TPAQ, we use the depth-first algorithm of [3] to retrieve the nearest neighbor of each computed centroid. In the case of FOR we have to set the parameters *numlocal* (number of restarts) and *maxneighbors* (sample size of the possible neighbor sets) of the CLARANS component. Ester et al. [8] suggest setting *numlocal* = 2 and *maxneighbors* = $k \cdot (M - k)/800$, where M is the number of leaf nodes in the R-tree of P. With these parameters, FOR does not terminate within reasonable time for our datasets. Therefore, we set *maxneighbors* = $k \cdot (M - k)/(8000 \cdot \log M)$ and keep *numlocal* = 2. These values speed up FOR considerably, while the deterioration of the resulting solutions (with respect to the suggested values of *numlocal* and *maxneighbors*) is negligible. One final remark concerning FOR, is that all results presented in this section are average values over 10 runs of the algorithm. This is necessary because the performance of FOR depends on the random choices of its CLARANS component. The algorithms are compared for different data cardinality $|P|$, number of medoids k and block size. Table 2 summarizes the parameters under investigation along with their ranges and default values. In each experiment we vary a single parameter, while setting the remaining ones to their default (median) values.

Table 2. Default parameter values

Parameter	Range	Default
$|P|$	256K – 4M	1M
k	2 – 512	32
Block size	1KB – 4KB	2KB

The first set of experiments measures the effect of $|P|$. In Figure 12(a), we show the running time of TPAQ and FOR, for SKW when $k = 32$ and $|P|$ varies between 256K and 4M. TPAQ is 2 to 4 orders of magnitude faster than FOR. Even for $|P| = 4M$ objects, our method terminates in less than 0.04 seconds (while FOR needs more than 3 minutes). Figure 12(b) shows the I/O cost (number of node accesses) for the same experiment. FOR is around 2 to 3 orders of magnitude more expensive than TPAQ since it reads the entire dataset once. Both the CPU and the I/O costs of TPAQ are relatively stable and small, because partitioning takes place at a high level of the R-tree.

The cost improvements of TPAQ come with no compromise of the answer quality. Figure 12(c) shows the average distance $C(R)$ achieved by the two algorithms. TPAQ outperforms FOR in all cases. An interesting observation is that the average distance for FOR drops when the cardinality of the dataset $|P|$ increases. This happens because higher $|P|$ implies more possible "paths" to a local minimum. To summarize, the results

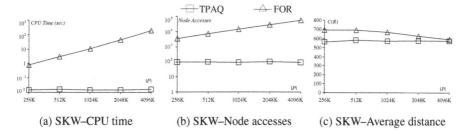

(a) SKW–CPU time   (b) SKW–Node accesses   (c) SKW–Average distance

Fig. 12. Performance versus $|P|$

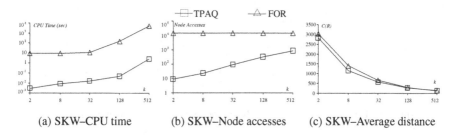

(a) SKW–CPU time   (b) SKW–Node accesses   (c) SKW–Average distance

Fig. 13. Performance versus k (synthetic data)

of Figure 12 verifies that TPAQ scales gracefully with dataset cardinality and incurs much lower cost than FOR, without sacrificing the medoid quality.

The next set of experiments studies the performance of TPAQ and FOR when k varies between 2 and 512, using a SKW dataset of cardinality $|P| = 1M$. Figure 13(a) compares the running time of the methods. In both cases, TPAQ is 3 orders of magnitude faster than FOR. It is worth mentioning that for $k = 512$ our method terminates in 2.5 seconds, while FOR requires around 1 hour and 20 minutes. For $k = 512$, both the partitioning into slots of TPAQ and the CLARANS component of FOR are applied on an input of size 14,184; the input of the TPAQ partitioning algorithm consists of the extended entries at the leaf level, while the input of CLARANS is the set of actual representatives retrieved in each leaf node. The large difference in CPU time verifies the efficiency of our partitioning algorithm.

Figure 13(b) shows the effect of k on the I/O cost. The node accesses of FOR are constant and equal to the total number of nodes in the R-tree of P (i.e., 14,391). On the other hand, TPAQ accesses more nodes as k increases. This happens because (i) it needs to descend more R-tree levels in order to find one with a sufficient number (i.e., k) of entries, and (ii) it performs more NN queries (i.e., k) at the final step. However, TPAQ is always more efficient than FOR; in the worst case TPAQ reads all R-tree nodes up to level 1 (this is the situation for $k = 512$), while FOR reads the entire dataset P for any value of k. Figure 13(c) compares the accuracy of the methods. TPAQ achieves lower $C(R)$ for all values of k.

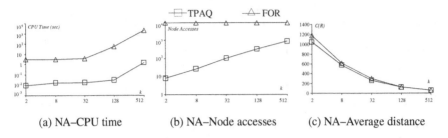

(a) NA–CPU time   (b) NA–Node accesses   (c) NA–Average distance

Fig. 14. Performance versus k (NA)

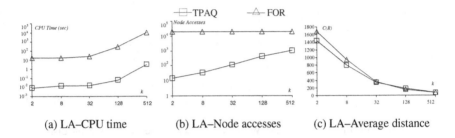

(a) LA–CPU time   (b) LA–Node accesses   (c) LA–Average distance

Fig. 15. Performance versus k (LA)

In order to confirm the generality of our observations, Figures 14 and 15 repeat the above experiment for real datasets NA and LA. TPAQ outperforms FOR by orders of magnitude in terms of both CPU time (Figures 14(a) and 15(a) for NA and LA, respectively) and number of node accesses (Figures 14(b) and 15(b)). Regarding the average distance $C(R)$, the methods achieve similar results, with TPAQ being the winner. Note that the CPU and I/O costs of the methods are higher for LA (than NA), since it is larger and its R-tree has more entries per level. The achieved $C(R)$ values are lower for NA, because it is more skewed than LA (i.e., the objects are concentrated in a smaller area of the workspace).

Figures 16(a) and 17(a) show the running time of TPAQ and FOR on 32-medoid queries as a function of the block size for datasets NA and LA. When the block size increases, the number of leaf nodes drops. Thus the CPU cost of FOR decreases because its expensive CLARANS step processes fewer representatives. TPAQ does not necessarily follow the same trend. For NA, the running time drops, since the number of entries at the partitioning level is 618, 143 and 33 for block size 1KB, 2KB and 4KB, respectively. For LA the populations of the partitioning levels are 43, 313 and 77, respectively, yielding higher running time in the 2KB case. Concerning the I/O cost, larger block size implies smaller R-tree height, and fewer nodes per level. Therefore, both methods are less costly (as illustrated in Figures 16(b) and 17(b)). Independently of the block size, TPAQ incurs much fewer node accesses than FOR. Finally, Figures 16(c) and 17(c) illustrate the effect of the block size in the quality of the retrieved medoid sets. In all cases, the average distance achieved by TPAQ is lower than that of FOR.

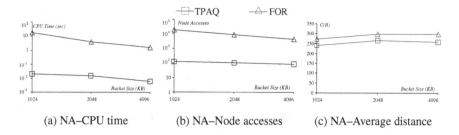

(a) NA–CPU time   (b) NA–Node accesses   (c) NA–Average distance

Fig. 16. Performance versus block size (NA)

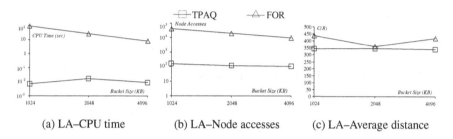

(a) LA–CPU time   (b) LA–Node accesses   (c) LA–Average distance

Fig. 17. Performance versus block size (LA)

6.2 Medoid-Aggregate Queries

In this section we study the performance of TPAQ-MA. We use datasets SKW (with 1M objects) and LA, and vary T in the range from 100 to 1500 (recall that our datasets cover a space with extent $10^4 \times 10^4$). Since there is no existing algorithm for processing such queries on large indexed datasets, we compare TPAQ-MA against an exhaustive algorithm (EXH) that works as follows. Let E be the set of entries at the partitioning level of TPAQ-MA; then, EXH computes and evaluates all the medoid sets for $|R| = 1$ up to $|R| = |E|$, by performing partitioning of E into slots with the technique presented in Section 4. EXH returns the medoid set that yields the closest average distance to T. Note that EXH is prohibitively expensive in practice because, for each examined value of $|R|$, it scans the entire dataset P in order to exactly evaluate $C(R)$. Therefore, we exclude EXH from the CPU and I/O cost charts.

Our evaluation starts with SKW. Figure 18(a) shows the $C(R)$ for TPAQ-MA versus T. Clearly, the average distance returned by TPAQ-MA approximates the desired distance (dotted line) very well. Figure 18(b) plots the deviation percentage between the average distances achieved by TPAQ-MA and EXH. The deviation is below 9% in all cases, except for $T = 300$ where it equals 13.4%. Interestingly, for $T = 1500$, TPAQ-MA returns exactly the same result as EXH with $|R| = 5$. Figures 18(c) and 18(d) illustrate the running time and the node accesses of our method, respectively. For $T = 100$, both costs are relatively high (100.8 seconds and 1839 node accesses) compared to larger values of T. The reason is that when $T = 100$, partitioning takes place

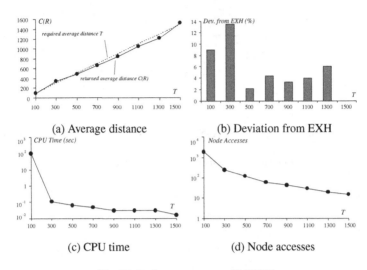

Fig. 18. Performance versus T (SKW)

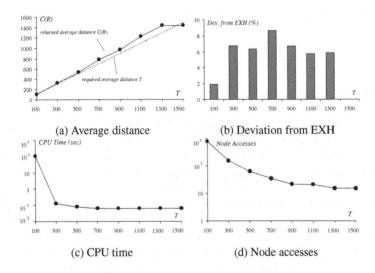

Fig. 19. Performance versus T (LA)

at level 1 (leaf level, which contains 14,184 entries) and returns $|R| = 1272$ medoids, incurring many computations and I/O operations. In all other cases, partitioning takes place at level 2 (containing 203 entries), and TPAQ-MA runs in less than 0.11 seconds and reads fewer than 251 pages.

Figure 19 repeats the above experiment for the LA dataset. Figures 19(a) and 19(b) compare the average distance achieved by TPAQ-MA with the input value T and the result of EXH, respectively. The deviation from EXH is always smaller than 8.6%, while for $T = 1500$ the answer of TPAQ-MA is the same as EXH. Concerning the efficiency of TPAQ-MA, we observe that the algorithm has, in general, very low CPU and I/O cost. The highest cost is again in the case of $T = 100$ for the reasons explained in the context of Figure 18; TPAQ-MA partitions 19,186 entries into slots and extracts $|R| = 296$ medoids, taking in total 105.6 seconds and performing 781 node accesses.

7 Conclusion

This paper studies k-medoids and related problems in large databases. In particular, we consider k-medoid and medoid-aggregate (MA) queries, and propose TPAQ (*Tree-based PArtition Querying*), a framework for their efficient processing. TPAQ provides high-quality answers almost instantaneously, thus facilitating data analysis, especially in time-critical resource allocation applications. Our techniques are the first ones to fully exploit the data partitioning properties of an already existing spatial access method on the dataset. TPAQ processes a query in three steps. Initially, it descends the index, and stops at the topmost level that provides sufficient information about the underlying data distribution. Next, it partitions the entries of the selected level into a number of slots. In the case of k-medoid queries, the number of slots is equal to k. For MA, this number is decided using binary search in conjunction with some average distance estimators. Finally, TPAQ retrieves one medoid for each slot with a NN query therein. An extensive experimental evaluation shows that TPAQ outperforms the state-of-the-art method for k-medoid queries by orders of magnitude, and achieves results of better or comparable quality. Our empirical study also illustrates the effectiveness and efficiency of TPAQ for processing MA queries.

The quality of the medoid sets returned by TPAQ is determined by the achieved average distance. An interesting direction for future work is to extend our index-based strategies to other aggregate distance functions, such as *max*. In the *max* case, we wish to minimize the maximum distance between the points in the input dataset and their closest medoid; i.e., $C(R) = \max_{p \in P} ||p - r(p)||$. Further, it is also interesting to solve constrained partitioning queries. For example, consider that each facility can serve up to a maximum number of clients. In this case the algorithms must be extended to take into account existing capacity (or processing capability) constraints.

References

1. Beckmann, N., Kriegel, H.P., Schneider, R., Seeger, B.: The R*-tree: An efficient and robust access method for points and rectangles. In: SIGMOD. (1990) 322–331
2. Theodoridis, Y., Stefanakis, E., Sellis, T.K.: Efficient cost models for spatial queries using r-trees. IEEE TKDE **12** (2000) 19–32
3. Roussopoulos, N., Kelley, S., Vincent, F.: Nearest neighbor queries. In: SIGMOD. (1995) 71–79
4. Hjaltason, G.R., Samet, H.: Distance browsing in spatial databases. ACM TODS **24** (1999) 265–318
5. Arora, S., Raghavan, P., Rao, S.: Approximation schemes for euclidean k-medians and related problems. In: STOC. (1998) 106–113

6. Kaufman, L., Rousseeuw, P.: Finding groups in data. Wiley-Interscience (1990)
7. Ng, R.T., Han, J.: Efficient and effective clustering methods for spatial data mining. In: VLDB. (1994) 144–155
8. Ester, M., Kriegel, H.P., Xu, X.: A database interface for clustering in large spatial databases. In: KDD. (1995) 94–99
9. Ester, M., Kriegel, H.P., Xu, X.: Knowledge discovery in large spatial databases: Focusing techniques for efficient class identification. In: SSD. (1995) 67–82
10. Ester, M., Kriegel, H.P., Sander, J., Xu, X.: A density-based algorithm for discovering clusters in large spatial databases with noise. In: KDD. (1996) 226–231
11. Zhang, T., Ramakrishnan, R., Livny, M.: BIRCH: An efficient data clustering method for very large databases. In: SIGMOD. (1996) 103–114
12. Guha, S., Rastogi, R., Shim, K.: CURE: An efficient clustering algorithm for large databases. In: SIGMOD. (1998) 73–84
13. Ankerst, M., Breunig, M.M., Kriegel, H.P., Sander, J.: OPTICS: Ordering points to identify the clustering structure. In: SIGMOD. (1999) 49–60
14. Hartigan, J.A.: Clustering algorithms. Wiley (1975)
15. Hastie, T., Tibshirani, R., Friedman, J.: The elements of statistical learning. Springer-Verlag (2001)
16. Pelleg, D., Moore, A.W.: Accelerating exact k-means algorithms with geometric reasoning. In: KDD. (1999) 277–281
17. Pelleg, D., Moore, A.W.: X-means: Extending K-means with efficient estimation of the number of clusters. In: ICML. (2000) 727–734
18. Hamerly, G., Elkan, C.: Learning the k in k-means. In: NIPS. (2003)
19. Fayyad, U., Piatetsjy-Shapiro, G., Smyth, P., Uthurusamy, R.: Advances in knowledge discovery and data mining. AAAI/MIT (1996)
20. Kamel, I., Faloutsos, C.: On packing r-trees. In: CIKM. (1993) 490–499
21. Moon, B., Jagadish, H.V., Faloutsos, C., Saltz, J.H.: Analysis of the clustering properties of the hilbert space-filling curve. IEEE TKDE 13 (2001) 124–141
22. Lo, M.L., Ravishankar, C.V.: Generating seeded trees from data sets. In: SSD. (1995) 328–347
23. Lo, M.L., Ravishankar, C.V.: The design and implementation of seeded trees: An efficient method for spatial joins. IEEE TKDE 10 (1998) 136–152
24. Mamoulis, N., Papadias, D.: Slot index spatial join. IEEE TKDE 15 (2003) 211–231

The Islands Approach to Nearest Neighbor Querying in Spatial Networks

Xuegang Huang, Christian S. Jensen, and Simonas Šaltenis

Department of Computer Science, Aalborg University,
Fredrik Bajers Vej 7E, DK-9220, Aalborg, Denmark
{xghuang, csj, simas}@cs.aau.dk

Abstract. Much research has recently been devoted to the data management foundations of location-based mobile services. In one important scenario, the service users are constrained to a transportation network. As a result, query processing in spatial road networks is of interest. We propose a versatile approach to k nearest neighbor computation in spatial networks, termed the Islands approach. By offering flexible yet simple means of balancing re-computation and pre-computation, this approach is able to manage the trade-off between query and update performance. The result is a single, efficient, and versatile approach to k nearest neighbor computation that obviates the need for using several k nearest neighbor approaches for supporting a single service scenario. The experimental comparison with the existing techniques uses real-world road network data and considers both I/O and CPU performance, for both queries and updates.

1 Introduction

An infrastructure is emerging that enables location-based mobile services, and we are witnessing substantial efforts in the research community to establish fundamental data management support for such services. Mobile services typically involve service users and so-called points of interest. We consider the scenario where these are located within a spatial network or, more specifically, a road network [12,13,14,17,21,29]. The movements of the users, often termed moving objects, are constrained by the network, and the points of interest can only be visited by traveling along the network. The relevant notion of distance is network distance based on shortest-path computation.

Existing approaches to k nearest neighbor (kNN) computation in spatial networks can be divided into two types: approaches that compute kNN queries by incrementally scanning the network until k neighbors are found, and approaches that apply some form of pre-computation and "compute" kNN queries by looking up data collected in pre-computed data structure. Both types of approaches assume that the spatial network is represented by graph-like data structures.

The first type of approach, denoted as "online computation," naturally captures the dynamic aspects of the network, e.g., the emergence or disappearance of points of interest, and applies some form of network expansion-based search. This type of approach is able to output the network distances and paths to each kNN, as these are computed as part of the process. The data structures used in online computation capture the connectivity of the network and are easily updated. When compared to online computation,

C. Bauzer Medeiros et al. (Eds.): SSTD 2005, LNCS 3633, pp. 73–90, 2005.

the second type of approach, termed "pre-computation," typically has better query performance, but has difficulty in coping with frequent updates of the road network and the points of interest.

We consider the performance of queries as well as updates, as both efficient querying and update are important for location-based mobile services. In particular, we propose a novel approach, termed the Islands approach, to kNN processing in spatial networks. This approach computes the kNNs along with the distance to each, but does not compute the corresponding shortest paths. The rationale for this design decision is that a mobile user is expected to only be interested in the actual path to one nearest neighbor selected from the kNN result, and so the path computation is better left to a subsequent processing step.

The Islands approach is designed with the assumption that the overall I/O cost of queries and updates is the main performance evaluation criterion, and the approach aims to be efficient for varying frequencies of queries and updates, which yields broad applicability. The versatility of the approach is demonstrated by an experimental comparison with two other approaches that covers the cases these two are optimized for.

The paper makes three main contributions.

- The Islands approach offers an attractive generalization of the existing kNN query processing techniques for spatial networks. It employs a relatively simple data structure and an intuitive search algorithm. And it is applicable to a broad range of mobile service scenarios, thus avoiding the need for using more specialized algorithms for different scenarios.
- The Islands approach offers a direct and elegant way of controlling the amount of pre-computation performed and thus also the trade-off between query and update performance. This enables the approach to accommodate varying densities of points of interest and varying query versus update frequencies.
- The paper presents an experimental evaluation that is significantly more comprehensive than previous evaluations. Specifically, this is the first evaluation that covers both online computation and pre-computation and hat considers both query and update performance in a setting with real road network data. The paper thus offers new insight into relative merits of the existing approaches.

In Section 2, we proceed to introduce related work. Section 3 presents the Islands approach and its variations. This is followed by a section that compares the Islands approach with existing kNN techniques. Section 5 then presents the empirical performance study that characterizes the Islands approach as well as compares it with the existing algorithms. The last section summarizes and offers directions for future research.

2   Related Work

Nearest neighbor computation is a classical topic. Many existing algorithms assume an indexing structure, e.g., an R-tree, and search in a branch-and-bound manner [10,18]. Many extensions and applications of kNN computation have also been proposed [1,5,11,15,20,23,24,27,28].

Query processing for objects moving in spatial networks, e.g., cars moving in road networks, has also received attention recently. However, most existing spatial query processing techniques cannot be applied directly in this setting, one reason being that the distance between two locations in a spatial network is the length of the shortest path in the network between these rather than being the Euclidean distance.

This paper assumes a specific data model and disk-based data structure for a spatial network and its associated data points as the foundation for its proposed algorithms. Among the several data models and data structures available [4,6,7,19,25], we adopt a fairly standard graph-based data model and structure [7,19] so that the algorithms are generally applicable.

We consider several existing disk-based data structures for shortest path computation and general query processing in spatial networks [8,17,22]. The CCAM structure [22] aims to support network computations such as route evaluation and aggregate queries. In this structure, a two-way partition algorithm [3] is adapted to partition the spatial network and then arrange network nodes into disk pages. Another algorithm for partitioning a road network is proposed by Huang et al. [8], and Papadias et al. [17] propose a network storage scheme for supporting both network-based and traditional Euclidean-distance-based spatial query processing. Our storage scheme enhances this scheme to capture additional aspects of real-world road networks.

To provide a thorough discussion of the existing techniques for kNN computation in spatial networks, and to compare them in detail to the Islands approach, we defer consideration of these works to Section 4.

3   The Islands Approach

Following a definition of the assumed transportation network model, concepts and observations related to the use of islands are presented. Section 3.3 presents an algorithm for kNN computation based on islands, and Section 3.4 covers several extensions to the algorithm.

3.1   Transportation Networks and Query and Data Points

We consider location-based mobile services in road networks as our application scenario. In this scenario, mobile service users are moving in a road network. A number of facilities or so-called points of interest, e.g., gas stations or supermarkets, are located within the road network. We define the network distance between a user and a point of interest as the length of the shortest path from the users' current location to the point of interest. A k nearest neighbor query issued by a service user will return the k nearest points of interest to the user based on the network distance. Using *query point* to denote a user and *data point* to denote a point of interest, we proceed to model the elements of the network scenario.

A *road network* is defined as a two tuple $RN = (G, co\mathcal{E})$, where G is a directed, labeled graph and $co\mathcal{E}$ is a binary, so-called co-edge, relationship on edges. Graph G is given by $G = (V, E)$, where V is a set of vertices and E is a set of edges. Vertices model intersections and the starts and ends of roads. An edge e models the road in-between

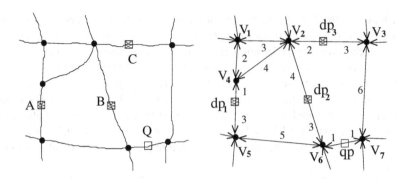

Fig. 1. Road Network Model

two vertices and is a three-tuple $e = (v_s, v_e, l)$, where $v_s, v_e \in V$ are, respectively, the start and the end vertex of the edge. The edge can be traversed only from v_s to v_e. The element l captures the travel length of the edge. Two edges e_i and e_j are in the co-edge relationship $((e_i, e_j) \in co\mathcal{E})$, if and only if they represent the same bi-directional part of a road for which U-turn is allowed.

Next, a *location* on the road network is a two tuple $loc = (e, pos)$ where e is the edge on which the location is located and *pos* represents the distance from the starting vertex of the edge to *loc*. A *data point* is modeled as a set of locations, i.e., $dp = \{loc_1, \cdots, loc_k\}$. Note that adding and removing data points or their locations does not affect the road network itself, which is important for maintainability in practice. A *query point qp* is modeled as a location.

An edge with start vertex v_i and end vertex v_j is denoted by $e_{i,j}$. Figure 1 illustrates the concepts defined above, e.g., edge $e_{1,4} = (v_1, v_4, 2)$, data point $dp_1 = \{(e_{4,5}, 1), (e_{5,4}, 3)\}$, and query point $qp = (e_{7,6}, 1)$.

The example road network in Figure 1 is assumed to have only bi-directional roads with no u-turn restrictions and each data point has two positions—one on each side of a bi-directional road. The remainder of the description of the Islands approach is carried out under these assumptions.

3.2 Observations

Intuitively, an incremental expansion process starting from the query point can be used to find the k nearest data points in Euclidean space. To optimize the search process, one can "enlarge" each data point into a big circle—see Figure 2(a)—so that the expansion process will terminate early. As shown in the figure, data point dp_3 will be found as the nearest neighbor, dp_1 is the second-nearest neighbor and dp_2 is the third-nearest neighbor. After touching the border of dp_2, the 3NN search process can stop.

In a road network, given a distance value r, the *island* of a data point dp is the subset of the road network covered by a network expansion from dp with the range r. We define r as a *radius* of this island. Intuitively, all vertices with distance to dp less than or equal to the radius belong to the island. We denote these vertices as *the island's vertices*. A vertex of an island is an *internal vertex* of the island if all its neighboring vertices are vertices of the same island. A vertex of an island is a *border vertex* of this island if at least one of its neighboring vertices does not belong to this island. All the edges connecting the island's vertices are *the island's edges*. A location (or, a query point) in the road network is *inside* an island if its network distance to the data point of this island is less than or equal to the radius of the island.

As illustrated in Figure 2(b), for the part of the road network belonging to the island of dp_1 with a radius of 5, v_4 is an internal vertex and v_1, v_2, and v_5 are border vertices. The location $loc = (e_{4,2}, 2)$ is inside this island.

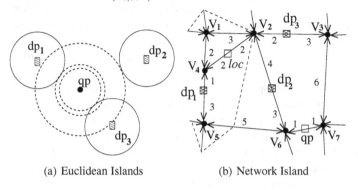

(a) Euclidean Islands   (b) Network Island

Fig. 2. Observations on Islands

To record information about islands, each vertex in the network stores references to all the data points that are centers of the islands covering the vertex. The distance from the vertex to the data point is stored with each such reference.

Then, similar to the Euclidean case in Figure 2(a), the network expansion process of a kNN query will be reduced, since a data point can be declared to be found when the expansion process visits a border vertex of this data point's island. If all islands have the same radius, and a query point is already inside l islands, the data points corresponding to these islands are the l nearest neighbors of the query point. The distances from the query point to these l neighbors are found from the above-mentioned pre-computed distances.

In general, the kNN search process includes two steps. First, we need to check the islands covering the query point. Second, if the number of such islands is smaller than k, a network expansion is needed to find additional islands.

If the islands have different radiuses, the islands approach uses the minimum radius, r_{min}, i.e., all data points are assumed to have islands with radius r_{min} (no larger than the islands they actually have). As will be explained later, having different island radiuses brings flexibility to the Islands approach. Specifically, the kNN query performance and the update efficiency can be controlled by changing the radiuses of the islands in different regions of the road network. We proceed to describe the Islands approach in more detail.

3.3   Islands-Based kNN Algorithm

The Islands approach consists of a pre-computation component and an online network-expansion component. The pre-computation component stores, for each vertex in the road network, references to the islands that cover the vertex and the network distances from the vertex to the data points that generate the islands. With this component, the network expansion, denoted as $IslandExpansion(qp, k)$, first checks the islands that the query point qp is inside and maintains the data points found in a priority queue, then starts a network expansion process from qp to find borders of new islands. The network expansion process terminates when the sum of the expansion radius and the minimum radius of all pre-computed islands exceeds the distance from the query point qp to the kth data point in the priority queue.

We proceed to describe $IslandExpansion(qp, k)$ algorithm in the following. It is similar to the INE algorithm [17], which in turn is a modified Dijkstra's single source shortest paths algorithm. Two priority queues, Q_{dp} and Q_v, are used in the algorithm to record the covered data points and vertices together with their distances to the query point, denoted as $d(qp, dp)$ and $d(qp, v)$. Both queues sort elements by the distance value and do not allow duplicate data points or vertices. The size of Q_{dp} is limited to k elements.

We introduce *update* and *deque* operations for the two queues. The $update(dp/v, dist)$ operation inserts a new data point or vertex and the corresponding distance into the queue. If this data point or vertex is already in the queue then, if $dist$ is smaller than the distance stored in the queue, the distance value in the queue is updated to $dist$. The *deque* operation removes and returns the vertex with the smallest distance. Suppose the minimum radius of all islands is r_{min}. The pseudo code of $IslandExpansion$ is given below. Queues Q_v and Q_{dp} are assumed to be empty initially.

(1)   **procedure** $IslandExpansion(qp, k)$
(2)   **for each** data point dp on edge $qp.e$: $Q_{dp}.update(dp, d(qp, dp))$
(3) $Q_v.update(qp.e.v_s, d(qp, dp.e.v_s)), Q_v.update(qp.e.v_e, d(qp, qp.e.v_e))$
(4)   **for each** dp, if its island covers $qp.e.v_s$ or $qp.e.v_e$: $Q_{dp}.update(dp, d(qp, dp))$
(5)   **if** $\exists a$, such that $(a, qp.e) \in co\mathcal{E}$, do lines (2)–(4) assuming $qp = (a, a.l - qp.pos)$
(6)   Let dp_k denote the k-th element in Q_{dp}, or $dp_k = \varnothing$, if there is no such element
(7) $d_k \leftarrow d(qp, dp_k)$ // $d_k \leftarrow \infty$ if $dp_k = \varnothing$
(8) $v \leftarrow Q_v.deque$, mark v visited
(9)   **while** $d(qp, v) + r_{min} < d_k$
(10)   **for each** non-visited adjacent vertex v_x of v
(11) $Q_v.update(v_x, d(qp, v_x))$ // $d(qp, v_x)$ assumes the path $qp \rightarrow \cdots \rightarrow v \rightarrow v_x$
(12)   **for each** dp, the center of an island covering v_x: $Q_{dp}.update(dp, d(qp, dp))$
(13) $d_k \leftarrow d(qp, dp_k)$
(14) $v \leftarrow Q_v.deque$, mark v visited
(15)   **return** Q_{dp}

Note that in line 12 of the algorithm, $d(qp, dp) = d(qp, v_x) + d(v_x, dp)$, where $d(qp, v_x)$ is taken from Q_v and $d(v_x, dp)$ is the pre-computed distance stored with v_x. Analogous computation of $d(qp, dp)$ is also performed in line 4.

To see how the algorithm works, consider Figure 2(b) and let all three data points have islands with radius 6. Starting from the query point $qp = (e_{7,6}, 1)$, the algorithm $IslandExpansion(qp, 2)$ first adds vertices v_6 and v_7 to Q_v ($Q_v = \langle(v_6, 1), (v_7, 1)\rangle$). Then it checks the islands covering v_6 and v_7, and data point dp_2 is found. Starting with v_6, the expansion process finds the islands of dp_1, dp_2 and dp_3 through the adjacent vertices v_5 and v_2. Thus, $Q_{dp} = \langle(dp_2, 4), (dp_1, 9)\rangle$ and $Q_v = \langle(v_7, 1), (v_5, 6), (v_2, 8)\rangle$. At the next step, since $r_{min} = 6$ and the distance d_2 from the query point to the 2nd

NN is 9, only vertex v_7 in Q_v needs to be checked (based on the while-loop criteria in line 9). Finally, dp_2 and dp_1 are the two data points returned. It can be observed that using the pre-computed information, the network expansion finds the data points dp_1 and dp_3 before reading the edges they are located at.

If $k = 1$, the algorithm starting from qp will find dp_2 in the first step. Since $r_{min} = 6$ and the distance from qp to dp_2 is $d_1 = 4$, the algorithm will finish without the network expansion process (since $d(qp, v_6) + r_{min} > d_1$). This, as mentioned in Section 3.2, is always the case if k or more islands cover the query point.

The *IslandExpansion* algorithm uses disk-based data structures for the network and pre-computed data. Section 4 provides a detailed description of the data structures, and it compares the Islands approach with the existing road network kNN algorithms using examples. We proceed to discuss several extensions of the Islands approach.

3.4 Extensions

An accompanying technical report [9] covers several extensions to the Islands approach, which we proceed to describe briefly. First, *shrink* and *expand* operations are provided that can be applied to any island to change its r_{min}. This offers a basis for balancing the overall query and update performance. The following section describes how these operations can be used to achieve different query and update performance trade-offs in different parts of a road network.

Second, we have seen that islands in different parts of a road network can have different radiuses. Intuitively, the radiuses of islands in urban areas should be relatively smaller than those of islands in rural areas since the densities of points of interest are relatively higher and there are relatively more frequent updates of road network data. When a kNN query is issued close to the border of areas with different r_{min} values, algorithm *IslandExpansion* can be improved to take into account the different r_{min} values. The accompanying technical report details how to modify the *IslandExpansion* algorithm to better handle such cross-area expansions.

Third, we extend the Islands approach, make changes to both the pre-computation component and the network expansion algorithm, to accommodate road networks with uni-directional edges, data points that have more than one network location, optional U-turn restrictions, and turn restrictions at intersections.

4   The Islands Approach in Comparison to Existing Techniques

The Islands approach consists of a pre-computation component and a network expansion algorithm. With the *shrink* and *expand* operations and the procedure for handling cross-area expansions, the trade-off between the performance of kNN queries and road-network updates can be controlled. For comparison purposes, we proceed to survey and exemplify the existing kNN algorithms. We also describe the disk-based data structure for the road network and the pre-computed data.

4.1   Online k Nearest Neighbor Computation

Intuitively, kNN computation can be done by employing a best-first search through adjacent edges until k neighbors are found. In contrast to the traditional shortest-path algorithms in graph theory, the kNN search in spatial networks has to employ a disk-based data structure for representing the network, the objective being to minimize I/O.

Papadias et al. [17] introduce two algorithms for kNN computation in spatial network, namely Incremental Euclidean Restriction (IER) and Incremental Network Expansion (INE). As they show that the INE algorithm outperforms the IER algorithm, we focus on the INE algorithm. This algorithm performs incremental network expansions from the query point and examines data points in the order they are encountered during the expansion process. The INE algorithm is an adaptation of Dijkstra's single source shortest paths algorithm on graphs. It terminates when the expansion's range exceeds the network distance to the kth nearest neighbor. It can be seen as a special case of the Islands approach where each data point's island has a radius of 0.

The performance of the INE approach depends on the density of the data points. Intuitively, for a large road network with only few data points, the expansion process of the INE algorithm will have to scan large parts of the road network until enough data points are collected.

4.2 kNN Pre-computation Approach

To reduce the cost of network expansion in queries, pre-computation techniques can be applied. Shahabi et al. [21] introduce a technique to transform a road network to a high dimensional space in which simpler distance functions can be used. However, this transformation requires pre-computation of the network distances between all pairs of vertices in the road network, which is often impractical.

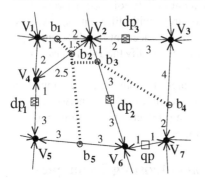

Fig. 3. Network Voronoi Diagram

Kolahdouzan and Shahabi [13] recently proposed the so-called VN3 technique for kNN computation in road networks. Starting from each data point, VN3 first creates a Network-Voronoi-Diagram [16], then pre-calculates the network distances within each Voronoi polygon. The network expansion within each Voronoi polygon can then be replaced by a look-up over the pre-computed distances.

Consider a Network-Voronoi-Diagram constructed for the example road network in Figure 1. As shown in Figure 3, the Voronoi polygon of dp_2 contains border points b_3, b_4, and b_5. Using the pre-computed information, a 2NN query from the query point qp first finds that qp is inside the Voronoi polygon of dp_2. Thus, dp_2 is its nearest neighbor. Then the network distance from qp to b_3, b_4, and b_5 can be found by look-up in a pre-computed distance table. To find the next nearest neighbor, the VN3 approach generates a candidate set consisting of "adjacent" data points, i.e., data points whose Voronoi polygons are adjacent to dp_2's polygon. Thus, dp_1 and dp_3 are included in the candidate set. Then a refinement step is used to find the actual network distance from qp to these candidate data points. Since the distances from border points of dp_1 and dp_3 to these data points are pre-computed, it requires just a look-up process to get the network distance from qp

to dp_1 and dp_3 via the border points b_3, b_4, and b_5. The VN3 approach continues this process until enough kNNs are found.

The VN3 approach excels in query performance when the density of data points is low, but is not efficient in situations when many data points are located in a small network area, e.g., points of interest in a city center. In addition, this approach does not provide a clear way of representing different "types" of data points in the road network. it is possible to construct a "multi-level" structure by constructing Voronoi diagrams for each type of data points, but such multi-level Voronoi-diagrams do not enable efficient processing of the kNN queries for multiple types of data points. An example of such query could be looking for k nearest tourist attractions such as museums, shopping malls, and parks.

4.3 Disk-Based Data Structure

The disk-based data structure for road network data used together with the INE, VN3, and Islands approaches, shown in Figure 4, is an adaptation of the data structures proposed for the INE [17] and VN3 [13] approaches. The structure has six components, the *Vertex-Edge*, *Edge-Data*, *Island-Precomputation*, *Voronoi-Polygon*, *Border-Adjacency*, and *Voronoi-Precomputation* component.

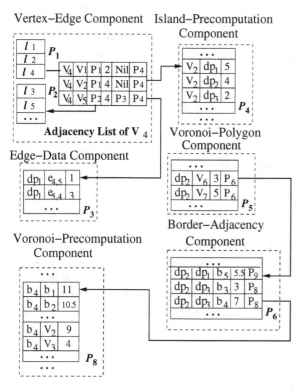

Fig. 4. Disk-Based Data Structure

The road network and data points are represented by the Vertex-Edge and Edge-Data components. Using the example road network in Figure 1, Figure 4 illustrates that the adjacency list l_4 for vertex v_4 is composed of entries standing for edges starting from v_4. The data point dp_1 located on edge $e_{4,5}$ is stored in an entry in the Edge-Data component.

Specifically, in the Vertex-Edge component, each entry denotes an edge and is of the form (*vsID*, *veID*, *pt NBVE*, *L*, *ptDP*, *ptI*). Here, *vsID* and *veID* are the id's of the start and end vertices, *ptNBVE* points to the disk page containing the end vertex, *L* is the length of this edge, *ptDP* points to the disk page containing the data points on this edge, and *ptI* points to the disk page containing Island-Precomputation data of the end vertex. Pointers are set to *Nil* if there is no linked page. Entries in the Vertex-Edge component are assigned to pages based on the Hilbert value of the start vertex.

Each entry in the Edge-Data component has the form $(dpID, eID, offset)$, where *dpID*, *eID* denote the data point and edge, and *offset* is the distance from the start vertex to the data point. We assume the *eID* value can be obtained from the *vsID* and *veID* values in the Vertex-Edge entry. Otherwise, these two attributes are used instead. Both the Vertex-Edge and the Edge-Data components are used in the INE algorithm.

For each vertex, the Island-Precomputation component stores a list containing its distance to related islands. Each entry in the list has the form $(vID, dpID, D)$ where *vID* and *dpID* denote the vertex and data point, and D is the network distance between them. These entries are arranged into pages based on the Hilbert values of the vertices. As illustrated in Figure 4, the list of vertex v_2 has three entries, describing its distances to the three data points. The *IslandExpansion* algorithm uses the Vertex-Edge component, the Island-Precomputation component and the Edge-Data component (only in the first step).

The VN3 approach uses the Voronoi-Polygon, Border-Adjacency and the Voronoi-Precomputation components. The Voronoi-Polygon component stores, for each data point, the vertices inside its Voronoi polygon. This component is used to decide the Voronoi polygon where the query point is located and provides the first nearest neighbor. Each entry has the form $(dpID, vID, D, ptB)$, where *dpID* denotes the data point generating this Voronoi polygon, *vID* is the id of a vertex inside the Voronoi polygon, D is their distance, and *ptB* points to the disk pages of the Border-Adjacency component containing the border points and adjacency information for all the Voronoi polygons. Each entry in the Border-Adjacency component has the form $(dpsID, dpeID, bID, D, ptP)$, where *dpsID* and *dpeID* denotes two data points whose Voronoi polygons are adjacent, *bID* denotes one border point of the two Voronoi polygons, D is the distance from the border point to the two data points, and *ptP* is the pointer to the disk page containing pre-computed distance values of this border point. The Voronoi-Precomputation component stores, for each border point, its distance to other border points and vertices of the same Voronoi polygons.

We assume that the edge where the query point is located is known before the query so that it can be visited directly. Otherwise, all the edges can be indexed using an R-tree, which can then be used for "map-matching." If the "id" or "name" of the edges can always be revealed for the query, a B^+-tree can be used to index these attributes and provide direct access to edges in the Vertex-Edge component. The whole disk-based data structure for the example road network in Figure 1, consisting of 9 disk pages, is presented in Figure 13 in the appendix. Each island is given a radius of 8. Each attribute value takes 1 unit size, and we set the page capacity to 54 units.

4.4 Example

Based on the example road network, we proceed to exemplify the workings of the INE, VN3, and Islands approaches. We employ an LRU buffer with a size of 2 pages and execute a 2NN query for query point $qp = (e_{7,6}, 1)$. We show the pages in the buffer and the total amount of I/Os for the three approaches. The I/O column denotes the

Approach	Steps	Q_v	Q_{dp}	d_2	Buffer	I/O
INE	1	$\langle(v_6,1),(v_7,1)\rangle$	$\emptyset$	∞	P_2	1/0
	2	$\langle(v_7,1),(v_5,6),(v_2,8)\rangle$	$\langle(dp_2,4)\rangle$	∞	P_2,P_3	2/0
	3	$\langle(v_5,6),(v_3,7),(v_2,8)\rangle$	$\langle(dp_2,4)\rangle$	∞	P_3,P_2	2/0
	4	$\langle(v_3,7),(v_2,8),(v_4,9)\rangle$	$\langle(dp_2,4),(dp_1,9)\rangle$	9	P_2,P_3	2/0
	5	$\langle(v_2,8),(v_4,9)\rangle$	$\langle(dp_2,4),(dp_1,9)\rangle$	9	P_2,P_3	2/0
	6	$\langle(v_4,9),(v_1,11)\rangle$	$\langle(dp_2,4),(dp_1,9)\rangle$	9	P_1,P_3	3/0
Island, $r_{min}=8$	1	$\langle(v_6,1),(v_7,1)\rangle$	$\langle(dp_2,4),(dp_1,9)\rangle$	9	P_2,P_4	2/0
Island, $r_{min}=7$	1	$\langle(v_6,1),(v_7,1)\rangle$	$\langle(dp_2,4)\rangle$	∞	P_2,P_4	2/0
	2	$\langle(v_7,1),(v_5,6),(v_2,8)\rangle$	$\langle(dp_2,4),(dp_1,9)\rangle$	9	P_2,P_4	2/0
	3	$\langle(v_5,6),(v_3,7),(v_2,8)\rangle$	$\langle(dp_2,4),(dp_1,9)\rangle$	9	P_2,P_4	2/0

(a) Example of the INE and Island ($r_{min}=7,8$) Approaches

Steps	Candidates	Distances	Results	Buffer	I/O
1	$\emptyset$	$\emptyset$	$\{(dp_2,4)\}$	P_2,P_5	2/0
2	$\{dp_1,dp_3\}$	$D(qp,b_4),D(b_4,dp_3),D(qp,b_5),$ $D(b_5,dp_1),D(qp,b_3),D(b_3,dp_3)$	$\{(dp_2,4)\}$	P_5,P_6	3/0
3	$\emptyset$	$\emptyset$	$\{(dp_2,4),(dp_1,9)\}$	P_8,P_9	5/0

(b) Example of the VN3 Approach

Fig. 5. Running Example of INE, Island, and VN3 Approach

amount of input/ouput (in pages). For the INE and Islands approaches, we also observe the content of the two queues Q_v and Q_{dp}, and the distance from qp to the second nearest data point, denoted as d_2. For the VN3 approach, we track the candidate set, the distance values used, and the final data points found.

It can be observed from Figure 5 that the query performance of the Islands approach is sensitive to the island radius used. When $r_{min}=8$, the query results are found by checking the islands within which the query point is located. When the radius is decreased to 7, the network expansion takes 2 more steps to finish.

Update of network and data points for the INE approach is obvious—updates only affect one or adjacent pages in the Vertex-Edge and Edge-Data component. For the Islands approach, updates cause the associated islands to be re-computed. As an example, to update data point dp_1, the re-computation will need pages P_1, P_2, and P_3 for network expansion and will then read page P_4 for updating data. Updating network and data point for the VN3 approach, as discussed in [13], requires adjacent Voronoi-Polygons to be re-generated. We use a network expansion process to update the Voronoi polygon of a data point. For example, to update the data point dp_1, a network expansion starting from dp_1 will stop after neighboring data points dp_2 and dp_3 are found. The re-computation process use disk pages P_1, P_2, and P_3. Then pages P_5, P_6, P_7, and P_9 and possibly page P_8 are accessed for updating.

5   Performance Evaluation

Two real-world datasets are used in the evaluation of the INE, VN3, and Islands approaches. The first, AAL, contains the road network of the Aalborg area in the Northern Jutland region of Denmark along with real points of interest. The network contains

11, 300 vertices, 13, 375 bi-directional edges, and 279 points of interest. The second, LA, represents the road network of Los Angeles, California. This data was obtained via the Internet [26] and converted into network files via the Tiger File Manager [2]. It contains 195, 010 vertices and 266, 335 bi-directional edges. We generate synthetic points of interest for this network.

We measure the performance of the three approaches in terms of CPU time and I/O cost. The CPU time checks, by loading the whole network and pre-computed data into physical memory, the actual running times of the experiments with the three approaches. To measure the I/O cost, we arrange the road network and pre-computation data into the data structures described in Section 4.3, we set the page size to 4k, and we employ an LRU buffer. The buffer size is set to 10% of the sum of the sizes of the Vertex-Edge and Edge-Data components. The AAL dataset contains 129 pages in the two components, and the LA dataset contains 4, 132 pages in the Vertex-Edge component. We disregard the space use that stems form the queues and variables used in the algorithms and thus do not consider them as part of the buffer.

Two series of experiments are conducted. The first series assumes that there are no updates to the road network and studies the effects on query performance of varying k, data point density, and islands radius. The density of data points is the ratio between the number of data points and the number of bi-directional edges in the road network. We define the maximum Euclidean distance between all vertices in the road network as D_{max}. The island radius used is represented as the fraction of D_{max}. In all experiments, islands of the same road network have the same radius.

The second series of experiments considers both query and update performance. We define the update ratio R_u as the ratio of updates being executed per query. The overall performance is the sum of the query and update cost. (To be consistent with the assumed application scenario, we assume an online-processing system where update operations have to be processed together with the query operations so as to provide correct query results). We use updates of edge lengths and updates of the positions of data points on an edge as standard update operations. Given an update ratio R_u and an amount of queries N, there are $N \cdot R_u$ updates on edges as well as data points. The experiments examine the effect on the overall performance of the three approaches of varying update ratio, data point density, and island radius.

In all experiments, the query points are randomly generated. For the first set of experiments, we execute a workload of 200 queries and report the average performance. For the second series of experiments, we increase the number of queries so as to get a proper amount of update operations (the update ratio is assumed to never exceed 0.1). Experiments with the same update ratio are conducted at least three times to obtain average performance figures.

The experiments are performed on a Pentium IV 1.3 GHZ processor with 512 MB of main memory and running Windows 2000. The C++ programming language is used.

5.1   Experiments on Query Performance

Query Performance Versus k. We set the island radius to 0.1 of D_{max} and k is varied from 5 to 200. The density of the (real) data points in AAL is 0.02, while the density of the data points (generated) in LA is 0.005. The results are shown in Figure 6. It can

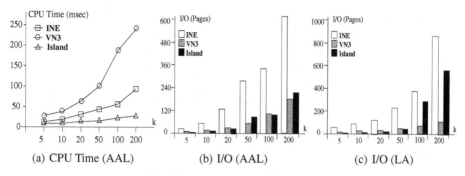

Fig. 6. Query Performance Versus k

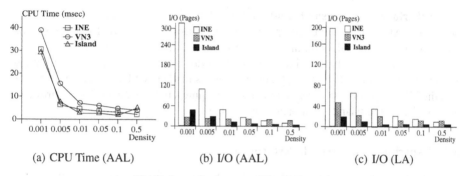

Fig. 7. Query Performance Versus Density

be observed that with the growth of k, the computational cost of all three approaches increases. The CPU time of the Islands approach is lower than those of the other two. Both the VN3 and Islands approaches show less I/O than the INE approach. The Islands approach is better than VN3 with respect to I/O cost until k exceeds 50.

Query Performance Versus Density of Data Points. We keep the island radius fixed at 0.1 of D_{max} and the value k is set to 10. We now use synthetic data points in both AAL and LA, varying the density from 0.001 to 0.5. It can be seen from Figure 7 that as the density increases, the INE approach improves substantially and becomes competitive. The Islands approach has similar behavior. It has worse performance for the AAL network and data than the VN3 approach (as shown in Figure 7(b)) when the density is less than 0.005, but becomes the best among the three approaches when the density exceeds 0.005. For the LA network and data, the Islands approach always shows better performance than the VN3 approach. The two networks differ in that the connectivity among vertices and the density of edges in the LA network are much higher than in the AAL network. This means that for the same density of data points, the network expansion process finishes earlier in the LA network than in the AAL network. The Islands approach works well in the LA network, as each island is related to more network vertices, which makes it fast for the network expansion process to discover an island. The VN3 approach, in the LA network with its high connectivity and density, possesses more border vertices and pre-computed distance data. It thus requires more I/O in its filter and refinement steps.

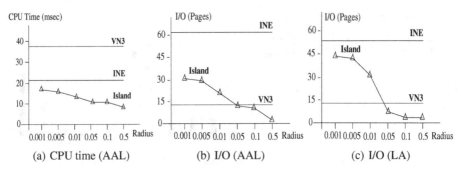

Fig. 8. Query Performance Versus Island Radius

Query Performance Versus Island Radius. We set $k = 10$ and use the real data points in AAL and synthetic data points in LA (density $= 0.005$). To determine the impact of the island radius on the query performance, the radius is varied from 0.001 to 0.5 of D_{max}. Note that in Figure 8, we also draw horizontal lines for the INE and VN3 approaches. The Islands approach always has the best CPU performance. As for I/O, the VN3 approach is best when the radius is quite small. When the radius grows to 0.05 of D_{max}, the Islands approach is preferable.

5.2 Experiments on Overall Performance

We proceed to consider the overall costs for different densities and update ratios of the INE, the VN3, and the Islands approach with two island sizes. The value of k is set to 10 in these experiments (this value is not related to the update operation). The performance costs reported are the sums of the cost of all queries and the cost of all updates of edges as well as data points, divide by the numbers of queries.

Overall Performance Versus Update Ratio. We fix the island radius at 0.01 of D_{max}. We use real data points for the AAL network and synthetic data points with density 0.005 for the LA network. The update ratio is varied from 0.0005 to 0.1 per query. It can be seen from Figure 9 that the INE approach has stable overall performance for the different update ratios, since the update operation only needs to read one or two disk pages. The VN3 approach is better than the other two approaches when the update ratio

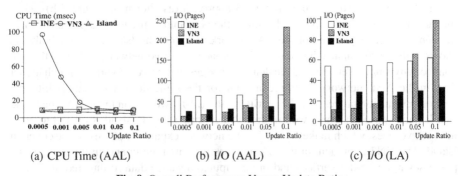

Fig. 9. Overall Performance Versus Update Ratio

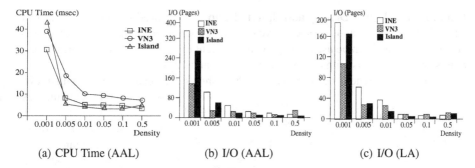

(a) CPU Time (AAL)   (b) I/O (AAL)   (c) I/O (LA)

Fig. 10. Overall Performance Versus Density

is less than 0.01. The Islands approach with a radius of 0.01 exhibits almost the same trend as the INE approach.

Overall Performance Versus Density of Data Points. The island radius remains at 0.01 of D_{max}, and the update ratio is set to 0.01. We use synthetic data points with both networks, varying the density from 0.001 to 0.5, to determine the effect on the overall performances. Figure 10 illustrates that as the density increases, the overall performances of the three approaches improve. At a low density, i.e., 0.001, the VN3 approach has the best performance. When the density grows to 0.01 and beyond, the Islands approach becomes dominant. The INE approach becomes superior when the density reaches 0.5.

Overall Performance Versus Island Radius. We retain the update ratio of 0.01 and use the real data points in AAL and synthetic data points in LA (with density = 0.005). The island radius is varied from 0.001 to 0.5 of D_{max}. Figure 11 has horizontal lines for the INE and VN3 approaches, for which the radius is not a parameter. It can be observed that the Islands approach has the best CPU performance when the radius is 0.05 or less. As for I/O, experiments on both the AAL and LA datasets show that the overall performance of the Islands approach is better than those of the INE and VN3 for certain radiuses (0.005 and 0.01 for AAL and 0.05 for LA). When the radius exceeds 0.05, the cost of re-computing the islands becomes substantial since islands grow large and overlap significantly.

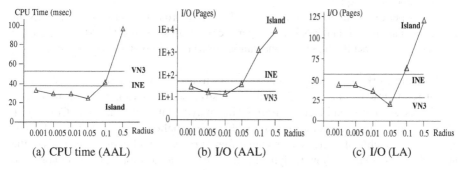

(a) CPU time (AAL)   (b) I/O (AAL)   (c) I/O (LA)

Fig. 11. Overall Performance Versus Island Radius

Island Radius Versus Density and Update Ratio. To obtain additional insight into the adaptability of the Islands approach, we conduct experiments on the LA data to check how this approach can be used to cope with different update ratios and densities of data points. We use islands with radiuses that are 0.01 and 0.05 of D_{max}.

Figure 12(a), where the density is 0.005, shows that using islands with radius 0.05 yields the best performance when the update ratio is less than 0.01. When the update ratio exceeds 0.01, the islands with radius 0.01 become the best.

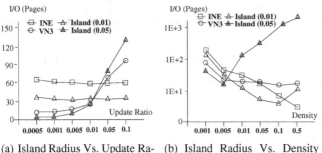

(a) Island Radius Vs. Update Ratio (LA)   (b) Island Radius Vs. Density (LA)

Fig. 12. Islands Versus Update and Density

In the experiment shown in Figure 12(b), the update ratio is fixed at 0.01. We still use islands with radiuses of 0.01 and 0.05. When the density is lower than 0.005, the Islands approach with an island radius of 0.005 achieves the best overall performance. For higher densities, the Islands approach with a radius of 0.01 is a good choice. When the density grows to 0.5, the INE approach shows the best overall performance.

6   Summary and Future Work

This paper presents a versatile approach to k nearest neighbor computation in spatial networks, termed the Islands approach, that generalizes existing re-computation and pre-computation approaches. In particular, pre-computation is performed inside so-called islands, and re-computation is performed in-between islands. An island intuitively is a sub-network with vertices and edges that are no further than a certain distance, termed the radius, away from a data point. Variation of the radiuses of islands enables the approach to accommodate networks with few as well as many data points and few as well as many updates. This enables flexible management of the trade-off between update and query cost.

The paper experimentally compares the Islands approach with two popular kNN algorithms, namely INE and VN3. The experiments result show that the Islands approach is indeed more versatile than these and can be tuned to yield better performance in most cases.

In future work, it would be of interest to try to take into account additional semantics of road networks and transportation infrastructures. For example, real-time road conditions, such as road blocks or traffic jams, may be taken into account. Computing kNN queries in such "dynamic" networks offers new challenges [4,6]. Next, the Islands approach is capable of using islands with different radiuses within different areas of the network. Techniques for how to dynamically maintain a partitioning of a network into different areas, each with its own, optimal island radius remains an open problem.

References

1. R. Benetis, C. S. Jensen, G. Karciauskas, S. Saltenis. Nearest Neighbor and Reverse Nearest Neighbor Queries for Moving Objects. In *Proc. IDEAS*, pp. 44–53, 2002.
2. T. Brinkhoff. The Tiger File Manager. http://www.fh-oow.de/institute/iapg/personen/brinkhoff/generator/.
3. C. K. Cheng, Y. C. Wei. An Improved Two-Way Partitioning Algorithm with Stable Performance. In *IEEE Trans. CAD*, 10(12), pp. 1502–1511, 1991.
4. Z. Ding, R. H. Güting. Modelling Temporally Variable Transportation Networks. In *Proc. DASFAA*, pp. 154–168, 2004.
5. H. Ferhatosmanoglu, I. Stanoi, D. Agrawal, A. E. Abbadi. Constrained Nearest Neighbor Queries. In *Proc. SSTD*, pp. 257–278, 2001.
6. R. H. Güting, V. T. de Almeida, and Z. Ding. Modeling and Querying Moving Objects in Networks. Fernuniversität Hagen, Informatik-Report 308, April 2004.
7. C. Hage, C. S. Jensen, T. B. Pedersen, L. Speičys, and I. Timko. Integrated Data Management for Mobile Services in the Real World. In *Proc. VLDB*, pp. 1019–1030, 2003.
8. Y. W. Huang, N. Jing, and E. Rundensteiner. Effective Graph Clustering for Path Queries in Digital Map Databases. In *Proc. CIKM*, pp. 215–222, 1996.
9. X. Huang, C. S. Jensen, and S. Šaltenis. The Islands Approach to Nearest Neighbor Querying in Spatial Networks. DB Tech Report TR-12. Department of Computer Science, Aalborg University, 2005.
10. G. R. Hjaltason and H. Samet. Distance Browsing in Spatial Databases. In *TODS*, 24(2), pp. 265–318, 1999.
11. G. S. Iwerks, H. Samet, K. Smith. Continuous K-Nearest Neighbor Queries for Continuously Moving Points with Updates. In *Proc. VLDB*, pp. 512–523, 2003.
12. C. S. Jensen, J. Kolář, T. B. Pedersen, I. Timko. Nearest Neighbor Queries in Road Networks. In *Proc. ACMGIS*, pp. 1–8, 2003.
13. M. Kolahdouzan and C. Shahabi. Voronoi-Based Nearest Neighbor Search for Spatial Network Databases. In *Proc. VLDB*, pp. 840–851, 2004.
14. M. Kolahdouzan and C. Shahabi. Continuous K-Nearest Neighbor Search for Spatial Network Databases. In *Proc. STDBM*, pp. 33–40, 2004.
15. F. Korn, N. Sidiropoulos, C. Faloutsos, E. Sieel, Z. Protopapas. Fast Nearest Neighbor Search in Medical Image Databases. In *Proc. VLDB*, pp. 215–226, 1996.
16. A. Okabe, B. Boots, K. Sugihara, and S. N. Chiu. Spatial Tessellations, Concepts and Applications of Voronoi Diagrams. John Wiley and Sons Ltd., 2nd edition, 2000.
17. D. Papadias, J. Zhang, N. Mamoulis, Y. Tao. Query Processing in Spatial Network Databases. In *Proc. VLDB*, pp. 802–813, 2003.
18. N. Roussopoulos, S. Kelley, F. Vincent. Nearest Neighbor Queries. In *Proc. SIGMOD*, pp. 71–79, 1995.
19. L. Speičys, C. S. Jensen, A. Kligys. Computational Data Modeling for Network Constrained Moving Objects. In *Proc. ACMGIS*, pp. 118–125, 2003.
20. T. Seidl, H. P. Kriegel. Optimal Multi-Step k-Nearest Neighbor Search. In *Proc. SIGMOD*, pp. 154–165, 1998.
21. C. Shahabi, M. R. Kolahdouzan, M. Sharifzadeh. A Road Network Embedding Technique for K-Nearest Neighbor Search in Moving Object Databases. In *GeoInformatica*, 7(3), pp. 255–273, 2003.
22. S. Shekhar, D. Liu. CCAM: A Connectivity-Clustered Access Method for Networks and Network Computations. In *TKDE*, 19(1), pp. 102-119, 1997.
23. Z. Song, N. Roussopoulos. K-Nearest Neighbor Search for Moving Query Point. In *Proc. SSTD*, pp. 79–96, 2001.

24. Y. Tao, D. Papadias, Q. Shen. Continuous Nearest Neighbor Search. In *Proc. VLDB,* pp. 287–298, 2002.
25. M. Vazirgiannis, O. Wolfson. A Spatio Temporal Model and Language for Moving Objects on Road Networks. In *Proc. SSTD,* pp. 20–35, 2001.
26. http://www.census.gov/geo/www/tiger/tgrcd108/tgr108cd.html.
27. X. Xiong, M. F. Mokbel, W. G. Aref. SEA-CNN: Scalable Processing of Continuous K-Nearest Neighbor Queries in Spatio-temporal Databases. In *ICDE,* 2005.
28. C. Yu, B. C. Ooi, K. L. Tan, H. V. Jagadish. Indexing the Distance: An Efficient Method to KNN Processing. In *Proc. VLDB,* pp. 421–430, 2001.
29. J. S. Yoo, S. Shekhar. In-Route Nearest Neighbor Queries. In *GeoInformatica,* 9(2), pp. 117–137, 2005.

Appendix

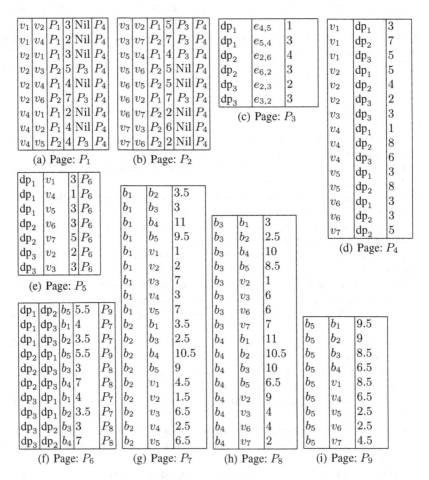

Fig. 13. Sample Data Pages

Estimating the Overlapping Area of Polygon Join

Leonardo Guerreiro Azevedo[1,*], Geraldo Zimbrão[1,2,**],
Jano Moreira de Souza[1,2], and Ralf Hartmut Güting[3]

[1] Computer Science Department, Graduate School of Engineering,
Federal University of Rio de Janeiro, PO Box 68511,
ZIP code: 21945-970, Rio de Janeiro, Brazil
[2] Computer Science Department, Institute of Mathematics,
Federal University of Rio de Janeiro, Rio de Janeiro, Brazil
[3] LG Datenbanksysteme für neue Anwendungen, FB Informatik,
Fernuniversität Hagen, D-58084 Hagen, Germany
{azevedo, zimbrao, jano}@cos.ufrj.br, rhg@fernuni-hagen.de

Abstract. Traditional query processing provides exact answers to queries trying to maximize throughput while minimizing response time. However, in many applications the response time of exact answers is often longer than what is acceptable. Approximate query processing has emerged as an alternative approach to give to the user an answer in a shorter time than the traditional approach. The goal is to provide an estimated result very close to the exact answer, along with a confidence interval, in a short time. There is a large set of techniques for approximate query processing available in different research areas. However most of them are only suitable for traditional data. This work is concerned with approximate query processing in spatial databases. We propose a new algorithm to estimate the overlapping area of polygon join using raster signatures. We executed experimental tests over real world data sets, and the results demonstrated our approach effectiveness.

1 Introduction

A main issue in the database area is to process queries efficiently so that the user does not have to wait a long time to get an answer. However, there are many cases where it is not easy to accomplish this requirement, for example: to process a huge volume of data requires a large number of I/O operations that can demand tens of minutes or hours; to access remote data can be reasonably time-consuming due to a slow network link or even temporary non-availability.

Environments for which providing an exact answer results in undesirable response times motivated the research for techniques in the approximate query processing field. The goal is to provide an estimated response in orders of magnitude less time than the time to compute an exact answer, by avoiding or minimizing the number of disk accesses to the base data [20].

* Leonardo Guerreiro Azevedo is supported by CNPq under grant number 200241/2004-4.
** Geraldo Zimbrão da Silva is supported by CAPES under grant number 3294-04-08.

C. Bauzer Medeiros et al. (Eds.): SSTD 2005, LNCS 3633, pp. 91–108, 2005.

There are many scenarios and applications where a slow exact answer can be replaced by a fast approximate one, provided that it has the desired accuracy. [13] emphasizes that in Decision Support Systems the intensification in business competitiveness that requires an information-based industry to make more use of its accumulated data, and thus techniques, of presenting useful data to decision makers in a timely manner, to be held as crucial. They also propose the use of approximate query processing during a drill-down query sequence in ad-hoc data mining, where the earlier queries in the sequence are used solely to determine what the interesting queries are. [14] and [21] present the need for performance and scalability when accessing very large volumes of data during the analysis process in data warehousing environments. [29] and [20] propose the use of approximate query processing techniques to define the most efficient access plan for a given query. [1] proposes their use in selectivity estimation in Spatial Database Management Systems (SDMS). An approximate answer can also be used as a tentative answer when the data is unavailable in warehousing environments and in distributed data recording as pointed by [20], [3] and [8] or in mobile computing as highlighted by [25]. [2] points to the use of approximate query processing in order to make decisions and infer interesting patterns online, such as over continuous data streams.

There is a large set of techniques for approximate query processing available in different research areas. However, most of them are only suitable for relational databases. Good surveys of techniques for approximate query processing are presented in [4] and [12]. On the other hand, providing a short time answer to users' queries becomes a bigger challenge in spatial database area, where the data usually have high complexity and is available in huge amounts. Furthermore, this subject is a hot research issue in spatial-temporal databases as pointed by [15]. Moreover, spatial query processing techniques assume that the positional attributes of spatial objects are precisely known. In practice, however, they are known only approximately, with the error depending on the nature of the measurement and the source of data, as pointed by [5] and [16]. So the "exact answer" is actually an approximation, although it is close to the real answer.

[23] defines a spatial database system as a full-fledged database system with additional capabilities for representing, querying, and manipulating geometric data. Such a system provides the underlying database technology needed to support applications such as geographical information systems and others. Spatial data types like point, line, and region provide a fundamental abstraction for modeling the structure of geometric entities, their relationships, properties, and operations.

Efficient evaluation of spatial queries is an important issue in spatial database. Among spatial operations, spatial join operations are very useful but costly to evaluate. Spatial joins have been well studied in the literature, and there are many approaches to process spatial join operations. [9] emphasizes that traditional approaches to performing spatial join processing in two steps ([11] and [24]), and proposes efficient algorithms to be used in the second step. In the two-step approach, the first step employs a Spatial Access Method (SAM) in order to reduce the search space. The Minimum Bounding Rectangle (MBR) is usually used by SAM methods. The second step is a refinement step where the objects resulting from the first step are read from disk and have their geometries processed. On the other hand, [26] proposes a Multi-Step Query Processor (MSQP) including another step between the first and the second step presented previously. In the proposed step the output resulting from the first step is processed against a geometric filter that uses a compact and approximate representation of the object, such as Convex Hull, 5C, RMBR and others found in [27]. The goal is to reduce the number of objects that will have their exact geometry processed in the last step. However, in both approaches (processing the spatial join in two or three steps) it is necessary to process the exact geometries of the objects, the most expensive step that consumes more CPU and I/O resources. To be the best of our knowledge, there is no approach that does not execute the last step, returning to the user an approximate answer along with a confidence interval, processing the join predicate on small approximations of data and not reading the real objects from the disk.

This work is concerned with approximate query processing in spatial databases. We extended the approach presented in [17] in which the use of Four-Colours Raster Signature [6] for approximate spatial query processing was introduced. We propose a new algorithm to compute the approximate intersection area of polygon × polygon, processing the query on 4CRS raster approximation, along with a confidence interval that is returned to the user allowing him to decide if the accuracy of the response is sufficient. Besides, we also present experimental results in order to show the effectiveness of our approach. One application that could benefit from our approach is the agriculture production estimation. According to the estimated values of agriculture production, several decisions must be taken, for example number and size of warehouses that will store the harvest, number of transports that must be available, roads and railroads that must be (re)constructed, etc. Several spatial joins involving the overlay of thematic planes such as soil, rural areas, rainfall indicators, pollution, areas that are open to pest attacks, etc., must be evaluated to estimate the agriculture production, something that can take a lot of time. On the other hand, a fast approximate answer could be enough for the agriculture production estimation.

The work has been divided in sections, as follows. Section 1 is the introduction. Section 2 presents the most important characteristics of Four-Colours Raster Signature for this work and our proposal of using Four-Colours Raster Signature for estimating the overlapping area of polygon join. Section 3 is dedicated to present the experimental results. Finally, Section 4 shows the conclusions and the future developments of this work.

2 Four-Colours Raster Signature and Estimating the Overlapping Area of Polygon Join

2.1 Four-Colours Raster Signature

The Four-Colours Raster Signature (4CRS) was introduced by [6] to be used as a polygon approximation in spatial join processing. The characteristics of 4CRS and its advantages over other methods motivated its use in approximate query processing area as well. The target of this new approach is to reduce the time required to process a query by avoiding accessing the real datasets which can lead to large amount of time, and processing an approximate query through the execution of a fast algorithm on approximate data, much smaller than the real one. On the other hand, the answer will be estimated and not exact. So, it is also necessary to return a confidence interval in order to have a precision measure of the approximate answer. In general, it is enough for the user to have an approximate answer to make his decision since it has a short execution time and the desired accuracy.

The 4CRS of one polygon is a raster approximation represented by a small four-colour bitmap upon a grid of cells. Each cell of the grid has a colour representing the percentage of the polygon's area within the cell, as shown in Table 1. In Figure 1, an example of 4CRS is presented. The grid can have its scale changed in order to obtain a more accurate representation (higher resolution) or a more compact one (lower resolution). Further details of 4CRS signature can be found in [6] and [17].

Table 1. Types of 4CRS cell

Bit value	Cell type	Description
00	Empty	The cell is not intersected by the polygon
01	Weak	The cell contains an intersection of 50% or less with the polygon
10	Strong	The cell contains an intersection of more than 50% with the polygon and less than 100%
11	Full	The cell is fully occupied by the polygon

Figure 2 presents two examples of grid of cells of the same size. It is easy to notice that it is harder to figure out a simple algorithm that executes on grids like the one presented in Figure 2.a than to figure out a simple algorithm that executes on perfectly overlapped grid, as shown in Figure 2.b. [6] presents an approach for computing the grid of raster approximations where the space is divided into cells independently of the object position through a universal grid so that the coordinate system determines the grid. By doing so, it is assured that if two cells overlap each other then their sides are perfectly superimposed (Figure 2.b). Also, the length of each cell side is always a power of two. So, if two 4CRS signatures have different lengths of cell side and they overlap each other, it is ensured that a small cell is entirely within a great one. This approach was employed in this work, and more details about it can be found in [6].

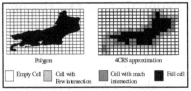

Fig. 1. Example of 4CRS signatures

Fig. 2. Grids of cells with same size (a) not overlapping perfectly and (b) overlapping perfectly

When executing query processing on two 4CRS approximations, it is essential that both of them have the same cell size. If that does not apply, it is imperative to perform a change of scale. This is accomplished through the grouping of cells of the approximation with smaller cell size. The algorithm to change the scale evaluates the average of the sum of numerical values assigned to each type of cell, which represents the percentage of the polygon's area within the cell. For *Empty* and *Full* cells the numerical values are 0% and 100%, respectively, since these values represent the exact percentage of intersection area of the cell and the polygon. Due to the fact that in approximate query processing an exact answer is not required, but a close approximate one, in this work we propose to use the average percentage of polygon's area inside the cell as the numerical values for *Weak* and *Strong* cells, which are of 25% and 75%, respectively. These values can be used because the grid and the polygon are independent from each other, and it is expected that the distribution of the percentage of the polygon's area within the cell is very close to the uniform distribution. In fact, we computed the distribution of the polygon area within the cell for the township dataset of Iowa (US) in intervals of 1%, and the result suggests that the uniform distribution assumption holds. Moreover, as shown in [17], the measure used for computing the confidence interval is the variance. Assuming the uniform distribution, the variance of area of weak cells in percentage is $(0.5-0)^2/12 = 1/48 = 0.020833$, since weak cells have distribution between $(0, 0.50]$. The strong cell has the same variance. In our test over township dataset of Iowa (US), the computed variances were 0.021978 and 0.021952 for weak and strong cells respectively, whose values are very close to the variance assuming the uniform distribution.

2.2 Expected Area

In this section, the calculus of the expected areas corresponding to the overlapping of two different types of cells with the same size is presented. These expected areas are employed by the algorithm for estimating the overlapping area of polygon join, which is presented in Sub-Section 2.3.

It is easy to notice that the expected area corresponding to a combination of an *Empty* cell with any other type of cell results in an expected area of 0% (zero percent). In the same way when two *Full* cells overlap, the expected area is 100%. Thus, we compute the expected intersection areas for the overlapping of the other type of cells. They were estimated as the mean value of the possible percentage occurrences of the intersection area between two types of cells.

As the datasets are reasonably independent (for example, there is no rule that all township boundaries must be defined by courses of rivers), we can assume that the expected area corresponding to the intersection of two cells with areas x_1 and x_2 is $x_1 \times x_2$. For instance, the expected area corresponding to the overlapping of two *Weak* cells with 10% and 15% of the area of the polygon within them is 1.5% (0.01×0.15). Besides, even though the area is a continuous value, in order to make easy the demonstration of the calculus, we are assuming that the cell area is computed as discrete values, in steps of size of $1/n$ for a large n ($n \rightarrow \infty$). Also, all the values are shown in percentage.

Let X be a random variable representing the computed intersection area of one cell of the grid against the polygon; $G(x_1, x_2)$ a function that gives the intersection area between two types of cells x_1 and x_2; and $p(x_1, x_2)$ the join probability function of two variables X_1 and X_2. The definition of mean (or expected value E) of two variables is presented in Equation 1 [7].

$$E[G(X_1, X_2)] = \sum_{x_1} \sum_{x_2} G(x_1, x_2) \times p(x_1, x_2) \cdot \tag{1}$$

Since the intersection area between a cell and a polygon is independent of the intersection area of another cell and the polygon, X_1 and X_2 are linearly independent and the joint probability function $p(x_1, x_2)$ can be expressed as $p(x_1, x_2) = p(x_1) \times p(x_2)$. In addition, let n be the possible observed values of the percentage of the area of the polygon within the cell. Thus $p(x_1)$ and $p(x_2)$ are equal to $1/n$, since that each value for the intersection area has the same probability of occurrence. Besides, $G(x_1, x_2)$ can be expressed as the multiplication of the intersection areas of the cells within the polygon. Therefore for n different kinds of cell intersections $E[G(x_1, x_2)]$ can be approximately given by Equation 2.

$$E[G(X_1, X_2)] = \mu = \sum_{i=1}^{n} \sum_{j=1}^{n} \delta(x_1) \times \delta(x_2) \times p(x_1) \times p(x_2) \cdot \tag{2}$$

Where $\delta(x)$ is a function that returns the percentages of the area of the polygon within the cell. This function can be expressed as equations 3 and 4.

$$\delta(k) = \frac{k}{2n}, \ 1 \le k \le n, \text{ for } Weak \text{ cell} . \tag{3}$$

$$\delta(k) = \left(\frac{k}{n} + \frac{1}{2} \right), \ 1 \le k \le n, \text{ for } Strong \text{ cell} . \tag{4}$$

In the case of *Weak* and *Strong* cells the percentages vary in the intervals $(0, 50\%]$ and $(50\%, 100\%)$, respectively. While the percentages for *Empty* cell is 0% and for *Full* cells is 100%.

From equations 2, 3 and 4 the expected area of the overlapping of two *Weak* cells employed by the algorithm for computing the approximate intersection area of two polygons can be calculated as follows.

- *Weak x Weak cells*

$$E[G(Weak_1, Weak_2)] = \mu = \sum_{i=1}^{n} \sum_{j=1}^{n} \delta_i(weak_1) \times \delta_j(weak_2) \times p(weak_1) \times p(weak_2) \tag{5}$$

$$\mu = \underset{n \to \infty}{Lim} \sum_{i=1}^{n} \sum_{j=1}^{n} \frac{i}{2n} \frac{j}{2n} \frac{1}{n} \frac{1}{n} = \underset{n \to \infty}{Lim} \frac{1}{4n^4} \sum_{i=1}^{n} i \sum_{j=1}^{n} j \cdot$$

Since the sum of the sequence $\sum_{k=1}^{n} k$ can be expressed as $\sum_{k=1}^{n} k = \frac{n(n+1)}{2}$, and using the L'Hôpital rule, Equation 5 can be rewritten as Equation 6.

$$\mu = \underset{n\to\infty}{Lim}\frac{1}{4n^4}\sum_{i=1}^{n}i\sum_{j=1}^{n}j = \underset{n\to\infty}{Lim}\frac{1}{4n^4}\frac{n(n+1)}{2}\frac{n(n+1)}{2} = \frac{1}{16}\underset{n\to\infty}{Lim}\frac{(n^2+n)(n^2+n)}{n^4} = \frac{1}{16}. \tag{6}$$

Following the same reasoning the expected area of the intersection of *Weak* x *Strong* cells, *Strong* × *Strong* cells, *Weak* × *Full* cells, and *Strong* × *Full* cells have the values 3/16, 9/16, 1/4 and 3/4, respectively. Table 2 presents the expected overlapping areas of different types of cells.

Table 2. Expected areas of the overlapping of different types of cells

Cell types	Empty	Weak	Strong	Full
Empty	0	0	0	0
Weak	0	0.0625	0.1875	0.25
Strong	0	0.1875	0.5625	0.75
Full	0	0.25	0.75	1

2.3 Algorithm for Estimating the Overlapping Area of Polygon Join

The algorithm for estimating the overlapping area of polygon join computes the sum of the expected area of their 4CRS signatures' cells that overlap each other, and multiplies the resulting value by the cell's area. Since there are four different types of cells, the superimposing possibilities are sixteen (Table 2), and the algorithm employs a matrix to store the expected areas. It is only necessary to consider the cells that are inside the intersection MBR of the two 4CRS signatures. The algorithm in C-like language is presented in Figure 3, and it handles 4CRS signatures with different or the same length of cell side. It is ensured that when two cells intersect, their sides overlap exactly, and when the lengths of cell sides are different it is always ensured that the smaller cell is whole contained by greater one, according to the approach used to compute the grid of cells presented in Sub-Section 2.1.

```
void approxIntersectionArea(signat4CRS1, signat4CRS2)
  approximateArea = 0;
  interMBR = intersectionMBR(signat4CRS1, signat4CRS2);
  if (signat4CRS1.lengthOfCellSide ==
      signat4CRS2.lengthOfCellSide) then
    s4CRS = signat4CRS1;
    b4CRS = signat4CRS2;
  else
    s4CRS = smallerCellSide(signat4CRS1, signat4CRS2);
    b4CRS = biggerCellSide (signat4CRS1, signat4CRS2);
  approximateArea = 0;
  For each b4CRS cell b that is inside interMBR Do
    For each s4CRS cell s that is inside cell b Do
      approximateArea += expectedArea[s.type,b.type];
  cellArea = s4CRS.lengthOfCellSide *
             s4CRS.lengthOfCellSide;
  return approximateArea * cellArea;
```

Fig. 3. Algorithm for computing the approximate intersection area of polygon × polygon

2.4 Confidence Interval Calculus

When executing a query whose result is an approximate answer, it is important to show to the user a confidence interval of the query's answer, so that the user can decide if the precision of the approximate answer is enough. The precision measure used in this work is based on the Central Limit Theorem [22], which holds almost regardless of the form of the density function. The Central Limit Theorem states that if a population has a mean μ and a variance σ^2, then the distribution of sample means derived from this distribution approaches the normal distribution with mean μ and variance σ^2/n as the sample size n increases. Thus, at some stage, means for large enough sample sizes, whether the random variable is discrete or continuous, will be approximately normally distributed. Clearly, the form of the parent density function will have some effect on the sample size required, and an asymmetric distribution will generally call for a large n than a symmetric one. However, a sample size of 30 is sufficiently large for many distributions. The confidence interval for approximate processing is computed as the sum of the confidence intervals of each combination of pair of cells. Consulting a statistical table of normal distribution, for a 95% confidence interval we have a range of $(\mu \pm 1.96 \times (\sigma^2/n)^{1/2})$, and for a 99% confidence interval we have $(\mu \pm 2.576 \times (\sigma^2/n)^{1/2})$. Equation 7 was used for computing the confidence interval of our experiments.

$$\sum_c n_c \times \left(\mu_c \pm p \times \sqrt{\frac{\sigma_c^2}{n_c}} \right). \tag{7}$$

- μ_c and σ_c^2 correspond to the mean and the variance of a combination of cells c in the set $\{Empty \times Empty,\ Empty \times Weak,\ ...,\ Weak \times Weak,\ ...,\ Full \times Full\}$;
- p is the value corresponding to the confidence interval chosen, i.e., 1.96 for a 95% confidence interval;
- n_c is the number of cells for the combination c.

In order to get the result in area units it is necessary to multiply the result by the cell's area.

For the confidence interval calculus it is necessary to have computed the mean and variance values of the expected areas corresponding to the overlapping of two different types of cells with the same size. Mean values are presented in Table 2 (Sub-Section 2.2) and the calculus of the variance for each combination is presented as follows.

The expected area corresponding to a combination of an *Empty* cell with any other type of cell results in an expected area of 0% (zero percent), because of the intersection area of such kinds of cells is zero. Consequently, the variance of the expected area is zero. In the same way, when two *Full* cells overlap, the expected area is always 100%, and the variance is also zero. Thus, we only need to compute the variances of the expected intersection areas for the overlapping of the other types of cells. We use the same assumptions that were used to calculate the expected areas corresponding to the overlapping of two different types of cells with the same size (Sub-Section 2.2).

Let X be a random variable representing the computed intersection area of one cell of the grid against the polygon; $G(x_1, x_2)$ a function that gives the intersection area between two types of cells x_1 and x_2; and $p(x_1, x_2)$ the join probability function of two variables X_1 and X_2, the variance of the intersection area of two different types of cells can be expressed as Equation 8.

$$\sigma^2 = \sum_{x_1} \sum_{x_2} \left(G(x_1, x_2) - \mu \right)^2 \times p(x_1, x_2) \cdot \tag{8}$$

In the same way as presented in Sub-Section 2.2, we assume that X_1 and X_2 are linearly independent and the joint probability function $p(x_1, x_2)$ can be expressed as $p(x_1, x_2) = p(x_1) \times p(x_2)$; $p(x_1)$ and $p(x_2)$ can be expressed as $p(x_1) = p(x_2) = 1/n$; and, $G(x_1, x_2)$ is the multiplication of the intersection areas of the cells within the polygon. By doing so, Equation 8 can be rewritten as Equation 9.

$$\sigma^2 = \sum_{i=1}^{n} \sum_{j=1}^{n} \left(\delta(x_1) \times \delta(x_2) - \mu \right)^2 \times p(x_1) \times p(x_2) \cdot \tag{9}$$

Where $\delta(x)$ is a function that returns the percentages of the area of the polygon within the cell. This function can be expressed as equations 3 and 4 (Sub-Section 2.2). Thus, from equations 3, 4 and 9 the variance of the percentage of the intersection area between two *Weak* cells can be calculated as follows (Equation 10).

- *Weak × Weak*

$$\sigma^2 = \sum_{i=1}^{n} \sum_{j=1}^{n} \left(\delta(weak_1) \times \delta(weak_2) - \mu_{weak \times weak} \right)^2 \times p(weak_1) \times p(weak_2)$$

$$\sigma^2 = Lim_{n \to \infty} \sum_{i=1}^{n} \sum_{j=1}^{n} \left(\frac{i}{2n} \times \frac{j}{2n} - \mu_{weak \times weak} \right)^2 \times \frac{1}{n} \times \frac{1}{n}$$

$$\sigma^2 = Lim_{n \to \infty} \sum_{i=1}^{n} \sum_{j=1}^{n} \left(\frac{i^2 j^2}{16n^4} - \frac{ij\mu_{weak \times weak}}{2n^2} + \mu_{weak \times weak}^2 \right) \times \frac{1}{n^2} \tag{10}$$

$$\sigma^2 = Lim_{n \to \infty} \sum_{i=1}^{n} \sum_{j=1}^{n} \frac{i^2 j^2}{16n^6} - Lim_{n \to \infty} \sum_{i=1}^{n} \sum_{j=1}^{n} \frac{ij\mu_{weak \times weak}}{2n^4} + Lim_{n \to \infty} \sum_{i=1}^{n} \sum_{j=1}^{n} \frac{\mu_{weak \times weak}^2}{n^2} \cdot$$

Since the sum of the sequences $\sum_{k=1}^{n} k$ and $\sum_{k=1}^{n} k^2$ can be expressed as

$\sum_{k=1}^{n} k = \dfrac{n(n+1)}{2}$ and $\sum_{k=1}^{n} k^2 = \dfrac{2n^3 + 3n^2 + n}{6}$, and using the L'Hôpital rule, the three

limits of Equation 10 can be solved as equations 11, 12 and 13.

$$Lim_{n \to \infty} \sum_{i=1}^{n} \sum_{j=1}^{n} \frac{i^2 j^2}{16n^6} = \frac{1}{16} Lim_{n \to \infty} \frac{\sum_{i=1}^{n} \sum_{j=1}^{n} i^2 j^2}{n^6} = \frac{1}{16} Lim_{n \to \infty} \frac{\frac{2n^3 + 3n^2 + n}{6} \times \frac{2n^3 + 3n^2 + n}{6}}{n^6} = \frac{1}{144} \cdot \tag{11}$$

$$Lim_{n\to\infty}\sum_{i=1}^{n}\sum_{j=1}^{n}\frac{ij\mu_{weak\times weak}}{2n^4}=\frac{\mu_{weak\times weak}}{2}Lim_{n\to\infty}\frac{\sum_{i=1}^{n}\sum_{j=1}^{n}ij}{n^4}=\frac{\mu_{weak\times weak}}{2}Lim_{n\to\infty}\frac{\sum_{i=1}^{n}i\frac{n(n+1)}{2}}{n^4}$$

(12)

$$=\frac{\mu_{weak\times weak}}{2}Lim_{n\to\infty}\frac{\frac{n(n+1)}{2}\times\frac{n(n+1)}{2}}{n^4}=\frac{\mu_{weak\times weak}}{8}=\frac{1}{16}\times\frac{1}{8}=\frac{1}{128} \cdot$$

$$Lim_{n\to\infty}\sum_{i=1}^{n}\sum_{j=1}^{n}\frac{\mu_{weak\times weak}^2}{n^2}=\mu_{weak\times weak}^2 Lim_{n\to\infty}\frac{n\times n}{n^2}=\mu_{weak\times weak}^2=\left(\frac{1}{16}\right)^2=\frac{1}{256}$$

(13)

Applying equations 11, 12 and 13 in Equation 10, the variance of the percentage of the intersection area between two *Weak* cells are presented in Equation 14.

$$\sigma^2=\sum_{i=1}^{n}\sum_{j=1}^{n}(\delta(weak_1)\times\delta(weak_2)-\mu_{weak\times weak})^2\times p(weak_1)\times p(weak_2)$$

(14)

$$\sigma^2=\frac{1}{144}-\frac{1}{128}+\frac{1}{256}=0.003038194 \cdot$$

The variances of the expected areas of the intersection of other types of cells can be calculated following the same reasoning. They are presented in Table 3, and we do not present their calculus due to space limitations.

Table 3. Variance of the expected areas of the overlapping of different types of cells

Cell types	Empty	Weak	Strong	Full
Empty	0	0	0	0
Weak	0	0.003038194	0.013454861	0.020833333
Strong	0	0.013454861	0.023871528	0.020833333
Full	0	0.020833333	0.020833333	0

Therefore it is possible to return to the user a confidence interval for the approximate query processing. For instance, let a query to produce the following pair of cells 100 *Weak* × *Weak* cells, 40 *Weak* × *Strong* cells, 70 *Weak* × *Full* cells, 60 *Strong* × *Strong* cells and 200 *Full* × *Full* cells we compute the 95% confidence interval as presented in Figure 4 (for simplicity we assume that each cell has the same area, equals to 1).

- WxW:100 × (0.0625 ± 1.96 × (0.0030382/100)$^{1/2}$) = 6.25 ± 1.0803
- WxS: 40 × (0.1875 ± 1.96 × (0.013454/40)$^{1/2}$) = 7.50 ± 1.4378
- WxF: 70 × (0.2500 ± 1.96 × (0.020833/70)$^{1/2}$) = 17.50 ± 2.3669
- SxS: 60 × (0.5625 ± 1.96 × (0.023872/60)$^{1/2}$)= 33.75 ± 2.3457
- FxF: 200 × 1 = 200 (full cells have the exact area!)
- Total: 265 ± 7.2308.

Fig. 4. Example of 95% confidence interval calculus

So, the confidence interval has a range of ±7.2308 that is 95% of the approximate answers with these numbers of cell combinations will have an error of at most ±2.7286%, a result with enough precision for most applications. For a 99% confidence interval, it is necessary to replace 1.96 to 2.576 in the calculus presented in Figure 4. In this case, the computed value is 265 ± 9.5034. The confidence interval has a range of ±9.5034, which means an error of at most ±3.5862% in 99% of the cases.

3 Experimental Results

This section is dedicated to present the experimental results found by using 4CRS signature for estimating the overlapping area of polygon join. In order to evaluate the effectiveness of our approach we compared the approximate processing against the exact processing according to the following metrics: response time (the time to provide an approximate answer for a query); accuracy (the precision of the answers, along with a confidence interval); and footprint (the storage requirements for the approximations).

3.1 Test Environment, Experimental Data Sets, 4CRS Signatures and R*-Tree Characteristics

Tests were executed on a PC Pentium IV 1.8 GHz with 512 MB of RAM. A page size of 2,048 bytes for I/O operations was defined. The polygon real data sets used in the experiments consist of township boundaries, census block-group, geologic map and hydrographic map from Iowa (US), available online at "http://www.igsb.uiowa.edu/rgis/gishome.htm", and Brazilian municipalities [10]. In order to simulate large datasets, the Iowa datasets were replicated six times, in the same way as suggested by [26]. The original polygons were shifted by random displacements of x and y coordinates. In the case of Brazilian municipalities, we performed only one replication (named Brazilian municipalities'), so that we could execute the test of Brazilian municipalities against Brazilian municipalities'. Some data characteristics are presented in Table 4.

Table 4. Test data sets characteristics

	Datasets	size (KB)	# pol.	# seg.	Avg # seg.
	Census block-group	38,824	17,844	1,764,588	98
	Topography	60,748	40,140	7,561,104	188
Iowa	Hydrologic map	6,904	2,670	475,812	178
	Township boundaries	25,288	12,216	1,059,438	86
	Geological map	21,856	9,984	640,428	64
Brazil	Municipalities	9,840	4,645	399,002	85
	Municipalities'	9,840	4,645	399,002	85
	Average	24,757	13,163	1,757,053	112

In order to generate the 4CRS signatures, we have to choose the maximum number of cells of the grid [6]. Intuitively, the larger the number of cells, the closer is the approximation to the original polygon. However, processing 4CRS signatures that have large sizes could produce high I/O and CPU costs. To evaluate the effects of different choices, we executed experimental tests with maximum number of cells of 250, 500, 1000, 1500 and 2000. We evaluated the approximate processing against the exact processing computing the intersection area of dataset 1 × dataset 2 presented in Table 7. Signatures with maximum number of cells equal to 250 have smaller storage requirements, but the precision of the approximate answers is not good enough. On the other hand, the answers are better estimated when the maximum number of cells was 2000; however the I/O and CPU costs are higher as well, because of the higher signature sizes. Figure 5 summarizes these experimental results showing: storage requirements (percentage of 4CRS signatures' sizes related to the datasets' sizes); error of the approximate answer (the percentage corresponding to the difference between the approximate value and the exact value related to exact value); percentage of the time required to execute the approximate processing related to the exact processing; and the percentage corresponding to the number of disk accesses needed to execute the approximate processing related to the exact one. We present in details in Sub-Section 3.2 the experimental results when 500 was used as the maximum number of the grid cells, which produced approximate answers with acceptable average error and confidence interval.

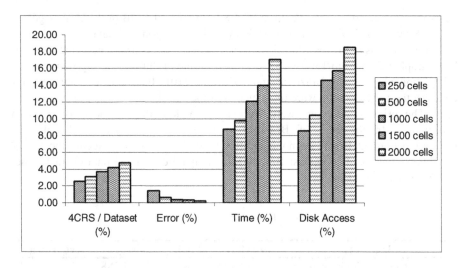

Fig. 5. Storage requirements, accuracy and number of disk access for maximum number of cells of the grid equal to 250, 500, 1000, 1500 and 2000

The 4CRS signature generation time was not shown because [6] evaluated its efficiency and presented good results. Table 5 presents the 4CRS signatures characteristics for the maximum number of cells equals to 500. We can notice that, in order to store 4CRS signatures of maximum number of cells equal to 500 it is only

Table 5. 4CRS signatures' characteristics with maximum number of cells equal to 500

	Datasets	Dataset size (KB)	4CRS size (KB)	4CRS / Dataset size (%)
Iowa	Census block-group	38,824	1,603	4.13
	Hydrologic map	6,904	177	2.56
	Township boundaries	25,288	838	3.31
	Geological map	21,856	676	3.09
Brazil	Municipalities	9,840	329	3.34
	Municipalities'	9,840	329	3.34
Average		18,759	659	3.30

needed, on average, 3.30% of the space needed to store the real datasets. In other words, it is necessary approximately 30 times more space to store the real datasets than to store the 4CRS signatures.

To perform the join, the R*-tree [19] was chosen as a spatial access method in order to reduce the search space. In other words, the R*-Tree was used to take account only the objects that have at least MBR intersection and not all of them. That choice was due to the wide use of R*-Tree, as well as, to the successful results found in the literature. The access methods traditionally used employ the object's Minimum Bounding Rectangle (MBR), and the access methods execution returns what is called a set of candidates, since it contains all the pairs of polygons that belong to the answer plus other pairs that have only MBR intersection. In the same way as [26] and [6] do, for our tests we generated R*-Trees that store the 4CRS signatures as part of the polygons' keys, and this means that they were stored in the leaf nodes in the R*-Tree index. It is a reasonable approach since in this way we have to compute the 4CRS just once.

Our tests can be described according to the concepts of Multi-Step Query Processor (MSQP) proposed by [26], presented in Section 1. In the approximate query processing, only the first two steps of the MSQP (SAM + Filter steps) were executed. Since it is not necessary to access the real objects when computing an approximate answer, the last step of MSQP was not executed. On the other hand, in the exact query processing, we executed the first and last step of MSQP (SAM + Refinement steps). In other words, after finding the objects that have MBR intersection, the exact representation of the objects was processed, and exact answers returned. To perform a fair test we generated R*-Trees without storing the 4CRS signatures on their leaf nodes to be used in the exact query processing. By doing so, the sizes of the R*-Trees without storing signatures are smaller than the sizes of the R*-Trees that store them, consequently the number of disk accesses in the first step is smaller as well. The R*-Trees characteristics are presented in Table 6. The column "R*-Tree type" shows that the characteristics presented are of R*-Tree that stores 4CRS signatures or R*-tree that do not store signatures.

In order to evaluate the 4CRS effectiveness in the approximate processing area, besides the storage requirements, we evaluated the approximate processing against the exact processing testing the accuracy of the approximate answer, execution time and disk accesses. The approximate query processing was done executing the algorithm proposed in the Sub-Section 2.3, while the exact query processing was performed using the General Polygon Clipping library that is available on the web at http://www.cs.man.ac.uk/aig/staff/alan/ software/#gpc.

Table 6. R*-Trees' characteristics

	Datasets	R*-Tree type	R*-Tree size (KB)	Time (sec)	Leaf node average use (%)	Height	# leafs
Iowa	Census block-group	4CRS	2,124	19.04	69.98	3	1045
		-	1,160	17.93	69.81	3	570
	Hydrologic map	4CRS	334	2.24	68.33	3	162
		-	162	2.14	75.35	2	79
	Township boundaries	4CRS	1,546	12.95	68.70	3	760
		-	800	11.97	69.50	3	392
	Geological map	4CRS	1,258	9.55	68.41	3	617
		-	644	9.32	70.46	3	316
Brazil	Municipalities	4CRS	586	4.66	71.15	3	286
		-	284	4.07	75.05	3	138
	Municipalities'	4CRS	582	4.92	71.63	3	284
		-	284	4.11	75.05	3	138
	Average	4CRS	1,289	8.89	69.70	3	525
		-	663	8.26	72.54	3	272

3.2 Results of Approximate Query Processing

This sub-section is dedicated to presenting, in detail, the experimental results when the maximum number of cells of the grid was 500. The results correspond to: precision of the approximate answer, including confidence intervals; processing time; and number of disk accesses. Storage requirements of 4CRS signatures were presented in Sub-Section 3.1 (Table 5). We executed queries computing the intersection area of dataset 1 against dataset 2 (presented in Table 7) comparing the approximate processing against the exact one. Each query was executed 20 times, and for each time we generated a random window so that only the considered pairs of objects were inside the window. In order to evaluate the effect of the number of objects returned by each query, we executed two different tests. In one test the random windows were generated with size of 4% of the size of the whole space of the datasets, and in the other test the windows were generated with size of 12.25%. The results are presented in Table 8 and Table 9. Since the values of both tests are quite similar, we will only analyze in more details the results corresponding to the second test (Table 9). The most relevant difference between the tests is that in the second test each window is intercepted by more objects than the number of intersections in the first test. As a result, the number of cells considered to compute the confidence interval is bigger and its value is closer to the computed approximate answer.

Experimental results show the effectiveness of the use of 4CRS signature in the approximate processing area due to the quite small error of the approximate answers, the short time of the approximate processing and the small number of disk accesses. The average error of the approximate answers is 0.59%, while the confidence intervals of 95% and 99% have average values of 0.97% and 1.28%, respectively (Table 9, column "Error and confidence interval"). In other words, the approximate answers have on average a difference of only 0.59% related to the exact ones. Besides, in order to show the accuracy of the approximate answers, a confidence interval is also returned to the user which means that for a precision of 95% the error is at most ±0.97%, while for a precision of 99% the error is at most ±1.28%.

The approximate query processing is on average approximately 9 times faster than the exact query processing, since it needs only approximately 11% of the time of the exact processing to execute the approximate one. Table 9 (columns "Processing Time") presents the processing time in seconds and the percentages corresponding to the approximate query processing related to exact one.

Table 7. Tests

Labels	Dataset 1	Dataset 2
Query-1	Brazilian municipalities	Brazilian municipalities'
Query-2	Township boundaries	Census block
Query-3	Township boundaries	Geological map
Query-4	Township boundaries	Hydrologic map
Query-5	Census Block	Hydrologic map
Query-6	Hydrologic map	Geological map

Table 8. Experimental results corresponding to the 20 executions of the intersection area of dataset 1 × dataset 2 × random window with size of 4% of the size of the whole space

	Error and Confidence Interval			Processing Time			Number of Disk Accesses			Average of objects per window
Queries	Error (%)	C. I. 95%	C. I. 99%	Approx. Proc.	Exact Proc.	%	Approx. Proc.	Exact Proc.	%	
Query-1	0.779	2.973	3.907	6.279	73.025	8.60	2138	26826	7.97	1813
Query-2	0.304	0.534	0.702	14.621	93.204	15.69	8691	50979	17.05	2590
Query-3	0.831	1.386	1.822	12.478	109.337	11.41	5440	44289	12.28	2551
Query-4	0.255	1.231	1.617	8.212	75.098	10.94	5747	27064	21.23	1591
Query-5	0.438	1.292	1.699	17.375	110.85	15.67	4470	33862	13.20	1935
Query-6	0.847	1.243	1.634	20.66	95.217	21.70	2736	22419	12.20	1267
Average	0.58	1.44	1.90	13.27	92.79	14.00	4870	34240	13.99	1958

The total execution time is not a good measure of performance gain as it is totally dependent on the algorithm used to execute the exact processing. Besides, the cache of the Operating System can influence processing time. Instead, the total number of disk accesses is a reliable performance gain measure, as the objects to be processed have to be, at least, read from disk. Table 9 (columns "Number of Disk Accesses") presents the gains of the approximate processing related to the exact one. The former needs in average only 16% of the number of disk access of the latter. In other words, the exact processing requires in average 6 times more disk accesses than the approximate processing.

Table 9. Experimental results corresponding to 20 executions of the intersection area of dataset 1 × dataset 2 × random window with size of 12.25% of the size of the whole space

| Queries | Error and Confidence Interval | | | Processing Time | | | Number of Disk Accesses | | | Average of objects per window |
	Error (%)	C. I. 95%	C. I. 99%	Approx. Proc.	Exact Proc.	%	Approx. Proc.	Exact Proc.	%	
Query-1	1.05	2.48	3.26	18.00	272	6.62	6166	62353	9.89	4907
Query-2	0.30	0.31	0.40	40.64	398	10.22	24512	133509	18.36	7394
Query-3	0.79	0.82	1.08	33.68	392	8.60	15737	108508	14.50	6940
Query-4	0.18	0.75	0.99	21.84	250	8.74	13378	65645	20.38	4305
Query-5	0.45	0.74	0.97	46.14	390	11.84	13055	87123	14.98	5581
Query-6	0.74	0.75	0.98	86.01	405	21.22	9504	59656	15.93	3886
Average	0.59	0.97	1.28	41.05	351	11.21	13725	86132	15.67	5502

4 Conclusions

This work proposes, implements, and evaluates a new approach for estimating the overlapping area of polygon join queries. The target is to provide an estimated result in orders of magnitude less time than the time to compute an exact answer, along with a confidence interval for the answer. We propose to compute the intersection area of pairs of polygons over 4CRS signatures of the polygons, processing compact and approximate representations of the objects, and avoiding accessing the whole data. By doing so, the exact geometries of the objects are not processed during the join execution, which is the most costly part of the spatial join since it requires the search and transfer of large objects from the disk to the main storage ([26] and [18]). Also, the exact processing algorithm needs to use complex CPU-time intensive algorithms for deciding whether the objects match the query condition [27]. There are many scenarios and applications where a slow exact answer can be replaced by a fast approximate one, provided that it has the desired accuracy, as presented in Section 1.

We evaluate our approach comparing the approximate processing against the exact processing according to storage requirements, accuracy, response time, and number of disk accesses. The results achieved were quite good, and demonstrated the effectiveness of our approach. The 4CRS signature has low storage requirements; the approximate answers have a quite small error; and, the processing time and the number of disk accesses required to execute the approximate processing are much smaller than the time and number of disk accesses of the exact processing. In Sub-Section 3.2, we presented details of the experimental results for small size signatures. These tests showed that an average of 30 times less space to store 4CRS signatures is needed than to store the real datasets. The approximate answers have an average error of 0.6%, while the confidence intervals of 95% and 99% have average values of 0.97% and 1.28% respectively, which is enough precision for most applications. Besides, the approximate processing varies from 5 to 15 times faster than the exact processing in response time and from 5 to 10 times relate to number of disk accesses.

As future work we plan to evaluate the use of more colours in the raster approximations, for example, eight colours. We believe that it can provide a better precision, and confidence intervals closer to the approximate answer. Besides, although this will have the extra cost of storing more bits for colour representation, storage requirements can be kept small since we apply compression methods on the 4CRS signatures. We also plan to investigate an algorithm to compute the number of cells that leads to a 4CRS signature that better represents the polygon, based on the complexity of the polygon [28]. A straightforward approach is to compute the 4CRS signature starting with the maximum number of cells equal to one, and then increase this number until the proportion between approximate area (*strong* and *weak* cells) and the exact area reaches a pre-defined threshold. Besides, the algorithm proposed in this work can be extended or new algorithms can be developed in order to process other kinds of operations.

References

1. Das, A., Gehrke, J., Riedwald, M.: Approximation Techniques for Spatial Data. In Proc. of ACM-SIGMOD Conference, Paris, France (2004) 695-706.
2. Dobra, A., Garofalakis, M., Gehrke, J. E., Rastogi, R.: Processing complex aggregate queries over data streams. In Proc. of SIGMOD (2002) 61-72.
3. Faloutsos, C., Jagadish, H. V., Sidiropoulos, N. D.: Recovering information from summary data. In Proc. of 23rd Int. Conf. on Very Large Data Bases, Athens, Greece (1997) 36-45.
4. Barbara, D., DuMouchel, W., Faloutsos, C., Hass, P., Hellerstein, J. M., Ioannidis, Y., Jagadish, H., Johnson, T., Ng, R., Poosala, V., Ross, K., Sevcik, K.: The New Jersey data reduction report. Bulletin of the Technical Committee on Data Engineering, IEEE Data Engineering Bulletin (1997) 20(4):3-45.
5. Heuvelink, G.: Error Propagation in Environmental Modeling with GIS, Taylor & Francis, London, UK (1998).
6. Zimbrao, G., Souza, J. M.: A Raster Approximation for Processing of Spatial Joins. In Proc. of the 24th VLDB Conference, New York City, New York (1998) 558-569.
7. Larson, H. J.: Introduction to probability theory and statistical inference. John Wiley & Sons (1982).
8. Jagadish, H. V., Mumick, I. S., Silberschatz, A.: View maintenance issues in the chronicle data model. In Proc. ACM PODS, San Jose, CA (1995) 113-124.
9. Zhu, H., Su, J., Ibarra, O. H.: Toward Spatial Joins for Polygons. In Proc. Int. Conf. of SSDBM, Berlin, Germany (2000) 2431-246.
10. IBGE (Brazilian Institute of Geography and Statistics): Malha Municipal Digital do Brasil, Rio de Janeiro, Brasil (1994).

11. Orenstein, J. A.: Spatial query processing in an object-oriented database system. In Proc. of ACM SIGMOD Int. Conf. on Management of Data, Washington, DC (1986) 326-336.
12. Han, J., Kamber, M.: Data Mining: concepts and techniques. Academic Press (2001).
13. Hellerstein, J. M., Haas, P. J., Wang, H. J.: Online aggregation. In Proc. of ACM SIGMOD Int. Conf on Management of Data, Tucson, Arizona (1997) 171-182.
14. Costa, J. P., Furtado, P.: Time-Stratified Sampling for Approximate Answers to Aggregate Queries. In Proc. of Int. Conf. on Database Systems for Advanced Applications, Kyoto, Japan (2003) 215-222.
15. Roddick, J., Egenhofer, M., Hoel, E., Papadias, D., Salzberg, B.: Spatial, Temporal and Spatiotemporal Databases Hot Issues and Directions for PhD Research. SIGMOD Record, (2004) 33(2):126-131.
16. Zhang, J., Goodchild, M.: Uncertainty in Geographical Information System. Taylor & Francis, Erewhon, NC (2002).
17. Azevedo, L. G., Monteiro, R. S., Zimbrao, G., Souza, J. M.: Approximate Spatial Query Processing Using Raster Signatures. In Proc. of VI Brazilian Symposium on GeoInformatics, Campos do Jordao, Brazil (2004).
18. Lo, M. L., Ravishankar, C. V.: Spatial Hash-Joins. In Proc. of the ACM-SIGMOD Conference, Montreal, Canada (1996) 247-258.
19. Beckmann, N., Kriegel, H. P., Schneider, R., Seeger, B.: The R*-tree: An Efficient and Robust Access Method for Points and Rectangles. In Proc. of ACM SIGMOD Int. Conf. on Management of Data, Atlantic City, NJ (1990) 322-331.
20. Gibbons, P. B., Matias, Y., Poosala, V.: Aqua project white paper. Technical report, Bell Laboratories, Murray Hill, NJ (1997).
21. Furtado, P., Costa, J. P.: Time-Interval Sampling for Improved Estimations in Data Warehouses. In Proc. of 4th Int. Conf. on Data Warehousing and Knowledge Discovery, Aix-en-Provence, France (2002) 327-338.
22. Steel, R. G. D., Torrie, J. H.: Introduction to statistics. McGraw-Hill Book Company (1976).
23. Güting, R. H., de Ridder, T., Schneider, M.: Implementation of the ROSE Algebra: Efficient Algorithms for Realm-Based Spatial Data Types. In Proc. of the 4th Int. Symposium on Large Spatial Databases, Portland, Maine (1995) 216-239.
24. Kothuri, R. K., Ravada, S.: Efficient Processing of Large Spatial Queries Using Interior Approximations. Proc. of the Int. Symposium on Spatial and Temporal Databases, Los Angeles, CA (2001) 404-424.
25. Madria, S. K., Mohania, M. K., Roddick, J. F.: A Query Processing Model for Mobile Computing using Concept Hierarchies and Summary Databases. Proc. of Int. Conference of Foundations of Data Organization, Kobe, Japan (1998) 147-157.
26. Brinkhoff, T., Kriegel, H. P., Schneider, R., Seeger, B.: Multi-step Processing of Spatial Joins. In Proc. of ACM-SIGMOD Int. Conference on Management of Data, Minneapolis, MN (1994) 197-208.
27. Brinkhoff, T., Kriegel, H. P., Schneider, R.: Comparison of Approximations of Complex Objects Used for Approximation-based Query Processing in Spatial Database Systems. In Procs. of Int. Conf. on Data Engineering, Vienna, Austria, (1993) 40-49.
28. Brinkhoff, T., Kriegel, H. P., Schneider, R., Braun, A.: Measuring the Complexity of Polygonal Objects. In Proc. of ACM Int. Workshop on Advances in Geographic Information Systems, Baltimore, MD, (1995) 109-118.
29. Ioannidis, Y. E., Poosala, V.: Balancing histogram optimality and practicality for query result size estimation. ACM SIGMOD, (1995) 233-244.

Density Estimation for Spatial Data Streams

Cecilia M. Procopiuc and Octavian Procopiuc

AT&T Shannon Labs, Florham Park, NJ 07950, USA
magda@research.att.com
oprocopiuc@gmail.com

Abstract. In this paper we study the problem of estimating several types of spatial queries in a streaming environment. We propose a new approach, which we call Local Kernels, for computing density estimators by using local rather than global statistics on the data. The approach is easy to extend to an on-line setting, by maintaining a small random sample with a kd-tree-like structure on top of it. Our structure dynamically adapts to changes in the locality of data and has small update time. Experimental results show that the proposed algorithm returns good approximate results for a variety of data and query distributions. We also show that it is useful in off-line computations, as well.

1 Introduction

We consider the problem of estimating the data distribution and query selectivity for spatial data streams. More precisely, we assume the so-called *cash register* model in which data points are inserted into the set, but they are never deleted. Each data point is a multidimensional real-valued tuple. In this paper, we propose new methods for maintaining an approximate density function on the data, by computing kernel density estimators in an on-line setting. We then use this information in order to approximate the selectivity of range queries.

As in all data stream applications, algorithms for computing the desired statistics must satisfy certain conditions. First, they must require only one pass over the data, and use only a small amount of space compared to the size of the dataset. In addition, processing an update should be fast, as in many applications new tuples arrive with high frequency. And finally, answering queries should be both fast and accurate.

The data stream model has become popular in recent years, motivated by applications that deal with massive information such as Internet and phone traffic log analysis, financial tickers, ATM and credit card operations, sensor networks, etc. Although in many such applications the data is stored and archived, its volume makes it prohibitively expensive to access. Thus, it is often necessary to monitor the contents of very large databases in an incremental, on-line fashion. In addition, for many application domains the ability to answer queries as the data arrives is crucial for mission-critical tasks such as fraud detection or financial transactions.

There is already a large body of literature that studies approximation algorithms for computing various statistics and aggregate queries over data streams. For example, [12,17] propose methods for computing approximate quantiles, and [4,3] estimate on-line self-join and multi-way join sizes. A recent result by Das et al. [9] addresses

C. Bauzer Medeiros et al. (Eds.): SSTD 2005, LNCS 3633, pp. 109–126, 2005.

the problem of computing on-line spatial joins and range queries. Summarization techniques such as sketches [7,21], wavelets [11,13], and histograms [10,14,22] have been proposed as a means of answering more complex queries, such as range queries. However, all these methods make the assumption that each tuple in the data stream has attribute values from some finite universe $\{1, \ldots, U\}$ (for simplicity, we henceforth refer to them as discrete methods). While this is a reasonable restriction for many types of data, it is not always feasible to assume such a priori knowledge of it. For example, in the case of spatial, temporal or multimedia datasets, objects are represented as feature vectors with real-valued attributes. Monitoring such data via discrete on-line methods requires an initial discretization of the objects. The accuracy of the result then depends not only on the guarantee provided by the algorithm, but also on the discretization grid. The higher the accuracy desired, the larger the size U of the discretized universe, and the slower the method (since its space and update time almost always depend on U). Moreover, the problem worsens as the dimensionality of the data increases.

In the experimental section we will discuss an application in which the goal is to maintain statistical information of network measurements over time. The data consists of real-valued tuples representing the current state of AT&T's backbone network, reflecting delay times between pairs of servers. Each tuple is an aggregate of measurements taken during a fifteen-minute interval, with multiple measurements generated each minute. Because of the size of the data, only a small amount of it is stored on disk, with older data being moved to tape, which makes accessing it a difficult task. However, it is often the case that only an approximate view of it will suffice, usually to be used for comparison purposes. Thus, our proposed algorithm offers a way to generate a low-storage data summary that is nonetheless powerful enough to answer queries with relatively small error.

Computing summary statistics for real-valued datasets has also been extensively studied in the literature. The Min-Skew [1], MHIST [19] and GENHIST [15] algorithms are histogram-based methods that estimate the selectivity of multi-dimensional range queries. In addition, kernel density algorithms have been proposed in [5,15]. However, all these methods either require multiple passes over the data, or a large enough memory (in relation to the overall data size). The multi-dimensional kernel estimator proposed in [15] is a one-pass algorithm that comes closest to being suitable for an on-line environment. However, its accuracy crucially depends on the computation of approximate values for the standard deviation along each dimension. Such computation requires storing a significant sample of the data. In addition, the proposed algorithm assumes that all queries are asked after the entire data has been seen once.

Our Contributions. In this paper we propose a new method for maintaining approximate data distributions on real-valued multi-dimensional data streams. We then use this information in order to estimate the selectivity of range queries arriving in an on-line fashion. To the best of our knowledge, this is the first algorithm that estimates range selectivity over real-valued data streams.

Our approach is to compute kernel density functions and maintain them over insertions and deletions from the sample set of kernel centers (thus making it suitable for on-line computations). We propose new methods for computing the *kernel bandwidths*, by estimating the standard deviations for each kernel only for points that fall in a locality of the kernel center. More precisely, we will maintain a kd-tree-like structure and we will compute separate density functions in each leaf of the tree. We then approximate the standard deviations only for the data points in the corresponding cell. As the kd-tree structure changes, we use previous density information from nearby cells, as well as newly arriving data in order to maintain the standard deviations in the new leaves.

This allows us to achieve good query accuracy while still using only a very limited amount of memory. Previous methods used global standard deviations to compute kernel bandwidths. As we discuss in the experimental section, such approaches usually require larger samples to achieve good accuracy; they may also suffer from an over-smoothing effect of the density estimators. Gunopulos et al. [15] suggested replacing global statistics by local ones, by first clustering the data. However, no experimental data was provided. Moreover, such an approach would be more difficult to adapt to an on-line setting, than the one we propose in this paper. We note that our Local Kernels algorithm is of independent interest and could also be used for off-line applications in order to compute locality-sensitive statistics. We also compare our method with a state-of-the-art on-line algorithm for discrete data, and conclude that it is significantly faster and generally more accurate. Hence, the Local Kernels method is a competitive alternative to current approaches for summarizing discrete on-line data.

Furthermore, the capability to compute on-line approximate density functions can lead to algorithms for other problems, such as maintaining an approximate visualization of the dataset, or detecting high-density areas. The latter is a generalization of the notion of heavy hitters, which were defined for discrete one-dimensional datasets. In general, any off-line application that uses the underlying distribution of the data in order to compute some fast summary statistics can also be translated into a streaming application using our proposed on-line density estimators.

2   Preliminaries

Data and Query Model. We assume that the dataset is a stream of tuples $\langle p_1, p_2, \ldots \rangle$, where each $p_i \in \mathbb{R}^d$ is a d-dimensional point. The points are indexed in the order in which they arrive. We focus on the so-called *cash register* model for data streams, in which points are inserted, but not deleted from the database. This is the natural model for the two real-life applications we consider, in which the data consists of network and weather measurements accumulated over certain periods of time. We will briefly discuss how our approach can be extended to the *turnstile* model, which allows points to be both inserted and deleted from the stream.

We focus on range selectivity queries, defined as pairs of type $Q_i = \langle i, R_i \rangle$, where R_i is a d-dimensional hyper-rectangle (see the discussion in Section 4 for extensions to other queries). The selectivity of Q_i is defined as $sel(Q_i) = |\{p_j | j \leq i, \ p_j \in R_i\}|$. In other words, it is the number of points that have arrived up to time step i and that lie inside the query range R_i. We study the problem of estimating $sel(Q_i)$, under the assumption that queries arrive in a continuous stream, which is interleaved with the data stream. A query Q_i must be answered before the point p_{i+1} is processed. This requirement arises from the fact that processing the point p_{i+1} changes a subset of the statistics we maintain, and thus has the potential to affect the outcome of Q_i. In practice, if the frequency with which points arrive is too high relative to query time, we can tolerate the processing of a small number of subsequent points, as the changes they induce are not significant. Moreover, as we explain below, processing a point usually affects only a constant number of statistical values, and thus can be handled 'out of sync' by using a small amount of extra memory.

One significant property of our algorithm is that it does not require a priori knowledge of the overall size of the data stream, nor of the range of values along each dimension. The latter is important in any approach that tries to discretize the data first, while the former is often required by discrete on-line methods.

Kernel Density Estimators. The problem of estimating an underlying data distribution is a central theme in statistics research [8,20]. Kernel estimators are statistical techniques for approximating the probability distribution, by generalizing random sampling. The first step is to compute a uniform random sample of the data, and to assign each sample point (also called *kernel center*) a weight of one. The second and crucial step is to distribute the weight of each point in the space around it according to a *kernel function*. In general, kernel functions distribute most of the weight over the area in the vicinity of the center, and taper off smoothly to zero as the distance from the center increases. However, in practice it is easier to use non-smooth kernel functions that are zero outside a given area. The study in [8] shows that the shape of the kernel function does not significantly affect the quality of the approximation.

In this paper we will use the Epanechnikov kernel function, which was also employed in [15]. More precisely, let $S = \{s_1, \ldots, s_m\}$ be a random subset of the data. Then the underlying probability distribution is approximated by the function

$$f(x) = \frac{1}{m} \sum_{i=1}^{m} k(x - s_i),$$

where $x = (x_1, \ldots, x_d)$ and $s_i = (s_{i1}, \ldots, s_{id})$ are d-dimensional points, and

$$k(x_1, \ldots, x_d) = 0.75^d \frac{1}{B_1 B_2 \ldots B_d} \prod_{j=1}^{d} \left(1 - \left(\frac{x_j}{B_j}\right)^2\right)$$

if $|\frac{x_j}{B_j} < 1|$ for all j, and 0 otherwise. See Figure 1 (a).

The parameters $B_1, \ldots, B_d$ are referred to as the *kernel bandwidth* along each dimension. Choosing the right values for these parameters is the crucial step in computing kernel estimators, as they determine the accuracy of selectivity computations. No efficient solution exists for finding optimal bandwidths. The problem has been addressed in [20], which proposes using the following rule: $B_j = \sqrt{5}\sigma_j m^{-1/(d+4)}$, where σ_j is the standard deviation of the sample along the jth dimension. Note that this rule implies that the same d parameter values are used for all kernel functions, in other words the local distribution around each center is assumed to be identical. Moreover, the accuracy of the method depends on how closely σ_j approximates the standard deviation of the entire data. Good approximations require large values for m (the sample size).

In this paper we will use a slightly different approach. We build a kd-tree-like structure on top of the sample S, and assume that each sample point s_i is the centroid of the

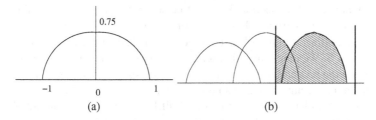

Fig. 1. One-dimensional kernels: (a) Kernel function, $B = 1$; (b) Contribution of multiple kernels to estimate of range query

dataset in its leaf. We then maintain a set of d values σ_{ij}, $1 \le j \le d$, that are generally distinct for each sample point, such that σ_{ij} approximates the standard distribution of the points in the cell of s_i, along dimension j. Note that we can update these values in an on-line fashion, by looking at all points in the cell (rather than just a sample). The detailed description of the procedure is given in the next section.

Computing range selectivities. Let $R = [a_1, b_1] \times \ldots \times [a_d, b_d]$ be a range selectivity query. Let T_i denote the subset of points in the tree leaf associated with sample point $s_i = (s_{i1}, \ldots, s_{id})$, and let $B_{i1}, \ldots, B_{id}$ be the kernel bandwidths for that leaf. Then the selectivity of R with respect to T_i is approximated in time $O(d)$ as follows:

$$sel(R, T_i) \approx |T_i|(0.75)^d \frac{1}{B_{i1} \ldots B_{id}} \times \prod_{j=1}^{d} \int_{[a_j, b_j]} \left(1 - \left(\frac{x_j - s_{ij}}{B_{ij}} \right)^2 \right) dx_d \ldots dx_1.$$

The overall selectivity is then computed in time $O(dm)$ as $sel(R) \approx \sum_{i=1}^{m} sel(R, T_i)$. See Figure 1 (b) for an illustration of the contribution of three kernel functions to the selectivity mass of a range query (in the one-dimensional case).

3   On-line Algorithm

In this section we describe our algorithm for maintaining a set of kernel density estimators that closely approximates the underlying data distribution of a spatial data stream. An important contribution of our approach is to design kernel estimators that use local statistics on the data in order to define the weight distribution functions. The advantage of using such estimators is that local statistics can be maintained very accurately with only a small amount of memory, and in an on-line fashion. By contrast, previous methods approximated global standard deviations and required a large sample size in order to ensure reasonable accuracy. In addition, our experimental results indicate that using local statistics improves the quality of the results. We will call our approach Local Kernels.

We propose using a kd-tree-like structure to partition the kernel centers, and define the neighborhood of a center to be its corresponding leaf. The algorithm will maintain the (approximate) standard deviations for each subset of data points in a leaf. Other partitioning schemes could also be employed to define such neighborhoods. However, as we show at the end of Section 3.1, the tree structure we consider exhibits certain properties that make it particularly suitable for our task.

Let S be a uniform sample of the data seen so far. The kernel centers are defined to be the points of S. We maintain a hierarchical decomposition tree on S, denoted $T(S)$, with the following properties: each leaf of $T(S)$ corresponds to a (possibly unbounded) axis-parallel hyper-rectangle in $\mathbb{R}^d$, so that there is exactly one point of S in each leaf; any two leaves are disjoint; and the union of all leaves is $\mathbb{R}^d$. For the initial random sample S, $T(S)$ is a kd-tree. As points are inserted and deleted from S, the structure $T(S)$ will no longer be a kd-tree, but it will maintain the above properties. For each sample point $s_i \in S$ we also maintain $d + 1$ values: τ_i, which is the (approximate) number of stream points that lie in the leaf corresponding to s_i, and $\sigma_{ij}, 1 \leq j \leq d$, which approximate the standard deviations of the points in the leaf of s_i along each dimension.

3.1 Random Sampling

The random sample S is chosen using the reservoir sampling method of [24]. More precisely, point p_i is chosen in S with probability $|S|/i$ (recall that the points are indexed in the order in which they arrive in the stream). If p_i is chosen, a random point of S is deleted. Vitter [24] proves that using this technique guarantees that S is always a uniform random sample of the data seen so far.

A powerful probabilistic result by Vapnik and Chervonenkis [23] allows us to compute the size of the random sample S for which certain estimation errors are guaranteed with high probability. More precisely, we can prove the following.

Theorem 1. *Let T be the data stream seen so far, and let $S \subseteq T$ be a random sample chosen via the reservoir sampling techqnique, such that $|S| = \Theta(\frac{d}{\epsilon^2} \log \frac{1}{\epsilon} + \log \frac{1}{\delta})$, where $0 < \epsilon, \delta < 1$. Then with probability $1 - \delta$, for any axis-parallel hyper-rectangle Q the following is true:*

$$|sel(Q) - sel(Q,S)\frac{|T|}{|S|}| \leq \epsilon|T| \tag{1}$$

where $sel(Q) = |Q \cap T|$ is the selectivity of Q with respect to the data stream seen so far, and $sel(Q,S) = |Q \cap S|$ is the selectivity of Q with respect to the random sample.

Proof. Consider the set system $(T, \mathcal{H}(T))$, where each $H \in \mathcal{H}(T)$ is a subset of T lying in an axis-parallel hyper-rectangle (for an introduction to the theory of set systems we refer the reader to, e.g., the book [18]). It is well known that the VC-dimension of this system is $2d$. By the result of [24], S is a uniform random sample of T. Then the main theorem of [23] states that, if S has the size specified above, with probability $1 - \delta$ equation 1 is true for any axis-parallel hyper-rectangle Q.

Theorem 1 implies that with high probability we can estimate the selectivity of any range query via the simple random sampling method, and achieve an *additive error*

$\epsilon|T|$. In the following, we will be interested in selectivity estimators with small *relative errors*, where the relative error for a query Q is defined as

$$\frac{|sel(Q) - estimated\_sel(Q)|}{\max\{sel(Q), 1\}}.$$

Hence, for the random sampling estimator, defined as

$$estimated\_sel(Q) = sel(Q, S)\frac{|T|}{|S|},$$

the relative error is $\epsilon\frac{|T|}{\max\{sel(Q),1\}}$. Note that this error is small for queries with high selectivity, but it can grow as large as $\Theta(|T|)$ if $sel(Q) = O(1)$. This behavior of the random sampling estimator is well known in practical applications, and has justified the study of more sophisticated estimators. As mentioned before, kernel density estimators can be viewed as a generalization of random sampling, in which the points in S distribute their weight over a local neighborhood, and the selectivity is estimated as an integral of the weight distribution functions. This smoothing technique improves the accuracy of the approximation, especially for ranges of small selectivity.

An immediate consequence of Theorem 1 is that, with high probability, no leaf of the decomposition tree $\mathcal{T}(S)$ contains more than $2\epsilon|T|$ data stream points. Indeed, let L be the axis-parallel hyper-rectangle associated with a leaf of $\mathcal{T}(S)$. By construction of the tree, $sel(L, S) = 1$. Then if we assume that $sel(L) \geq 2\epsilon|T|$, equation 1 implies $|S| \leq 1/\epsilon$, a contradiction. Hence, $\mathcal{T}(S)$ induces a good partitioning of the data stream, in the sense that no leaf is too dense. This is particularly important if the underlying data is clustered in a reasonably large number of dense subsets, as it ensures that points from different clusters fall in different leaves. This in turn means that our estimates σ_{ij} of the standard deviations for points in each leaf are close to the real values for that region.

The above observations justify our choice of using a kd-tree-like structure for partitioning the data stream. Note that other means of defining local neighborhoods for the sample points can also be employed. For example, one could use the Voronoi diagram of S, and compute the values τ_i and σ_{ij} for each subset of points lying in a Voronoi cell. However, in order to ensure that no cell of the partition is too dense (in the sense discussed above), we have to significantly increase the sample S. More precisely, $|S|$ must have a linear dependence on the VC-dimension of the corresponding set-system; this is asymptotically larger for any other reasonable decomposition schemes.

In addition to this well-balanced property in terms of data density in each cell, $\mathcal{T}(S)$ also has the advantage of being easy to maintain. As we show below, we can update $\mathcal{T}(S)$ in time $O(|S|)$ under insertions and deletions from S.

3.2   Updating $\mathcal{T}(S)$

For ease of presentation, we introduce the following notations. Let $box(v)$ be the axis-parallel hyper-rectangle associated with a node v of $\mathcal{T}(S)$. If v is an internal node, let $h(v)$ denote the hyper-plane orthogonal to a coordinate axis that divides $box(v)$ into the two smaller boxes associated with the children of v. Let $leaf(s_i)$ be the leaf containing sample point s_i. We maintain $\tau_i \approx$ number of stream points contained in $leaf(s_i)$ and

$$\Sigma_{ij} \approx \sum_{p \in leaf(s_i)} (p_j - s_{ij})^2,\ 1 \leq j \leq d.$$

Then $\sigma_{ij} = \sqrt{\Sigma_{ij}/(\tau_i - 1)}$ is the approximate standard deviation of the points in $leaf(s_i)$ along dimension j (assuming s_i is the centroid of the distribution).

Let p be the current point in the data stream. If p is not selected in the sample S, we find the leaf that contains p - say this is $leaf(s_i)$. Then we increment τ_i by one, and we add $(p_j - s_{ij})^2$ to Σ_{ij}, $1 \leq j \leq d$.

We now consider the case when p is selected in the random sample S. Let q denote the point that gets deleted from S. The updating procedure first deletes $leaf(q)$ from the tree, and then adds a new leaf corresponding to p. We detail each step below.

Deleting a leaf. Let u denote the parent node of $leaf(q)$, and let v be the sibling of $leaf(q)$. Without loss of generality, assume that $leaf(q)$ lies to the left of $h(u)$ and v lies to the right of $h(u)$; see Figure 2. The structure of $T(S)$ will be modified as illustrated in Figure 3: make v a descendant of the parent node of u; delete nodes u and $leaf(q)$. Let $\mathcal{N}(q)$ denote the leaves in the subtree of v that have one boundary contained in $h(u)$; we will call these the *neighbors* of $leaf(q)$. The deletion procedure can be viewed as extending the bounding box of each neighbor of $leaf(q)$ past the hyper-plane $h(u)$, until it hits the left boundary of $leaf(q)$. The points that were previously contained in $leaf(q)$ must thus be redistributed among the leaves in $\mathcal{N}(q)$, and the corresponding τ and Σ values must be updated for all these leaves. The procedure

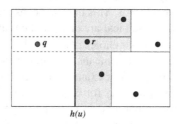

Fig. 2. Deleting $leaf(q)$ means extending the bounding boxes of leaves in $\mathcal{N}(q)$ (represented as gray rectangles)

for updating the τ values is simple: for each leaf in $\mathcal{N}(q)$, we increment its τ value by the selectivity of its (expanded) bounding box with respect to the points contained in $leaf(q)$. As long as the kernel function for $leaf(q)$ is a good model for the distribution of points inside the leaf, τ remains a good approximation for the actual number of points inside each leaf of $\mathcal{N}(q)$. However, updating the Σ values requires more information than we store. This is because points in $leaf(q)$ will contribute differently to the standard deviations of the (expanded) leaves in $\mathcal{N}(q)$, based on their relative position to the centroids of those leaves. Let r be a point so that $leaf(r) \in \mathcal{N}(q)$, and for a fixed

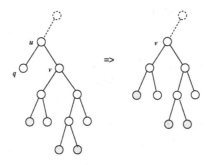

Fig. 3. Deleting $leaf(q)$ from the tree. Gray nodes correspond to $\mathcal{N}(q)$

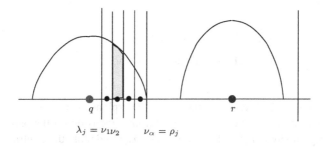

Fig. 4. Updating Σ_{rj} by discretizing the intersection of $box_e(r)$ and the kernel of q along dimension j (the gray area represents wt_2)

dimension j, let $[\lambda_j, \rho_j]$ be the intersection of the expanded box of r, denoted $box_e(r)$ and the kernel function of q along dimension j (see Figure 4). We discretize the interval $[\lambda_j, \rho_j]$ by choosing a set of equidistant points $\lambda_j = \nu_1, \nu_2, \ldots, \nu_\alpha = \rho_j$ (where α is a sufficiently large constant), and update the value of Σ_{rj} as follows:

$$\Sigma_{rj} = \Sigma_{rj} + \sum_{i=1}^{\alpha-1}((\nu_i + \nu_{i+1})/2 - r_j)^2 wt_i,$$

where

$$wt_i = 0.75 \cdot \tau_q \frac{1}{B_{qj}} \int_{\nu_i}^{\nu_{i+1}} \left(1 - (\frac{x - q_j}{B_{qj}})^2\right) dx$$

is the (approximate) number of points of $leaf(q)$ whose j'th coordinate lies in the interval $[\nu_i, \nu_{i+1}]$. In other words, we approximate all points in this interval by its midpoint, and use this approximation to update the Σ value for r.

The update procedures are summarized in Figure 5. The overall procedure for deleting a leaf is given in Figure 6 (a).

Inserting a leaf. Let p be the point newly selected in the sample S, and q be an existing sample point such that $p \in leaf(q)$. We split $leaf(q)$ by a hyperplane passing through

PROCEDURE τ-$Update(q, r, box(r))$
  τ_r += it $sel(box(r), leaf(q))$;
end

PROCEDURE Σ-$Update(q, r, box(r))$
  $box_j(r)$: projection of $box(r)$ along dimension j;
  α: constant;
  for $j = 1, \ldots, d$
    $[\lambda_j, \rho_j] = box_j(r) \cap [q_j - B_{qj}, q_j + B_{qj}]$;
    discretize $\lambda_j = \nu_1, \nu_2, \ldots, \nu_\alpha = \rho_j$;
    for $i = 1, \ldots, \alpha - 1$
      $$wt_i = 0.75 \cdot \tau_q \frac{1}{B_{qj}} \int_{\nu_i}^{\nu_{i+1}} \left(1 - \left(\frac{x - q_j}{B_{qj}}\right)^2\right) dx;$$
      Σ_{rj} += $((\nu_i + \nu_{i+1})/2 - r_j)^2 wt_i$;
    end for
  end for
end

Fig. 5. Updating values τ and Σ

the midpoint $(p + q)/2$. The direction of the hyperplane is chosen according to the alternating rule of a kd-tree, i.e., if i is the splitting dimension for the parent of q, then we split q along dimension $(i + 1) \bmod d$. Let $box_c(q)$ denote the contracted box that bounds the new $leaf(q)$. We set τ_q and τ_p to be the selectivity of $box_c(q)$, respectively $box(p)$, with respect to the set of points previously contained in $leaf(q)$ (in fact, τ_p is one more than the selectivity, to account for the new point p). We then compute the Σ values for p and q using the procedure described in Figure 5, using $box_c(q)$ and $box(p)$ as the third argument. The overall procedure is summarized in Figure 6 (b).

Clearly, the update time is dominated by the leaf deletion procedure. Its time complexity is proportional to the number of leaves in the subtree rooted at the sibling of the leaf, which is $O(|S|)$ in the worst case. Hence, we conclude with the following.

Theorem 2. *The update and query time for our online range searching procedure is* $O(|S|)$, *where* S *is the sample size.*
Extension to turnstile model. The overall approach can also be extended to handle the case when points are allowed to be deleted from the data stream. If the point p to be deleted is not a kernel center, we compute s_i such that $p \in leaf(s_i)$. We then subtract 1 from τ_i, and $(p_j - s_{ij})^2$ from Σ_{ij}, $1 \leq j \leq d$. If p is a kernel center, then we delete $leaf(p)$ and choose the next point inserted in the stream to replace p in the sample. This approach suffers from the same problems as maintaining a data sample over deletions, which is that we can no longer guarantee (with high probability) that the sample is uniform with respect to the data currently present in the stream. However, if the deletions do not exhibit strong spatial or temporal locality, then the approach is likely to work well in practice.

PROCEDURE *Delete*(*leaf*(*q*))
  Compute $\mathcal{N}(q)$;
  for each $r \in \mathcal{N}(q)$
    $box_e(r)$: expanded box of r;
    τ-*Update*$(q, r, box_e(r))$;
    Σ-*Update*$(q, r, box_e(r))$;
  end for
  Replace parent of *leaf*(*q*) by sibling of *leaf*(*q*)
  in $\mathcal{T}(\mathcal{S})$;
end

PROCEDURE *Insert*(*leaf*(*p*))
  Compute q so that $p \in leaf(q)$;
  Split *leaf*(*q*) through $(p + q)/2$ along appropriate axis;
  $box_c(q)$: contracted box of new *leaf*(*q*);
  oldq: all values assoc. with q before splitting;
  $\tau_p = \tau_q = 0$;
  $\Sigma_{qj} = \Sigma_{pj} = 0, 1 \le j \le d$;
  τ-*Update*$(oldq, q, box_c(q))$;
  τ-*Update*$(oldq, p, box(p))$; $\tau_p+ = 1$;
  Σ-*Update*$(oldq, q, box_c(q))$;
  Σ-*Update*$(oldq, p, box(p))$;
  end

(a)　　　　　　　　　　　　　　　　　　　(b)

Fig. 6. (a) Deleting *leaf*(*q*). (b) Inserting *leaf*(*p*)

4　Experimental Results

We evaluate the performance of our on-line algorithm on both synthetic and real data sets, in different number of dimensions and under varying query loads. We focus on range selectivity queries in the experiments described in this section, as they are one of the most common type of queries asked against multi-dimensional data, and have been extensively studied in previous literature. Note, however, that other types of queries can also be handled using the statistical information maintained by Local Kernels, and the range selectivity computation as a basic procedure. For example, the *rank* of a point $p = (p_1, \ldots, p_d)$, defined as the number of points dominated by p on all coordinates, can be computed as the range selectivity of the hyper-rectangle $(-\infty, p_1) \times \cdots \times (-\infty, p_d)$. *Hot spots*, defined as unit cubes containing at least αn points in the stream, for some user-defined $0 < \alpha < 1$ (see [16]), could also be computed by answering range selectivity queries on an appropriate set of candidate cubes.

As we discuss below, the data distribution for one of the real datasets we used was significantly different from the synthetic data. However, our method returned good results on both distributions. There are two main issues we are interested in evaluating in our experimental set-up: the accuracy of our method with respect to existing techniques, both on-line and off-line, and the trade-off between accuracy and space usage.

We first provide experimental evidence that our proposed local-statistics kernels computation is competitive in an off-line setting with previous density-based methods. In fact, as we show below, it has better accuracy in most of the cases we study. This is an important issue, because the off-line accuracy of local kernels is an upper bound on the accuracy that we can achieve when moving to the on-line setting. Hence, we need to validate our strategy as a multi-pass method first. By this, we mean that instead of maintaining the approximate values for the number of points and standard distribution in a leaf, we compute them exactly in one pass over the current data, every time the set of kernel centers changes. Note that off-line local-statistics kernels are of independent interest, as they can be used for summarization and mining for large-scale data warehousing. As we discuss below, in the context of data warehousing, off-line local kernels require only two data scans. Our experiments also show that they yield highly accurate results with small space requirements, even when query selectivity is very low.

Next, we study the performance of our on-line algorithm. We first compare it to the discrete one-pass histogram method proposed in [22], on a set of integer-valued two-dimensional data. Although the method can be used to answer range queries that arrive interspersed with the data, the expensive histogram computation makes it impractical for such a setting. Therefore, we restricted our experiments to the case when all queries arrive after the entire data stream has been processed. Finally, we evaluate our method in its most general setting, i.e. over real-valued data that arrives interspersed with the queries. We compare it with random sampling, which we consider as a base method for on-line range searching, due to its simplicity and effectiveness in practice. The version that we implemented uses reservoir sampling [24] to maintain a sample of the points in a streaming environment. We also consider the off-line version of local kernels computation, which we use as a basis of comparison for accuracy results. As mentioned above, this represents the best approximation that can be obtained via local kernels.

Datasets and queries. We used four datasets: a 2-dimensional synthetic set SD2, a 4-dimensional synthetic set SD4, and a 2-dimensional set NM2 containing network measurements. Each of the two synthetic datasets contains 1 million points, of which 90% are contained in clusters, and 10% are uniformly distributed. The data generator we used is similar to the one described in [2], which was introduced in order to model local dependencies in the data. There are 100 clusters in each dataset, and the points in each cluster are drawn from a normal distribution around a randomly chosen center. The variance in each cluster is determined randomly in the following manner: Fix a *spread parameter* r and choose a scale factor $s_{ij} \in [1, s]$ uniformly at random, where s is user defined. Then the variance of the normal distribution in cluster i and along dimension j is $(s_{ij} \cdot r)^2$. The number of points in each cluster is proportional to the realization of an exponential random variable. Once all clusters are generated, we compute a random permutation of the points, and choose that to be the order in which the data arrives. Thus, the stream does not exhibit temporal correlation, i.e. consecutive points are likely to belong to different clusters.

The network dataset NM2 contains 1 million two-dimensional data points with real-valued attributes. Each point is an aggregate of measurements taken during a fifteen-minute interval, reflecting minimum and maximum delay times between pairs of servers on AT&T's backbone network.[1] The dataset is only a small snapshot of the entire information stored in the course of a month, which we have chosen for the purpose of experimental evaluation. As mentioned in Section 1, summarization methods with small storage are highly desirable in this case, as older data is no longer stored on disk, making it difficult to access.

For every dataset, we create two query workloads. The queries are chosen randomly in the attribute space, but each query in a workload has approximately the same selectivity: 0.5% for the first workload, and 10% for the second. Hence, the first workload

---

[1] The proprietary nature of the data prohibits us from disclosing more details.

corresponds to low selectivity ranges, and reflects how much our kernel density approach manages to improve over random sampling. The second workload corresponds to high selectivity ranges, and is used in order to verify that our method does not result in an over-smoothing of the distribution function, which would then imply significantly under-estimating the query counts. All workloads contain 200 queries each. The timestamps of the queries are randomly interspersed in the data stream.

Accuracy measure. For the remainder of this paper, we report the accuracy of each method as the average 1-norm relative error, defined below. Because of the random nature of all the algorithms we discuss, each point on a graph is the average value over five runs.

For a query $Q_i = \langle i, R_i \rangle$, the relative estimation error of Q_i is defined as

$$err_i = \frac{|sel(Q_i) - estimated\_sel(Q_i)|}{\max\{sel(Q_i), 1\}}.$$

Let $\{Q_{i1}, \ldots, Q_{ik}\}$ be the query workload for a given experiment. Then the average 1-norm error for this workload is defined as $avg\_err = (\sum_{j=1}^{k} err_{ij})/k$. Note that in the off-line setting, $N \leq i_1, \ldots, i_k$, where N is the size of the data stream; i.e., all queries are asked after the entire data is seen.

Validating local kernels in an off-line setting. In the context of data warehousing, the computation of summaries or density functions does not take place concurrently with query processing. The assumption is that the entire data is available for pre-processing before queries can be answered. Hence, we do not need to maintain our local kernels under insertions and deletions from the sample set. Rather, we scan the data once to select our random sample of centers, compute the associated kd-tree, and then use a second scan to compute the exact values for the number of points and standard distribution in each leaf. All queries are answered after the two data scans are completed. We denote this algorithm by MPLKernels, from **Multi-Pass Local Kernels**.

We will compare this approach with the kernel-density method proposed in [15], which we denote by GKernels (**Global Kernels**). The main difference between MPLKernels and GKernels is that the latter defines kernel bandwidths as functions of the global standard deviations of the data along each dimension. In their paper, the authors note that their algorithm can be implemented as a one-pass method, by approximating the global standard deviations with the standard deviations of the random sample. However, since we are interested in comparing the relative accuracy of the two methods, we report results for the two-pass GKernels, which provides better query estimates.

For the sake of completion, we also include accuracy results for the random sampling estimator, as well as for our on-line local kernels method, which we denote by LKernels. For the latter, we emphasize that, although we maintain the kernels in the on-line manner described in the previous section, the fact that all queries are processed after the entire data has been seen implies that only the last set of kernels is important. Hence, LKernels should be regarded, in this context, as a one-pass warehousing method, rather than an on-line one.

Figures 7 and 8 indicate that both MPLKernels and LKernels are competitive with GKernels in terms of accuracy. In fact, they out-perform it in almost all the cases, except

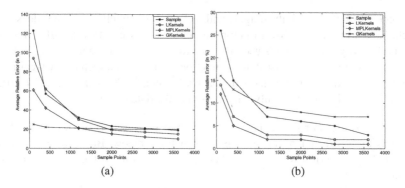

Fig. 7. Off-line query workloads on dataset SD2: (a) selectivity 0.5%; (b) selectivity 10%

for small (below 1500) sample sizes for SD2. This is not surprising, since intuitively, using local statistics should provide a better estimate on the data density. Gunopulos et al. [15] have also suggested that the performance of their density-based estimator would improve by using a clustering algorithm first, and then replacing global statistics by local statistics in each cluster. However, they did not provide any experimental results, and such an approach would be more difficult to adapt to an on-line setting. As noted in the previous section, one of the main advantages of using a kd-tree-like structure is that updating it is easy and fast.

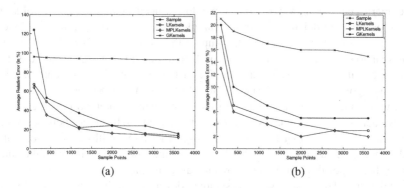

Fig. 8. Off-line query workloads on dataset SD4: (a) selectivity 0.5%; (b) selectivity 10%

Note that for the low selectivity workload, the accuracy of GKernels is almost independent of the sample size. This can be explained by the fact that using global statistics is too coarse an estimate of local density, and the accuracy degrades as the selectivity decreases. Perhaps even more surprising is the fact that the random sample estimator outperforms GKernels on both SD2 and SD4, for sample sizes bigger than 1000. In looking more closely at the results, we noticed that in these cases GKernels under-estimates almost all of the queries. This is due to an over-smoothing of the data, a problem that has been studied intensely in the statistics literature [20] in relation to density estimators. Our approach of computing local kernels significantly alleviates this drawback.

Table 1. The performance of the Histogram algorithm for various values of parameter ρ, compared with the performance of the LKernels algorithm

	Error	Update time	Query time
LKernels	35%	0.0039ms	2.5ms
Histogram (optimal)	33%	0.007ms	2011ms
Histogram ($\rho = 3$)	50%	0.007ms	286ms
Histogram ($\rho = 2$)	60%	0.007ms	252ms
Histogram ($\rho = 1$)	63%	0.007ms	211ms
Histogram ($\rho = 0.5$)	81%	0.007ms	167ms

Comparison with histogram methods. Histograms are a widely used method for computing summary statistics and answering selectivity queries. While off-line histogram computation has been studied for a long time, it is only more recently that the approach has been adapted to on-line settings. Two recent approaches have been proposed in [6,22]. The first algorithm, called STHoles [6], computes the histogram via a training phase, which poses queries and uses the exact answers to build the resulting histogram. The second approach [22] maintains a sketch of the actual data distribution, and uses it to extract histogram buckets. As the sketch can be maintained on-line, this method is closer in spirit to the problem we consider. Moreover, the experimental evaluation in [22] shows that it has comparable or sometimes better accuracy than STHoles. We therefore restrict our attention only to this algorithm.

We evaluate it on a synthetic two-dimensional dataset of 100x100, generated similarly to the one above, except that the points are restricted to the integer grid. For a fair comparison, we restrict both methods to use the same amount of memory cache, which is 11000 integers. This translates into 1200 sample points for Local Kernels (we use 9 integers per sample point). As for the dynamic histogram, a careful analysis of the results in [22] shows that the space utilization is $\min(n^2, sn) + 2s$, where s is the sketch size, and n is the attribute range (in our case, 100). Based also on the evaluations presented in the original paper, we use a sketch size of 500, and extract 50 buckets. We have implemented the faster heuristic (EGreedy), and present the accuracy and computation time in table 1. Update time represents the average time spent per data point, and query time is the average time to answer a query (it includes the histogram computation time for EGreedy). The parameter ρ represents the ratio α/k in the original paper, which is used to speed up histogram computation, at the cost of reducing accuracy. The higher the value of ρ, the better the accuracy. The notation 'optimal' denotes the algorithm EGreedy in which the optimal bucket is computed at each iteration.

As is apparent from Table 1, our approach is at least two orders of magnitude faster than the on-line histograms method, as well as significantly more accurate in most cases. The only situation in which EGreedy is slightly more accurate is when optimal buckets are computed, but in that case EGreedy is three orders of magnitude slower. Note also that the update time per point for our method is about half the time required by EGreedy. Thus, in an on-line environment in which points arrive with reasonably high frequency, and/or the queries are interspersed with the points, Local Kernels is clearly a better choice. Even in a setting in which all queries are computed after the entire dataset has been seen, Local Kernels may still be the preferred solution, as the small accuracy gain of EGreedy comes at a much higher cost in terms of processing time. We conclude that, although designed with enough flexibility to handle real-valued data, Local Kernels is competitive as an on-line algorithm over discrete data, as well.

General on-line setting. We now evaluate the accuracy of our method for real-valued datasets in an on-line context, in which queries arrive interleaved with points. Each query must be processed with respect to the data seen so far, rather than at the end. We first present the performance of LKernels for the 2 and 4-dimensional synthetic datasets, and then discuss the NM2 set. As mentioned above, we use both the random sample estimator and MPLKernels as basis of comparison for accuracy results. In this context, MPLKernels performs a data scan every time it must answer a query: it first computes the exact number of points and the standard deviations in each leaf of $T(S)$, and then processes the query. Note that in this case MPLKernels is an impractical approach. We include it here only as a benchmark, to better understand the limits of our method.

As expected, MPLKernels has the smallest relative error for all sample sizes. However, it is important to note that, just as in the off-line setting, the error of LKernels is always smaller than that of random sampling, which shows that we are indeed able to minimize the problem of over-smoothing in a consistent manner.

It is interesting to look at the graphs corresponding to LKernels in Figures 9 and 7 (respectively, Figures 10 and 8) side by side. The relative errors are very similar,

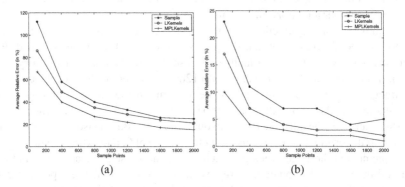

Fig. 9. Query workloads on dataset SD2: (a) selectivity 0.5%; (b) selectivity 10%

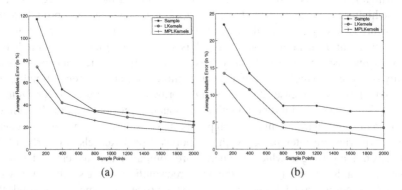

Fig. 10. Query workloads on dataset SD4: (a) selectivity 0.5%; (b) selectivity 10%

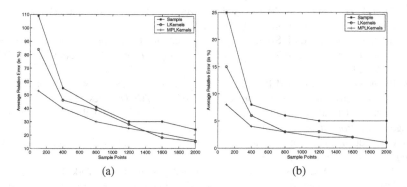

Fig. 11. Query workloads on dataset NM2: (a) selectivity 0.5%; (b) selectivity 10%

whether the queries are answered on the fly, or all at the end. This indicates that the performance of our method does not decrease with time, as we update our data structure and statistics estimates.

The experiments on the network dataset NM2 are shown in Figure 11. Unlike the synthetic data, for which we randomly chose the order in which points arrive, in this case points have a well established order given by the timestamp of the measurement. However, since each point refers to a different network connection, there is little locality in the data. The graphs exhibit the same patterns as before, with LKernels proving the robustness of the kernel-based methods.

5 Conclusions

In this paper we have proposed a new approach for computing density estimators over spatial data, and showed how it can be adapted to on-line environments. Our method maintains a random sample with a kd-tree-like structure on top of it, which permits the estimators to easily adapt to changes in the locality of the data. Given these density estimators, we can approximate the selectivity of range queries that arrive interspersed in the data stream. Our algorithm requires no a priori knowledge of the range of attribute values, nor of the number of tuples in the data stream, and is thus easy to use in a large number of practical applications. We have also provided extensive experimental evaluations that prove that the method is competitive (in terms of both accuracy and running time) with off-line summarization approaches and with one-pass histograms. Finally, we note that the idea of maintaining an indexing structure over spatial data streams, together with density functions, may be of independent interest, such as for visualizing the distribution of low-dimensional streaming data (e.g., network measurements).

References

1. S. Acharya, V. Poosala, and S. Ramaswamy. Selectivity estimation n spatial databases. In *Proceedings of ACM SIGMOD*, pages 13–24, 1999.
2. C. C. Aggarwal, C. M. Procopiuc, J. L. Wolf, P. S. Yu, and J. S. Park. Fast algorithms for projected clustering. In *Proceedings of ACM SIGMOD*, pages 61–72, 1999.

3. N. Alon, P. B. Gibbons, Y. Matias, and M. Szegedy. Tracking join and self-join sizes in limited storage. In *Proceedings of ACM PODS*, pages 10–20, 1999.
4. N. Alon, Y. Matias, and M. Szegedy. The space complexity of approximating the frequency moments. In *Proceedings of the 28th Annu. ACM Symp. on the Theory of Computing (STOC)*, pages 20–29, 1996.
5. B. Blohsfeld, D. Korus, and B. Seeger. A comparison of selectivity estimators for range queries on metric attributes. In *Proceedings of ACM SIGMOD*, pages 239–250, 1999.
6. N. Bruno, S. Chaudhuri, and L. Gravano. Stholes: a multidimensional workload-aware histogram. In *Proceedings of ACM SIGMOD*, pages 211–222, 2001.
7. G. Cormode and S. Muthukrishnan. The count-min sketch and its applications. In *Proceedings of LATIN*, pages 29–38, 2004.
8. N. A. C. Cressie. *Statistics for Spatial Data*. J. Wiley & Sons, New York, 1993.
9. A. Das, J. Gehrke, and M. Riedewald. Approximation techniques for spatial data. In *Proceedings of ACM SIGMOD*, pages 695–706, 2004.
10. A. C. Gilbert, S. Guha, P. Indyk, Y. Kotidis, S. Muthukrishnan, and M. Strauss. Fast, small-space algorithms for approximate histogram maintenance. In *Proceedings of the 34th Annu. ACM Symp. on the Theory of Computing (STOC)*, pages 389–398, 2002.
11. A. C. Gilbert, Y. Kotidis, S. Muthukrishnan, and M. Strauss. One-pass wavelet decompositions of data streams. *IEEE Trans. Knowl. Data Eng.*, 15(3):541–554, 2003.
12. M. Greenwald and S. Khanna. Efficient online computation of quantile summaries. In *Proceedings of ACM SIGMOD*, pages 58–66, 2001.
13. S. Guha, C. Kim, and K. Shim. Xwave: Approximate extended wavelets for streaming data. In *Proceedings of the 30th VLDB Conference*, 2004.
14. S. Guha and N. Koudas. Approximating a data stream for querying and estimation: Algorithms and performance evaluation. In *Proceedings of ICDE*, pages 567–578, 2002.
15. D. Gunopulos, G. Kollios, V. J. Tsotras, and C. Domeniconi. Approximating multidimensional aggregate range queries over real attributes. In *Proceedings of ACM SIGMOD*, pages 463–474, 2000.
16. J. Hershberger, N. Shrivastava, S. Suri, and C. D. Toth. Adaptive spatial partitioning for multidimensional data streams. In *Proceedings of ISAAC 2004 (LNCS 3341)*, pages 522–533, 2004.
17. G. S. Manku, S. Rajagopalan, and B. G. Lindsay. Random sampling techniques for space efficient online computation of order statistics of large datasets. In *Proceedings of ACM SIGMOD*, pages 251–262, 1999.
18. J. Pach and P. K. Agarwal. *Combinatorial Geometry*. J. Wiley & Sons, New York, 1995.
19. V. Poosala and Y. E. Ioannidis. Selectivity estimation without the attribute value independence assumption. In *Proceedings of the 23rd VLDB Conference*, pages 486–495, 1997.
20. D. W. Scott. *Multivariate Density Estimation*. Wiley-Interscience, 1992.
21. S. Suri, C. D. Toth, and Y. Zhou. Range counting over multidimensional data streams. In *Proceedings of Symp. on Computational Geometry (SCG)*, 2004.
22. N. Thaper, S. Guha, P. Indyk, and N. Koudas. Dynamic multidimensional histograms. In *Proceedings of ACM SIGMOD*, pages 428–439, 2002.
23. V. N. Vapnik and A. Y. Chervonenkis. On the uniform convergence of relative frequencies of events to their probabilities. *Theory of Probability and its Applications*, 16:264–280, 1971.
24. J. S. Vitter. Random sampling with a reservoir. *ACM Transactions on Mathematical Software*, 11(1):37–57, 1985.

Change Detection in Time Series Data Using Wavelet Footprints

Mehdi Sharifzadeh, Farnaz Azmoodeh, and Cyrus Shahabi

Computer Science Department,
University of Southern California, Los Angeles, CA 90089-0781
{sharifza, azmoodeh, shahabi}@usc.edu

Abstract. In this paper, we propose a novel approach to address the problem of change detection in time series data. Our approach is based on *wavelet footprints* proposed originally by the signal processing community for signal compression. We, however, exploit the properties of footprints to capture discontinuities in a signal. We show that transforming data using footprints generates nonzero coefficients only at the change points. Exploiting this property, we propose a change detection query processing scheme which employs footprint-transformed data to identify change points, their amplitudes, and degrees of change efficiently and accurately. Our analytical and empirical results show that our approach outperforms the best known change detection approach in terms of both performance and accuracy. Furthermore, unlike the state of the art approaches, our query response time is independent of the number of change points and the user-defined change threshold.

1 Introduction

Time series data are generated, maintained, and processed within a broad range of application domains in different fields such as economics, meteorology, or sociology. Moreover, recent advances in the manufacturing of modern sensory devices have caused several applications to utilize these sensors towards better understanding of the physical world. These sensors when deployed in an environment generate large amount of measurement data streams which can be stored as time series data.

Mining such time series data becomes vital as the applications demand for understanding of the underlying processes/phenomena that generate the data. There has been an explosion of interest within the data mining community in indexing, segmenting, clustering, and classifying time series [13,14,15,16]. A specific interesting mining task is to detect change points in a given time series [7,8,12,21,24]. These are the time positions in the original data where the local trend in the data values has changed. They may indicate the points in time when external events have caused the underlying process to behave differently.

The problem of detecting change in time series has been mostly studied in the class of segmentation problems [14,6] where each portion of the data is modelled by a known function. Subsequently, change points are defined as the points in data where two adjacent segments of the time series are connected.

C. Bauzer Medeiros et al. (Eds.): SSTD 2005, LNCS 3633, pp. 127–144, 2005.

For the past year, we have been working with Chevron on mining real data generated during oil well tests. This is a real-world petroleum engineering application studied within the USC's Center of Excellence for Research and Academic Training on Interactive Smart Oilfield Technologies (CiSoft). Petroleum engineers deploy sensors in oil wells to monitor different characteristics of the underlying reservoir. Here, the underneath pressure values measured by sensors form a time series. When the second derivative in the pressure vs. time plot becomes fixed (i.e., a radial flow event), they estimate the "permeability" of the reservoir [11]. At the same time, they would like to know if the first derivative is changing. To us, these points are the positions in the pressure time series where a change of degree 1 or 2 occurs. In this paper we focus on identifying both the *change points* and the *degrees* of change in time series data. While the definition of change is highly application-specific, we focus on points in data where discontinuities occur in the data or any of its ith derivatives. Moreover, we consider the notion of *degree* of change as the degree of the changing derivatives at the change point. This general definition of change has been broadly used in many scientific application areas such as petroleum engineering [11]. However, its significance has been ignored within the data mining community.

We propose a novel efficient approach to find change points in time series data. Our approach utilizes *wavelet footprints*, a new family of wavelets recently introduced by the signal processing community for signal compression and denoising [5]. While footprints are defined to address a different problem in a different context, we exploit their interesting properties that make them a powerful data analysis tool for our change detection problem. Our contribution starts with employing the idea of wavelet footprints in the context of a data mining problem. This is yet another example of adapting signal processing techniques for the purpose of data mining which started by Vitter et. al proposing the use of wavelets in answering OLAP range-sum queries [19], and Chakrabarti et. al using multi-dimensional wavelets for general approximate query processing [1].

We show that footprints efficiently capture discontinuities of any degree in the time series data by gathering the change information in the corresponding coefficients. Motivated by this property, we make the following additional contributions:

- We propose two database-friendly methods to transform the time series data using footprints up to degree D. These methods enable us to detect all the change points of degree 0 to degree D, their corresponding amplitude and degree of change. To the best of our knowledge, this is the first change detection approach that captures all the above parameters at the same time.
- While we transform the data using footprints, our methods can work with any user-defined threshold value. That is, there is no need to rerun our algorithms each time the user-defined threshold value changes; we answer any new query via a single scan over the transformed data to return the coefficients greater than the user threshold. This is a considerable improvement over the best change detection algorithms which are highly dependent on this threshold value.

- Both analytically and empirically, we show that our query processing schemes significantly outperform the state of the art change detection methods in terms of performance. Our query response time is independent of the number of change points in the data. This is while both our methods demonstrate a significant increase in accuracy.

The remainder of this paper is organized as follows. Section 2 reviews the current data mining research on change detection in time series data. Section 3 provides the background on linear algebra and wavelet theory. In Section 4, we first illustrate the idea of using footprints to capture discontinuities by focusing only on piecewise linear time series. We then generalize our change detection approach and propose our lazy and absolute methods for footprint transformation in Section 5. In Section 6, we show how our footprint-based approach can be incorporated within systems where time series data is stored in wavelet domain. Section 7 includes our experimental results, and Section 8 discusses the conclusion and our future plans.

2   Related Work

Change detection in time series has recently received considerable attention in the field of data mining. Change detection has also been studied for a long time in statistics literature where the main purpose is to find the number of change-points first and identify the stationary model to fit the dataset based on the number of change points.

In the data mining literature, change detection has mainly been studied in the time series segmentation problems. Most of these studies use *linear* interpolation to approximate the signal with a series of best fitting lines and return the endpoints of the segments as change points. However, there are many examples of real-world time series which fitting a linear model is inappropriate. For example, Puttagunta et al. [21] use incremental LSR to detect the change points and outlier points with the assumption that the data can be fit with linear models. Also Keogh et al. [12] use probabilistic methods to fit the data with linear segments in order to find patterns in time series.

Yamanish et al. [24] reduce the problem of change point detection in time series into that of outlier detection from time series of moving-averaged scores. Ge et al. [7] extend hidden semi markov model for change detection. Both these solutions are applicable to different data distributions using different regression functions; however, they are not scalable to large size datasets due to their time complexity.

Guralnik et al. [8] suggest using maximum likelihood technique to find the best function to fit the data in each segment. Their method is mainly based on the trade-off between the data fit quality and the number of estimated changes. They also consider a wider group of curve fitting functions; however, they do not consider the possible disagreement among different human observers on the actual change points. Also their approach lacks enough flexibility in the sense that they have to rerun the algorithm for different change thresholds asserted by the user.

The method described in this paper is mostly similar to the work done by Guralnik et al. in [8]; however, we return all the possible change points for several polynomial curve fitting functions since then users have the flexibility to focus only on the interesting change points. For example, the user can focus only on the change points detected by the quadratic and linear models. Moreover, after we find the change points once, there is no need to rerun the algorithm for different change thresholds asserted by the user.

3   Preliminaries

We consider time series $\mathbf{X}_n$ of size n as a vector $(x_1, ..., x_n)$ where each x_i is a real number (i.e., $x_i \in \mathbb{R}$). Given F, a class of functions (e.g., polynomials), one can find the piecewise segmented function $\mathcal{X} : [1, n] \to \mathbb{R}$ that models time series $\mathbf{X}_n$ as follows:

$$\mathcal{X}(t) = \begin{cases} P_1(t) + e_1(t) & 1 < t \leq \theta_1 \\ P_2(t) + e_2(t) & \theta_1 < t \leq \theta_2 \\ \vdots \\ P_{K+1}(t) + e_{K+1}(t) & \theta_K < t \leq n \,. \end{cases} \quad (1)$$

Each function P_i is a member of class F that best fits the corresponding segment of the data in $\mathbf{X}_n$ and each $e_i(t)$ is the amount of error introduced when fitting the segment with P_i. Our ultimate goal is to identify $\theta_1, ..., \theta_K$ when P_i's are not known a priori. We refer to these points as the *change points* in data where discontinuities occur in the data or its derivatives. We use change point and discontinuity interchangeably in this paper.

Throughout the paper, F is the class of polynomial functions of maximum degree D. That is, each $P_i(t)$ in Equation 1 can be represented as

$$P_i(t) = p_{i,D}t^D + p_{i,D-1}t^{D-1} + ... + p_{i,2}t^2 + p_{i,1}t + p_{i,0} \,. \quad (2)$$

We call a change point θ_i, a *change point of degree j* if the corresponding coefficients $p_{i,j}$ and $p_{i+1,j}$ differ in the polynomial representations of $P_i(t)$ and $P_{i+1}(t)$. Notice that θ_i is a change point of all degrees j where we have $p_{i,j} \neq p_{i+1,j}$.

3.1   Linear Algebra

In this section, we present some background linear algebraic definitions. We use these definitions in Section 4 when we discuss transforming time series with wavelet footprints.

Definition 1. *A finite basis B for a vector space $\mathbb{R}^d$ is a set of vectors $B_i \in \mathbb{R}^d$ (i.e., $B = \{B_1, B_2, ..., B_n\}$ where $n = d$) such that any vector $V \in \mathbb{R}^d$ can be written as a linear combination of B_i's, i.e., $V = \sum_{i=1}^n c_i B_i$. Note that given a basis B, the set of coefficients c_i is unique for a vector V. However, if the number of vectors in B is greater than d (i.e., $n > d$) then the vector V can be represented* *as the linear combination of B_i's in infinite number of ways. Such collection B where $n > d$ is an overcomplete collection for $\mathbb{R}^d$. For ease of use we call it an "overcomplete basis" for the rest of the paper.*

Definition 2. *Suppose $B = \{B_1, B_2, ..., B_n\}$ is a finite basis for vector space $\mathbb{R}^d$ and there exists a basis $\tilde{B} = \{\tilde{B}_1, \tilde{B}_2, ..., \tilde{B}_n\}$ such that*

$$\langle B_i, \tilde{B}_j \rangle = \begin{cases} 1 \ if \ i = j \\ 0 \ if \ i \neq j. \end{cases} \tag{3}$$

where $\langle X, Y \rangle$ denotes the inner product of vectors X and Y. The "unique" basis $\tilde{B}$ is known as the dual basis *of B.*

Definition 3. *A basis $B = \{B_1, B_2, ..., B_n\}$ is a* biorthogonal *basis if we have $\langle B_i, \tilde{B}_j \rangle = 0$ for any $B_i \neq B_j$ and $\langle B_i, \tilde{B}_j \rangle = 1$ otherwise.*

Definition 4. *A basis $B = \{B_1, B_2, ..., B_n\}$ is an* orthogonal *basis if for any $B_i \neq B_j$, we have $\langle B_i, B_j \rangle = 0$. According to Definition 2, an orthogonal basis is the dual basis of itself (i.e., it is* self-dual).

To find each coefficient c_i where $1 \leq i \leq n$ for a vector V given a basis B (as in Definition 1), we simply compute $\langle V, \tilde{B}_i \rangle$. For orthogonal bases, due to their self-duality, c_i is computed by the inner product of V to the basis itself, i.e., $\langle V, B_i \rangle$.

The basic idea of compression is to find the basis B and then for each given vector V, only store the coefficients c_i's. The main question is what is the best basis for a given application and dataset, such that several of c_i's become zero or take negligible values. In our case, wavelet footprints would result in c_i's that would take non-zero values only if a change occurs in the vector V. The value of c_i then corresponds to the amount (or amplitude) of change.

3.2 Wavelets

We develop the background on the wavelet transformation using an example. We use Haar wavelets to transform our example time series into the wavelet domain. Consider the time series $\mathbf{X}_8 = (0, 0, 0, 0, 0, 1, 1, 1)$. The transformation starts by computing the pairwise averages and differences of data (also multiplying by a normalization factor at each level) to produce two vectors of *summary* coefficients $H_1 = (0, 0, 0.7, 1.4)$ and *detail* coefficients $G_1 = (0, 0, -0.7, 0)$, respectively. This process repeats by applying the same computation on the vector of summary coefficients. The last summary coefficient followed by all $n - 1$ detail coefficients form the transformed data, i.e., $(1.06, -1.06, 0, -0.5, 0, 0, -0.7, 0)$.

We can conceptualize the process of wavelet transformation as the projection of the time series vector of size n to n different vectors ψ_i termed as *wavelet basis* vectors. Suppose $|X|$ denotes the length of a vector X, the wavelet transformation[1] of $\mathbf{X}_n$ is $\hat{\mathbf{X}}_n = (\hat{x}_1, ..., \hat{x}_n)$ where $\hat{x}_i = \langle \mathbf{X}_n, \psi_i \rangle \cdot \frac{1}{|\psi_i|}$ and the term $\frac{1}{|\psi_i|}$ is a

---

[1] Throughout the paper, we assume that the size of the time series we work with is always a power of 2. This can be achieved in practice by padding the time series with zeroes.

normalization factor (notice that Haar wavelet basis is orthogonal, and hence self-dual). Moreover, time series $\mathbf{X}_n$ can be represented as $\mathbf{X}_n = \sum_{i=1}^{n} \hat{x}_i \psi_i$. Figure 1 shows Haar wavelet basis vectors of size 8 as different rows of an 8×8 matrix.

In general, we identify Haar wavelet basis vectors of size n as ψ_i where $1 \leq i \leq n$. The first vector ψ_1 consists of n 1's. The remainder $n - 1$ vectors corresponding to the detail coefficients are defined as follows:

$$\psi_{2^j+k+1}(l) = \begin{cases} 1 & k.\frac{N}{2^j} \leq l \leq k.\frac{N}{2^j} + \frac{N}{2^{j+1}} - 1 \\ -1 & k.\frac{N}{2^j} + \frac{N}{2^{j+1}} \leq l \leq k.\frac{N}{2^j} + \frac{N}{2^j} - 1 \\ 0 & \text{otherwise} \end{cases} \tag{4}$$

where $0 \leq j \leq \log n$, $k = 0, ..., 2^j - 1$, and $1 \leq l \leq n$. We now define the term *support interval* that we will use throughout the paper.

Definition 5. *Let $\hat{\mathbf{X}}_n = (\hat{x}_1, ..., \hat{x}_n)$ be the wavelet transformation of the time series $\mathbf{X}_n$. The support interval of a wavelet coefficient $\hat{x}_i$ is the range of indices $j \in [1, n]$ such that $\hat{x}_i$ is derived from x_j's.*

For example, the support interval of the first coefficient $\hat{x}_1$ is the entire time series (i.e., $[1, n]$), while that of the last coefficient $\hat{x}_n$ is the last two elements of $\mathbf{X}_n$ (i.e., $[n-1, n]$). We use $Sup(\hat{x}_i)$ to denote the support interval of coefficient $\hat{x}_i$. Similarly, we use $Sup^{-1}(j)$ to refer to the set of all wavelet coefficients which are derived from x_j, i.e., all $\hat{x}_i$'s such that $x_j \in Sup(\hat{x}_i)$.

$$\begin{array}{cccccccc}
1 & 1 & 1 & 1 & 1 & 1 & 1 & 1 \\
1 & 1 & 1 & 1 & -1 & -1 & -1 & -1 \\
1 & 1 & -1 & -1 & 0 & 0 & 0 & 0 \\
0 & 0 & 0 & 0 & 1 & 1 & -1 & -1 \\
1 & -1 & 0 & 0 & 0 & 0 & 0 & 0 \\
0 & 0 & 1 & -1 & 0 & 0 & 0 & 0 \\
0 & 0 & 0 & 0 & 1 & -1 & 0 & 0 \\
0 & 0 & 0 & 0 & 0 & 0 & 1 & -1
\end{array}$$

Fig. 1. Haar wavelet basis of size 8

4   Footprints

Wavelets have been widely used in different data mining applications due to their power in capturing the trend of the data as well as their approximation property [17]. However, wavelets in their general form do not efficiently model discontinuities in the data. To illustrate the problem, consider the example time series $\mathbf{X}_8 = (0, 0, 0, 0, 0, 1, 1, 1)$. Although there exists only one discontinuity point at fifth position of $\mathbf{X}_8$, we get three nonzero coefficients (other than the average) in the final transformed vector $\hat{\mathbf{X}}_8$. The reason is that there is a great amount of overlap among the support intervals of different coefficients at different levels. Therefore, to benefit from the approximation power of wavelets and efficiently model the change points in the underlying data at the same time, a new form of basis is required.

Dragotti et al. [3] introduce a new basis which removes the overlap among the support intervals of corresponding wavelet coefficients at different levels. They call this basis *wavelet footprints* or *footprints* for short. We now explain the idea behind the footprints, assuming for simplicity piecewise constant data only. Consider $\mathbf{X}_n$ with only one discontinuity at position θ:

$$\mathbf{X}_n(i) = \begin{cases} a & 1 \leq i \leq \theta \\ b & \theta < i \leq n. \end{cases} \tag{5}$$

where a and b are two real values. In our example $\mathbf{X}_8$, we have $a = 0$, $b = 1$, and $\theta = 5$. We transform $\mathbf{X}_n$ using Haar wavelet basis vectors ψ_i as:

$$\mathbf{X}_n = \hat{x_1}\psi_1 + \sum_{i=2}^{n} \hat{x_i}\psi_i. \tag{6}$$

where $\hat{x_i} = \langle \mathbf{X}_n, \psi_i \rangle$.

Considering the procedure of transformation discussed in Section 3.2, only those coefficients whose support interval include the point of discontinuity θ (i.e., x_θ) will be nonzero (i.e., $\hat{x_i}$ if $i \in Sup^{-1}(\theta)$). In other word, we get

$$\mathbf{X}_n = \hat{x_1}\psi_1 + \sum_{\substack{i \in Sup^{-1}(\theta) \\ }}^{i \neq 1} \hat{x_i}\psi_i. \tag{7}$$

Now, if we define a new vector f_θ as the linear combination of the multiplication of nonzero coefficients to their corresponding wavelet basis vectors (i.e., the second term in Equation 7), we obtain a representation of $\mathbf{X}_n$ as follows:

$$\mathbf{X}_n = \hat{x_1}\psi_1 + \alpha_\theta f_\theta. \tag{8}$$

where $\alpha_\theta = 1$.

We refer to f_θ as the footprint for discontinuity point x_θ of degree zero as it captures a change point of degree zero.

Here, we apply the same scenario to our example time series $\mathbf{X}_8$. If we rewrite $\mathbf{X}_8$ in terms of ψ_i's we get $\mathbf{X}_8 = 1.0607\frac{\psi_1}{|\psi_1|} - 1.0607\frac{\psi_2}{|\psi_2|} - 0.5\frac{\psi_4}{|\psi_4|} - 0.7071\frac{\psi_7}{|\psi_7|}$. Let $f_5 = -1.0607\frac{\psi_2}{|\psi_2|} - 0.5\frac{\psi_4}{|\psi_4|} - 0.7071\frac{\psi_7}{|\psi_7|}$. Similar to Equation 8 we get $\mathbf{X}_8 = 1.0607 \times \psi_1 + 1 \times f_5$. Therefore, we can represent $\mathbf{X}_8$ with one summary coefficient and only one nonzero detail coefficient at position 5.

Now, we need to show that the above scenario is extendable to time series data with m discontinuities of degree zero (piecewise constant). It is easy to see that any piecewise constant time series with m discontinuities can be represented as linear combination of m time series each with only one discontinuity. We use Parseval's theorem [20] to extend the scenario we developed so far:

Theorem 1. *Let $\hat{X}$ denote wavelet transformation of vector $\mathbf{X}$ using orthogonal wavelets (e.g., Haar). If $\mathbf{X}_n = \mathbf{X}_{1n} + ... + \mathbf{X}_{mn}$, we have $\hat{\mathbf{X}}_n = \hat{\mathbf{X}}_{1n} + ... + \hat{\mathbf{X}}_{mn}$.*

The direct conclusion of Parseval's theorem is that $\hat{\mathbf{X}}_n$ is represented in terms of the set of summary vector and footprints f_i each used in representing $\mathbf{X}_{in}$. That is, any piecewise constant time series with m discontinuities can be represented with the summary vector together with m footprints.

As the previous example shows, we get a much sparser representation of the data if we use wavelet footprints as our basis. This idea can easily be generalized to generate all vectors f_i of degree $d = 0, ..., D$ each with only one discontinuity point i ($i = 1, ..., n$) of degree d. Using these vectors as a basis enables us to capture polynomial changes up to degree D in time series data. Interested readers can find the details in [23]. Dragotti et al. [3] prove that each discontinuity of degree D at x_i can be represented with the summary vector together with maximum $D + 1$ footprints (i.e., $f_i^{(0)}, ..., f_i^{(D)}$).

4.1   Properties

In this section, we enumerate different properties of footprints.

- The set of footprints together with the *summary vector* constitutes a basis.
- The footprint basis is an overcomplete basis (except for the case of constant piecewise where we have footprints of degree 0 only). Notice that when the length of the data is n the number of footprint vectors in $f_i^{(D)}$ is $n \times (D+1)$, and hence the resulting basis is overcomplete.
- The footprints efficiently model discontinuities in time series data; a piecewise polynomial time series with K discontinuities, can be represented with only $K \times (D + 1)$ footprints together with the summary coefficients. Each footprint coefficient also contains
  1. The amplitude of the discontinuity that it represents.
  2. The characteristics of the two polynomials right before and after the discontinuity point.

5   Change Detection with Footprints

We showed that nonzero coefficients in footprint transformed time series data are representatives of the change points in the data. Therefore, a novel change detection approach emerges by employing footprints. Throughout this section, we assume that we have pre-computed the biorthogonal footprint basis F_D. Note that this is a one time process independent of either the data or the queries. We would like to answer two major categories of change detection queries:

- Q_d: Return change points of all degrees.
- Q_{da}: Return change points of all degrees, their corresponding degrees and change amplitudes.

Similar to any general SQL query, user can enforce restrictions on degree of change point or its change amplitude. For example, user can ask for change points of degree d where the change amplitude is greater than a threshold T.

Our approach stores the time series data in the wavelet domain. We use footprint basis F_D as our wavelet basis. Therefore, our approach answers change detection queries by returning the nonzero coefficients stored in its database. Figure 2 illustrates the process flow of our approach. We describe each part in details.

5.1   Insert/Update

Upon receiving the new data (i.e., time series), we transform it using F_D and then store it in the database. The transformation will further be explained in

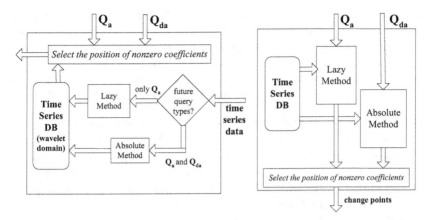

Fig. 2. Query processing in wavelet domain *(on the left)* and Ad hoc query processing *(on the right)*

Section 5.3. To update the transformed data, approaches such as Shift-Split [9] can be used to update the transform data stored in the wavelet domain efficiently.

5.2 Query Processing

On receiving a change detection query on a portion of the data, we retrieve the nonzero coefficients corresponding to that portion of the data from the database. For each nonzero coefficient $f_i^{(d)}$, we return a change of degree d at point i. If the user is interested in changes greater than a given threshold, we return only those coefficients that are greater than the threshold. The time complexity of this approach is $O(n)$, since we only need a single scan over the data.[2]

The pre-transformation of the data eliminates the need to restart the entire process whenever user specifies a new degree and/or threshold for the change values. This makes our approach faster and more practical than the other change detection approaches where the algorithms are highly dependent on either threshold value or degree.

5.3 Footprint Transformation

The challenge here is to transform the data using the footprint basis. We propose two different methods for footprint transformation. In both these methods, we assume that F_D and $\tilde{F}_D$ are pre-computed. Note that the computation of these vectors is completely data-independent.

The first *lazy* method is mainly based on *approximating* the footprint coefficients by projecting the time series on the dual basis of the footprint basis. This method is highly efficient in terms of performance.

---

[2] This can be improved further (perhaps to $O(\log n)$) by using an index-structure on the coefficients.

The second *absolute* method is based on a greedy iterative algorithm termed *matching pursuit* [18] which is a proven approach in signal processing for representing signals in terms of an overcomplete basis. The outputs of both methods enable us to answer change detection queries by retrieving the nonzero coefficients and reporting their positions as change points. Because of possible noise in the data, both methods may employ thresholds to select nonzero coefficients.

The Lazy Footprint Transformation. The lazy method approximates the coefficients of $\mathbf{X}_n$ by simply computing the $\alpha_i^{(d)} = \langle \mathbf{X}_n, \tilde{f}_i^{(d)} \rangle$ for all $f_i^{(d)}$'s in the basis. During the query processing, it returns i as a change point if $\alpha_i^{(d)}$ is greater than the user defined threshold in each footprint basis of degree d (see Section 5.2). The universal threshold $u = \sigma\sqrt{2 \ln N}$ suggested in [4] is an appropriate candidate for the threshold value.

Notice that the coefficients computed by the lazy method are not the exact footprint coefficients due to the overcompleteness of the basis. They only approximate the discontinuity points. However, in Section 7 we show that the lazy transformation performs very effectively for detecting the change points. For each time series of size n, it is easy to see that the time complexity of the lazy transformation is $O(n^2)$.[3]

The Absolute Footprint Transformation. As mentioned before, the footprint vectors constitute an overcomplete basis. This overcomplete basis gives us more power and flexibility in modelling changes in the data. As a drawback, transformation of the time series $\mathbf{X}_n$ (i.e., coefficients $\alpha_i^{(d)}$) becomes a more challenging task. Here, in order to compute the exact values for $\alpha_i^{(d)}$'s we use the *matching pursuit technique* to find the nonzero $\alpha_i^{(d)}$ coefficients. Due to lack of space, we omit the details of the technique and corresponding algorithm. Interested readers can find the details in [23].

The set of coefficients returned using the matching pursuit technique are the change points in the time series $\mathbf{X_n}$. Notice that the coefficients computed by the absolute approach are the exact coefficients of the footprint transformation. We can modify matching pursuit such that the algorithm terminates after maximum $\lceil \frac{K}{2} \rceil$ iterations where K is the number of change points in $\mathbf{X_n}$. Since each iteration of matching pursuit algorithm takes $O(n^2)$, the overall time complexity of the absolute transformation becomes $O(\lceil \frac{K}{2} \rceil .n^2)$. [4]

5.4   Ad Hoc Query Processing

If the data cannot be stored in the wavelet domain, we must transform it in real-time when we receive a query. We choose between the lazy or absolute method based on the type of query. The time complexity of this ad hoc change detection approach is equal to the time complexity of the transformation using footprints.

---

[3] This can be improved further by utilizing the fact that the $f_i^{(d)}$ matrices are very sparse.

[4] The same as what we had for the Lazy transformation the time complexity can be improved.

6   Customizing Footprints for Wavelet-Based Applications

In this section, we show that our approach can be incorporated within systems where the time series data is maintained in the wavelet domain (e.g., ProDA [10]). An example of an approach dealing with the data directly in the wavelet domain is ProPolyne introduced in [22]. ProPolyne is a wavelet-based technique for answering polynomial range-aggregate queries. It uses the transformed data from the wavelet domain to generate the result. We show that ProPolyne's approach to answer polynomial range-aggregate queries is still feasible when the data is transformed using footprint basis.

With ProPolyne, a polynomial range-aggregate query (e.g., SUM, AVERAGE, or VARIANCE) is represented as a query vector Q_n. Then, the answer to the query is $\langle \mathbf{X}_n, Q_n \rangle$. The family of wavelet basis used by ProPolyne each constitutes an orthogonal basis. Thus, according to *Parseval's theorem*, they preserve the energy of the data after the transformation and hence we have:

$$\langle \mathbf{X}_n, Q_n \rangle = \langle \hat{\mathbf{X}}_n, \hat{Q}_n \rangle \ . \tag{9}$$

Therefore, ProPolyne evaluates $\langle \hat{\mathbf{X}}_n, \hat{Q}_n \rangle$ as the answer to the query Q_n.

However, it is easy to see that Equation 9 does not hold for wavelet footprints since the footprint basis is not an orthogonal basis. Here, we extend Equation 9 to hold for the biorthogonal bases. Assume that $\hat{\mathbf{X}}_n$ and $\tilde{\mathbf{X}}_n$ are the transformations of $\mathbf{X}_n$ using the footprint wavelet basis, and its dual basis, respectively. Now, according to Definitions 3 and 4, and Equation 9 it is easy to see that

$$\langle \mathbf{X}_n, Q_n \rangle = \langle \hat{\mathbf{X}}_n, \tilde{Q}_n \rangle \tag{10}$$

$$\langle \mathbf{X}_n, Q_n \rangle = \langle \tilde{\mathbf{X}}_n, \hat{Q}_n \rangle \ . \tag{11}$$

In practice, we use Equation 10 where the data is transformed to wavelet in advance and the dual of the query will be computed on the fly to perform the dot product at the query time. Hence, we are still able to answer polynomial range-aggregate queries proposed by ProPolyne. Therefore, ProPolyne can transform data using footprint basis and still benefit from its unique properties.

7   Experimental Results

We conducted several experiments to evaluate the performance of our proposed approach for change detection in time series data. We compared the query response time of our ad hoc query processing approach described in Section 5.4 with the maximum likelihood-based algorithm proposed by Guralnik et. al [8]. We chose their approach for comparison because it is the fastest change detection algorithm that considers different degrees of change. Throughout this section, we refer to their method as the *Likelihood* method. We studied how the size of the time series (n) and the total number of its change points (K) affects the performance of each method.

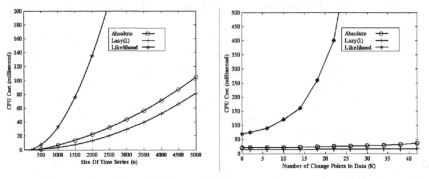

a. Query cost vs. size of data   b. Query cost vs. number of change points K

Fig. 3. Performance Comparison

We also evaluated our lazy and absolute methods by investigating the effect of the following parameters on their accuracy: 1) the minimum distance between two consecutive change points (MinDist), 2) the maximum degree of change points in the data (MaxDeg), 3) the maximum degree D of footprint basis (MaxDegF), and 4) the amount of noise in the data (Noise). We represent the accuracy of each method in terms of the average number of false negatives (AFN) and the average number of detected false hits (AFH).

We used both synthetic and real-world datasets. We generated a synthetic dataset $D3$ of 80 time series each with a size in the range of 100 to 5,000. Each time series of the dataset $D3$ is a concatenation of several segments each modelled by a polynomial of degree up to 3. Our real-world dataset include oil and gas time series for different oil wells in California.

Notice that for the absolute method, we used a modified version of matching pursuit introduced in [3] which terminates after $\lceil \frac{K}{2} \rceil$ iterations. For the likelihood method, we use the threshold value with which the method computes the most accurate result. Sections 7.1 and 7.2 focus on our synthetic dataset as we already know the exact characteristics of the change in their time series. This enables us to measure the accuracy of our approach. Section 7.3 discusses our experiments with the real-world data.

7.1   Performance

In our first set of experiments, we compared the performance of our ad hoc change detection query processing. As the Likelihood method uses the original time series data as input, it is only fair to compare it with our ad hoc approach. That is, the CPU time reported for the lazy and absolute methods include both the time for the footprint transformation of time series and that of detecting the change points. We used footprint basis of up to degree 3 to transform the data in the lazy and absolute methods. That is, MaxDeg = MaxDegF. Also, we used polynomials of up to degree 3 in the Likelihood method.

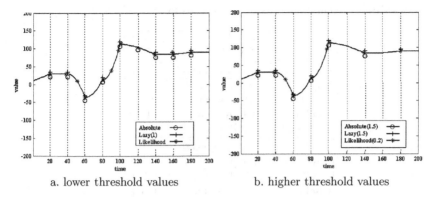

a. lower threshold values   b. higher threshold values

Fig. 4. Detected change points by the absolute, lazy and Likelihood methods with the actual change points in the data *(the vertical lines)*

We varied the size of time series data from 100 to 5,000 and measured the CPU cost of each method. In Figure 3a, Lazy(i) denotes the measurements of the lazy method in which the threshold value is $i \times u$ (u is the universal threshold and i is simply a factor multiplied by u). As shown in the figure, our lazy and absolute methods outperform the Likelihood method by a factor of 2 to 8 when the size of data increases from 100 to 5,000. We also compared the performance of all the mentioned methods by running the algorithms on overlapping chunks of size 256 for the subset of time series with size larger than 1000. However, since all the methods benefit from the input with smaller sizes the result diagram shows approximately the same trend.

The theoretical time complexities of the absolute and Likelihood methods depend on K. On the other hand, the lazy method is a series of simple projections which are independent of the characteristics of the data. To study the effect of K on the performance of each method, we varied K and measured the CPU cost. Figure 3b illustrates that while the performance of both our methods remains almost fixed for different number of change points, the CPU cost of Likelihood method dramatically increases.

7.2 Accuracy

Our second set of experiments were [2] aimed to evaluate the accuracy of each method in terms of number of missed change points and the detected spurious change points (i.e., precision and recall). Figure 4 shows a small time series of size 200 generated with polynomial segments of maximum degree 2 as well as the true change points and those detected by each method. We used footprint basis of up to degree 2 to transform the data and for the Likelihood method we used polynomials of up to degree 2.[5]

---

[5] Notice that the performance of the likelihood method can be improved further by adapting the technique described in [2]. As part of our future work, we plan to compare the performance of the improved version of the likelihood method with our improved versions of Lazy and Absolute methods.

Table 1. Accuracy results of all methods for cases $F2$ and $F3$

F3		*F2*	
Method	*AFN*	*Method*	*AFN*
MinDist= 5		MinDist= 5	
Lazy(1.5)	3	Lazy(1.5)	3.5
Lazy(1)	2.1	Lazy(1)	3.1
Absolute	0.9	Absolute	1
Likelihood	4.5	Likelihood	4.1
MinDist= 10		MinDist= 10	
Lazy(1.5)	1.9	Lazy(1.5)	2.9
Lazy(1)	0.9	Lazy(1)	2.7
Absolute	0.2	Absolute	0.5
Likelihood	1.8	Likelihood	3.0
MinDist= 20		MinDist= 20	
Lazy(1.5)	0.5	Lazy(1.5)	1.1
Lazy(1)	0.2	Lazy(1)	0.6
Absolute	0	Absolute	0.1
Likelihood	0.2	Likelihood	1
MinDist= 50		MinDist= 50	
Lazy(1.5)	0.3	Lazy(1.5)	0.9
Lazy(1)	0.2	Lazy(1)	0.4
Absolute	0	Absolute	0.1
Likelihood	0.2	Likelihood	0.8

There are ten actual change points as shown in Figure 4 and the minimum distance between each two change points is 20. In Figure 4a, the likelihood and lazy(1) methods both miss the change point at $t = 120$. Also, Likelihood method detects two false hits at points $t = 51$ and $t = 90$. And for the change that occurs at point $t = 100$, it detects two change points at $t = 98$ and $t = 101$. The lazy method returns no false hit. The absolute method returns all 10 actual change points at their exact positions without any false hit. It is interesting to note that the increased threshold values result in ignoring the minor change points at $t = 120$ and $t = 160$ by all the methods as shown in Figure 4b. However the likelihood method still has a false hit at $t = 51$ and for the change that occurs at point $t = 100$, it also detects two change points at $t = 98$ and $t = 101$.

Notice that using our footprint-based approach, we also acquire valuable information about the degree and amplitude of each change point. For example, at point $t = 40$, we have a discontinuity caused by a quadratic segment following a constant segment. Also at point $t = 140$, we have a discontinuity caused by a constant segment following a linear segment.

We repeated the previous experiment on time series of the dataset $D3$ for which $K = 10$. We varied the minimum distance between each two consecutive change points in the data (MinDist) from 5 to 50. Table 1 shows the average number of false negatives (AFN) for all the methods. Column $F3$ shows the results for the experiment where we used footprints of degrees 0, 1, 2, and 3.

Note that in this case MaxDegF is identical to MaxDeg. Likewise, column $F2$ shows the case where we used footprints of degrees 0, 1, and 2. The Likelihood method also used polynomials of degrees up to 2 (resp. 3) for case $F2$ (resp. $F3$).

The Effect of MinDist. As Table 1 depicts, our footprint methods always outperform the Likelihood approach in terms of accuracy with *absolute* being the superior method. With small values of MinDist, the accuracies of all methods dramatically downgrade. However, even for close change points, the absolute method misses only one of 10 change points on average. This yields that the absolute method is resilient to the effect of closeness of change points.

The Effect of Noise. In our third set of experiments, we fixed the minimum distance between change points to 10, threshold value equal to $1.5 \times u$, and $F_D = F_2$. Using polynomials of degree up to 2, we generated two noisy datasets. We added noise with the standard deviation of about 1/15 to 1/30 and 1/150 to 1/300 of the average of values in time series to generate two noisy datasets $Noisy(1)$ and $Noisy(0.1)$, respectively. Table 2 shows the accuracy results of applying all three methods on both datasets.

Table 2. Accuracy results of all methods for datasets $Noisy(1)$ and $Noisy(0.1)$

Noisy(1) Method	AFN	AFH	Noisy(0.1) Method	AFN	AFH
Lazy(1)	5	1.2	Lazy(1)	3.5	1
Absolute	2	0.9	Absolute	1.5	0.8
Likelihood	2.8	3	likelihood	2.6	3

7.3   Experiment with Real-World Datasets

Finally, the last set of experiments focuses on real-world time series data. We tested our methods on different time series generated within the oil industry. Here, we only report the results on three time series $OIL1$, $OIL2$, and GAS obtained from Petroleum Technology Transfer Council[6] due to lack of space. These time series are collected from wells in active oil fields in California. $OIL1$ and $OIL2$ include oil production during 1985-1995 and 1974-2002, respectively. GAS includes the gas production rate measured in a 2300 day period, sampling once every 15 days.

Unlike synthetic time series, here we do not know where the *exact* change points are. Therefore, we evaluated our methods visually based on the position of their detected change points. Figure 5a, 5b and 6 depicts the change points detected in time series $OIL1$, $OIL2$ and GAS, respectively.

Notice that our absolute method does not identify any change at points such as $t = 235$ in Figure 5b. The reason here is that the segment corresponding to the range $[228, 240]$ can be modelled by a polynomial of degree 3. Therefore, $t = 235$ is not a discontinuity of degree 3 in $OIL2$.

---

[6] http://www.westcoastpttc.org/.

Figure 6b illustrates the detected change points in GAS data by each of the three methods when they use higher threshold values as compared to those used in Figure 6a. Comparing Figures 6a and 6b shows that the former detects all small changes while the later identifies only major changes in the data. Notice that once our methods detect changes using a given threshold value, changes above different thresholds can be identified only by performing a simple scan over all the coefficients. However, with the likelihood method, we need to rerun the whole process when the user changes her threshold value. For example, we ran the likelihood method separately for the results of each of Figures 6a and 6b.

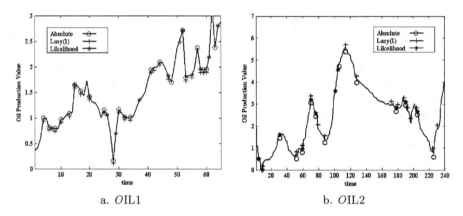

a. $OIL1$ b. $OIL2$

Fig. 5. Detected change points by the absolute, lazy and Likelihood methods

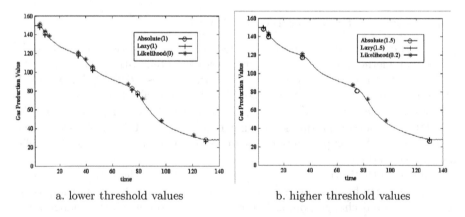

a. lower threshold values   b. higher threshold values

Fig. 6. Detected change points by the absolute, lazy and Likelihood methods

8   Conclusions and Future Work

We studied the problem of detecting changes in time series data. We formally defined the degree of change for change points in the time series data. Our definition is closely related to the difference in two polynomial functions fitting two adjacent segments of data. We then described our novel approach which employs wavelet footprints for defining discontinuities of different degrees. We proposed lazy and absolute methods to transform the data using footprint basis. Finally, we compared the performance and accuracy of our footprint-based approach with the maximum likelihood method [8] through exhaustive sets of experiments with both synthetic and real-world data. The results show that our approaches are faster, more accurate and return more information about the changes. In addition their performances are less sensitive to user defined parameters such as threshold values and number of changes.

In this paper, for the first time we exploited the interesting characteristics of footprints for change detection in time series data. Motivated by our results, we plan to develop a footprint-based tool for real-time change detection on data streams.

Acknowledgement

We would like to acknowledge Dr. Antonio Ortega and his student En-Shuo Tsau for their valuable help in our understanding of the theory of wavelet footprints.

This research has been funded in part by NSF grants EEC-9529152 (IMSC ERC) and IIS-0238560 (PECASE), unrestricted cash gifts from Microsoft, an on-going collaboration under NASA's GENESIS-II REASON project and partly funded by the Center of Excellence for Research and Academic Training on Interactive Smart Oilfield Technologies (CiSoft); CiSoft is a joint University of Southern California - Chevron initiative.

References

1. K. Chakrabarti, M. N. Garofalakis, R. Rastogi, and K. Shim. Approximate query processing using wavelets. In A. E. Abbadi, M. L. Brodie, S. Chakravarthy, U. Dayal, N. Kamel, G. Schlageter, and K.-Y. Whang, editors, *VLDB 2000, Proceedings of 26th International Conference on Very Large Data Bases, September 10-14, 2000, Cairo, Egypt*, pages 111–122. Morgan Kaufmann, 2000.
2. Y. Chen, G. Dong, J. Han, B. W. Wah, and J. Wang. Multi-dimensional regression analysis of time-series data streams. In *VLDB*, pages 323–334, 2002.
3. P. L. Dragotti and M. Vetterli. Wavelet transform footprints: Catching singularities for compression and denoising. In *ICIP*, 2000.
4. P. L. Dragotti and M. Vetterli. Deconvolution with wavelet footprints for ill-posed inverse problems. In *IEEE Conference on Acoustics, Speech and Signal Processing*, volume 2, pages 1257–1260, Orlando, Florida, USA, May 2002.
5. P. L. Dragotti and M. Vetterli. Wavelet footprints: Theory, algorithms and applications. *IEEE Transactions on Signal Processing*, 51(5):1306–1323, May 2003.
6. L. Firoiu and P. R. Cohen. Segmenting time series with a hybrid neural networks - hidden markov model. In *Eighteenth national conference on Artificial intelligence*, pages 247–252. American Association for Artificial Intelligence, 2002.
7. X. Ge. *Segmental semi-markov models and applications to sequence analysis*. PhD thesis, 2002. Chair-Padhraic Smyth.

8. V. Guralnik and J. Srivastava. Event detection from time series data. In *KDD '99: Proceedings of the fifth ACM SIGKDD international conference on Knowledge discovery and data mining*, pages 33–42. ACM Press, 1999.

9. M. Jahangiri, D. Sacharidis, and C. Shahabi. SHIFT-SPLIT: I/O Efficient Maintenance of Wavelet-Transformed Multidimensional Data. In *Proceedings of the 24th ACM SIGMOD International Conference on Management of Data*, 2005.

10. M. Jahangiri and C. Shahabi. ProDA: A Suit of WebServices for Progressive Data Analysis. In *Proceedings of 24th ACM SIGMOD International Conference on Management of Data*, 2005. (demostration).

11. R. Jr. *Advances in Well Test Analysis*, volume 5. Society of Petroleum Engineers, 1977.

12. E. Keogh and P. Smyth. A probabilistic approach to fast pattern matching in time series databases. In D. Heckerman, H. Mannila, D. Pregibon, and R. Uthurusamy, editors, *Third International Conference on Knowledge Discovery and Data Mining*, pages 24–30, Newport Beach, CA, USA, 1997. AAAI Press, Menlo Park, California.

13. E. J. Keogh, K. Chakrabarti, S. Mehrotra, and M. J. Pazzani. Locally adaptive dimensionality reduction for indexing large time series databases. In *SIGMOD Conference*, 2001.

14. E. J. Keogh, S. Chu, D. Hart, and M. J. Pazzani. An online algorithm for segmenting time series. In *ICDM '01: Proceedings of the 2001 IEEE International Conference on Data Mining*, pages 289–296. IEEE Computer Society, 2001.

15. J. Lin, E. Keogh, and W. Truppel. Clustering of streaming time series is meaningless. In *DMKD '03: Proceedings of the 8th ACM SIGMOD workshop on Research issues in data mining and knowledge discovery*, pages 56–65. ACM Press, 2003.

16. J. Lin, M. Vlachos, E. J. Keogh, and D. Gunopulos. Iterative incremental clustering of time series. In *EDBT '04: Proceedings of the 8th International Conference on Extending Database Technology*, pages 106–122, 2004.

17. S. Mallat and W. L. Hwang. Singularity detection and processing with wavelets. *IEEE Trans. Inf. Th*, 38:617–643, 1992.

18. S. Mallat and Z. Zhang. Matching pursuits with time-frequency dictionaries. *IEEE Transactions on Signal Processing*, 41(12):3397–3415, 1993.

19. Y. Matias, J. S. Vitter, and M. Wang. Wavelet-based histograms for selectivity estimation. In *SIGMOD '98: Proceedings of the 1998 ACM SIGMOD international conference on Management of data*, pages 448–459. ACM Press, 1998.

20. A. V. Oppenheim and R. W. Schafer. *Digital Signal Processing*. Prentice Hall, Englewood Cliffs, NJ, USA, 1975.

21. V. Puttagunta and K. Kalpakis. Adaptive methods for activity monitoring of streaming data. In *ICMLA*, pages 197–203, 2002.

22. R. R. Schmidt and C. Shahabi. Propolyne: A fast wavelet-based algorithm for progressive evaluation of polynomial range-sum queries. In *EDBT '02: Proceedings of the 8th International Conference on Extending Database Technology*, pages 664–681. Springer-Verlag, 2002.

23. M. Sharifzadeh, F. Azmoodeh, and C. Shahabi. Change detection in time series data using wavelet footprints. Technical Report 05-855, Department of Computer Science, University of Southern California, 2005.

24. K. Yamanishi and J. ichi Takeuchi. A unifying framework for detecting outliers and change points from non-stationary time series data. In *KDD '02: Proceedings of the eighth ACM SIGKDD international conference on Knowledge discovery and data mining*, pages 676–681. ACM Press, 2002.

Evaluation of a Dynamic Tree Structure for Indexing Query Regions on Streaming Geospatial Data

Quinn Hart[1], Michael Gertz[2], and Jie Zhang[2]

[1] CalSpace, University of California, Davis, CA 95616, USA
[2] Dept. of Computer Science, University of California, Davis, CA 95616, USA

Abstract. Most recent research on querying and managing data streams has concentrated on traditional data models where the data come in the form of tuples or XML data. Complex types of streaming data, in particular spatio-temporal data, have primarily been investigated in the context of moving objects and location-aware services. In this paper, we study query processing and optimization aspects for streaming Remotely-Sensed Imagery (RSI) data. Streaming RSI is typical for the vast amount of imaging satellites orbiting the Earth, and it exhibits certain characteristics that make it very attractive to tailored query optimization techniques. Our approach uses a Dynamic Cascade Tree (*DCT*) to (1) index spatio-temporal query regions associated with continuous user queries and (2) efficiently determine what incoming RSI data is relevant to what queries. The *DCT* supports the processing of different types of RSI data, ranging from point data to more general spatial extents in which the incoming imagery can be single pixels, rows of pixels, or discrete parts of images. The *DCT* exploits spatial trends in incoming RSI data to efficiently filter the data of interest to the individual query regions. Experimental results using random input and Geostationary Operational Environmental Satellite (GOES) data give a good insight into processing streaming RSI and verify the efficiency and utility of the *DCT*.

1 Introduction

The current interest in data stream management [1,2,6] has driven various new processing methods and paradigms for streaming data, such as adaptivity [11,15] and operator scheduling [3,5]. Some proposed approaches have extended to the realm of spatio-temporal data, where new methods defining the spatial relationships between queries and data streams are being investigated, in particular in the context of continuous queries over moving objects (e.g., [12,18,19]). A new area of research where these streaming data paradigms can have a great impact is in applications surrounding remotely-sensed geospatial image data originating from the various satellites orbiting the Earth.

Figure 1(a) gives an overview of a data stream management system (DSMS) for Remotely-Sensed Imagery (RSI) data. Such data has a number of characteristics that are different from traditional streaming relational or spatio-temporal

C. Bauzer Medeiros et al. (Eds.): SSTD 2005, LNCS 3633, pp. 145–162, 2005.

data typically described in a DSMS context. One aspect is that the incoming RSI data stream is very large, usually arriving as discrete parts of binary image data. Another important point is that streaming RSI data is more highly organized with respect to its spatial and temporal components than is usually assumed for more generic types of spatio-temporal data. This organization has some profound effects on how data and queries can be processed in a RSI DSMS. The organization of the incoming RSI data affects the query evaluation and the index structures used. In this paper, we discuss in detail an indexing scheme for queries in such a system. This structure, the Dynamic Cascade Tree (*DCT*) utilizes the characteristics of the RSI data and facilitates the efficient routing of relevant portions of the streaming data to query operators.

Exactly how data arrives varies for different RSI data streams, but generally image data arrives as discrete contiguous packets of data. The packets may be individual pixels or organized sets of pixel data, depending on the instrument. Different types of RSI data have different orderings and structures. Figure 1(b) illustrates the common structures, including image-by-image, row-by-row, and pixel-by-pixel. Many satellite instruments, such as the Geostationary Operational Environmental Satellite (GOES) weather satellite [7], obtain RSI data in a row-by-row fashion. Although conceptually the data collected by GOES is represented as a set of complete images, the images are incrementally received in *row-scan order* where pixels are delivered a few lines at a time. Other types of sensors gather data on a pixel-by-pixel basis. These include imaging sensors with large pixel sizes, or active sensors such as Light Detection and Ranging (LIDAR) sounding instruments.

Within a packet of data, the spatial organization, represented as image pixels, is well defined. The ordering of the packets is also well defined, and consecutive data packets from an RSI data stream are in close spatial and temporal proximity. Data from an RSI stream usually arrive only at one or a small number of spatial locations for a given instrument.

The above scenarios illustrate an important characteristic exploited in the following approach. Consecutive packets in an RSI data stream have close spatial and temporal proximity. In the *DCT*, this spatial trend is used to influence how multiple queries against a stream of RSI data are processed.

1.1 Problem Description and Objectives

Most continuous queries against an RSI data stream include operations to restrict the spatio-temporal data to specified regions of interest. Such *Continuous Query (CQ) regions* are part of more complex queries users issue against a DSMS. Other, more complex operators of a query against an RSI data stream, such as map projections or spatial and value transforms, typically follow a spatial restriction as they are much less selective. Clearly, query regions specified by different users may overlap. This is typical for RSI streams that have geographic *hot spots* or regions that are of interest to many users. An RSI stream management system needs to (1) efficiently intersect incoming image data with a possibly large number of query regions, and (2) it should provide a means that allows queries to share the incoming data for further processing. These aspects are illustrated

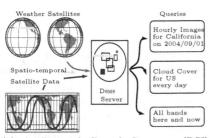

(a) A Remotely-Sensed Imagery (RSI) DSMS. High bandwidth input image streams are handled by the system. The figure shows imagery from both polar orbiting and geostationary sources. Typical queries request image data or other products for specific times and spatial extents.

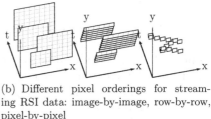

(b) Different pixel orderings for streaming RSI data: image-by-image, row-by-row, pixel-by-pixel

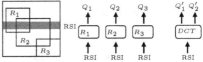

(c) RSI data intersecting with query regions

Fig. 1. Geospatial data stream system and characteristics of streaming RSI data

in Fig. 1(c), where one region query R_i is associated with each of the user queries Q_i. In the figure, incoming RSI data intersects with two query regions R_1 and R_2 (left). Instead of filtering incoming data for each individual query Q_1, Q_2, and Q_3 (middle), a mechanism is needed to determine what incoming image data is relevant to what queries and pipeline the relevant data to subsequent query operators. The DCT is such a mechanism that filters and streams relevant data to subsequent operators of individual queries, here only Q_1' and Q_2' (right). To efficiently process these spatial restrictions, the organization of the incoming data stream needs to be considered.

The Dynamic Cascade Tree (DCT) is a space efficient structure that indexes query regions that are part of more complex queries against RSI data streams. In particular, the DCT supports the efficient processing of a *moving data stream*. A data stream query is one that, for a window describing the spatial extent of the incoming image data (pixel, row, or image), will identify all CQ regions that spatially and temporally overlap the data window. This allows the pipelining of image data to those queries to which the data is of interest, and it facilitates the sharing of image data among queries in the case of non-disjoint query regions. The design of the data structures and algorithms underlying the DCT is guided by some important requirements, which are typical for RSI data streams: (1) The DCT indexes CQ regions and is sufficiently small to be kept in main memory. It also has to support efficient insertion and deletion of CQ regions associated with user queries. (2) The geospatial data stream comes from a single source corresponding to the real-time streaming satellite data. Sequential stream data are usually in close proximity and might have a regular trajectory through space. The DCT has to account for both size and spatio-temporal trends of the incoming data. (3) Because of the size of the CQ regions and the size and shape of the incoming stream data, selectivity is high. For example, incoming RSI data packets intersect about 20% of CQ regions for typical GOES applications.

1.2   Structure of the Paper

Section 2 describes the *DCT* and shows how it filters incoming image data for multiple CQ regions. Section 3 details the performance of the *DCT* in general terms and discusses the parameters affecting performance. In Section 4, several experimental results are presented. These include experiments on random data to study the performance under changing input parameters, and experiments closer to real world scenarios using GOES data as a practical example. Section 5 describes related research for similar problem domains. Section 6 concludes the paper and highlights future research directions.

2   The Dynamic Cascade Tree (DCT)

The *DCT* data structure is designed to use a single rectangular window of pixel data from a RSI data stream as input to a *window query* on a number of CQ regions associated with user queries, as illustrated in Fig. 1(c). The *DCT* uses the spatial extent of the most recent *input window* to quickly answer multiple queries on a stream of RSI data. That is, *for a given input window from the stream, the DCT returns all CQ regions that overlap this input.* For streaming RSI data, input windows correspond to individual packets of contiguous image data. The regions correspond to the spatial extents of the individual CQ queries registered in the DSMS.

The most important aspect for RSI data in terms of designing the *DCT* is that the incoming stream data comes as contiguous data packets that are typically in close spatial and temporal proximity to the previous data. Our goal is to take advantage of this proximity and to develop an index structure that improves the search performance for subsequent parts of the stream.

The *DCT* structure realizes an index that is dynamically tuned to the current location of an input RSI window. For the spatial extent of the most recent input window, the *DCT* maintains the CQ regions around that window where the result set will change. New RSI input can be processed very quickly if the new window has the same result set of CQ regions as the previous window; and will incrementally update the result set otherwise. The structure is designed to be small, and allow for fast insertions and deletions of new CQ regions registered by the DSMS. It assumes some particular characteristics of the input stream, notably that the stream changes in such a way that many successive data extents from the RSI stream will share the same result set of CQ regions. Therefore, the cost of maintaining a dynamic structure can be amortized over the incoming data stream. Sections 3 and 4 examine in more detail the actual performance with different CQ regions and data stream parameters. First a general overview of the components of the *DCT* is given, followed with details on how the *DCT* is built, maintained, and queried. Hart [10] gives a more complete algorithmic description of the *DCT*, including a simpler point-based version of the *DCT* and algorithms for insertion, deletion, and queries.

2.1 DCT Components

Figure 2 gives an overview of the *DCT*. The figure shows a set of rectangular CQ regions $a, b, \ldots, f$, and a rectangular input window corresponding to the most recent data in the RSI stream. Also shown are the associated structures that make up and maintain the *DCT*. This figure describes a *DCT* that indexes two dimensions, although the *DCT* can be extended to more. The figure shows the dimensions being cascaded first in the x dimension and then in the y dimension. The *DCT* maintains separate trees for both the minimum and maximum end-

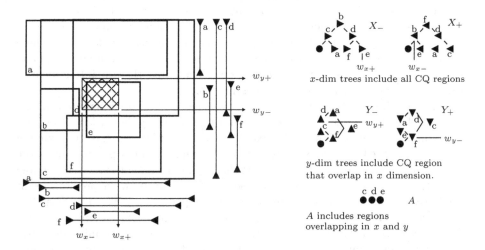

Fig. 2. Dynamic Cascade Tree (DCT) with six CQ regions, $a, b, \ldots, f$. Shown are the indexed regions, the current data stream input window, and the cascading trees, X_-, X_+, Y_-, Y_+, A, and the window node pointers, w_{x-}, w_{x+}, w_{y-}, w_{y+} that make up the data structure. $\triangle$ and $\triangleright$ represent the minimum endpoints of the regions in x and y dimension, $\triangledown$ and $\triangleleft$ the maximum endpoints of the regions in x and y dimension

points of each of the CQ regions for each dimension. In Fig. 2, these are denoted as X_- and X_+ for the x-dimension, and Y_- and Y_+ for the y-dimension and are termed *endpoint trees* in the *DCT*. A final tree, A, maintains an index of all the regions that overlap with the current data in the input stream. The figure shows these trees notionally, emphasizing their ordering and structure; the endpoint trees using one endpoint and the unique identifier r_{id} for each indexed region, and the A tree only using r_{id}. The values of the nodes contain pointers back to the regions, or the associated CQ query in the DSMS.

Which of the CQ regions are included in the trees varies with the dimension. The trees for the first dimension, x, contain the minimum (in X_-) and maximum (in X_+) endpoints for *every* CQ region. Y_- and Y_+ do not contain the endpoints of all the regions in *DCT*, but only *the regions with x extent that overlap the current data stream's x extent*. This is easily expanded to more dimensions, where each additional dimension adds another set of trees to the *DCT* that hold the minimum/maximum endpoints of the CQ regions in this new dimension. Again, each of these new trees only indexes those regions that overlap the current window up to that dimension.

A is the final tree and contains all the regions that overlap with the current window in all dimensions. Just as the trees in the y-dimension contain only a subset of the regions that are indexed in the x-dimension, A contains the subset of the regions where the y-dimension of the window and the CQ regions overlap. These tree structures in each dimension make a cascade of indices, each a subset of the previous index.

In addition, pointers are maintained identifying where the current window is located. This is accomplished by pointing to nodes within each of the endpoint trees for each of the dimensions in the DCT. The pointers match the closest endpoint with a value less than or equal to the current window location. These are pointers to existing nodes and not the actual values of the window endpoints themselves. The nodes correspond to line segments and these pointers identify regions where the current result set is still valid.

Figure 2 shows these pointers in each dimension. They are designated as w_{x-} and w_{x+} for the x-dimension, and w_{y-} and w_{y+} for the y-dimension. Note that the minimum window pointers, for example w_{x-}, point into the tree of maximum endpoints for the CQ regions. Similarly, the maximum window pointers are in the trees of minimum endpoints. By moving these pointers along their corresponding trees, the DCT constructs a new result set of CQ regions from the previous result set for a new input window. A more detailed explanation is given in Sect. 2.4. In short, imagine as a spatial extent grows in size, new CQ regions are added to the result set as the maximum edge of the extent crosses the minimum edge of new regions. Similarly, regions would be added as the data extent minimum crosses new region maximum edges. The same idea applies for shrinking edges. Tracking the movement of the spatial extent of the incoming data stream from input window to input window is how the DCT incrementally maintains a result set of intersecting CQ regions.

The endpoint trees of the DCT are implemented using simple binary tree structures that support insertion, deletion, and iteration of nodes in both forward and reverse directions. The keys for each node in the endpoint trees are made up of two values, one corresponding to an endpoint value for each CQ region in one dimension, and one corresponding to a unique identifier for each region, r_{id}. The two values are combined so that spatial order is maintained. Different CQ regions that share an endpoint value are differentiated by the r_{id}'s. Individual nodes correspond to the half-open line segments between two endpoints in a single dimension. The trees all have an additional node with a minimum endpoint that is less than any possible region, shown as a dot in the trees of Fig. 2. These allow half-open line segments to completely cover any potential location of the incoming data stream.

2.2   DCT Initialization

The DCT is initialized by creating the minimum/maximum endpoint trees for each dimension of the DCT. Initial nodes for each endpoint tree are created by adding a node with a value that is less than any possible value for a region in each dimension. These are shown as dots in the trees of Fig. 2. The current window location is then created by assigning the pointers w_{x-}, w_{x+}, w_{y-}, and w_{y+} to the appropriate minimum node for each of the endpoint trees in each dimension. The A structure is initially empty.

2.3   Inserting and Deleting CQ Regions

For inserting a new region r, first the x-dimension endpoints are inserted into X_- and X_+. The new region is checked for overlap with the current stream window extent, that is both $w_{x-} < r_{x+}$ and $w_{x+} >= r_{x-}$ hold, where r_{x+} and r_{x-} are the maximum and minimum values of the new region in the x dimension. If the new region does overlap the current window, then the region is also added into trees Y_- and Y_+. If the new region overlaps in this dimension, $w_{y-} < r_{y+}$ and $w_{y+} >= r_{y-}$, the region is added into A using r_{id} as the index. Insertion maintains the validity of the result set A with respect to the current stream window.

Deletion is similar, following the cascade of the DCT, with one modification. The current window pointers need to be checked to verify that they are not pointing to a node that is being removed. If they are, then they need to be modified. For example, to delete region r, if w_{x+} points to this region, then decrement w_{x+} to the previous node in X_-. If w_{x-} points to the maximum endpoint of r, then decrement w_{x-} to the previous point in X_+. These changes maintain the validity for Y_-, Y_+, and A. The endpoints of r are then deleted from X_- and X_+. If the region intersects the current window, it needs to be deleted in a similar manner from Y_- and Y_+, and potentially A as well.

2.4   Queries

Figure 3 describes the changes to the DCT for new stream data (windows) with new extent. As X_- and X_+ are not changed, only the values of Y_-, Y_+, and A are shown. Figure 3(a) shows the query change due to the movement of the window in the x dimension, and 3(b) shows the change due to the movement in the y dimension. Just as the trees for each dimension and A need to be maintained when new regions are inserted and deleted, each new stream window requires maintenance of these trees as the window changes its location. These changes are made incrementally on the existing structure of the DCT.

Since Y_- and Y_+ contain regions overlapping the current window's x extent, when new data arrives with a different x-extent, Y_-, Y_+, and A need to be modified. These modifications occur when the window region endpoints cross a boundary of a CQ region. For each x region boundary in the DCT that is crossed, the y-dimensional trees need to be modified to account for inclusion or deletion of this region. A similar method needs to be associated with boundary crossings in the y-dimension, requiring modification of A.

The algorithm for reporting overlapping CQ regions for a new window $w = \{(w_{x-}, w_{y-}), (w_{x+}, w_{y+})\}$ begins by traversing the endpoint trees in the x dimension. Figure 3(a) shows an example where the new window region has increasing

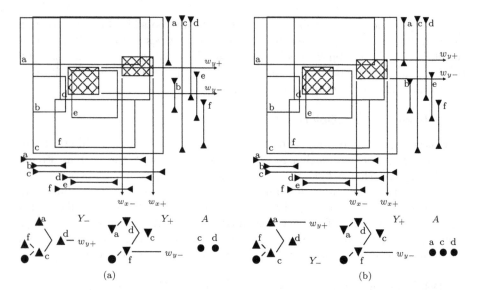

Fig. 3. Query on a new input stream window. The original window position (as in Fig. 2) is the hatched box on the left, and the region of the new window is the hatched box on the right. (a) shows the updated Y_-,Y_+, and A structures after moving the window in x dimension; (b) shows the final structures after moving in y dimension

minimum and maximum x edges. An increasing minimum edge implies that the movement can result in CQ regions being deleted from the result set A. The movement of w_{x-} is tracked by incrementing along the nodes of X_+. Moving w_{x-} to a new node corresponds to a boundary crossing of the window minimum with a region maximum, and in this case corresponds to one region, e, being removed from Y_- and Y_+. This deletion is cascaded to A as well. The increasing maximum, w_{x+}, corresponds to a movement that would add CQ regions. However, in this case, although the window maximum is greater, no minimum nodes are crossed, and w_{x+} points to the node that contains the minimum endpoint of the region d. The net result of moving the window in the x dimension is that region e is removed from Y_-, Y_+, and A.

Next, the movement of the window in the y dimension is accounted for as shown in Fig. 3(b). In this case, although the y minimum of the window has changed, no maximum boundaries are crossed, and w_{y-} remains pointing to the maximum value of region f. Moving the window maximum in y, however, results in the minimum edge of region a to be crossed. w_{y+} now points to the minimum edge of a, and a is added to the result set A.

This example shows one type of movement of the window region, but the same algorithm works for all changes of the window region. The window can grow and contract on all sides, or move in any direction. The algorithm works the same way; each edge of the window query is handled separately either adding to or deleting from the subsequent trees and the final result, based on the direction of movement of the edge. There is one caveat to this algorithm for updating the *DCT* based on the movement of edges individually. Without modification of the above algorithm, the expansive movements of the new window need to be performed before the shrinking movements. Large movements of the window can produce movements across both edges of a CQ region. Expanding before shrinking prevents a region from being deleted first on a shrinking movement, and then erroneously inserted back on expansion. Executing deletions first, however, is preferable as they decrease the size of the trees in the *DCT*. To allow shrinking movements first, during the expansive movements, range checks need to verify that the region indeed belongs to the overlapping set before insertion into subsequent trees.

3   DCT Performance

The window query performance of the *DCT*, i.e., determining which CQ regions intersect with a new incoming stream window, is highly dependant on the location of the CQ regions, the properties of the input window, and the interaction of these parameters. For window queries over n CQ regions, with k being the result count, the execution time for window query can range from $O(k)$ in the best case, when the result set is the same as the previous set to $O(n \lg n + k)$ in the worst case, when every region is entered in a single movement of the spatial extent of the incoming window. Before looking at some experimental results, there are some simple guidelines to consider for the application of the *DCT*.

3.1   Region Insertion and Deletion Cost

The *DCT* data structure is small and robust to many insertions and deletions of CQ regions. Insertions and deletions take $O(\lg n)$ time as the query region is potentially added to the endpoint trees in some constant dimension and A. These trees are simple to maintain dynamically in $O(n)$ space.

3.2   Query Cost Versus the Number of Boundaries Crossed

The *DCT* is designed for trending data, which can be measured by the number of region boundary crossings from one window query to the next. The structure works best when the number of region boundaries crossed by successive input windows is not large. When no boundaries are crossed, then no internal lists are modified, and reporting CQ regions runs in $O(k)$ time. When a boundary is crossed in movement of the window, then each region with a crossed boundary needs to be inserted into or deleted from subsequent endpoint trees. This is true for at least one set of endpoint trees or A, even for CQ regions whose domains in other dimensions do not overlap the new window and thus do not contribute to the final result. The cost of a window query in this case can be as high as $O(n \lg n + k)$, since all CQ regions could potentially be inserted into dimensional trees during one window query.

The *DCT* data structure indexes somewhat lazily in the sense that for insertions of new regions that do not overlap the current window, only the first dimension is indexed, and not the other dimensions. The indexing costs on window queries can be thought of as finally incurring those indexing costs. However, the problem with the DCT is that these costs can occur many times in the motion of the input stream data. Rather than indexing region values once, the DCT re-indexes a subset of points multiple times as boundaries are crossed. The hope is that many successive window queries will be in the same region with the same result, and that the low cost for those queries will make up for the extra cost of maintaining a dynamic index.

3.3   Trajectory of the Trending Windows

Another aspect affecting performance is the movement trajectory of the input windows. For example, consider a DCT in two dimensions as shown in Fig. 2 with trajectories that are monotonic in the x and y dimensions. In this case, CQ regions are inserted into the y endpoint trees and A at most one time, and once more potentially for deletion. The total time for maintaining the DCT then is at most $O(n \lg n)$, fixing a bound on the dynamic maintenance costs. For m window queries over that trajectory, the cost of all queries would be $O(n \lg n + mk)$ where k is the average number of results per window query. Similar savings exist for trajectories that are only generally monotonic, such as most RSI data streams.

On the other hand, more erratic window trajectories can result in poor performance. Consider a window movement that repeatedly crosses all n region boundaries. Each window query iteration would require $O(n \lg n)$ time, as the y dimensional trees are repeatedly made up and torn down.

Also, the DCT as described in the above figures, which indexes on x and then y, favors windows that trend in the y direction over windows that trend in the x direction. This is because boundary crossings in the x dimension have to modify more trees, and the trees tend to be bigger, so they take more time. In our examples, movements in the y dimension only modify A, the smallest tree in the DCT. Also, movements in the x direction result in insertions into Y_- and Y_+ that can be somewhat wasted in the sense that these CQ regions may never contribute to a final result set, whereas boundary crossings in the y direction will always affect the result set. Although, the time it takes to update the DCT due to a boundary crossing is $O(\lg n)$, for either x or y dimension, y modifications will always be faster.

This shows that order in the cascade is very important, and dimensions that see more boundary crossings for successive window queries into the DCT should be pushed deeper into the structure. Boundary crossings are of course dependant on the trajectory of incoming windows and the organization of the regions in the DCT. Section 4 tests and quantifies some of these performance features.

4   Experiments

To test the performance of the DCT, two experimental setups were used. The first tests are on randomly moving stream data (windows) and random CQ regions, with a number of variations on specific input parameters. The second experiment was developed to more closely replicate the queries made on DSMS with GOES RSI data as input, and more realistic region parameters.

Comparisons were made to an existing in memory R*-tree implementation from the *Spatial Index Library* [9]. For better comparison, the *DCT* implementation included some components from this same library. All trees in this implementation of the *DCT* were simple set objects from the Standard Template Library (STL) [22]. The total size of this experimental *DCT* implementation was about 600 lines of C++ code. All experiments were run on a single 1.6 GHz Pentium M CPU with a 1MB L2 cache and 512MB of main memory.

4.1   Random Continuous Queries with a Random Data Stream

The more extensive tests were run using a set of CQ regions that were randomly located throughout a two dimensional space, with some variation of parameters as shown in Fig. 4(a). For most tests, the input stream window moved randomly through the region, with the default parameters shown in Fig. 4(a). The CQ regions were distributed uniformly throughout a square region. Some region parameters are modified for certain tests, these tables show the default values. Aspect is the ratio of lengths $\frac{x}{y}$ of the spatial extent of the input window. For most tests, input windows moved in a random direction from one window to the next. Sometimes, the extent of the input windows would trend off the region of interest. When this occurred, a new starting point within the region was chosen to reset the window location.

These experiments do not correspond too closely to any real world application, but are instructive in testing the performance of the *DCT* with respect to various parameters. All experiments plot the average response time for a window query as a function of a single parameter. All experiments were run using 4K and 16K CQ regions.

Figure 4(b) shows the average window query time while varying the average distance moved from on window to another. The direction of the move was randomly determined. This is one of the most important parameters to consider when using the *DCT*. The windows must come close to one another for the index to work effectively. As expected, while the R*-tree is insensitive to this parameter, the *DCT* performance is very dependant on the distance moved for a window. As the distance between windows increases, the amount of information that can be shared in the trees of the *DCT* becomes less and less. For data in the input stream at a size of 1% of the total area, a move over a distance of 2% virtually eliminates all sharing of information between window queries. At what point the distance between queries becomes limiting is dependant on other application parameters. For example, as more regions are added into the *DCT*, more boundaries are likely to be crossed over the same distance moved.

The size of the stream input window spatial extent also affects the performance of the *DCT*. In general, windows with a larger spatial extent would be expected to share more results from query to query and therefore improve the performance of the *DCT*. Figure 5(a) compares the *DCT* to the R*-tree for dif-

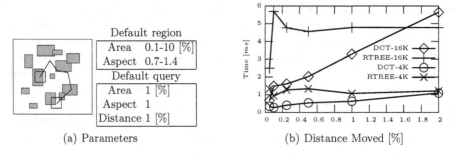

(a) Parameters

(b) Distance Moved [%]

Fig. 4. (a) Random test parameters for stream input windows and query regions; (b) average query time as function of the distance between subsequent windows and window queries

ferent incoming input window extent sizes. The average response time increases for both implementations, but less so for the DCT. For this experiment, the result set for the window queries increases with window query size as well, so it is expected that the response time increases with stream data extent. The DCT results are somewhat artificially high for another reason. As was described, the query window is relocated to a new random location when the window trends off the experiment's region. Since this is more likely to occur with a larger extent, more of these relocations occur in those experiments, and since each relocation corresponds to larger query distance movements in the DCT, this increases the average distance between data in the stream.

As described in the introduction, the intent of the DCT is for use in streaming RSI, and these usually entail incoming data packets of a row, or small number of rows of data. Since these correspond to the extents of the individual data in the incoming geospatial data stream, DCT queries can have a high aspect ratios, that is, $\frac{x}{y} \gg 1$. Figure 5(b) shows the change in response time as a function of aspect ratio. The R*-tree performance is slightly degraded as the aspect ratio increases, while the DCT performance remains insensitive to this parameter.

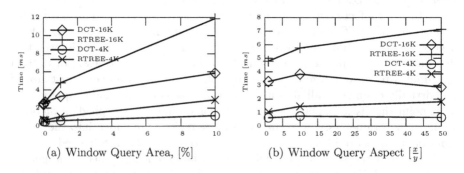

(a) Window Query Area, [%]

(b) Window Query Aspect $[\frac{x}{y}]$

Fig. 5. Average window query times as a function of area of the window (a) and aspect ratio of spatial extent of the stream windows (b)

In Sect. 3, we described how the *DCT* can be affected by the average trajectory of the successive data (windows) from the stream. Performance is expected to improve as the window trajectory aligns with the dimensions farther down in the *DCT* cascade. Figure 6(a) illustrates this gain. As expected, the R*-tree is insensitive to this parameter, whereas the *DCT*'s performance increases by a factor of two based on the trajectory of the window.

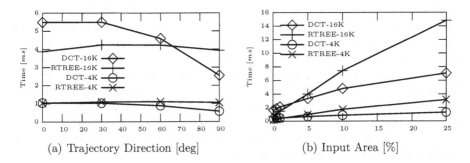

(a) Trajectory Direction [deg]   (b) Input Area [%]

Fig. 6. Average query times as a function of trajectory of the input window (a), and size of CQ regions (b)

Figure 6(b) shows the affect of the average size of the CQ regions on the *DCT* performance. As the size of the regions increases, the average response time increases for both search structures, due in part to the fact that more regions overlap with each individual window. The R*-tree slows down more than the *DCT*, since more of the results from successive window queries can be shared from query to query. This is an important consideration for applications where the CQ regions are large.

In summary, the *DCT* performance is affected by the number of region boundaries crossed for successive window queries, and the trajectory of the window. Also, the size of the CQ regions and input data extent affect performance. Often the performance changes are different than seen in the more typical R*-tree. How the *DCT* performs under more realistic situations is described next.

4.2   GOES Experiment

From the discussion of the *DCT* performance and the experimental results for random windows and regions, the parameters associated with higher performance of the *DCT* can be anticipated. These include: (1) relatively large extents of data in the stream, possibly with a high aspect ratio, (2) small distances from data to data in the stream, and (3) a regular trajectory of the data in the stream, with the *DCT* tuned to that direction. The *DCT* also works well with large CQ regions, having a relatively high selectivity. These are all aspects of a typical satellite RSI data stream.

For a more realistic scenario for the DCT, an experiment using queries for weather satellite data from the National Oceanic and Atmospheric Administration National Oceanic and Atmospheric Administration (NOAA) GOES satellite [7] was developed. Figure 7(a) gives an overview of the GOES sensors. The GOES satellite continuously scans a hemisphere of the Earth with two sensors, the Imager and the Sounder. Imager data arrives row-by-row, whereas Sounder data arrives more in a pixel-by-pixel manner. The experiment uses Imager data as input. GOES offers a continuous stream of data for regions ranging from the continental United States to a hemisphere centered near Hawaii. Each complete image at some time is termed a frame. Data from the GOES visible channel comes in blocks that contain 8 rows of data. Other channels from the GOES imager come in blocks of 1 or 2 rows. The number of columns in a row varies from frame to frame. An entire frame of data is reported 8 rows at a time from North to South. New frames start from their most northern extent. On average there are about 5125 rows per frame, which means that for every frame start that requires a long traverse through the DCT structure, there are about 640 small steps downward in the y direction. Query regions also tend to be approximately square regions covering relatively large regions.

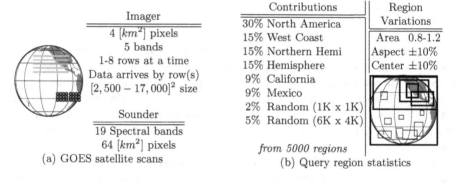

Fig. 7. GOES based DCT experiment

GOES streaming RSI data is well suited to the DCT. In each frame, the data trends only in the downward direction and incrementally. Therefore, endpoints are only added into the X_-, X_+ structures one time, limiting the maintenance time to $O(n \lg n)$ for each frame. For normal GOES data, the starting column of each row does not change within a frame, and the Y_-, Y_+, and A structures are the only structures that change in determining which regions overlap any given set of incoming streaming rows of data.

For the experiment, 5000 regions were indexed. Rather than randomly locating these regions throughout the image domain, they were preferentially located around a small number of "hot spots" in the hemisphere. This more closely relates to real world scenarios, where specific parts of the RSI data are requested by a large numbers of users. Although the general area of the queries was fixed to a small number of locations, the individual region centers, total sizes and aspect ratios were varied for each region. This corresponds to many users requesting slightly different particulars for a general region of interest. A small fraction of random regions was also included in the experimental setup. Figure 7(b) describes the parameters for the CQ regions. The table shows the various regions and their contributions to the total regions. The other table shows the variation of the individual regions. The figure graphically depicts the experimental setup.

10,000 window queries were performed on the CQ regions. The RSI data corresponding to the input data stream were taken from a sample of the GOES West Imager instrument, each query corresponding to 8 rows of data. Depending on some assumptions about other aspects of a DSMS indexing GOES data, this corresponds to somewhere between 7.5 to 60 minutes of streaming time.

Again, the *DCT* was compared to the in-memory R*-tree implementation. The R*-tree implementation required 48 seconds to perform all the queries, for an average query time of 4800 [μs]. The *DCT* performed the queries in 9 seconds, 900 [μs] per query. It is also interesting to note that of those nine seconds for the *DCT*, only 0.88 seconds where used in the maintenance of the *DCT* and its structures. Nearly all the other 8.22 seconds involved copying the final result A structure into new lists for output. This was done to match the output of the R*-tree, but scenarios can be envisioned where applications access the A structure directly after a query, reducing this overhead cost.

5   Related Work

This work is an extension of a more simple Dynamic Cascade Tree (*DCT*), first proposed to efficiently restrict streaming spatial data *points* to specified regions of interest [10]. That structure defined the problem as solving many *stabbing point* queries, where the incoming data stream was a single moving point used for each query. The *DCT* proposed in this paper aims to allow many spatial regions to be evaluated simultaneously as the stream of input spatial data arrives in the DSMS. The obvious application of the *DCT* is in multi-query evaluation, where a single operator uses the *DCT* structure to perform the spatial restrictions for a large number of queries, see Fig. 1(c). Similar multi-query optimizations have been discussed in streaming databases where they have been described as *group filters* [15,16] and in spatio-temporal databases where this optimization has been described as *query indexing* [19].

Many data structures have been developed for one and two dimensional window queries including among others, interval trees, priority search trees, and segment trees [4,20]. Space partitioning methods of answering window queries include quadtrees, hashes, and numerous variants of R-trees.

The common methods for solving window queries in two dimensions are multi-level segment trees [4], interval trees [20], and R-trees [8]. It is difficult to modify these methods to take advantage of trending data. Using multi-level segment trees, one dimension of the region is stored in a segment tree, while the second dimension is indexed with an associated interval structure for each node in the first segment tree. Storage for these structures can be $O(n \lg n)$, with query times of $O(\lg 2n)$. Dynamic maintenance of such a structure is more complicated and requires larger storage costs [14]. It is difficult to modify the multi-level segment tree to improve results for trending data. If the input window moves a small distance, which does not change the query result, it still would take $O(\lg n)$ time to respond. That is because even if every node in the multi-level segment tree maintains knowledge of the previous point, it would still take $\lg(n)$ time to traverse the primary segment tree to discover that no changes to the query result occurred.

Interval trees yield an optimal worst-case space and query time solution to the window query problem, $O(n)$ in space and $O(n \lg n + k)$ in query time, where k is the number of regions intersected with the given region. However, interval trees are static structures, because all the region intervals need to be known in advance to construct the trees. Also, like segment trees, it is difficult to modify interval trees to take advantage of data trendiness.

R-trees solve window queries by traversing through minimum bounding rectangles that include the extent of all regions in the sub-tree, generally with good performance. Since these rectangle regions generally overlap, there can be no savings from knowing the previous window, as there is no way to know if an entirely new path through the segment tree needs to be traversed. R^+-trees [21] may be modified for trending windows, since the minimum bounding rectangles are not allowed to overlap. Thus, maintaining the previous window can help to verify a window query that has not left a particular region. R^+-trees, however, have problems with redundant storage and dynamic updates [17].

Another approach similar to the *DCT* described in this paper is the Query Index [12,19]. The Query Index builds a space partitioning index on a set of static query regions and at each time interval, allows a number of moving objects to probe the index to determine overlapping queries. Experiments with main memory implementations show that grid-based hashing of query regions generally outperform R-tree or quad-tree based methods. The query index is most effective for small regions and query points, and experiments show performance degradation with larger regions or high selectivity query windows [13].

The *DCT* is also an incremental approach to answering queries where updates are made to a current query result. SINA [18] describes an incremental method to solving the problem of intersecting moving objects, though the approach focuses on efficient integration of incremental query changes with disk-based static queries, and a complete main-memory implementation would be more similar to the query index approach.

In another sense, the *DCT* is a method of dynamically maintaining boundaries around a current window where the current result set is valid, and identifying where this result set will change. Another method of dynamically describing a neighborhood of validity for a query using R-trees was proposed in [23], which builds an explicit region of validity around a current query point. This method is meant to minimize transmission costs to a client, and the technique makes many additional queries to an R-tree style index to build these regions of validity.

6   Conclusions and Future Work

Remotely-sensed imagery clearly provides a great opportunity to study concepts and paradigms for the management and processing of streaming data, given the existence of various satellites that are used constantly for numerous data products in environmental sciences, disaster management, climatology etc.

In this paper, we have presented the Dynamic Cascade Tree (DCT), which is suitable for window queries that are most common to most of RSI data streams. This simple data structure is designed to work especially well for streaming data with spatial extents that are in close proximity and follow certain trendiness in their movement. These are characteristics of solving multi-query optimizations for streaming RSI data, for which the DCT was developed. Performance evaluation and experiments over random data identified some strengths and weaknesses of the DCT in more general situations, while an experiment using GOES weather imagery demonstrated its suitability for streaming RSI database applications.

The DCT is part of query processing architecture being developed to support complex continuous queries over streams of remotely-sensed geospatial image data and will be an important component to facilitate the optimization of multiple restriction queries against such a stream.

Acknowledgment. This work was partially supported by the National Science Foundation under Grant No. IIS-0326517.

References

1. Abadi, D., Carney, D., Cetintemel, U., Cherniack, M., Convey, C., Lee, S., Stonebraker, M., Tatbul, N., Zdonik, S.: Aurora: A new model and architecture for data stream management. The VLDB Journal **12**(2) (2003) 120–139
2. Babcock, B., Babu, S., Datar, M., Motwani, R., Widom:, J.: Models and issues in data stream systems. In: Proc. 21th ACM Symposium on Principles of Database Systems (PODS). (2002) 1–16
3. Babcock, B., Babu, S., Datar, M., Motwani, R., Thomas, D.: Operator scheduling in data stream systems. The VLDB Journal **13**(4) (2004) 335–353
4. de Berg, M., van Kreveld, M., Overmars, M., Schwarzkopf, O.: Computational Geometry: Algorithms and Applications. Springer Verlag (2000)
5. Carney, D., Cetintemel, U., Rasin, A., Zdonik, S., Cherniack, M., Stonebraker, M.: Operator scheduling in a data stream manager. In: Proceedings of 29th VLDB Conference (2003) 838–849
6. Chandrasekaran, S., Cooper, O., Deshpande, A., Franklin, M.J., Hellerstein, J.M., Hong, W., Krishnamurthy, S., Madden, S., Raman, V., Reiss, F., Shah, M.A.: TelegraphCQ: Continuous dataflow processing for an uncertain world. In: First Biennial Conference on Innovative Data Systems Research (CIDR 2003) (2003)
7. GOES I-M DataBook. Space Systems-Loral
   `rsd.gsfc.nasa.gov/goes/text/goes.databook.html` (2001)
8. Guttman, A.: R-trees: a dynamic index structure for spatial searching. In: Proceedings of the ACM SIGMOD Int. Conf. on Management of Data (1984) 47–57
9. Hadjieleftheriou, M.: Spatial Index Library. Department of Computer Science and Engineering, University of California, Riverside. version 0.80b edn. (2004)

162   Q. Hart, M. Gertz, and J. Zhang

10. Hart, Q., Gertz, M.: Indexing query regions for streaming geospatial data. In: 2nd Workshop on Spatio-temporal Database Management (STDBM'04) (2004)
11. Hellerstein, J.M., Franklin, M.J., Chandrasekaran, S., Deshpande, A., Hildrum, K., Madden, S., Raman, V., Shah, M.A.: Adaptive query processing: Technology in evolution. IEEE Data Eng. Bulletin **23** (2000) 7–18
12. Kalashnikov, D.V., Prabhakar, S., Hambrusch, S.E.: Main memory evaluation of monitoring queries over moving objects. Distributed and Parallel Databases **15**(2) (2004) 117–135
13. Kim, K., Cha, S.K., Kwon, K.: Optimizing multidimensional index trees for main memory access. In: Proc. of the ACM SIGMOD International Conference on Management of Data (2001) 139–150
14. van Kreveld, M.J., Overmars, M.K.: Concatenable segment trees. Technical report, Rijksuniversiteit Utrecht (1988)
15. Madden, S., Shah, M., Hellerstein, J.M., Raman, V.: Continuously adaptive continuous queries over streams. In: Proc. of the ACM SIGMOD International Conference on Management of Data (2002) 49–60
16. Madden, S., Franklin, M.J.: Fjording the stream: An architecture for queries over streaming sensor data. In: Intern. Conf. on Data Engineering (2002) 555–566
17. Manolopoulos, Y., Nanopoulos, A., Papadopoulos, A.N., Theodoridis, Y.: R-trees have grown everywhere. Unpublished Technical Report, www.rtreeportal.org/pubs/MNPT03.pdf (2003)
18. Mokbel, M.F., Xiong, X., Aref, W.G.: SINA: Scalable incremental processing of continuous queries in spatio-temporal databases. In: Proc. of the ACM SIGMOD International Conference on Management of Data (2004) 623–634
19. Prabhakar, S., Xia, Y., Kalashnikov, D., Aref, W., Hambrusch, S.: Query indexing and velocity constrained indexing: Scalable techniques for continuous queries on moving objects. IEEE Trans. on Computers **51**(10) (2002) 1124–1140
20. Samet, H.: Hierarchical representations of collections of small rectangles. ACM Comput. Surv. **20**(4) (1988) 271–309
21. Sellis, T.K., Roussopoulos, N., Faloutsos, C.: The R+-tree: A dynamic index for multi-dimensional objects. In Proceedings of 13th International Conference on Very Large Data Bases (1987) 507–518
22. SGI: Standard Template Library Programmer's Guide. (1999)
23. Zhang, J., Zhu, M., Papadias, D., Tao, Y., Lee, D.L.: Location-based spatial queries. In: Proceedings of the ACM SIGMOD International Conference on on Management of Data (2003) 443–454

The Optimal-Location Query

Yang Du, Donghui Zhang*, and Tian Xia

College of Computer & Information Science,
Northeastern University, Boston, MA 02115
{duy, donghui, tianxia}@ccs.neu.edu

Abstract. We propose and solve the optimal-location query in spatial databases. Given a set S of sites, a set O of weighted objects, and a spatial region Q, the *optimal-location query* returns a location in Q with maximum influence. Here the *influence* of a location l is the total weight of its RNNs, i.e. the total weight of objects in O that are closer to l than to any site in S. This new query has practical applications, but is very challenging to solve. Existing work on computing RNNs assumes a single query location, and thus cannot be used to compute optimal locations. The reason is that there are infinite candidate locations in Q. If we check a finite set of candidate locations, the result can be inaccurate, i.e. the revealed location may not have maximum influence. This paper proposes three methods that *accurately* compute optimal locations. The first method uses a standard R*-tree. To compute an optimal location, the method retrieves certain objects from the R*-tree and sends them as a stream to a plane-sweep algorithm, which uses a new data structure called the aSB-tree to ensure query efficiency. The second method is based on a new index structure called the *OL-tree*, which novelly extends the k-d-B-tree to store segmented rectangular records. The OL-tree is only of theoretical usage for it is not space efficient. The most practical approach is based on a new index structure called the *Virtual OL-tree*. These methods are theoretically and experimentally evaluated.

1 Introduction

Spatial databases play more and more important roles in applications such as corporation decision-support systems. For instance, an interesting query that the McDonald's Corporation may ask again and again is: "what is the optimal location in a given region to put a new McDonald's store?" Here an optimal location can be defined as a location which geographically benefits the most number of customers. This example motivates the optimal-location query. In general, let S be the set of sites (e.g. existing McDonald's stores) and let O be the set of weighted objects (e.g. residential buildings, where the weight for a building is the number of residents in it). Given a spatial region Q, the *optimal-location query* computes a location l in Q which maximizes the total weight of objects that are closer to l than to any site.

* Supported by NSF CAREER Award IIS-0347600.

C. Bauzer Medeiros et al. (Eds.): SSTD 2005, LNCS 3633, pp. 163–180, 2005.

We focus our discussions on the L_1 distance (also known as Manhattan distance) for it more accurately models the driving distance in a city road network [SKC93]. Given two locations (x_1, y_1) and (x_2, y_2), their L_1 distance is $|x_1 - x_2| + |y_1 - y_2|$. If a road network consists of a set of north-south roads and a set of east-west roads (e.g. in Manhattan), the L_1 distance is the shortest driving distance. When we say "the closest site of o", we mean the site whose L_1 distance to o is the smallest.

A closely related problem is the *bichromatic Reverse Nearest Neighbor (RNN) query* [KM00, YL01, SRAE01]. There, a query location is given, and the RNN query computes the set of objects in O that are closer to l than to any site in S. There are three differences between the bichromatic RNN query and our newly proposed optimal-location query. A small difference is that the RNN query considers L_2 distance (also known as Euclidean distance). The most important difference is that the optimal-location query involves a query region Q, which consists of infinite number of candidate locations. One can approximate Q as a grid, and limit the candidates to the finite set of grid intersections. But this approach cannot accurately compute optimal locations, for the optimal location may be off the grid. The third difference is that the optimal-location query is interested in the influence of a candidate location, or the total weight of objects in the RNN set, instead of the RNN objects themselves.

This paper proposes three methods that accurately compute optimal locations. The first solution (Section 4) assumes we have an R*-tree indexing the set O of objects. Similar to how the Rdnn-tree [YL01] extends the R-tree, we assume the R-tree stores some extra information. Every object stores the L_1 distance to its closest site in S, and every index entry stores the maximum L_1 distance of objects in the sub-tree. In particular, we propose a concept called the *nn_buffer*, for an object o. It is a spatial contour such that a location l is inside $o.nn\_buffer$, if and only if o is closer to l than to any site. As we will see later, each such contour, based on L_1 distance, has the shape of a diamond which has four right angles. If we rotate the coordinate by 45° counter-clockwise, every *nn_buffer* is an axis-parallel square in the rotated coordinate. The solution follows two steps. The first step is to retrieve from the R*-tree those objects which may affect the influence of some locations in Q. The objects are identified in certain order which enables a plane-sweep algorithm (as the second step) to go through the stream of objects once and identify an optimal location. The only objects that may affect the influence of locations in the query region Q are the ones whose *nn_buffers* intersect with Q. Our approach retrieves such objects in increasing order of *nn_buffer.x_low* in the rotated coordinate, even though the R*-tree was built in the original coordinate. This enables the run-time plane sweep. A naive plane-sweep solution has $O(n^2)$ cost, where n is the number of objects in the stream. We propose the *aggregation SB-tree (aSB-tree)*, extended from the SB-tree [YW01], to reduce the query cost to $O(n \log n)$.

Our second solution (Section 5.1) to the optimal-location query is based on a new specialized aggregation index called the *OL-tree*. It is disk-based, balanced and dynamically updateable. The index is built in the rotated coordinate. It is a novel extension to the k-d-B-tree. While the k-d-B-tree maintains point objects, the OL-tree keeps axis-parallel squares. In the OL-tree, each index entry maintains a value called *fullcover* to count how many squares fully contain the range of it. Besides, Each index entry stores *maxoverlap*: the maximum local influence in the sub-tree, and *maxrange*, a rectangular region where any location in it has maximum local influence.

The two solutions have interesting tradeoffs. The R*-tree based solution has efficient (linear) space cost. However, as objects are not pre-aggregated, a query needs to examine all objects whose *nn_buffers* intersect with Q. If Q has large size, the query performance is poor. On the other hand, the OL-tree is a specialized aggregation index, whose space overhead is higher since an object may have many copies. But it may have faster query support. For instance, if Q intersects with the *maxrange* stored at the root, the algorithm instantly returns.

The third solution (Section 5) combines the benefits of the previous two approaches. As in solution 1, we use an R*-tree to store the objects. But to guide the search, we use a small, in-memory OL-tree-like structure. This index is named the *Virtual OL-tree (VOL-tree)*. It looks like an OL-tree, but it does not store any *nn_buffer*. A leaf entry has the same meaning of an index entry. It corresponds to a spatial range, and it logically references a node that stores (pieces of) *nn_buffers* in that range. These *nn_buffers* can be retrieved from the R*-tree dynamically. Because the VOL-tree is small, each leaf entry may correspond to many *nn_buffers* (as a comparison, each OL-tree leaf node has at most B *nn_buffers*). Thus it is more costly to maintain *maxoverlap* in a VOL-tree, which may require to retrieve all *nn_buffers* corresponding to some leaf entry. For this reason, instead of maintaining *maxoverlap*, the VOL-tree maintains *lower_max* and *upper_max*, which are a lower bound and an upper bound of *maxoverlap*. In particular, *maxrange* is associated with *lower_max*.

This paper contributes in several ways to the understanding of the emerging class of located-based applications.

1. We propose the optimal-location query. It has practical applications such as corporate decision-support systems.
2. We present an R*-tree-based solution. In particular, our solution retrieves objects of interest in some given order and then uses a plane-sweep algorithm to identify an optimal location. The plane-sweep algorithm uses a new data structure called the aSB-tree to improve the query performance from $O(n^2)$ to $O(n \log n)$.
3. We introduce a theoretical solution based on a new index structure called the OL-tree. It is a specialized index with higher space cost but possibly more efficient query performance.
4. We provide a practical solution based on the Virtual OL-tree, which is both space efficient and query efficient.
5. We show experimental results on real datasets which reveal the tradeoffs of the proposed methods.

The rest of the paper is organized as follows. Section 2 reviews related work. Section 3 provides problem transformation and introduces the rotated space.

Section 4 presents the R*-tree-based solution. Section 5 presents the OL-tree-based and the Virtual-OL-tree-based solutions. Section 7 shows performance results. And Section 8 concludes the paper.

2   Related Work

The nearest neighbor (NN) query, since its introduction in [RKV95], has received vast attention in spatial database research community. One recent variation introduced by [KM00] was the RNN query. That is, given a query location l, find the objects in a given set O that consider l as their nearest neighbor. Note that existing work assumes the Euclidean distance while we focus on the Manhattan distance.

There are two variations of the RNN query: the monochromatic case and the bichromatic case. In the monochromatic case [SAE00, TPL04], the distance between an object $o \in O$ and the query location l is compared with the distances between o and other objects in O. In the bichromatic case [SRAE01], there is another dataset: a set S of sites. And the distance between o and l is compared with the distances between o and sites in S. Many real-life applications correspond to the bichromatic case. For instance, given a new location, compute the set of residence buildings that are closer to this location than to any existing McDonald's store. In [Smi97], it was proved that for the monochromatic case, the number of RNNs is bounded. For instance, there are at most 6 RNNs in the 2D case and at most 12 RNNs in the 3D case. But for the bichromatic case, the number of RNNs is unbounded even for the 2D case.

One solution to the bichromatic RNN query is based on precomputation [KM00, YL01]. (It also works for the monochromatic case.) The idea of [KM00] is to build an R-tree that stores circles instead of points. Every circle is centered at some object o, with radius being the distance from o to its nearest site. Precomputation is required to get these distances. Given a query location l, its RNNs are retrieved by locating the circles that enclose l.

Yang and Lin [YL01] proposed the Rdnn-tree which combines the R-tree of circles with the R-tree of objects. It is an R-tree of objects, where every object stores the distance to its closest site, while every index entry stores the maximum distance of all objects in the sub-tree. The structure logically maintains the R-tree of circles. It remains to determine, given a location l and an index entry e, whether the sub-tree referenced by e may contain some object whose "circle" (not stored) encloses l. The solution is to expand the index entry's MBR outward by the associated maximum distance. If the expanded region does not enclose l, there is no need to check the sub-tree.

The R-tree that we use, in the first solution to the optimal-location query, is the Rdnn-tree [YL01] which stores L_1 distance instead of L_2 distance. Our concept of *nn_buffer* corresponds to their concept of *circle* for each object. However, in this paper both the addressed problem and the R-tree-based solution are different from [YL01]. Our problem takes as input a spatial region and aims at identifying an optimal location (with maximum influence), while [YL01] takes as input a single location and finds its RNNs. As for the solution, while [YL01] finds the objects whose circles enclose the given location and terminates, we find the objects whose *nn_buffers* intersects with the given region as the first stage of a pipeline process. The second pipeline stage takes this stream of objects and identifies an optimal location via plane sweep.

Another bichromatic RNN query solution was proposed by [SRAE01]. The idea is to dynamically construct the *influence region* of the query location l. Here, the influence region is defined as a polygon in space which encloses and only encloses all possible RNNs of l. This is equivalent to the *Voronoi cell* enclosing l [BKOS97]. Conceptually, if we draw a bisector line between l and a site s, any object located on the l side of the bisector will have smaller Euclidean distance to l than to s. The l side of the bisector is a half plane. If we compare l against all sites and take the intersection of these l-side half planes, we get the Voronoi cell containing l.

Of course, to compare with all sites is expensive. [SRAE01] provides a clever way to compute the rectangle that is guaranteed to contain all sites needed for computing the exact Voronoi cell of l. Then the Voronoi cell of l can be computed by only examining these sites with in the rectangle. So to find the RNNs of l, a range query using the Voronoi cell is performed on the R-tree of objects.

If we had limited candidate locations, the approach could be extended to solve the optimal location problem, with two modifications. First, we now need to construct a Voronoi cell with regards to the L_1 distance [LW80]. Second, for each location, we need to know its influence rather than the actual RNN objects. So we can index the set of objects using the aggregation R-tree (aR-tree) [PKZT01] where each index entry stores the total weight of objects in the sub-tree. If the MBR of an index entry is contained in a Voronoi cell, the stored total weight contributes to the computation of influence, without browsing the sub-tree.

Unfortunately, in the optimal-location query, there are infinite number of candidate locations. So the approach of [SRAE01] (with the above modifications) does not work. Before presenting our solutions, let's first study a problem reduction.

3   Problem Transformation

An illustration of the optimal-location query, defined in Section 1, appears in Figure 1(a). There are four objects and two sites. In particular, the object o_3 with weight 5 has s_1 as the closest site, where $d(o_3, s_1)=22$. And the object o_4 with weight 6 has s_2 as the closest site, where $d(o_4, s_2)=12$. The influence of a location is the total weight of objects that are closer to this location than to their closest sites. For instance, the influence of l is the total weight of o_3 and o_4, which is 5+6=11. Given a query region Q, the optimal-location query finds an location inside Q with maximum influence. In this example, l is an optimal location. There may be more than one optimal location. The query asks for one of them.

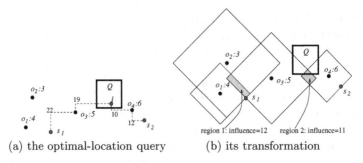

(a) the optimal-location query   (b) its transformation

Fig. 1. In (a), l is an optimal location, with influence 11. The transformation in (b) shows that any location in the intersection between Q and region 2 is an optimal location

To efficiently compute an optimal location, we first define the concept of *nn_buffer*, and then transform the optimal-location query into the problem of finding a location with maximum overlap among objects' *nn_buffers*.

Definition 1. *Let s be the closest site of an object o. The nn_buffer of o is a contour such that $\forall l$ on the contour, $d(l, o) = d(o, s)$. Here $d()$ is L_1 distance. Given a MBR of objects, let t be the maximum distance between any such object to its closest site, the nn_buffer of the MBR is a outside contour such that $\forall l$ on the contour, the minimum distance between l and the MBR is t.*

Fig. 2. The *nn_buffer* of an object and the *nn_buffer* of an MBR

As shown in Figure 2(a), the *nn_buffer* of an object o is a diamond with four right angles. It is easy to check that the L_1 distance between o and any location on the boundary of the diamond is fixed.

The weight of object o contributes to the influence of a location l, if and only if l is inside the *nn_buffer* of o. So as shown in Figure 1(b), an optimal location is a location l inside Q which maximizes the total weight of objects whose *nn_buffers* contain l. The concept of *nn_buffer* can also be defined for an *minimum bounding rectangle (MBR)* of a set of objects. The *nn_buffer* of an MBR is the tightest contour which is guaranteed to contain the *nn_buffers* of all objects in the MBR, without knowing the locations of the objects. The *nn_buffer* of an MBR is a polygon with eight edges, as illustrated in Figure 2(b).

Consider the coordinate which has the same origin as in the original coordinate, but whose X and Y axes are rotated 45° counter-clockwise. We call it the $\searchbox45^\circ$ (reads rotate-45-degree) coordinate (Figure 3). In this paper, the R*-tree indexes in the original coordinate (to satisfy the possible need for other applications), while the aSB-tree, the OL-tree and the VOL-tree indexes in the $\searchbox45^\circ$ coordinate.

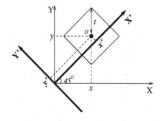

Fig. 3. Illustration of the rotated coordinate

Our analysis shows that an object o located at (x, y) in the original coordinate is mapped to $(\frac{x+y}{\sqrt{2}}, \frac{-x+y}{\sqrt{2}})$ in the $\searchbox45^\circ$ coordinate. Furthermore, let t be the L_1 distance from o to its closest site. The *nn_buffer* of o is an axis-parallel square in the $\searchbox45^\circ$ coordinate, whose lower-left corner and upper-right corner are: $(\frac{x+y-t}{\sqrt{2}}, \frac{-x+y-t}{\sqrt{2}})$ and $(\frac{x+y+t}{\sqrt{2}}, \frac{-x+y+t}{\sqrt{2}})$.

4   The R*-Tree-Based Solution

Our first solution to the optimal-location query assumes the objects are indexed by an R*-tree. Similar to how the Rdnn-tree [YL01] extends the R*-tree, we assume every object stores the L_1 distance to its closest site, and every index entry stores the maximum L_1 distance of objects in the sub-tree.

The R*-tree indexes objects in the original coordinate (not the $\searchbox45^\circ$ coordinate), since there may be other applications that need to access the data in the original coordinate. However the plane-sweep algorithm works in the $\searchbox45^\circ$ coordinate. In order to do the plain sweep, we have to retrieve the objects in increasing order of their *nn_buffer*'s *x_low* in the $\searchbox45^\circ$ coordinate. Section 4.1 shows how to retrieve objects, Section 4.2 describe a naive plane sweep algorithm with $O(n^2)$ cost, Section 4.3 propose the aSB-tree structure which can reduce the worst-case query cost to $O(n \log n)$, and Section 4.4 extends the algorithms to incorporate a rotated query region.

4.1   Retrieving Objects from the R*-Tree

To retrieve the objects whose *nn_buffers* intersects with Q, we can browse the R*-tree in a top-down fashion, similar to the range query. The difference is that to determine whether to expand an sub-tree, instead of checking whether its MBR intersects with Q, we check whether the MBR's *nn_buffer* intersects with Q.

The remaining issue of object retrieval is how to return objects in increasing order of their *nn_buffer*'s *x_low* in the $\searchbox45^\circ$ coordinate. This is achieved by using a best-first search. That is, we keep a heap of the R*-tree's index entries as well as objects. The entries are ordered in increasing *nn_buffer.x_low* in the $\searchbox45^\circ$ coordinate. Initially, the heap contains the index entry referencing the whole tree. In each iteration, the entry e with minimum *nn_buffer.x_low* is extracted.

If e is a object, output it (to be sent, as the next element of an input stream, to the plane-sweep algorithm discussed in the next section). Otherwise, e is an index entry. We examine every entry se in the node referenced by e, and push se into the heap if its nn_buffer intersects with Q.

4.2   The Naive Plane Sweep

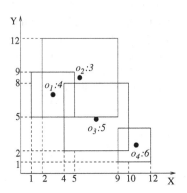

Fig. 4. $nn\_buffers$ in the rotated coordinate

In the rotated coordinate, the $nn\_buffers$ are axis-parallel squares. To find the optimal location, the basic idea is to perform a plane sweep in increasing order of X. For each particular X, the Y axis is partitioned into a set of intervals, each associated with an influence value. For instance, at $X=4$, the Y axis is partitioned into six intervals: $(-\infty,2):0$, $(2,5):5$, $(5,8):12$, $(8,9):7$, $(9,12):3$, and $(12,\infty):0$. Whenever a change happens, update the set of intervals. During the process, always maintain a location with maximum influence. In fact we can maintain a rectangular region with maximum influence, instead of a single location.

At the end, any location in the maintained region is an optimal location. As an example, in Figure 4, any location in the X range of (4,5) and Y range of (5,8) is an optimal location, with influence 12.

4.3   The aSB-Tree

The naive plane sweep has $O(n^2)$ worst-case performance. The reason is that there are $O(n)$ events to handle, and each event needs to scan through $O(n)$ intervals that partition the Y axis. We hereby propose a data structure called the *Aggregation SB-tree (aSB-tree)*, derived from the SB-tree [YW01]. The new structure enables any event to be processed in $O(\log n)$ time, and therefore reduces the overall cost to $O(n \log n)$.

The idea is to organize the intervals (that partition the Y axis) into a balanced B-tree-like structure. To insert a new Y range which may affect many intervals in the naive approach, with the aSB-tree we only need to update two paths from root to leaf. The key idea that enables this is: if the Y range to be inserted fully contains the interval of an index entry, we do not insert into the sub-tree. Instead, we update a value $fullcover$ maintained along with the index entry. The aSB-tree extends the SB-tree by storing the max influence and the corresponding spatial region in the sub-tree. Figure 5 shows a two-level aSB-tree, which corresponds to Figure 4 right after processing the event at $X = 4$ and the event at $X = 5$. Let's examine it in more details.

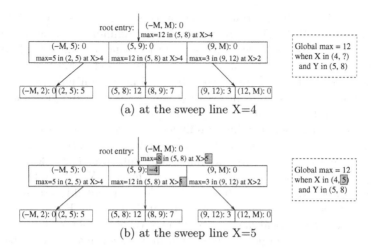

(a) at the sweep line X=4

(b) at the sweep line X=5

Fig. 5. An example of aSB-tree

Properties inherited from the SB-tree:

– The aSB-tree is a balanced tree structure. The maximum number of entries in a (leaf or index) node is fixed. Except the root, every node must be at least half full.
– Every entry corresponds to an interval (a Y range). For any index entry e, all intervals of entries in $Node(e)$ form an exact partition of $e.interval$. E.g. in Figure 5(a), the root entry has an interval as the whole Y space.
– Every leaf entry has a value *influence*. In Figure 5(a), the leaf entry (5,8):12 means that right after the current $X = 4$, any location with $Y \in (5,8)$ has influence 12.
– Every index entry has a value *fullcover*, which corresponds to the total weight of inserted Y ranges which fully cover the entry's interval. E.g. the second index entry in the root of Figure 5(a), which has (5,9):0, means its fullcover is 0, while the interval is (5,9).
– To insert a range I with weight w, we update (at most) two paths from the root to the leaf level. These are the nodes whose referencing entry's interval partially intersects with I. E.g. Figure 5(b) shows the result after processing the event at $X = 5$, i.e. inserting $I=(5,9)$ and $w=-4$. In particular, the insertion stopped at the root node, since no entry in the root node has an interval partially intersecting with I. For any entry (e.g. the second index entry in root) whose interval is contained in I, w is added to its *fullcover*. An overflow/underflow, if happens, is treated like in the B-tree.

Properties extended from the SB-tree:

– There is a gap between what the SB-tree provides and what we need. The ultimate goal we need is: after all the *nn_buffers* are seen, report an optimal location with its influence. To do so, separate from the aSB-tree we maintain a globally maximum influence and its spatial range.

- This global max is maintained after processing each insertion. Here every index entry in the aSB-tree stores the local maximum influence of some location in the sub-tree. It is local because the actual influence should consists of the *fullcover* of index entries for all ancestor nodes. For instance, in the second index entry in Figure 5(b), the local maximum influence is 12. But the actual maximum influence is 12+(-4)=8. Since the old global max is no smaller than the new one, it is not changed.
- Along with each local maximum influence stored at an index entry, or with the global max, we also store the corresponding spatial region. That is, any location in this region has this maximum influence. For a local max, its corresponding region only needs to store the left X border for the right border is not known yet. With a new insertion, it is possible to close the right border of the previous max region and start a new one. E.g. in Figure 5(b), two index entries' local max region change their left border from $X = 4$ to $X = 5$. The right border of the global max region may need to be closed correspondingly. All the additional information can be maintained along with the insertion process, by following the insertion paths backwards. So the update cost remains $O(\log n)$ as in the SB-tree. Therefore, the plane-sweep algorithm integrated with the aSB-tree has $O(n \log n)$ query cost.

4.4   Extension to Involve a Query Region

In the original coordinate, the query region Q is an axis-parallel rectangle. Thus, in the ⤵45° coordinate, the query region Q becomes a rectangle rotated 45° clockwise (as shown in Figure 6). To perform the query correctly, our aSB-tree based plane-sweep algorithm needs to be extended as follows:

- The Y space of the aSB-tree is not the whole space $(-M,M)$, but the Y projection of Q. This is because we only care about the locations in Q. In Figure 6, the Y space of the aSB-tree should be (y_l, y_h).
- For each *nn_buffer*, we calculate the smallest X (called *start*) when it 'enters' Q and the largest X (called *end*) when it 'leaves' Q. The insertion/removal events occur at these calculated X values, instead of the *x_low* and *x_high* of the *nn_buffers*.

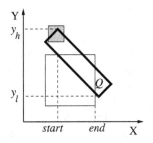

Fig. 6. Illustration of a rotated query region

- Finally, it is no longer true that whenever the aSB-tree is updated due to an event, the current maximum influence is known only by checking the root entry. The reason is that the maintained maximum influence may be in a region outside Q. To address this issue, we perform a *range-max query* on the aSB-tree. That is to find the maximum influence within the actual Y range. The range-max query can be performed in $O(\log n)$ time, since it only needs to examine two paths of the aSB-tree. The reason is that, similar to the insertion algorithm, a sub-tree whose interval is contained in the query interval does not need to be expanded.

A side note is: even though we use the aSB-tree as an in-memory structure to improve the plain sweep, if needed the structure can be implemented as a disk-based index like the SB-tree.

5 The Virtual OL-Tree

The R*-tree based solution examines all objects whose *nn_buffers* intersect with the query region Q, and thus is not efficient when a large Q results in the examination of many objects. This section first proposes an theoretical solution to the optimal-location query based on a new index structure called the *Optimal-Location Tree (OL-tree)*. Then we extend it to a more practical and efficient solution based on the *Virtual Optimal-Location Tree (VOL-tree)*.

5.1 The OL-Tree-Based Solution

The OL-tree is a k-d-B-tree-like structure which is balanced, disk-based and dynamically-updateable. Roughly speaking, it stores the *nn_buffers* in the 45° coordinate. Like the k-d-B-tree, the OL-tree is a space-partitioning method (versus a data-partitioning method like the R*-tree). Unlike the k-d-B-tree, the OL-tree stores rectangular records in its leaf nodes. If a square partially intersects with the ranges of multiple index entries, it is split and multiple copies are inserted. However, if the square fully contains the range of some index entry, we only update a value called *fullcover* stored along with the index entry, without further inserting into the sub-tree. Each index entry stores *maxoverlap*: the maximum local influence in the sub-tree. That is, the maximum influence in the sub-tree, subtracted by the *fullcover* values of all ancestor index entries. A rectangular region *maxrange* is also stored, where any location in it has maximum local influence. Due to the space limitation, we skip the details of the update and query processing.

The OL-tree may cause cascading split of child nodes if splitting an index node. One may wonder how bad the space complexity can be. We argue that the space complexity of the OL-tree (with an additional requirement) is $O(n^2/B)$ for the following reasons. First, the total number of leaf entries is $O(n^2)$. With n axis-parallel squares, there are $O(n)$ different X positions and $O(n)$ different Y positions, which form $O(n^2)$ cells. In the worse case each cell is stored in the tree separately. Thus there are at most $O(n^2)$ leaf entries. Second, the total number of nodes in $O(n^2/B)$. The linear storage of the k-d-B-tree can be guaranteed by re-organization of sub-trees which contain too few leaf entries. Similarly, the OL-tree with $O(n^2)$ leaf entries needs $O(n^2/B)$ nodes.

This bound reveals that the OL-tree is not a practical spatial index structure. In the next subsection we introduce a practical structure, named the VOL-tree, to solve the optimal-location query.

5.2   The VOL-Tree Structure

The OL-tree has more than linear space because if an *nn_buffer* is split into multiple pieces, each of them is physically stored in some leaf node(s). What if we do not physically store any leaf node of the OL-tree? We can use an R*-tree to store the original objects, and whenever the content of a leaf node is needed, we perform a range query on the R*-tree. This is the key idea to the *Virtual OL-tree (VOL-tree)*.

It is challenging to implement this idea. As we already spend the space to store the R*-tree, it is ideal to have a small VOL-tree that fits in memory. On the other hand, as there are $O(n^2/B)$ leaf nodes in an OL-tree, there are $O(n^2/B^2)$ index nodes, which would be the size of the VOL-tree if we treat it as an OL-tree without leaf nodes. There is a big gap. Thus we claim that the VOL-tree is NOT merely an OL-tree without the leaf level. It has to be much much smaller, possibly only consisting of one index level besides the root. A consequence is that each leaf entry of the VOL-tree corresponds to a virtual node (content stored in the R*-tree) with much more than B *nn_buffers*. So a crucial issue jumps out: it is expensive to maintain *maxrange* and *maxoverlap* because an update requires us to perform plane sweeps on the virtual nodes.

To address this issue, we propose another change from the OL-tree: along with each index entry, instead of keeping the accurate *maxoverlap*, keep two values *lowermax* and *uppermax*, which are a lower bound and an upper bound of *maxoverlap*.

In more detail, the entries in the tree are as follows:

- An index entry e has the following format: (*range*, *nodeID*, *fullcover*, *lowermax*, *maxrange*, *uppermax*). Here *range* is the spatial range of the corresponding sub-tree, and *nodeID* points to the referenced node. The value *fullcover* is the total weight of *nn_buffers* whose insertion stopped at e (such a *nn_buffer* contains *e.range*, but not the *range* of e's parent).
- The values *lowermax* and *uppermax* are some lower and upper bounds of the maximum local influence in *e.range*. And *maxrange* is a rectangle fully contained in *e.range* where every location in *maxrange* has local influence = *lowermax*.
- A leaf entry and an index entry have the same content, with a minor difference that a leaf entry's *nodeID* is empty.

5.3   The VOL-Tree Query Algorithm

Figure 7 shows the optimal-location-query algorithm in the VOL-tree. We start from the root node. In the VOL-tree, even if *root.maxrange* intersects with Q, it is possible that some location in $Q - root.maxrange$ has an influence larger than *root.lowermax* (when *root.lowermax* < *root.uppermax*). So as Step 1 shows, we can safely return a location in *root.maxrange* $\cap Q$ only if *root.lowermax* = *root.uppermax* or Q is completely inside *root.maxrange*.

Step 2 inserts the root entry into a heap. Every entry in the heap has, besides an index entry, two values *min* and *upper*. These are the actual (not local) lower

Algorithm. *VOLTreeQuery*

Input: Query region Q, VOL-tree *root*.

Return: An optimal location in Q.

1. **if** *root.maxrange* $\cap Q \neq \emptyset$ and (*root.lowermax* $=$ *root.uppermax* or $Q \subseteq$ *root.maxrange*), **return** any location in *root.maxrange* $\cap Q$.
2. *heap*.Insert(*root*, 0, *root.uppermax*)
3. Set *opt_loc* as an arbitrary location in Q, and *opt_inf* $= 0$,
4. **while** *heap* is not empty
   - (a) $(e, min, upper) =$ *heap*.ExtractMaxUpper().
   - (b) **if** *upper* $\leq$ *opt_inf*, **return** *opt_loc*.
   - (c) **if** e references an index node
      - **for** every entry *se* in $Node(e.nodeID)$ s.t. *se.range* $\cap Q \neq \emptyset$
         - A. Set $m = min +$ *se.fullcover*, and $u = min +$ *se.fullcover* $+$ *se.uppermax*.
         - B. **if** $u \leq$ *opt_inf*, **goto** next entry.
         - C. **if** *opt_inf* $< m$, set *opt_inf* $= m$ and *opt_loc* be any location in *se.range* $\cap Q$.
         - D. **if** *se.maxrange* $\cap Q \neq \emptyset$,
            - (i) $l = min +$ *se.fullcover* $+$ *se.lowermax*
            - (ii) **if** *opt_inf* $< l$, set *opt_inf* $= l$ and *opt_loc* be any location in *se.maxrange* $\cap Q$.
            - (iii) **if** $u \neq l$ and $(Q \cap se.range) \nsubseteq$ *se.maxrange*, *heap*.Insert(*se*, m, u)
         - E. **else**
            - *heap*.Insert(*se*, m, u)
         - F. **end if**
      - **end for**
   - (d) **else**
      - A. Using *e.range* $\cap Q$ as a new query region, retrieve *nn_buffers* from the R*-tree of objects. Use plane sweep to find an optimal location (inf, loc) within the new query region.
      - B. **if** *opt_inf* $< min + inf$, set *opt_loc* $= loc$, and *opt_inf* $= min + inf$.
   - (e) **end if**
5. **end while**
6. **return** *opt_loc*.

Fig. 7. Finding an optimal location using the VOL-tree

bound and upper bound of influence for locations in the sub-tree. Meanwhile, we maintain the currently seen optimal location *opt_loc* along with its influence *opt_inf*, initialized to be an arbitrary location with influence 0 (Step 3).

While the heap is not empty, we process each element at a time. In each iteration, the heap entry with maximum *upper* is extracted. As Step 4(b) of the algorithm shows, if this extracted *upper* is no larger than *opt_inf*, we can determine that *opt_loc* is an optimal location and thus the algorithm returns. The crucial steps are Step 4(c) which expands an index node and Step 4(d) which expands a leaf node.

To expand an index node, we examine every child entry *se* whose *range* intersects with Q, and try to push *se.nodeID* into the heap. Here the new lower bound is $m = min +$ *se.fullcover*, and the new upper bound is $u = min +$ *se.fullcover* $+$ *se.lowermax*. There are two pruning opportunities. First, if the new upper bound u is no larger than *opt_inf*, there is no need to expand the sub-tree (Step 4(c)B). Second, if *se.maxrange* intersects with Q, we already know the influence of the locations within the intersection, and thus we may have the chance to update the maintained optimal location before expanding the sub-tree (Step 4(c)D). It likely causes other entries to be pruned earlier.

To expand a leaf node (Step 4(d)), we go to the R*-tree to retrieve the *nn_buffers* that intersect with *e.range* ∩ *Q* and then perform a plane sweep technique of Section 4 to compute a location inside *Q* with maximum global influence *inf*. If this influence is bigger than *opt_inf*, we update the maintained *opt_inf* and *opt_loc*.

5.4   The Update Algorithm

The VOL-tree can be bulk-loaded. Due to space limitations, the algorithm is omitted. We only point out that immediately after bulk-loading, every entry in the VOL-tree has accurate local maximum information, i.e. *lowermax* = *uppermax*. With dynamic update, this may not be true. Let us examine the update algorithm below.

The insertion algorithm is shown in Figure 8, while a deletion is treated as an insertion with negative weight. To insert into an index node (Step 1), we consider every child entry *se* whose *range* intersects with the parameter *R*. If *se.range* is contained in *R*, we simply add *w* to *se.fullcover*. If *se.range* partially intersects with *R*, we recursively insert into the sub-tree referenced by *se*. After insertion, we need to re-aggregate the *lowermax*, *uppermax* and *maxrange* if necessary.

When *e* refers to a virtual leaf node which is not stored, the actual object is maintained in a separate R*-tree. So we only need to modify *e.lowermax*, *e.uppermax* and *e.maxrange*. The update of *e.uppermax* is simple. As Step 2(a) shows, for a positive weight, *e.uppermax* is increased by *w*. For a negative weight, *e.uppermax* remains unchanged. We may modify *e.lowermax* and/or *e.maxrange* only if *e.maxrange* intersects with *R*. For a positive weight (Step 2(c)), the intersection part of *e.maxrange* and *R* is the new *e.maxrange*, with weight increased by *w*. For a negative weight (Step 2(d)), there are two cases. If *e.maxrange* is fully covered by *R*, we decrease *e.lowermax*. Otherwise, we shrink *e.maxrange* to *e.maxrange* − *R* but keep *e.lowermax* unchanged.

6   Performance

In this section, we report experimental results on the R*-tree approach and the VOL-tree approach. In our experiments we used real datasets: the Digital Chart of the World from the R-tree Portal [The03]. It contains two type of point data: the populated places and cultural landmarks in North America, a total of 24,493 and 9,203 points respectively. We use the populated places as the objects and cultural landmarks as the sites. From the dataset, we generated an object R*-tree for all populated place, which is augmented by the L_1 distance from each object to its nearest site. We set the page size to 1k and the default buffer size to 256 pages. All the programs were written in Java and run on a Pentium IV Dell PC equipped with 3.2GHz CPU.

Algorithm. *VOLTreeInsert*

Input: Range R, Weight w, VOL-tree index entry e.

Pre-condition: R intersects with, but does not fully contain, $e.range$.

Action: Insert range R with weight w to the sub-tree referenced by e.

1. **if** e refers to a node in the VOL-tree
   (a) **for** every se in $Node(e)$ s.t. R contains $se.range$, $se.fullcover \mathrel{+}= w$.
   (b) **for** every se in $Node(e)$ s.t. R partially intersects with $se.range$, VOLTreeInsert(R, w, se).
   (c) Let se_0 be the entry in $Node(e)$ with maximum $se.fullcover + se.lowermax$.
   (d) Set $e.maxrange = se_0.maxrange$ and $e.lowermax = se_0.fullcover + se_0.lowermax$.
   (e) $e.uppermax = \max\{se.fullcover + se.uppermax\}$ for all entry se in $Node(e)$.
2. **else** /* e refers to a virtual leaf node */
   (a) **if** $w > 0$, $e.uppermax \mathrel{+}= w$.
   (b) **if** $e.maxrange \cap R = \emptyset$, **return**.
   (c) **if** $w > 0$
       i. $e.maxrange = e.maxrange \cap R$
       ii. $e.lowermax\mathord{+} = w$
   (d) **else**
       i. **if** $e.maxrange \subseteq R$, $e.lowermax\mathord{+} = w$.
       ii. **else** $e.maxrange = e.maxrange - R$.
   (e) **end if**
3. **end if**

Fig. 8. The Insertion algorithm of the VOL-tree

From our preliminary experimental results, we found that it does not help to make the VOL-tree disk based. So the VOL-tree is in memory, and only the I/O of R*-tree will be measured. However, it consumes part of the buffer. For example, if the size of the VOL-tree is 50 pages, the buffer available to the R*-tree retrieval in the VOL-tree based method should be $256 - 50 = 206$. For each experiment we start with a clean buffer, run 100 random queries, and measure the total I/Os. Buffer will not be flushed during the execution of 100 queries. Our preliminary experimental results show that, with fixed area of the query range, the shape of the query range has little impact on the query performance of the VOL-tree based methods. Thus we always use square query ranges.

In many applications, the datasets are known in advance. For instance, the set of McDonald's stores and the set of residential buildings can be given in advance when building the index, although changes may happen later on. Therefore, we use the VOL-tree based method with high bulk-loading percentage (80% and 100%). In the experiments, we compare the performance of three methods listed in Table 1.

Table 1. There different settings for experiments

Name of method	Explanation
R*	R*-tree based method (without using VOL-tree)
VOL80	VOL-tree based method with 80% of the objects being bulk loaded
VOL100	VOL-tree based method with 100% of the objects being bulk loaded

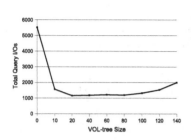

Fig. 9. The I/O performance of the VOL-tree for various size

To utilize the VOL-tree, the first question to answer is how large the VOL-tree should be. Figure 9 shows the I/O of the R*-tree of various sizes (in the unit of the page size). When the size of the VOL-tree is small (< 20 pages), the I/Os become close to the R*-tree based method (which corresponds to VOL-tree size = 0). When the size of the VOL-tree is large (> 80 pages), the I/Os also increase. That is because the larger VOL-tree does not help much to prune the search space, but it uses a large proportion of the buffer, which results in the worse I/O of the R*-tree.

From the results, we draw the conclusion that a small VOL-tree is sufficient. Thus, in the later experiments, we set the size of VOL-tree to 20 pages.

To study the effect of the size of query range on the I/Os, we change the *area* of the query range. Figure 10 (a) and (b) shows the results when the query range is small and is large respectively. When the query area is smaller than 1% of the whole space, their performances are very close although VOL-tree methods outperform. When the query area is larger than 1%, the R*-tree based method has I/Os of more than 10,000 so we do not even show it in figure. An expected fact is that when the query range becomes very large, the performance of both VOL80 and VOL100 improve. That is because, the large query ranges is more likely to intersect with *maxranges* stored in the VOL-tree, and thus is more likely to prune some subtrees.

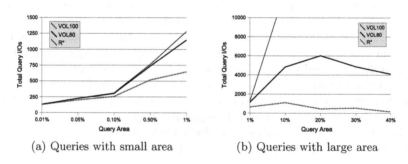

(a) Queries with small area   (b) Queries with large area

Fig. 10. The I/O performance of of the VOL-tree for various query area

The updates increase the difference between *lowermax* (*uppermax*) and the local maximal influence, thus decreasing the pruning capability. Figure 11 shows how updates affect the I/O performance. We bulk load some objects and insert the others. The X-axis presents the percentage of the number of inserted objects to the number of bulk loaded objects. For example, $X = 50\%$ corresponds to the case when we bulk load 2/3 of the objects and insert the remaining 1/3. With the increase of the percentage, the I/O performance decreases. After about 50% insertions, the performance becomes comparable to the R*-tree based method. There are two reasons for that. First, the VOL-tree uses some buffer of the R*-tree. Second, the VOL-tree may cause multiple scans of same page. We need to point out that even if an application is update intensive, the VOL-tree based method is still a good choice since the tree can be rebuilt in part or in full. And the rebuilding cost is amortized. Furthermore, the rebuilding can be integrated with the query processing.

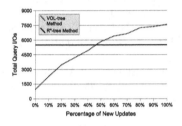

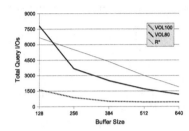

Fig. 11. The Effect of the Updates   **Fig. 12.** The Effect of the Buffer Size

Figure 12 shows how the buffer size affects the I/O performance of the VOL-tree. When buffer size is 128, the R* outperforms VOL80. That is because the VOL-tree has size of 20 and occupies about 20% the buffer. After the buffer size is doubled to 256, the I/O of VOL80 dramatically drops to below R*. With the increase of the buffer size, the performance of all the three methods get improved, while the VOL-tree based method is again better.

7   Conclusions

In this paper we proposed and solved the optimal-location query. The query has real applications, e.g. in corporate decision-support systems. We presented three solutions to accurately answer such a query. In particular, the VOL-tree approach is the most efficient. The approach uses an R*-tree to index the objects, while a small, in-memory VOL-tree is used to prune the search space. The query performance is much better than the plain R*-tree approach, especially when the query size is large. (Notice that the R*-tree approach is already optimized via a new index called the aSB-tree.) For instance, if the query area is 5% of the space, the VOL-tree approach computes an optimal location 6 times faster than the R*-tree approach. If the query size increases, the improvement increases as well, which can be multiple orders of magnitude better. Also, the size of the VOL-tree is small. In our experiments, while the R*-tree of objects is over 700 disk pages, the VOL-tree is only 20 pages. The VOL-tree has very efficient updates, as the index is small and updating it does not need to touch the R*-tree (except for ordinary object insertion/removal). One set of experiments showed that within 50% new updates, the VOL-tree approach remained to have better query performance. Of course, if there are too many updates, the VOL-tree can be re-built and the cost is amortized across all the new updates. In summary, the VOL-tree approach is the most efficient solution to the optimal-location query.

References

[BKOS97] M. de Berg, M. van Kreveld, M. Overmars, and O. Schwarzkopf. *Computational Geometry: Algorithms and Applications.* Sprinter Verlag, 1997.

[KM00]   F. Korn and S. Muthukrishnan. Influence Sets Based on Reverse Nearest Neighbor Queries. In *SIGMOD*, pages 201–212, 2000.

[LW80]   D. T. Lee and C. K. Wong. Voronoi Diagram in L_1 (L_∞) Metrics with 2-Dimensional Storage Applications. *SIAM Journal on Computing*, 9:200–211, 1980.

[PKZT01] D. Papadias, P. Kalnis, J. Zhang, and Y. Tao. Efficient OLAP Operations in Spatial Data Warehouses. In *SSTD*, pages 443–459, 2001.

[RKV95]   N. Roussopoulos, S. Kelley, and F. Vincent. Nearest Neighbor Queries. In *SIGMOD*, pages 71–79, 1995.

[SAE00]   I. Stanoi, D. Agrawal, and A. El Abbadi. Reverse Nearest Neighbor Queries for Dynamic Databases. In *ACM/SIGMOD Int. Workshop on Research Issues on Data Mining and Knowledge Discovery (DMKD)*, pages 44–53, 2000.

[SKC93]   S. Shekhar, A. Kohli, and M. Coyle. Path Computation Algorithms for Advanced Traveller Information System (ATIS). In *ICDE*, pages 31–39, 1993.

[Smi97]   M. Smid. Closest Point Problems in Computational Geometry. In J.-R. Sack and J. Urrutia, editors, *Handbook on Computational Geometry.* Elsevier Science Publishing, 1997.

[SRAE01] I. Stanoi, M. Riedewald, D. Agrawal, and A. El Abbadi. Discovery of Influence Sets in Frequently Updated Databases. In *VLDB*, pages 99–108, 2001.

[The03]   Yannis Theodoridis. The R-tree-portal. *http://www.rtreeportal.org*, 2003.

[TPL04]   Y. Tao, D. Papadias, and X. Lian. Reverse kNN Search in Arbitrary Dimensionality. In *VLDB*, pages 744–755, 2004.

[YL01]   C. Yang and K.-I. Lin. An Index Structure for Efficient Reverse Nearest Neighbor Queries. In *ICDE*, pages 485–492, 2001.

[YW01]   J. Yang and J. Widom. Incremental Computation and Maintenance of Temporal Aggregates. In *ICDE*, pages 51–60, 2001.

Constrained Shortest Path Computation*

Manolis Terrovitis[1], Spiridon Bakiras[2], Dimitris Papadias[2], and Kyriakos Mouratidis[2]

[1] Electrical and Computer Engineering Department,
National Technical University of Athens, Greece
`mter@dblab.ece.ntua.gr`
[2] Computer Science Department, Hong Kong University of Science and Technology,
Clear Water Bay, Hong Kong
{`sbakiras, dimitris, kyriakos`}`@cs.ust.hk`

Abstract. This paper proposes and solves *a-autonomy* and *k-stops* shortest path problems in large spatial databases. Given a source s and a destination d, an *a*-autonomy query retrieves a sequence of data points connecting s and d, such that the distance between any two consecutive points in the path is not greater than a. A k-stops query retrieves a sequence that contains exactly k intermediate data points. In both cases our aim is to compute the shortest path subject to these constraints. Assuming that the dataset is indexed by a data-partitioning method, the proposed techniques initially compute a sub-optimal path by utilizing the Euclidean distance information provided by the index. The length of the retrieved path is used to prune the search space, filtering out large parts of the input dataset. In a final step, the optimal (*a*-autonomy or *k*-stops) path is computed (using only the non-eliminated data points) by an exact algorithm. We discuss several processing methods for both problems, and evaluate their efficiency through extensive experiments.

1 Introduction

Shortest path computation has been studied extensively in graph theory and computer networks, assuming in-memory processing. However, the emergence of time-critical applications that require processing of voluminous spatial datasets necessitates the design of efficient shortest path algorithms for disk-resident data. In this paper we study two variations of the problem, and demonstrate how spatial access methods can be exploited to speed up processing. In particular, we consider the existence of a large collection of data in a Euclidean space, where each point is accessible from any other point in the database, with a cost equal to their distance. In this context, identifying the shortest path between two points is trivial; it is always the straight line connecting them. Nevertheless, real-world applications impose constraints that complicate the computation of the answer. We propose solutions to the following variations of the problem: (i) the *a-autonomy shortest path*, and (ii) the *k-stops shortest path*.

Definition 1 (*a*-**autonomy path**). *Let DB be a collection of points in the Euclidean space and a be a constant. An a-autonomy path from source s to destination d is a sequence of points (path) $s \rightarrow p_1 \rightarrow p_2 \rightarrow \ldots \rightarrow d$, where each intermediate point*

---

* Supported by grant HKUST 6180/03E from Hong Kong RGC.

C. Bauzer Medeiros et al. (Eds.): SSTD 2005, LNCS 3633, pp. 181–199, 2005.

belongs to DB, and the distance between any two consecutive points in the path is not greater than a.

The parameter a is called the autonomy constraint. Informally, it expresses the maximum distance that one can travel without a stop. Assume, for example, that an airplane must perform a flight from city A to city B, whose distance is D. If a is the autonomy of the plane and $D > a$, which intermediate airports should we choose to use as refueling bases, in order to minimize the overall flight distance? The a-autonomy problem also arises in the area of mobile communications. A signal can be successfully received when the distance between the sender and the receiver is no greater than a constant a. Given a source s and a destination d, we want to determine the intermediate communication centers that a message has to pass from, so that the latency (which is proportional to the overall covered distance) is minimized.

Definition 2 (*k*-stops path). *Let DB be a collection of points in the Euclidean space and k be a constant. A k-stops path from source s to destination d is a sequence $s \rightarrow p_1 \rightarrow \ldots \rightarrow p_k \rightarrow d$, where each intermediate point belongs to DB, and the number of intermediate points is exactly k.*

As an instance of the k-stops shortest path problem, assume that a delivery vehicle loads some goods at point s, and has to drive to its terminus at point d. In its course it has to deliver the goods to k of the company's customers, where k depends on its cargo capacity. Which k customers should it choose to serve in order to minimize the total traveled distance? The k-stops shortest path is also related to the prize collecting traveling salesman problem [1], where the salesman must choose k out of the total N cities to visit.

To the best of our knowledge, there is no previous work on the aforementioned problems in the context of spatial databases. On the other hand, naïve solutions, such as exhaustive search, are inapplicable to large datasets, due to their prohibitive CPU and I/O cost. In this paper, we propose algorithms for both the a-autonomy and the k-stops shortest paths. Specifically, given s and d, we initially compute an approximate solution (i.e., a sub-optimal path connecting s and d) that satisfies the input constraint. To obtain this approximate answer we design fast heuristics that utilize an existing data-partition index on the input dataset DB. The length of the retrieved path is used to prune the search space, filtering out large parts of the input dataset. Finally, an exact algorithm computes the optimal a-autonomy or k-stops shortest path among the remaining points. The proposed methodology reads only a fraction of DB from the disk, and has very low cost.

The rest of the paper is organized as follows. Section 2 surveys related work. Section 3 describes the general framework and states our basic pruning criterion. Sections 4 and 5 focus on the a-autonomy and the k-stops problem, respectively. Section 6 experimentally evaluates our techniques with real datasets, and Section 7 concludes the paper with a discussion on future work.

2 Related Work

Section 2.1 describes R-trees and algorithms for nearest neighbor (NN) search. Section 2.2 presents existing methods for shortest path computation and related problems.

2.1   R-Trees and Nearest Neighbor Queries

Although our techniques can be used with any data-partition method, here we assume R-trees [2,3] due to their popularity. Figure 1 shows an R-tree for point set $DB = \{p_1,p_2,\ldots,p_{12}\}$ with a capacity of three entries per node. Points that are close in space (e.g., p_1, p_2, p_3) are clustered in the same leaf node (N_3). Nodes are then recursively grouped together with the same principle up to the top level, which consists of a single root. Given a node N and a query point q, the $mindist(N,q)$ corresponds to the closest possible distance between q and any point in the sub-tree of node N. Figure 1(a) shows the $mindist$ between point q and node N_1.

The first NN algorithm for R-trees [4] searches the tree in a depth-first (DF) manner, by recursively visiting the node with the minimum $mindist$ from q. In Figure 1, for example, DF accesses the root, followed by N_1 and N_4, where the first potential nearest neighbor is found (p_5). During backtracking to the upper level (node N_1), the algorithm prunes entries whose $mindist$ is equal to or larger than the distance ($best\_dist$) of the nearest neighbor already retrieved. In the example of Figure 1, after discovering p_5, DF backtracks to the root level (without visiting N_3), and then follows the path N_2, N_6 where the actual NN p_{11} is found.

The DF algorithm is sub-optimal, i.e., it accesses more nodes than necessary. On the other hand, the best-first (BF) algorithm of [5] achieves the optimal I/O performance, visiting only nodes intersecting the circle centered at the query point q with radius equal to the distance between q and its nearest neighbor. These nodes have to be examined anyway in order to avoid false misses. In Figure 1(a), for instance, BF visits only the root, N_1, N_2, and N_6 (whereas DF also visits N_4). BF maintains a heap H with the entries encountered so far, sorted by their $mindist$. Starting from the root, it inserts all the entries into H (together with their $mindist$), e.g., in Figure 1(a), $H = \{< N_1, mindist(N_1,q) >, < N_2, mindist(N_2,q) >\}$. Then, at each step, it visits the node in H with the smallest $mindist$. Continuing the example, the algorithm retrieves the contents of N_1 and inserts all its entries in H, after which $H = \{<$

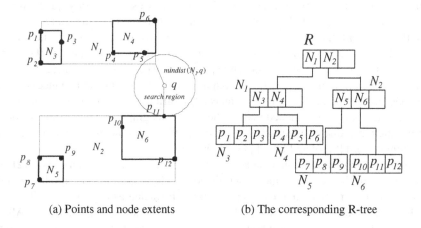

(a) Points and node extents   (b) The corresponding R-tree

Fig. 1. Example of an R-tree and a NN query

$N_2, mindist(N_2, q) >, < N_4, mindist(N_4, q) >, < N_3, mindist(N_3, q) >\}$. Similarly, the next two nodes accessed are N_2 and N_6 (inserted in H after visiting N_2), in which p_{11} is discovered as the current NN. At this time, the algorithm terminates (with p_{11} as the final result) since the next entry (N_4) in H is farther (from q) than p_{11}. BF (as well as DF) can be easily extended to kNN queries, where $k > 1$. Additionally, BF is incremental, implying that it can output the NNs in ascending order of their distance to the query without a pre-defined termination condition.

An interesting variation of the NN search is the *aggregate nearest neighbor* (ANN) query. Given a set of query points $Q = \{q_1, q_2, \ldots, q_m\}$ and an object p, the aggregate distance $adist(p, Q)$ is defined as a function f over the individual distances $|p, q_i|$ between p and each point $q_i \in Q$. Assuming, for example, n users at locations $q_1, \ldots, q_n$ and $f = sum$, an ANN query outputs the data object p that minimizes $adist(p, Q) = \sum_{q_i \in Q} |p, q_i|$, i.e., the *sum* of distances that the users have to travel in order to meet at the position of p. Similarly, if $f = max$, the ANN query reports the object p that minimizes the maximum distance that any user has to travel to reach p. In turn, this leads to the earliest time that all users will arrive at the location of p (assuming that they move with the same speed). Finally, if $f = min$, the result is the object p which is closest to any user, i.e., p has the smallest $adist(p, Q) = \min_{q_i \in Q} |p, q_i|$. Assuming that the data set is indexed by an R-tree, the *minimum bounding method* [6] applies best-first NN search, with the difference that each encountered node N is inserted into the heap H with key equal to $f(mindist(N, q_i), mindist(N, q_2), \ldots, mindist(N, q_m))$. We use this technique, as a module of the proposed algorithms, in Sections 4 and 5.

2.2 Shortest Path Computation and Related Problems

To the best of our knowledge, a-autonomy and k-stops shortest paths have not been studied before. On the other hand, there is extensive work on shortest path algorithms for main memory and disk-resident graphs. The most popular algorithm of the former category is proposed by Dijkstra [7]. This technique expands the input graph starting from the source node until it reaches the destination. It uses a priority queue to store the encountered nodes with key equal to their graph distance from the source. In every step, the node with the smallest key is de-queued, and its adjacent (non-visited) nodes are en-queued. The procedure terminates when a complete path (connecting the source and the destination) is found. A* search [8] uses heuristics in order to direct the graph expansion and prune the search space, assuming that the Euclidean distance between two nodes lower bounds their graph distance. The difference from Dijkstra's algorithm is that the key of each en-queued node is the sum of its graph distance from the source and its Euclidean distance from the destination. Other main memory methods include the Bellman-Ford [9,10] and Floyd [11] algorithms.

Shortest path computation techniques for disk-resident data, such as HiTi [12] and HEPV [13], are based on partial materialization. They partition the graph into subgraphs that fit in memory, and each sub-graph is abstracted as a graph node. The subgraphs are grouped recursively into higher level nodes, thus forming a hierarchy. All the distances between the sub-graph boundary nodes are computed and stored in the upper level. To answer a shortest path query, the algorithms (i) determine the lowest-level subgraph containing the source and destination, and (ii) utilize the materialized information along the two search paths to retrieve the result. [14,8] analyze the performance of several secondary memory adaptations of shortest path algorithms.

Papadias et al. [15] propose a storage scheme for large graphs and algorithms for nearest neighbors, range search and distance joins. Their methods combine connectivity and location information about the data objects (indexed by R-trees) to guide the search. Kolahdouzan and Shahabi [16] use the concept of network Voronoi cells and materialization to speed-up query processing. Shahabi et al. [17] find approximate nearest neighbors in road networks by transforming the problem to high dimensional space. Jensen et al. [18] discuss nearest neighbor queries for points moving in a network. Shekhar and Yoo [19] find all the nearest neighbors along a given route. Yiu and Mamoulis [20] study clustering problems in spatial networks.

The most related paper to our work is [21] that uses thematic spatial constraints to restrict the permitted paths (e.g., "find the shortest path that passes only through rural areas"). Although the problem is similar, in the sense that it also deals with constrained shortest path computation in spatial databases, the thematic restrictions are very different from our autonomy and cardinality constraints. Summarizing, all the existing techniques are inapplicable to the proposed problems. In the sequel, we discuss algorithms for a-autonomy and k-stops shortest paths, starting with the general framework.

3   General Framework and Pruning Criterion

The proposed techniques follow the methodology of Figure 2. The first step applies heuristics to efficiently retrieve a path, not necessarily optimal, that satisfies the given constraint (on a or k). The second step uses the length of this path to prune the search space and eliminate the majority of the data points. Finally, the third step computes the actual shortest path (subject to the constraints) using only the non-eliminated points.

Algorithm **Find Shortest Path**

// Input: the source s, the destination d, the dataset DB, the parameter a or k

// Output: the constrained (i.e., a-autonomy or k-stops) shortest path

1. Find a sub-optimal solution with a fast algorithm
2. Use the length of the obtained path to prune parts of the workspace
3. Compute the exact (a-autonomy or k-stops) shortest path using only the non-eliminated data points

---

Fig. 2. The general processing methodology

Whereas steps 1 and 3 are problem-dependent, the pruning criterion is common for both *a-autonomy* and *k-stops* shortest paths. Consider that in Figure 3, we already have a path $s \rightarrow p_1 \rightarrow p_2 \rightarrow d$ with length l satisfying the given constraint (a or k). Our goal is to use this path in order to restrict the search space. For example, point p_3 cannot belong to a better path because $|s, p_3| + |p_3, d| > l$, where $|s, p_3|$ and $|p_3, d|$ are the distances of p_3 from s and d, respectively. In general, any data point p that may be part of a path with length equal to or shorter than l must satisfy the condition $|s, p| + |p, d| \leq l$, i.e., it must lie in the ellipse with foci at points s and d, and sum of

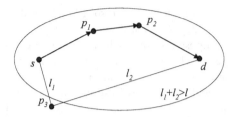

Fig. 3. Pruning example

distances from the foci equal to l. Such points are efficiently retrieved by a range query (in the shape of the ellipse) on the data R-tree. Based on the above, the goal of the first step of our framework is to retrieve a nearly-optimal path, so that the area of the ellipse, and the number of data points therein, is minimized. Then, the third step computes the actual shortest path using only these points. In the following sections we discuss steps 1 and 3 for *a-autonomy* and *k-stops* shortest paths.

4 a-Autonomy Shortest Paths

Section 4.1 describes alternative ways to obtain a good approximate solution for the a-autonomy problem, while Section 4.2 deals with the optimal path computation.

4.1   Fast Sub-optimal Path Computation

The shortest route between the source s and the destination d in the Euclidean space is obviously the line segment $\overline{sd}$ connecting them. If the distance $|s, d|$ between s and d is greater than the autonomy of the problem a, we have to introduce intermediate stops. An intuitive strategy for choosing the stops is to select points that cause the least diversion from the optimal route (i.e., the line segment $\overline{sd}$). In particular, the point p in DB that lies closest to $\overline{sd}$ is chosen as a part of the path. The process continues recursively with segments $\overline{sp}$ and $\overline{pd}$ if their individual lengths exceed a. This Least Diversion Method (LDM) terminates when (i) a complete path fulfilling the autonomy constraint is found, or (ii) when all possible solutions are examined. In the latter case, there is no solution for the given value of a and the required path is *infeasible*.

We illustrate the functionality of LDM using the example of Figure 4(a). Initially, LDM retrieves the NN of the line segment $\overline{sd}$, which is point p_1. Assuming that the distance $|s, p_1|$ is less than a, $s \rightarrow p_1$ is accepted as a component of the path. On the other hand, if $|p_1, d|$ is greater than a, the process is repeated for the line segment $\overline{p_1 d}$. Continuing the example, in Figure 4(b) the NN of $\overline{p_1 d}$ is point p_2. Since both distances $|p_1, p_2|$ and $|p_2, d|$ are smaller than the autonomy a, the components $p_1 \rightarrow p_2$ and $p_2 \rightarrow d$ are inserted into the path, and LDM terminates with $s \rightarrow p_1 \rightarrow p_2 \rightarrow d$ as the result.

Figure 5 contains a divide-and-conquer version of LDM. The first call has input parameters s, d, and an empty list $path$. The algorithm recursively examines points according to their distance from $\overline{sd}$. If some point cannot lead to a feasible solution (lines 11, 12), LDM backtracks, and continues with the next NN of the line segment

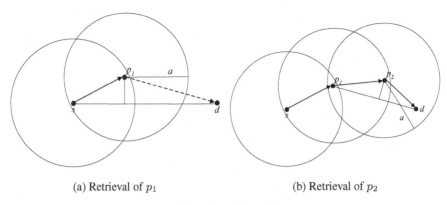

(a) Retrieval of p_1 (b) Retrieval of p_2

Fig. 4. LDM example

Algorithm **LDM**$(p, p', path)$
// p and p' are two intermediate points, and $path$ is the path constructed so far
1.   If $|p, p'| > a$
2.   Find the next nearest neighbor p_{NN} of line segment $\overline{pp'}$ in $DB - \{path\}$
3.   If no p_{NN} is found // i.e., all possible p_{NN} have been unsuccessfully examined
4.   Return *false* // Backtracking
5.   Else
6.   **LDM**$(p, p_{NN}, path)$ // Recursion
7.   **LDM**$(p_{NN}, p', path)$
8.   If both above calls of LDM return *true*
9.   Add p_{NN} to $path$
10.   Return *true*
11.   Else // Selection of p_{NN} cannot lead to a valid path
12.   Go to line 2 and continue with the next NN of $\overline{pp'}$
13.   Else // i.e., $|p, p'| \leq a$
14.   Add p to $path$
15.   Return *true*

Fig. 5. The least diversion method for the a-autonomy problem

in line 2. Upon termination, if the returned result is *false*, the problem is infeasible. Otherwise, the obtained path is stored in $path$. The nearest neighbor of a line segment (in line 2) is retrieved in a way similar to the best-first NN search discussed in Section 2.1. The difference is that the $mindist$ between the query line segment and an MBR is computed according to the method of [22].

LDM may incur relatively high cost because each NN query (to a line segment) may visit numerous nodes (that intersect, or are near, the segment). Furthermore, since LDM does not aim at minimizing the intermediate points in the path, the number of such queries may be large. Motivated by this we propose a Greedy Heuristic Method (GHM) that (i) applies point (instead of line segment) NN queries and (ii) tries to minimize the number of intermediate points. GHM is based on the observation that an optimal set of intermediate points would lie on $\overline{sd}$, and that the distance between any consecutive

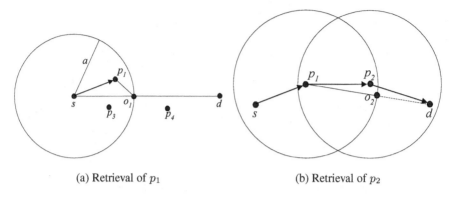

(a) Retrieval of p_1 (b) Retrieval of p_2

Fig. 6. GHM example

Algorithm **GHM**$(p, path)$
// p is an intermediate point, and $path$ is the path constructed so far
1. If $|p, d| > a$
2.   Set o to be the point in the line segment $\overline{pd}$ with distance from p equal to a
3.   Find the next NN of o (p_{NN}) in the circle with center at p and radius a
4.   If no p_{NN} is found // i.e., all possible p_{NN} have been unsuccessfully examined
5.   Return *false* // Backtracking
6.   Else
7.   **GHM**$(p_{NN}, path)$ // Recursion
8.   If the above call of GHM returns *true*
9.   Add p_{NN} to $path$
10.   Return *true*
11.   Else // Selection of p_{NN} cannot lead to a valid path
12.   Go to line 3 and continue with the next NN of o
13. Else // i.e., $|p, p'| \leq a$
14.   Add p to $path$
15.   Return *true*

Fig. 7. The greedy heuristic method

pair would be equal to the autonomy a. Since such points do not necessarily exist in the database, it tries to use their NNs.

Figure 6(a) illustrates GHM with an example. Ideally, the first intermediate point o_1 would lie on the line segment $\overline{sd}$ at distance a from s. GHM retrieves the NN p_1 of o_1 among the points that are directly reachable by s (i.e., the points falling in the circle centered at s with radius a) and inserts it into the path. To compute the second intermediate point, it determines the ideal point o_2 that lies on $\overline{p_1 d}$ at distance a from p_1 (see Figure 6(b)). Then, it retrieves the NN of o_2 (i.e., p_2) among the points that are directly reachable from p_1, and inserts it into the path. The distance $|p_2, d|$ is smaller than a, and GHM terminates with $s \to p_1 \to p_2 \to d$ as the result.

Figure 7 shows the pseudo-code of GHM. The NN computation in line 3 is an instance of a constrained NN query [23], since the retrieved points must fall in a specified region (i.e., the circle centered at p with radius a). The search algorithm follows the best-first paradigm and, thus, it is incremental. In particular, it inserts into a search heap only R-tree entries N where $mindist(N, p) \leq a$, with sorting key $mindist(N, o)$. It reports only points p_{NN} with $|p, p_{NN}| \leq a$ in ascending order of their distance from o.

Note that the paths obtained by LDM and GHM are possibly different and sub-optimal. For instance, LDM first exploits paths containing the first NN p of the segment $\overline{sd}$, but the best path does not necessarily contain this point (even if there is a path passing from p that satisfies the autonomy constraint). Similarly, in Figure 6(a), GHM will not discover the best path $s \rightarrow p_3 \rightarrow p_4 \rightarrow d$ because p_3 is not the NN of o_1. Therefore, both LDM and GHM only constitute fast filter steps before the exact computation, which is discussed next.

4.2 Optimal Path Computation

After obtaining a sub-optimal path, we perform a query on the R-tree to retrieve the data points that may lead to better solutions. As discussed in Section 3, if l is the length of the sub-optimal path (returned by LDM or GHM), potential candidates lie in the ellipse defined by points s, d, and the value of l. Assume that the corresponding set of points is $DB_{eps} \subseteq DB$. To identify the optimal path, we process DB_{eps} with a modified version of A* search, which takes into account the autonomy constraint and the approximate solution available. In order to apply the algorithm, we consider that the retrieved points form a graph, such that (i) two points (nodes) are connected, if their Euclidean distance

---

Algorithm **Optimal Path Computation**
// s: source point, d: destination point, a: autonomy
// DB_{eps}: the subset of DB after the pruning step
1. Initialize a min-priority queue Q
2. Insert s into Q with key $dist(s) = |s, d|$
3. While Q is not empty
4.   Get the next entry $< e, dist(e) >$ in Q
5.   If $e \neq d$ // expand the graph around point e
6.   For each point p inside the circle centered at e with radius a
7.   If $dist(e) - |e, d| + |e, p| + |p, d| \geq l$
8.   Go back to line 6 and continue with the next point
9.   If p has not been de-queued before
10.   If p is not currently in Q // i.e., p is visited for the first time
11.   En-queue $< p, dist(e) - |e, d| + |e, p| + |p, d| >$ in Q
12.   Else // i.e., p was visited before and it is contained in Q
13.   Let $dist(p)$ be the key of p in Q
14.   If $dist(p) > dist(e) - |e, d| + |e, p| + |p, d|$
15.   Update the key of p in Q to be $dist(e) - |e, d| + |e, p| + |p, d|$
16.   Else // i.e., $e = d$
17.   Return the corresponding path as the result, and terminate
18. Return the path at hand as the result // i.e., no better path was found

---

Fig. 8. The optimal a-autonomy path computation over the un-pruned part of the dataset

does not exceed the autonomy a, (ii) the cost of an edge connecting two points equals their distance.

Figure 8 illustrates the pseudo-code for the optimal path computation module. Line 6 guarantees that the path returned is valid by considering as reachable only points within distance a from the de-queued entry. The knowledge of a path with length l is used to reduce the search space in line 7. Note that the pruning condition $(dist(e) - |e, d| + |e, p| + |p, d| \geq l)$ for a considered point p also takes into account $|p, d|$, which constitutes a lower bound for the length of the shortest path between p and d, in accordance with the A* algorithm.

5 k-Stops Shortest Paths

Section 5.1 presents two heuristics for the efficient retrieval of sub-optimal k-stops paths. Section 5.2 describes an algorithm for computing the optimal answer.

5.1   Fast Sub-optimal Path Computation

A naïve heuristic for computing a good initial path is to select the k closest points to the line segment that connects s and d. In certain cases, however, this may lead to a poor solution. Consider, for instance, that in Figure 9(a) we want to compute the shortest path that passes through three intermediate stops. The four NNs of $\overline{sd}$ are p_1, p_2, p_3, p_4 (in this order). The path $s \rightarrow p_1 \rightarrow p_2 \rightarrow p_3 \rightarrow d$ containing the first 3NNs is relatively long since p_2 is on the opposite side (of $\overline{sd}$) with respect to p_1 and p_3. In order to avoid this problem we follow the least diversion paradigm. Figure 9(b) illustrates the adaptation of LDM (called k-LDM) to the k-stops problem on the example of Figure 9(a). First, k-LDM adds to the path the NN (p_1) of line segment $\overline{sd}$. Then, it retrieves the point with the minimum distance from line segments $\overline{sp_1}$ and $\overline{p_1d}$. Among the NNs (p_3 and p_4) of $\overline{sp_1}$ and $\overline{p_1d}$, p_3 is inserted into the path. The process is repeated for the NN of $\overline{sp_1}$, $\overline{p_1p_3}$, $\overline{p_3d}$, and LDM terminates with $s \rightarrow p_4 \rightarrow p_1 \rightarrow p_3 \rightarrow d$ as the result.

The k-LDM algorithm is illustrated in Figure 10. It is worth mentioning that the best NN computation in line 3 is not performed by individual NN queries for each edge of the path, because this approach would lead to multiple traversals of the R-tree of DB.

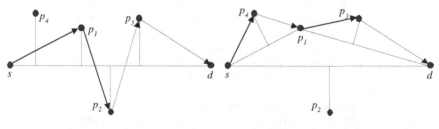

(a) 3NNs as 3 intermediate stops   (b) k-LDM shortest path

Fig. 9. k-LDM motivation and example

Algorithm k-**LDM**(s, d)
// s and d are the source and destination, and $path$ is the path constructed so far
1.   Initialize $path$ to $s \rightarrow d$
2.   For $i = 1$ to k
3.   Find the point p in DB with the min distance from any line segment in $path$
4.   Add p to $path$
5.   Return $path$

Fig. 10. The least diversion method for the k-stops problem

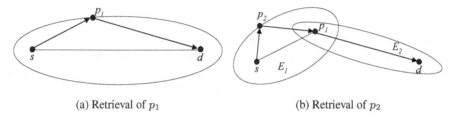

(a) Retrieval of p_1 (b) Retrieval of p_2

Fig. 11. LOM example

To compute an intermediate point with a single traversal, line 3 is implemented using an adaptation of the aggregate nearest neighbor search discussed in Section 2.1. In particular, the query set Q now consists of the edges in the path constructed so far, and the aggregate distance of a point is defined as the distance from its closest line segment in Q. The algorithm follows the best-first paradigm, by inserting each encountered node N in the R-tree of DB into the search heap H with key equal to $\min_{p \rightarrow p' \in path} mindist(N, \overline{pp'})$, i.e., the minimum $mindist$ between N and any of the edges in the path.

An alternative to k-LDM is the Local Optimum Method (LOM). Given any pair of points p and p', if we want to select one intermediate point $o \in DB$ so that the length of the path $p \rightarrow o \rightarrow p'$ is minimized, then o is by definition the point that minimizes the sum of distances from p and p' (i.e., $|p, o| + |o, p'|$). Based on this observation, given a path with fewer than k points, LOM chooses as the next point o the one that minimizes the sum of distances from any pair of consecutive points in the current path. In other words, it selects $o \in \{DB - path\}$ that minimizes $\min_{p \rightarrow p' \in path} |p, o| + |o, p'|$. Figure 11 gives an example for a 2-stops query. The first intermediate point is p_1, since it minimizes the sum of distances from s and d. Geometrically, this implies that the ellipse in Figure 11(a) does not contain any other point. The second intermediate point p_2 is computed as the point in $DB - \{s, d, p_1\}$ that minimizes the quantity $\min(|s, p_2| + |p_2, p_1|, |p_1, p_2| + |p_2, d|)$. Consequently, in Figure 11(b) the ellipse E_1 does not contain any point other than p_2, while ellipse E_2 (defined by foci p_1 and d, and length $|p_1, p_2| + |p_2, d|$) is empty.

Figure 12 shows the LOM algorithm. Similar to the implementation of k-LDM, step 3 is performed with an ANN algorithm, in order to avoid multiple traversals of the R-tree of DB. The query set Q contains all the edges in the current path, and the aggregate distance of a point is defined as the minimum sum of distances from the

---

Algorithm **LOM**(s, d)
// s and d are the source and destination, and $path$ is the path constructed so far
1.   Initialize $path$ to $s \to d$
2.   For $i = 1$ to k
3.   Find the point p in DB with the min sum of distances from the endpoints
4.   of any line segment in $path$
5.   Add p to $path$
6.   Return $path$

Fig. 12. The local optimum heuristic method for the k-stops problem

endpoints of any line segment in Q. The ANN search traverses the R-tree of DB in a best-first manner, inserting each encountered node N into the search heap H with key equal to $\min_{p \to p' \in path} mindist(N, p) + mindist(N, p')$, i.e., the minimum sum of $mindist$ between N and the endpoints of any of the edges in the path.

5.2 Optimal Path Computation

The optimal path computation involves an implementation of Bellman-Ford's algorithm. This algorithm works iteratively, and calculates the shortest paths in increasing number of hops. In particular, during step i it computes the shortest paths consisting of exactly i hops between the source node and every other node of the graph. The complexity of Bellman-Ford's algorithm is $O(mE)$ for a graph with E edges and a maximum path length of m hops, rendering it very expensive for dense graphs. Notice that in the k-stops problem formulation there is no autonomy constraint, and therefore, the number of edges is $O(N^2)$ (where N is the number of points after the pruning step). Consequently, the running time of the algorithm is expected to be $O(kN^2)$, and a good first solution is crucial for achieving low cost.

6   Experimental Evaluation

In this section we experimentally evaluate the performance of our methods, in terms of I/O and CPU cost. We use the two real spatial datasets TCB and LA (available at *www.rtreeportal.org*), containing 450K and 1.3M points, respectively. Both datasets are normalized to fit in a $[0, 10000]^2$ workspace. The block size of the R-trees is set to 2 KBytes. For each simulation, we select two random points from the dataset and compute the constrained shortest path using the proposed methods. In order to reduce randomness, each result is obtained by averaging over the measurements of 10 simulations. For all experiments we use a Pentium 3.2 GHz CPU with 1 GByte memory. Section 6.1 focuses on the a-autonomy problem, while Section 6.2 on k-stops shortest paths.

6.1   Evaluation of a-Autonomy

We first study the effect of the autonomy value a using the GHM and LDM heuristics. We fix the distance between the source and destination points to 3000 and vary a from

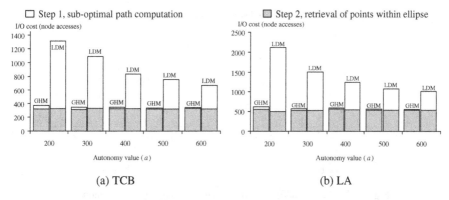

Fig. 13. Total I/O cost vs. autonomy a

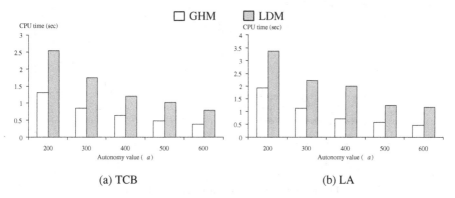

Fig. 14. Total CPU time vs. autonomy a

200 to 600. Figure 13 shows the total I/O cost for datasets TCB and LA, and its break-down into the sub-optimal path computation (step 1) and the elliptical range search (step 2) that collects the points passing the criterion of Section 3. As a increases, the cost drops because the path consists of fewer intermediate points. Regarding the final range search, the methods incur similar overhead. On the other hand, obtaining the initial path with GHM incurs significantly fewer node accesses compared to LDM. This happens because GHM performs (cheap) point NN queries, as opposed to the (expensive) linear NN queries of LDM. Furthermore, since GHM aims at reaching the destination with the minimum number of steps, it performs fewer NN searches than LDM.

Figure 14 depicts the total CPU time for the previous experiment. The running time decreases with a because both the sub-optimal and the optimal path consist of fewer points. The performance gain of GHM is similar to the I/O gain for the reasons ex-plained in the context of Figure 13. An important remark concerning both methods is that the initial path computation dominates the total CPU time because, as discussed shortly the number of non-eliminated points (that participate in the selection of the op-timal path) is small.

We now present some interesting measurements regarding the cost and accuracy of the sub-optimal path computation. Figure 15 depicts the CPU time for calculating

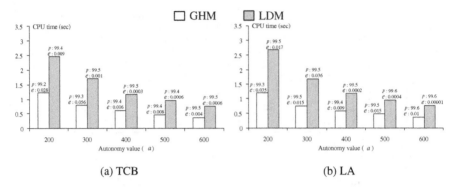

Fig. 15. CPU time and quality of the sub-optimal path vs. autonomy a

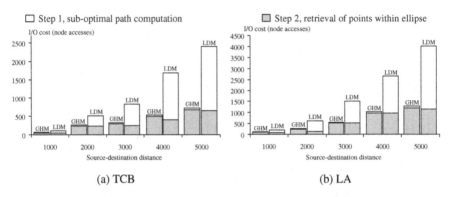

Fig. 16. Total I/O cost vs. distance of endpoints

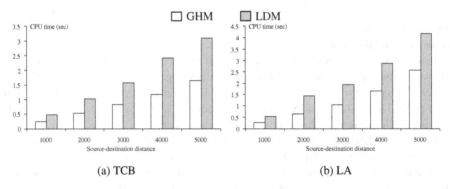

Fig. 17. Total CPU time vs. distance of endpoints

the initial path, the deviation percentage e of the achieved path length compared to the optimal one, and the percentage p of the dataset that is pruned according to the criterion of Section 3. LDM provides a better quality path than GHM, at the expense of higher CPU cost. Both methods, however, produce a very accurate result (with less than 0.04% deviation from the optimal path length) and are able to prune over 99% of the database.

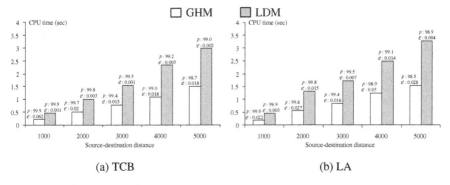

Fig. 18. CPU time and quality of the sub-optimal path vs. distance of endpoints

Next, we investigate the effect of the distance between the source and destination. We set the autonomy variable a to 300, and vary the (Euclidean) distance between the endpoints of the path from 1000 to 5000. The total number of node accesses and the overall CPU time are illustrated in Figures 16 and 17, respectively. As expected, a larger distance implies higher I/O and CPU cost. LDM is affected more because of the numerous linear NN queries. Figure 18 focuses on the sub-optimal path computation step. As the distance between the endpoints of the path increases, the pruning percentage p decreases, since the ellipse of the criterion of Section 3 grows. On the other hand, the deviation e from the optimal solution is very small in all cases.

6.2   Evaluation of k-Stops

In the following experiments we evaluate the performance of the k-stops shortest path algorithms. First, we study the effect of k on the LOM and k-LDM techniques. We fix the distance between the source and destination points to 3000 and vary the number k of intermediate stops from 3 to 7. Figure 19 shows the total I/O cost, and its breakdown into the initial sub-optimal path computation and the retrieval of points passing the pruning criterion. The node accesses of LOM remain relatively stable, while for k-

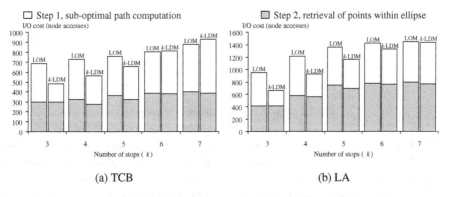

Fig. 19. Total I/O cost vs. number k of required stops

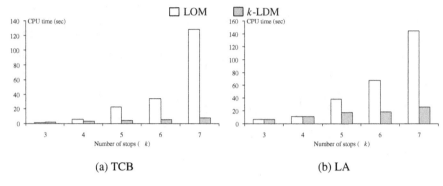

Fig. 20. Total CPU time vs. number k of required stops

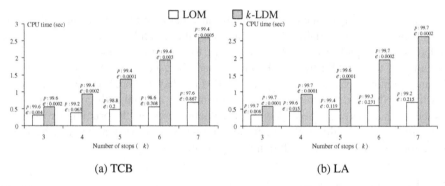

Fig. 21. CPU time and quality of the sub-optimal path computation vs. number k of required stops

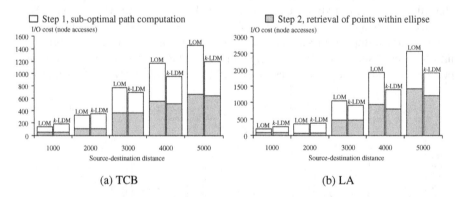

Fig. 22. Total I/O cost vs. distance of endpoints

LDM they increase linearly with k. Initially, k-LDM incurs fewer node accesses than LOM, but this changes for larger values of k. Regarding the elliptical range search for points passing the pruning criterion, its I/O cost is similar for both methods.

Figure 20 illustrates the total running time for the previous experiment. k-LDM is considerably faster than LOM, and the performance gap increases sharply with k. This

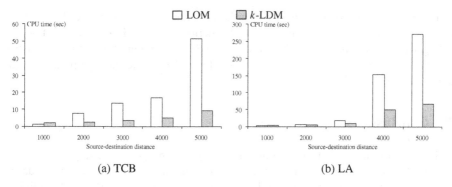

Fig. 23. Total CPU time vs. distance of endpoints

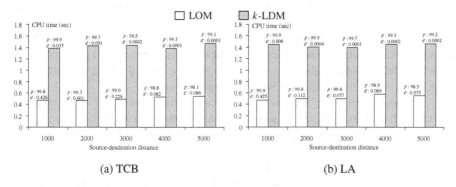

Fig. 24. CPU time and quality of the sub-optimal path vs. distance of endpoints

fact indicates that the k-LDM technique retrieves a better sub-optimal solution, and thus achieves more effective pruning than LOM. Recall that the time complexity of the optimal path computation is $O(kN^2)$, where N is the number of points that pass the criterion of Section 3. It follows that even a slightly worse initial solution can increase significantly the overall running time. Regarding the detailed behavior of each method, the CPU time for LOM is dominated by the optimal path computation, while for k-LDM approximately 10%-50% of the CPU time is spent on the initial path computation.

Figure 21 verifies the above observation, by comparing the accuracy and the CPU time of the sub-optimal path computation for k-LDM and LOM. The path returned by k-LDM is at most 0.003% longer than the optimal one in all cases. LOM is not as effective, especially for large values of k. Its solution deviates from the optimal by up to 0.86%. On the other hand, LOM is considerably faster than k-LDM, and it is a good choice for providing approximate results to time-critical applications that can tolerate a certain amount of inaccuracy in return for a fast response.

Finally, in Figures 22, 23 and 24 we investigate the impact of the distance between the two endpoints of the path. We set the number k of stops to 5, and vary the distance between the source and destination from 1000 to 5000. As expected, when the distance increases, both the I/O and CPU costs are higher. The reason is that the ellipse covers a larger area of the workspace, and prunes fewer nodes and points (also verified by the pruning percentages p in Figure 24). Furthermore, the I/O cost increases because the edges of the path are longer, and the ANN queries of both k-LDM and LOM access a larger part of the index. As shown in Figure 24, the running time of the heuristics remains relatively stable because the required number of intermediate points is constant (i.e., k).

7 Conclusion

This paper formulates and solves a-autonomy and k-stops shortest path queries, in spatial databases. Assuming a large collection of points in the Euclidean space indexed by a data-partitioning access method, we propose several techniques for the efficient computation of the constrained shortest paths. Our methods exploit the spatial information provided by the index, in order to produce very fast an initial sub-optimal path. The length of this path is then used to prune the workspace, according to a geometric criterion. The optimal path is retrieved by utilizing an exact shortest path algorithm on the non-eliminated data points. Our experimental results on real spatial datasets demonstrate that the proposed techniques are able to prune over 98% of the database for all examined settings, thus leading to very low response times.

A promising direction for future work concerns a top-K version of the constrained shortest path problem, where instead of a single path, we are asked to compute the best K paths according to some input constraint (e.g., a-autonomy or k-stops). Furthermore, in this paper we consider that all points are equivalent. It would be interesting to study cases where the data points have different properties. For instance, in autonomy problems it may be beneficial to visit a point that incurs relatively high diversion, if it can provide a large benefit (e.g., in terms of refueling capacity).

References

1. Balas, E.: The prize collecting travelling salesman problem. Networks **19** (1989) 621–636
2. Beckmann, N., Kriegel, H.P., Schneider, R., Seeger, B.: The R*-tree: An efficient and robust access method for points and rectangles. In: SIGMOD. (1990) 322–331
3. Guttman, A.: R-trees: A dynamic index structure for spatial searching. In: SIGMOD. (1984) 47–57
4. Roussopoulos, N., Kelley, S., Vincent, F.: Nearest neighbor queries. In: SIGMOD. (1995) 71–79
5. Hjaltason, G.R., Samet, H.: Distance browsing in spatial databases. ACM TODS **24** (1999) 265–318
6. Papadias, D., Tao, Y., Mouratidis, K., Hui, C.: Aggregate nearest neighbor queries in spatial databases. ACM TODS (2005, to appear)
7. Dijkstra, E.: A note on two problems in connection with graphs. Numerische Mathematik **1** (1959) 269–271
8. Shekhar, S., Kohli, A., Coyle, M.: Path computation algorithms for advanced traveller information system (ATIS). In: ICDE. (1993) 31–39
9. Bellman, R.: On a routing problem. Quarterly of Applied Mathematics **16** (1958) 87–90
10. Ford, L., Fukelson, D.: Flows in networks. Princeton University Press (1962)
11. Floyd, R.: Shortest path. Communications of the ACM **5** (1962) 345

12. Jung, S., Pramanik, S.: HiTi graph model of topographical roadmaps in navigation systems. In: ICDE. (1996) 76–84

13. Jing, N., Huang, Y., Rundensteiner, E.A.: Hierarchical encoded path views for path query processing: An optimal model and its performance evaluation. IEEE TKDE **10** (1998) 409–432

14. Jiang, B.: I/O-efficiency of shortest path algorithms: An analysis. In: ICDE. (1992) 12–19

15. Papadias, D., Zhang, J., Mamoulis, N., Tao, Y.: Query processing in spatial network databases. In: VLDB. (2003) 802–813

16. Kolahdouzan, M.R., Shahabi, C.: Voronoi-based K nearest neighbor search for spatial network databases. In: VLDB. (2004) 840–851

17. Shahabi, C., Kolahdouzan, M.R., Sharifzadeh, M.: A road network embedding technique for k-nearest neighbor search in moving object databases. In: GIS. (2002) 94–100

18. Jensen, C.S., Kolárvr, J., Pedersen, T.B., Timko, I.: Nearest neighbor queries in road networks. In: GIS. (2003) 1–8

19. Shekhar, S., Yoo, J.S.: Processing in-route nearest neighbor queries: a comparison of alternative approaches. In: GIS. (2003) 9–16

20. Yiu, M.L., Mamoulis, N.: Clustering objects on a spatial network. In: SIGMOD. (2004) 443–454

21. Huang, Y., Jing, N., Rundensteiner, E.A.: Integrated query processing strategies for spatial path queries. In: ICDE. (1997) 477–486

22. Tao, Y., Papadias, D., Shen, Q.: Continuous nearest neighbor search. In: VLDB. (2002) 287–298

23. Ferhatosmanoglu, H., Stanoi, I., Agrawal, D., Abbadi, A.: Constrained nearest neighbor queries. In: SSTD. (2001) 257–278

Accurate and Efficient Similarity Search on 3D Objects Using Point Sampling, Redundancy, and Proportionality

Johannes Aßfalg[1,2], Hans-Peter Kriegel[1], Peer Kröger[1], and Marco Pötke[2]

[1] Institute for Computer Science, University of Munich, Germany
[2] sd&m AG, Munich, Germany
{assfalg, kriegel, kroegerp}@dbs.ifi.lmu.de, marco.poetke@sdm.de

Abstract. With fast evolving resources for 3D objects such as the Protein Data Bank (PDB) or the World Wide Web, new techniques, so-called *similarity models* to efficiently and effectively search for these 3D objects become indispensible. Invariances w.r.t. specific geometric transformations such as scaling, translation, and rotation are important features of similarity models. In this paper, we focus on rotation invariance. We first propose a new method of representing objects more accurately in the context of rotation invariance than the well-known voxelization technique.In addition, we extend existing feature-based similarity models by proposing a new spherical partitioning of the data objects based on proportionality and redundancy[1], and generalizing an existing method for feature extraction. A broad experimental evaluation compares our method with existing methods in terms of accuracy and efficiency. In particular, we experimentally confirm that our point sampling method is better suited to represent 3D objects in the context of rotation invariance than voxelized representations. In addition, we empirically show that our new similarity model significantly outperfoms competitive rotation invariant models in terms of accuracy as well as efficiency.

1 Introduction

During the last years, more and more 3D models became available, e.g. through the fast-growing protein database PDB [1] or through the World Wide Web [2]. This trend will probably continue and thus, new techniques are required to efficiently and effectively search within such 3D databases.

This paper covers feature-based methods to describe 3D objects, so called *shape descriptors*. Shape descriptors extract numerical features of a 3D object so that the object is mapped to a metric space called *feature space*. The similarity of two spatial objects is then measured by the proximity of their feature vectors.

In this paper, we are particularly interested in shape descriptors that are invariant with respect to rotation, translation, and scaling.

We will base our work on an existing shape descriptor (the *volume model*[3]) by applying a combination of new techniques. The original descriptor is already

---

[1] Patent pending.

C. Bauzer Medeiros et al. (Eds.): SSTD 2005, LNCS 3633, pp. 200–217, 2005.

invariant w.r.t. the mentioned transformations because it uses a spherical partitioning of data objects after moving the balance point of an object to the origin of the coordinate system. For a given object, the model extracts the volume of the object from each partition as feature.

The retrieval quality of this very intuitive but simple model can be significantly enhanced by our newly proposed techniques. The first improvement we propose in this paper is a new representation of 3D objects. Instead of the commonly used voxel representation, we use uniformly distributed surface points.

Our second enhancement is a new way of decomposing an object into several partitions. Instead of constructing equidistant shells like in [3], our method creates partitions that are dynamically adapted to the specific shape of an object. In addition, we propose a new technique for the assignment of parts of an object to different partitions. Instead of assigning each part to exactly one partition, our new *redundant assignment* method associates parts of an object to several overlapping partitions.

A third enhancement we are proposing is to use an advanced feature extraction method instead of the rotation invariant but rather simple feature extraction based on the volume of the object in a partition. In particular, we show how the *eigenvalue model* which is presented in [4] for voxelized data working with cubic partitionings, can be generalized for arbitrarily shaped partitionings. We further adopt this generalization to our new object partitioning method.

We empirically show how these newly introduced techniques can be combined, to significantly improve quality of similarity search on 3D models.

The rest of the paper is organized as follows. We review related work on rotation invariant shape descriptors in Section 2. We then discuss the limitations of voxel representation in the context of rotation invariance and present our solution to this problem based on point sampling in Section 3. Section 4 presents our new object partitioning methods and Section 5 describes the generalization of the eigenvalue model. Section 6 summerizes our new method to compute rotation invariant shape descriptors. The experimental evaluation of our methods is presented in Section 7. Section 8 provides conclusions.

2   Related Work

Various methods for the characterization of 3D shapes have been reported. In this section we focus on techniques that result in a rotation invariant representation of an object without requiring a normalization step.

In [5] Osada et al. presented a technique called "D2" to describe 3D shapes. At first the distances of pairs of randomly selected points on the surface of an object are measured. These distances are used to create a histogram that is finally used as a feature vector. This method is intended to distinguish large differences in geometric shape. It is not able to discriminate between objects that differ only in small details. In addition, a high number of pairwise distances has to be calculated to obtain a robust description of the object.

The idea of the D2 descriptor was refined in [6]. The authors not only calculated the distance between two randomly distributed surface points, but also classified it according to the position of the two points. The line between two points either can lie completely inside the object, completely outside of the object, or inside and outside of the object. This separation yields three histograms instead of one. The presented results show that very similar shape distributions are still derived for dissimilar parts. The complexity of the method is the same as above while the time that has to be spent for each pair of points is even longer since the connecting line has to be classified.

In [7], Kazhdan et al. presented a rotation invariant shape descriptor by applying the spherical Fourier analysis to a number of spherical functions defined on a voxel grid. The concentric spheres constructed around the center of the voxel grid are used to define corresponding spherical functions.

In [8] the authors describe a shape descriptor based on the work in [7]. A spherical function is defined by measuring the distance between the surface of an object and its balance point. This function is afterwards analyzed with the spherical Fourier transform and the Fourier coefficients are used to characterize the object.

As the computation of the Fourier coefficients by means of the SFT algorithm requires a lot of function values we decided to use the method in [7] for our experimental comparisons because it is easier to determine whether or not a voxel is filled than to intersect a large number of rays with a triangle mesh.

Hilaga et al. in [9] introduced a technique to characterize 3D shapes based on so-called Reeb graphs. The better these graphs match, the higher is the similarity between the corresponding objects. However, this method compares the topology of objects rather than their geometry.

An intuitive and rotation invariant technique to describe the 3D shapes of proteins was presented by Ankerst et al. in [3]. Since our method uses a similar technique for object partitioning, we will present details on this method in Section 4. The method also will be included in our experimental comparisons.

3   Representation of 3D Objects

In this section, we discuss how 3D objects can be represented in order to efficiently derive spatial features. Often, 3D objects are given by triangle meshes, i.e. sets of connected triangles. The algorithm in [10] for example calculates a triangulated surface of a protein. Although well suited for the graphical display, this representation is still too complex for the efficient computation of feature vectors. In the following, we discuss voxelization as a method to represent 3D objects and its problems in terms of rotation invariance. Thereafter, we propose a new representation method called *point sampling* that overcomes the shortcomings of voxelization regarding rotation.

3.1   Voxelization

A well-known method to represent 3D data for extracting spatial features is voxelization. For voxelization, an object is usually placed into a standardized cube.

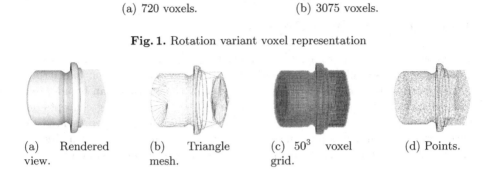

(a) 720 voxels.   (b) 3075 voxels.

Fig. 1. Rotation variant voxel representation

(a)   Rendered   (b)   Triangle   (c) 50^3 voxel   (d) Points.
view.   mesh.   grid.

Fig. 2. Different representations of a 3D object

This cube is partitioned into n^3 small cubes (so-called voxels) using an equidistant grid. A voxel is considered as filled if it intersects a triangle of the triangle mesh. The algorithm proposed by Kaufman in [11] yields a conservative approximation of the object surface by means of voxels. A further advantage of this representation is the possibility to model filled objects by adding interior voxels.

However, this representation is not invariant with respect to rotations, i.e. the number of voxels changes when the object is rotated. This effect is illustrated in Figure 1. As we will demonstrate experimentally, this rotation variant representation of objects reduces the quality of shape descriptors working with voxel input.

3.2   Point Sampling

In this section we introduce another approach to represent a given mesh of triangles. Let $\mathcal{M} = \{T_1, \ldots, T_m\}$ be the set of triangles the mesh consists of. For $1 \leq i \leq m$ let A_i denote the area of the triangle T_i and let $P_i, Q_i, R_i \in \mathbb{R}^3$ be the three vertices of T_i. Then in a first step the total area $A = \sum_{i=1}^{m} A_i$ of all triangles is calculated and the desired number $n \in \mathbb{N}$ of points to be sampled is chosen. Afterwards, the following steps are repeated n times:

Selection of a Triangle. With probability $p = \frac{A_i}{A}$ triangle T_i is selected. This selection is implemented by a preprocessing step assigning to each triangle the sum of the areas of all its predecessors plus its own area, i.e. $A_{sum_i} = \sum_{j=1}^{i} A_j$. Afterwards, a random number $r \in \mathbb{R}, 0 \leq r \leq A$ is created and the triangle T_j is selected so that $A_{sum_j} \geq r$ and $\not\exists i : 1 \leq i < j : A_{sum_i} \geq r$.

Sampling of a Random Point. According to [12] a random point is created inside triangle T_j.

In contrast to the voxelization method this method is not able to model filled volumes. Nevertheless, our experiments show an improved quality when using shape descriptors working on models that have not been normalized before the shape descriptor is applied.

Figure 2 shows different representations of a 3D object.

4 Partitioning Data Objects

The general idea of our rotation invariant shape descriptors is to partition an object into shells similar to the method proposed in [3] and then extracting spatial features from each partition. We show how this method can be adopted and enhanced in terms of acurracy. Let us note that the following concepts do not rely on a certain kind of object representation, i.e. objects represented as voxels or using sampled points. In the following, we assume that the data objects are represented as points, either center points of voxels or sampled points.

4.1 Construction of the Shells

The shell model to partition the data was originally proposed in [3]. After the computation of the balance point M (cf. Section 5) of a given object O and the radius of the bounding sphere around O the space enclosed by the bounding sphere is divided up into $k \in \mathbb{N}$ shells.

Equidistant Shell Construction. The original approach in [3] constructs k shells whose distance to each other is the same for each pair of adjacent shells. So we will refer to this method as the *equidistant* construction method. Although the innermost shell actually is a sphere, we nonetheless will denote it as a shell. Let $r_{BS} \in \mathbb{R}$ be the radius of the bounding sphere. Then the thickness δ of each shell,i.e. the radius of the inner sphere, can be computed by $\delta = r_{BS}/k$.

For each $i \in \{1, \ldots, k\}$ the shell S_i is characterized by its inner radius r_{i-1} and its outer radius r_i where $r_i = \delta \cdot i$.

This method is illustrated in Figure 3. An object is inscribed into its minimal bounding sphere and the bounding sphere is afterwards partitioned into $k = 3$ equidistant shells. This method corresponds to the calculation of so-called *shape histograms* [3].

(a) Original object.   (b) Partitioned bounding sphere.   (c) Equidistant shell construction.   (d) Proportional shell construction.

Fig. 3. Shell construction

Proportional Shell Construction.[1] The shell construction method of [3] described above keeps the distance between adjacent shells constant. However, this method does not partition the bounding sphere into areas of equal complexity and thus of equal interest. Shells of very low complexity (i.e. with few points) are most likely of very low interest. The granularity of the partitioning is too fine to extract meaningful spatial features in that case. In the worst case, the equidistant construction method may result in empty partitions that are of no interest. On the other hand, other shells contain many more points. In this case, the granularity of the partitioning may be too coarse to extract meaningful spatial features. The solution we are proposing in the following is to apply a proportional shell construction such that each partitioning has the same level of complexity and interest.

Let $O = \{p_1, \ldots, p_n\}$ be an object represented by a set of points (either representing voxels or sampled points as discussed above). Let M be the balance point of O, let r_{BS} be the radius of the bounding sphere around O, and let $k \in \mathbb{N}$ be the desired number of shells to be constructed. Then the number $a \in \mathbb{N}$ of points to be placed into each shell is given by $a = \lfloor n/k \rfloor$.

Now the points of O are sorted in ascending order with respect to the Euclidean distance d_{eucl} of its elements to M such that O can be rewritten as $O = \{p_{s(1)}, \ldots, p_{s(n)}\}$, where $d_{eucl}(M, p_{s(1)}) \leq d_{eucl}(M, p_{s(2)}) \leq \ldots \leq d_{eucl}(M, p_{s(n)})$.

Let $k \neq 1$. Then the shell S_i, $i \in \{1, \ldots, k-1\}$, is characterized by its inner radius r_{i-1} and its outer radius r_i where $r_i = d_{eucl}(M, p_{s(i \cdot a)})$.

The outer radius r_k of the outmost shell S_k is set to r_{BS}. In case of objects represented using point sampling, this radius is not necessarily equal to $d_{eucl}(M, p_{s(n)})$. This is due to the random process with which the sampled points of O have been created. The inner radius r_{k-1} of S_k is calculated as described above.

The radii of the shells are important for our method of feature extraction from the partitions which will be presented in Section 5.

If $n \neq k \cdot a$, the remaining elements of O $(p_{s(k \cdot a+1)}, \ldots, p_{s(n)})$ are assigned to the outmost shell.

Obviously, the proportional shell construction method partitions the bounding sphere into areas of equal complexity and thus of equal interest. The granularity of the partitioning adopts to the shape of the data objects. The method is illustrated in Figure 3(d). While the equidistant shell construction method corresponds to the calculation of shape histograms, the proportional method generates *shape quantiles*.

4.2   Assignment to Shells

After the space enclosed by the bounding sphere has been partitioned, i.e. after the shells have been constructed in an equidistant or proportional way, points of $O = \{v_1, \ldots, v_n\}$ (either representing voxels or sampled points) have to be assigned to these partitions.

Let M be the balance point of O and let the bounding sphere be segmented into shells $S_1, \ldots, S_k$, $k \in \mathbb{N}$. Each shell S_i is characterized by its inner radius r_{i-1} and its outer radius r_i, where $r_0 := 0$.

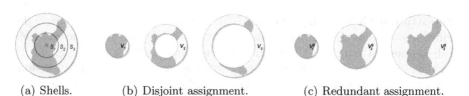

(a) Shells.   (b) Disjoint assignment.   (c) Redundant assignment.

Fig. 4. Disjoint vs. redundant assignment. Cross section of the bounding sphere

Disjoint Assignment. In [3] a disjoint assignment is proposed. For each shell S_i, a set V_i is created. An element v_j of O is assigned to V_i if and only if $r_{i-1} < d_{eucl}(M, v_j) \leq r_i$.

As every element of O is assigned to exactly one shell, this assignment results in a disjoint decomposition of O into $V_1, \ldots, V_k$. Each set corresponds to the space enclosed by a certain shell. This method is illustrated in Figure 4(b).

Redundant Assignment.[1] The disjoint assignment proposed in [3] has a severe limitation. If two (or more) shells (and their associated elements) are rotated against each other, the resulting feature vector will be exactly the same as the one derived from the unrotated elements. Although the so changed set of points may represent a totally different object, this object will be regarded as very similar to the one described by the unrotated points.

To overcome this weakness we introduce a new method for the assignment of elements to spherical partitions. This method can be applied to both, equidistant and proportional shells.

A set V_i^S is created for each shell S_i. An element v_j of O is assigned to V_i^S if and only if $d_{eucl}(M, v_j) \leq r_i$.

To illustrate the method, imagine $x \in O, r_{i-1} < d_{eucl}(M, x) \leq r_i$, i.e. x lies inside the shell S_i. The disjoint assignment method would assign x to only one specific set V_i, whereas the redundant assignment method assigns x to the sets $V_i^S, \ldots, V_k^S$.

It is obvious that rotating a shell against other shells now results in a different feature vector. If for example shell S_i is rotated, the sets $V_i^S, \ldots, V_k^S$ will yield different features.

The set V_i^S corresponds to the space enclosed by a sphere with radius r_i that is centered at M. V_k^S represents the space enclosed by the bounding sphere and thus $V_k^S = O$. In Figure 4(c), an example of the partitions resulting from the redundant assignment method is depicted.

4.3   Problems with Thin Shells

No matter how the elements are assigned to the spherical partitions, a problem can occur when the underlying shells are constructed proportionally. If all or almost all points representing an object have the same distance to their balance point, the constructed shells will be very thin.

Let object O be a perfectly triangulated globe, i.e. the triangle mesh of O consists of infinitesimally small triangles and thus the points that are sampled on the surface of O all have the same Euclidean distance to their balance point

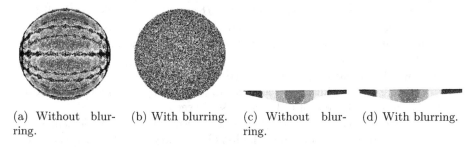

(a) Without blurring. (b) With blurring. (c) Without blurring. (d) With blurring.

Fig. 5. Impact of blurring on (a),(b) thin shells and (c),(d) thick shells

M, the center of the globe. Sorting the points according to their distance to M will then result in a list whose ordering is in the best case random and in the worst case determined by the creation time of a certain point. The elements of a certain partition will therefore not reflect the globular shape of O. It is more likely that a partition only describes a small (non-globular) area of O.

In reality, due to a triangulation process that is far from being perfect, not all sampled points representing O will have the same Euclidean distance to M. Nonetheless the constructed shells will be very thin and the distribution of points in the partitions will most probably not reflect the globular distribution of the complete set of sampled points. The points assigned to a certain shell will more likely reflect geometric deviations from the ideal surface of the globe due to a non-perfect triangulation of O.

An example of this problem is illustrated in Figure 5(a). The points are marked with four different colors. The colors correspond to the disjoint assignment of the sampled points to different shells. The shape of the object is globular and thus the mentioned problem becomes visible. Instead of being equally distributed, points of the same color are grouped together in certain areas.

A possible solution to this problem is to shuffle the points. Therefore, we introduce a value we call the *blur-distance* β. Let $O = (p_{s(1)}, \ldots, p_{s(n)})$ be represented by a list of sampled points. Let M be the balance point of O and let $d_{eucl}(M, p_{s(1)}) \leq \ldots \leq d_{eucl}(M, p_{s(n)})$, i.e. the points are sorted in ascending order with respect to the Euclidean distance to M. Thus the blur-distance β is defined by: $\beta = c \cdot d_{eucl}(M, p_{s(n)}), c \in [0, 1]$.

The blur-distance is specified as a portion of the maximum distance to M over all elements of O. Our experiments suggest that 0.01 is a good choice for c, so we will set $c = 0.01$ throughout the rest of the paper. The blur-distance specifies regions adjacent to borders between shells. The elements of these regions are afterwards shuffled with regard to their position in the sorted list O. The exact procedure is as follows:

1. Let $k \in \mathbb{N}$ be the number of shells to be constructed.
2. The number a of elements to be assigned to each shell is calculated. Then the set $B = \{s(a), s(2a), \ldots, s((k-1)a)\}$ contains the indices of the outmost elements (with respect to M) of the shells (except for the outmost shell). These elements mark the border between two adjacent shells.

3. Now we successively use each index $b \in B$ as a starting point in the list O. We proceed downwards in the list until we reach for the first time an index l such that $d_{eucl}(p_b, p_{s(l)}) > \beta$ or until we reach the innermost element of O. In this case, we set $l = 1$.

4. Afterwards, for each $b \in B$ we proceed upwards in the list O until we for the first time reach an index u, such that $d_{eucl}(p_b, p_{s(u)}) > \beta$ or until we reach the outermost element of O In this case we set $u = n$.

5. As a result, a pair of indices (l, u) was determined for each element of B. Each pair specifies a portion of the list O that now is shuffled (the elements from $p_{s(l+1)}$ to $p_{s(u-1)}$). If two or more of these regions overlap, they are all shuffled at once such that an element of one part of O can be swapped with an element of another part.

This shuffling ensures that the distribution of points of a very thin shell describes all parts of an object that lie in adjacent thin shells. In Figure 5(b) this effect is demonstrated with the globular object mentioned above. Now all shells reflect the spherical shape of the original object. In Figure 5(c) we depicted an object partitioned into thick shells. The clearly defined borders become only slightly blurred when applying the method described above (cf. Figure 5(d)).

5   Extracting Spatial Features

Having partitioned the objects into shells, we have to extract spatial features to build a feature vector. In [3], the volume, i.e. the number of representing points, of the object in a given partition is extracted as feature. The resulting feature vector consists of one volume measurement in each of the k shells. We refer to this method of extracting spatial features as the *volume model*.

In [4] the eigenvalue model has been introduced which clearly outperfoms the volume model in terms of effectiveness. Thus, we argue that it is more accurate to use this eigenvalue model to extract spatial features. The eigenvalue model was originally introduced in [4] but was only applied to a cubic partitioning of a voxelized object. The method is based on PCA, the principal component analysis [13], to analyze a given set of data and extract shape descriptors from a given partitioning. In the following, we describe the basic idea of the eigenvalue model, and generalize a procedure for obtaining feature vectors from the data of a given partition. The sets to be analyzed are the sets of 3D points that lie inside a certain partition.

The eigenvalue model. The basic idea of the eigenvalue model is to obtain the eigenvalues of a set of data within a cubic partitioning as shape descriptors [4]. Originally, it was applied to voxelized data, however, it is quite simple to apply it on objects being represented by sampled points.

Let $V = \{v_1, \ldots, v_n\}$ be a set of 3D points, i.e. $v_i \in \mathbb{R}^3, 1 \le i \le n$, where $v_i = (v_{i_1}, v_{i_2}, v_{i_3})^{\mathrm{T}}$. If the object is voxelized, the v_i represent the center of voxel v_i, whereas if the object is represented using point sampling, v_i is simply one of the sampled points.

Each $v_i \in V$ is translated such that M afterwards coincides with the origin and afterwards the covariance matrix $\mathbf{C}$ for V is computed as follows:

$$\mathbf{C} = \frac{1}{|V|-1} \sum_{j=1}^{n} (v_j - M) \cdot (v_j - M)^T, \text{ where } M = \frac{1}{n} \sum_{j=1}^{n} v_j \text{ is the balance point.}$$

The covariance matrix can be decomposed as $\mathbf{C} = \mathbf{V}\mathbf{E}\mathbf{V}^T$, where $\mathbf{V}$ is an orthonormal matrix containing the eigenvectors of $\mathbf{C}$ and $\mathbf{E}$ is a diagonal matrix containing the eigenvalues of $\mathbf{C}$. The eigenvectors are called *principal axes* of V. They describe the three orthogonal axes where the scattering of the elements is greatest. The eigenvalues describe the variance along the three principal axes and thus can be used to characterize the shape of the elements of V. As stated above, in [4], the authors apply this idea to voxelized data. The data objects are partitioned into axis-parallel units. From each unit, the eigenvalues of the voxels (represented as vectors of their center point) are obtained as shape descriptors. Let us note, that this method is not rotation invariant due to the cubic partitioning of the data. The resulting feature vector consists of $3k$ values since for each partition, 3 features (eigenvalues) are extracted.

Generalization and Adoption to Spherical Partitionings. The eigenvalue model as proposed in [4] is applicable on cubic partitionings only and thus is not rotation invariant. In the following, we present a way to normalize the calculated eigenvalues such that eigenvalues of different partitions or even different similarity models can be compared to each other or can be combined with each other. In particular, we show how this model can be applied to spherical partitions like the shells constructed in Section 4. The idea is to express each eigenvalue as a portion of the maximum possible eigenvalue.

Lemma 1. *Let V be a 3D space and let d_{max} be the maximum Euclidean distance two points in V can be apart from each other. The maximum variance Var_{max} that can occur in V is given by $Var_{max} = \frac{1}{2}d_{max}^2$.*

Proof. The highest variance in a certain direction is given by two points that have the highest possible distance from each other and that lie on a line indicating the specific direction. Let therefore $T = \{(a_1, a_2, a_3), (b_1, b_2, b_3)\} =: \{a, b\}$ be a set with two points in a 3D space V and let $d_{eucl}(a, b)$ be the largest Euclidean distance two elements in V can be apart from each other.

The mean value M of T is equal to $(\frac{a_1+b_1}{2}, \frac{a_2+b_2}{2}, \frac{a_3+b_3}{2})$. The variance of T is then given by:

$$
\begin{aligned}
Var(T) &= \frac{\sum_{t \in T}(d_{eucl}(t, M))^2}{|T| - 1} \\
&= \frac{(a_1 - \frac{a_1+b_1}{2})^2 + (a_2 - \frac{a_2+b_2}{2})^2 + (a_3 - \frac{a_3+b_3}{2})^2}{1} \\
&\quad + \frac{(b_1 - \frac{a_1+b_1}{2})^2 + (b_2 - \frac{a_2+b_2}{2})^2 + (b_3 - \frac{a_3+b_3}{2})^2}{1} \\
&= \frac{1}{2}((a_1 - b_1)^2 + (a_2 - b_2)^2 + (a_3 - b_3)^2) = \frac{1}{2}(d_{eucl}(a, b))^2
\end{aligned}
$$

Lemma 1 states, that in order to normalize the calculated eigenvalues of V, it is sufficient to determine the largest possible distance d_{max} that can occur in V. Using d_{max}, the maximum possible variance, i.e. the maximum possible eigenvalue, can be computed as indicated in Lemma 1. For example, the maximum distance d_{max} can be the diameter of a shell or the diagonal of a cube. An eigenvalue λ is normalized by the mapping $\lambda \mapsto \lambda/Var_{max}$.

In the following, if we speak of the eigenvalue model, we mean the generalized version of the model, i.e. we will assume that the normalized eigenvalues are obtained as shape descriptors.

The eigenvalue model can now be applied to spherical partitionings such as the shells constructed in Section 4. In both cases of a disjoint or a redundant assignment of points to the shells, the maximum Euclidean distance d_{max} of two elements within a shell S_i or S_i^S, respectively, can be determined by $2r_i$. The maximum variance, i.e. the maximum eigenvalue for scaling, can then be computed according to Lemma 1.

6   Computing Invariant Shape Descriptors

Now we can summarize our proposed methods of computing shape descriptors and discuss whether they are invariant w.r.t. several transformations, including translation, rotation, reflection, and scaling.

The general technique of how invariance with respect to translation, rotation, reflection, and scaling is achieved is the same for all descriptors. Let O be the object whose feature vector is to be calculated. The following steps depend on the type of representation.

Voxel Representation: Let $O = \{v_1, \ldots, v_n\}$, $v_i \in \mathbb{R}^3$ be represented by a set of voxels (cf. Section 3.1). Then the balance point $M = \frac{1}{n} \sum_{i=1}^{n} v_i$ of O is computed. Afterwards, O is moved so that M coincides with the origin. Finally, the minimum bounding sphere centered at M is constructed. The radius r_{BS} of the bounding sphere is determined by $r_{BS} = \max_{v_i \in O}\{d_{eucl}(M, v_i)\}$.

Sampled Points Representation: Let $O = \{p_1, \ldots, p_n\}$, $v_i \in \mathbb{R}^3$ be represented by a set of sampled points. In principle the method for the calculation of the center and the radius of the bounding sphere described above can be applied to any set of 3D coordinates. But with regard to the random process, the sampled point set was created with (cf. Section 3.2), we decided to calculate these values with the help of the original triangle mesh. The triangle vertices used to calculate the balance point are weighted by the area of the corresponding triangle. This procedure was proposed in [14]. To determine the radius of the bounding sphere we again use the vertices of the triangles rather than the set of interpolated points. This is due to a problem that may occur when using the sampled points. Consider a very small but long, needle-shaped triangle with one vertex being the most remote point with respect to the balance point. Due to its tininess only one point might be sampled into the triangle. Thus the radius of the bounding sphere may vary significantly depending on the position of this point. Using the vertices of the triangles will result in a radius that is invariant with respect to the randomly generated points.

After these steps, invariance with respect to *translation* is achieved as the balance point of O has been moved to the origin.

Our method partitions the space enclosed by the bounding sphere of a given object into several shells centered at the balance point of the object. In Section 4, we presented two methods to construct these shells, in particular equidistant partitioning as proposed in [3] and proportional partitioning. In addition, we presented two methods to assign the points representing a given object to the shells. The first method uses a disjoint assignment of the points to shells (cf. [3]) whereas the second method uses a redundant assignment.

Afterwards, we apply the eigenvalue model [4] to the elements (voxels or sampled points) of each shell. We presented a generalization of this model in Section 5. As the variance of the elements inside a certain shell does not change even if the object is rotated the resulting eigenvalues are invariant with respect to *rotation*. At the same time, invariance with respect to *reflection* is achieved. The amount of scattering of the elements of a certain shell remains unchanged, if the object is reflected. Finally, the eigenvalues derived for the single partitions are scaled (cf. Section 5) and thus, invariance with respect to *scaling* is achieved.

7   Experiments

In this section, we present our experimental evaluation of the proposed methods. In particular, we will first evaluate the usability of the sampled points representation in comparison to a voxelized representation in the context of rotation invariance. Next, we confirm the superiority of our adoption of the eigenvalue model over the volume model. Last but not least, we evaluate the performance of the different shell construction techniques and point assignment strategies described in Section 4 and compare the best shape descriptor with existing work on rotation invariant similarity models.

7.1   Data Sets

For our experiments we used two different real-world data sets that are described in the following.

Princeton Shape Benchmark. The models of this set originate from the Princeton Shape Benchmark Set [2] consisting of 1814 models collected from the World Wide Web. Along with the models, a hierarchical classification is provided that can be used to evaluate the quality of different shape descriptors with precision/recall plots. We decided to only regard the leaves of the classification system. Thus, the set is partitioned into 161 disjoint classes. We will refer to this set of objects as the "PSB set".

Proteins. A huge amount of 3D protein structure data is available at the online repository of the Protein Data Bank (PDB) [1]. We used the MSMS program [10]

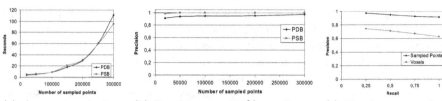

(a) Average real time for feature extraction per model.

(b) Precision at 100% recall vs. number of sampled points.

(c) Voxel vs. sampled points representation.

Fig. 6. Evaluation of sample points vs. voxel representation

to calculate the solvent excluded surface [15] of each molecule. Afterwards the surface was triangulated and so all experiments could be performed on the same type of data. For the classification of the proteins we used the FSSP (Families of Structurally Similar Proteins) classification [16], a well-known classification system for proteins. To create classes of approximately the same size without selecting too many proteins for the test set, we scanned the PDB for classes whose size ranged from 50 to 100 members. This resulted in a set of 3279 proteins in 47 classes which we will refer to as the "PDB set".

We rotated all models in both sets around a random axis by a randomly determined angle in a preprocessing step to ensure a non-canonical orientation. All experiments were run on an Intel Pentium 4 processor featuring 2.53 GHz and 1 GB RAM.

7.2   Sampled Points vs. Voxel Representation

Our first experiments evaluated the representation of 3D objects. In Section 3 we discussed the limitation of voxelization in the context of rotation invariance and proposed the technique of point sampling to overcome this limitation. To evaluate the usability of voxelization and point sampling for rotation invariance, we randomly selected 100 models from the PDB set and duplicated each selected model four times. In addition, we rotated each copy randomly. Finally the resulting 500 models together with the rest of the PDB set were mapped to a feature space using the eigenvalue model on 8 disjoint and equidistant shells. We applied the same preprocessing to the PSB set.

First we analyzed the influence of the number of sampled points on retrieval quality and computational cost. The task was for each of the 500 preprocessed models to retrieve the 4 corresponding (arbitrarily rotated) models. The experiment was performed for different numbers of sampled points. We measured the average time needed for the feature extraction for one model. The results shown in Figure 6(a) and Figure 6(b) suggest a sensible trade-off between retrieval quality and computational cost is a number of 50,000 sampled points.

We then compared the accuracy of both the sampled points representation and the voxel representation, using precision/recall plots. To confirm our assumption that the representation by voxels is inferior to the sampled points

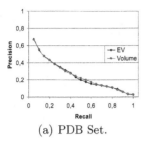 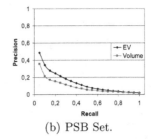

(a) PDB Set.   (b) PSB Set.

Fig. 7. Volume model vs. eigenvalue model

representation we repeated the above described experiments for the 500 duplicated models in both data sets. The models were mapped to a feature space
using 20,000 sampled points, and for the 4 possible recall values the average precision was determined. Then the objects of both data sets were represented by a
60^3 voxel grid. The resulting precision/recall plots (cf. Figure 6(c)) confirm our
presumption. Although we used a relatively low number of sampled points and
a relatively high voxel resolution, the sampled points representation generated
significantly higher precision values.

Thus, throughout the rest of our experiments we used the sampled points
representation for the 3D objects. Due to our experiments presented above, we
set the number of sampled points to 50,000.

7.3   Volume Model vs. Eigenvalue Model

The following experiments evaluate the applicability of our generalization of the
eigenvalue model (cf. Section 5). Therefore, we compared the volume model as
proposed in [3] with our generalized eigenvalue model. Both models are based
on an equidistant shell construction with disjoint point assignment. In case of
the eigenvalue model, we used 10 shells whereas in case of the volume model we
used the parameter setting that performed best according to [3], i.e. 120 shells.

Figure 7 illustrates the result of the comparison. The precision/recall plot of
the volume model is labelled with "Volume" and the results for the eigenvalue
model are marked with "EV".

While the two models yield almost the same results on the PDB set (cf.
Figure 7(a)), the eigenvalue model yields significantly better results on the PSB
set (cf. Figure 7(b)). These results underline that the eigenvalue model is more
accurate than the volume model. Let us note that in case of the volume model
we need 12 times more partitions than using the eigenvalue model resulting in
significantly higher waste of resources for the volume model.

Thus, we use the eigenvalue model throughout the rest of the experiments as
the method for extracting spatial features.

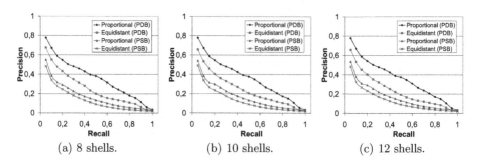

(a) 8 shells.   (b) 10 shells.   (c) 12 shells.

Fig. 8. Proportional vs. equidistant shell construction

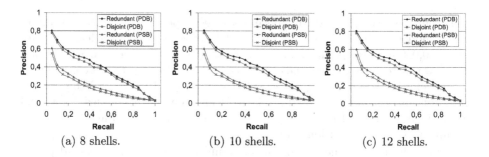

(a) 8 shells.   (b) 10 shells.   (c) 12 shells.

Fig. 9. Redundant vs. disjoint assignment

7.4 Shell Construction and Assignment to the Shells

In this section, we evaluated the different methods of shell construction and shell assignment presented in Section 4.

First we analyzed the impact of the different methods for the shell construction. Figure 8 shows that for both data sets and for different numbers of constructed shells the proportional shell construction significantly outperforms the equidistant method. Let us note that the presented results were achieved with a disjoint assignment step. The same effect occurs when assigning the sampled points redundantly (results are not shown here due to space limitations).

A further significant improvement in the quality of 3D model retrieval can be observed when using the redundant assignment to the shells. This is shown for different numbers of proportional shells and for both data sets in Figure 9. We empirically verified that this is also true for all considered numbers of equidistant shells (results are not shown here due to space limitations).

7.5 Comparison with Existing Approaches

Last but not least, we compared the best of our proposed shape descriptors with existing rotation invariant methods. In particular, we compared the accuracy of proportional and redundant shells and equidistant and disjoint shells (both

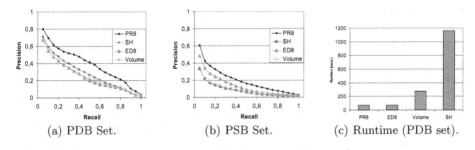

Fig. 10. Results for different shape descriptors

combined with the eigenvalue model) with the work presented in [7] and the method proposed in [3]. The latter model corresponds to 120 equidistant and disjoint shells combined with the volume model.

In Figure 10, we show the resulting precision/recall plots for the competing shape descriptors. The plot of the method of [3] is marked with "Volume", the result of our generalized eigenvalue model applied to 8 equidistant and disjoint shells is labelled with "ED8", and the results of our generalized eigenvalue model applied to our 8 proportional and redundant shells are labelled with "PR8". We furthermore implemented the shape descriptor based on spherical harmonics (marked with "SH" in Figure 10) as described in [7]. The dimensionality of "Volume" is 120, the dimensionality of "ED8" and "PR8" is $3 \cdot 8 = 24$, and the dimensionality of "SH" is 512. As can be seen from Figure 10, the combination of the newly introduced proportional shell construction and the newly introduced redundant assignement in combination with the generalized eigenvalue model and the point sampling representation leads to a significantly higher retrieval quality.

We also determined the efficiency of the competitive methods by measuring the time that was necessary to create a precision/recall plot for the PDB data set. We started the measurement after all features were loaded into main memory. As we employed no index structures, the measured time depends only on the dimensionality of the feature space.

The results depicted in Figure 10(c) show that the presented combination of proportional shell construction and redundant assignment with the generalized eigenvalue model does not outperform the other methods only in terms of accuracy, but also in terms of efficiency. The reason for this performance is the comparatively low dimensionality of the resulting feature space. This is also an advantage if a spatial index structure is applied for further speeding-up similarity queries since the performance of spatial index structures usually detoriates with increasing data dimensionality.

In summary, we can observe that our newly proposed method of proportional shell construction and redundant shell assignment in combination with our generalized eigenvalue model and our new object representation based on point sampling significantly outperfoms existing rotation invariant approaches in terms of accuracy and efficiency.

8   Conclusions

In this paper, we presented different similarity models featuring many important invariances with a focus on rotation invariance. In particular, we discussed the limitation of voxel representations (a well-known standard technique to represent 3D objects) in the context of rotation invariance and proposed a solution to this limitation based on point sampling. Furthermore, we introduced a new similarity model that is based on two key ingredients: (1) a new partitioning of the data objects that extends spherical partitionings using the ideas of redundancy and proportionality; (2) a generalization of the existing eigenvalue model and an adoption of this generalization to the newly introduced sherical partitionings.

Our broad experimental evaluation shows that the sampled point representation is better suited in the context of rotation invariance than the well-known voxel representation. In addition, we showed that our new similarity model using proportional shells with redundant point assignment as object partitioning method and the adoption of the eigenvalue model as feature extraction technique clearly outperfoms existing rotation invariant models in terms of accuracy as well as efficiency.

References

1. Berman, H., Westbrook, J., Feng, Z., Gilliland, G., Bhat, T., Weissig, H., Shindyalov, I., Bourne, P.: The Protein Data Bank. Nucleic Acids Research **28** (2000) 235–242
2. Shilane, P., Min, P., Kazhdan, M., Funkhouser, T.: The Princeton Shape Benchmark. In: Shape Modeling International, Genova, Italy. (2004)
3. Ankerst, M., Kastenmüller, G., Kriegel, H.P., Seidl, T.: 3D Shape Histograms for Similarity Search and Classification in Spatial Databases. In: Proc. 6th International Symposium on Spatial Databases. (1999) 207–226
4. Kriegel, H.P., Kröger, P., Mashael, Z., Pfeifle, M., Pötke, M., Seidl, T.: Effective Similarity Search on Voxelized CAD Objects. In: Proc. 8th DASFAA. (2003) 27–36
5. Osada, R., Funkhouser, T., Chazelle, B., Dobkin, D.: Matching 3D Models with Shape Distributions. In: International Conference on Shape Modeling and Applications. (2001) 154–166
6. Ip, C., Lapadat, D., Sieger, L., Regli, W.: Using Shape Distributions to Compare Solid Models. In: 7th ACM/SIGGRAPH Symposium on Solid Modeling and Applications. (2002) 273–280
7. Kazhdan, M., Funkhouser, T., Rusinkiewicz, S.: Rotation Invariant Spherical Harmonic Representation of 3D Shape Descriptors. In: Symposium on Geometry Processing. (2003) 167–175
8. Vranić, D.V.: An Improvement of Rotation Invariant 3D-Shape Descriptor Based on Functions on Concentric Spheres. In: Proc. IEEE International Conference on Image Processing. (2003) 757– 760
9. Hilaga, M., Shinagawa, Y., Kohmura, T., Kunii, T.L.: Topology Matching for Fully Automatic Similarity Estimation of 3D Shapes. In: Proc. ACM SIGGRAPH 2001. (2001) 203–212

10. Sanner, M., Olson, A., Spehner, J.C.: Fast and Robust Computation of Molecular Surfaces. In: Proc. 11th ACM Symposium on Computational Geometry. (1995) C6–C7
11. Kaufmann, A.: An Algorithm for 3D Scan-Conversion of Polygons. In: Proc. Eurographics. (1987) 197–208
12. Glassner, A., ed. In: Graphics Gems. Academic Press (1990) 24–28
13. Jolliffe, I.: Principal Component Analysis. Springer (1986)
14. Vranić, D.V., Saupe, D.: 3D Model Retrieval. In: Proc. SCCG2000, Budmerice, Slovakia. (2000) 89–93
15. Sanner, M.: Modeling and Applications of Molecular Surfaces. PhD thesis, University of Haute-Alsace, France (1992)
16. Holm, L., Sander, C.: Mapping the Protein Universe. Science **273** (1996) 595–602

Spatio-textual Indexing for
Geographical Search on the Web

Subodh Vaid[1], Christopher B. Jones[1], Hideo Joho[2], and Mark Sanderson[2]

[1] School of Computer Science, Cardiff University, UK
{c.b.jones, subodh.vaid},@cs.cf.ac.uk
[2] Department of Information Studies, University of Sheffield, UK
{h.joho, m.sanderson}@sheffield.ac.uk

Abstract. Many web documents refer to specific geographic localities and many people include geographic context in queries to web search engines. Standard web search engines treat the geographical terms in the same way as other terms. This can result in failure to find relevant documents that refer to the place of interest using alternative related names, such as those of included or nearby places. This can be overcome by associating text indexing with spatial indexing methods that exploit geo-tagging procedures to categorise documents with respect to geographic space. We describe three methods for spatio-textual indexing based on multiple spatially indexed text indexes, attaching spatial indexes to the document occurrences of a text index, and merging text index access results with results of access to a spatial index of documents. These schemes are compared experimentally with a conventional text index search engine, using a collection of geo-tagged web documents, and are shown to be able to compete in speed and storage performance with pure text indexing.

1 Introduction

The main focus of developments in spatial database design has been in support of the maintenance of highly structured map-based geometric data and their attributes. The World Wide Web introduces a challenge to spatial databases in that it consists of a vast repository of largely unstructured information that is dominantly in the form of text documents. A large amount of information on the web is geographically specific, in the sense that it refers to particular geographical locations, but the geographic references are as a rule embedded within the textual content of the documents, in the form of place names, addresses, postcodes and the associated geographical terminology of spatial relationships. Users of the web often submit geographical enquiries requesting information about, for example, services relating to retailing, tourist attractions, accommodation, sport, entertainment, transport, public services and cultural heritage. In a study of a log of the Excite search engine, it was found that about one fifth of all queries were geographical, as determined by the presence of a geographical term such as a place name, a post code, a type of place or a directional qualifier such as north [18].

C. Bauzer Medeiros et al. (Eds.): SSTD 2005, LNCS 3633, pp. 218–235, 2005.

When a user submits a geographically-specific web search they usually use a place name to provide the geographic reference. This name will then be treated the same as the other search terms and documents containing the query terms will be retrieved. For purposes of geographic search, this approach has major limitations in that it will ignore potentially relevant documents that refer to the place of interest but do not include the specified place name. Thus relevant documents could refer to places that are inside or near the specified place or they could use an alternative version of the specified place name. It is also the case that there are many places in different locations sharing the same name, resulting in the return of irrelevant documents. Another problem with using place names for geographic search with conventional search engines is that place names are commonly used in the names of organisations, people and buildings, resulting in the retrieval of documents that may have no geographical relevance despite the inclusion of the place name. In theory, the limitations above that result in missing relevant documents can be overcome by creating an expanded list of query terms. The expanded list could include alternative names and the names of places inside and nearby the target geographical location. In practice this would lead to the possibility of intractable query expressions containing many thousands of geographical terms. This would occur for target places that were spatially extensive to the extent that they contained many other named places. The approach would also inevitably result in the return of irrelevant documents that used the target place names to refer to the names of organizations, people or other phenomena for which the name does not provide geographical context.

There is a need therefore to develop geographically-aware web search technology that can index and retrieve effectively documents according to their geographical context. Indexing documents according to their geographical context would not only overcome the problems referred to above. It would also facilitate intelligent interpretation of spatial relationships that the user may employ to qualify the query place names. This includes terms such as *near, north of*, and *within 5 kilometers of*. Several working and experimental systems for geographical web search have appeared in recent years (some examples of which are reviewed below) but there is much work to be done to create effective systems. There are several important aspects of geographically-aware search that introduce challenges in their own right. Categorisation of web documents according their geographical content (geo-tagging) requires geo-parsing and geocoding procedures to detect and interpret geographical terminology in web documents and to "ground" (geocode) the resulting references with coordinates. This process of document categorisation requires a source of place name knowledge in the form of a gazetteer or geographical ontology that maintains information about place names in association with, for example, alternative names, geometric footprints that give coordinates for places, place types and the hierarchical structure of geographic space. Once documents have been categorised geographically they must be indexed with respect both to the textual content and to geographic space. Retrieval of documents must be accompanied by relevance ranking that needs to take account both of the geographical context and of the non-spatial concept terms that the user has employed in a query. Effective geographical search also requires a user interface that can help the user to disambiguate place names that refer to multiple places and to assist in formulating geographically-specific queries and reporting the results.

In this paper we focus on the issue of combining text indexing with spatial indexing. We present two spatio-textual indexing schemes which may be regarded as spatial-primary and text-primary respectively and compare them with each other and with using a pure text index in conjunction with a separate spatial index of documents and with a pure text index by itself. Experiments are conducted in the context of the SPIRIT prototype spatially-aware web search engine [20], using a collection of actual web documents. The performance of the various schemes are compared with respect to index costs and to query times and the numbers of documents retrieved. In the remainder of the paper we summarise briefly previous work on geographical web search, before providing an overview of the architecture of the SPIRIT system. In Section 4 we describe the indexing schemes that have been implemented for these experiments, before reporting on results of applying them using several types of query in Section 5. The paper concludes in Section 6 highlighting the relative merits of the implemented techniques and indicating future research directions.

2 Related Work

A geosearch tool from the Vicinity company was implemented in association with the Northern Light search engine [14], but no longer operates. It provided the facility to search for web documents on a specified topic relating to an address in the USA or Canada, allowing the user to specify a radius of search, The techniques employed were not openly published. Google introduced a geographical search facility in the Google Local version of their search engine [6]. Indexing of web pages is associated with a Yellow Pages directory, but again no technical details are published. In Europe the Mirago web site [13] provides a geographically specific search facility that allows the user to perform web searches based on administrative regions, which are also displayed on a map on the user interface. Sagara and Kitsuregawa [17] have described briefly a system for retrieving and scoring geographically specific documents from the web with a prototype spatial search engine. They used Yellow Pages to generate key words to find documents on the web relating to listed businesses. These were then scored, according to measures of popularity and reliability but the indexing methods were not described. An experimental system for geographical navigation of the web has been described by McCurley [12]. A variety of techniques was proposed for extraction of the geographical context of a web page, on the basis of the occurrence of text addresses and post codes, place names and telephone numbers. This information was then transformed to one of a limited set of point-referenced map locations. An early example of developing methods to determine the geographical scope of web pages was described by Buyukokkten [3]. This involved associating IP addresses with telephone area codes of the associated network administrators, and hence, via zip code databases, to place names and geographical coordinates. The approach facilitates the analysis of the geographical distribution of web sites. But it appears to require that the content of a web page is related to the place where the web page was created. Ding et al [5] attempted to determine the geographic scope of pages using a gazetteer to recognise place names which were then analysed with respect to their frequency of occurrence. They also considered the geography of the sources that linked to the web document. Silva et al [19] described methods for determining of the scope of web documents in the Portugese tumba! web search engine. After transforming web documents to a structured XML/RDF format they were progressively augmented with geographical descriptors through a sequence of lexical analysis, geographical entity recognition and semantic and web inference procedures.

Recent work on establishing the geographic scope of web pages has been presented by Amitay et al [1]. They identify the presence of candidate place names using a gazetteer, before assigning confidence levels to the interpretation of the name based on associated evidence. A technique for indexing web documents geographically using spatio-textual keys was presented briefly in [7] and evaluated using synthetic data. In the context of a synthetic web document collection, the approach was shown to be beneficial, but no evidence was provided for its accuracy when applied to real data. A large proportion of recent published research relating to geographic web search has been concerned with the problems of geotagging (see also e.g. [10] [15]).

3   Overview of SPIRIT Search Engine

The spatio-textual indexing methods described in this paper were implemented in the experimental SPIRIT search engine [9] [7]. Here we describe briefly the overall architecture of the SPIRIT search engine in order to place the indexing methods in context. The main components are the user interface, document analysis and metadata extraction, core search engine, indexes, the geographical ontology, and relevance ranking. The user interface allows users to specify a concept, a geographical place name and a spatial relationship to the named place. Spatial relationships may be proximal (distance), topological or directional. Examples of types these types of queries are illustrated in Table 1.

Table 1. Query types for a SPIRIT query

Query Type	Example
1. Distance	1. schools *within 10 km* of Zurich city centre
	2. hotels *near* Cardiff University
2. Topological	1. hospitals *in* London
3. Directional	1. holiday resorts *north of* Milan

SPIRIT employs disambiguation functionality, to allow the user to select the appropriate instance of a place name that has multiple occurrences, and presents the search results as a list of URLs and on an interactive map linked to the retrieved document list. The geographical ontology stores knowledge of instances of place names with alternative names, place types, qualitative spatial relationships to other places and one or more geometric footprints [8]. Place footprints may be in the form of a representative point (centroid), a minimum bounding box, a polygon or a line. The user interface uses the geographical ontology for disambiguation of the part of a user's query that specifies place. This results in a query footprint F^Q representing a geometric interpretation of the specified spatial relationship to the named place. For many geographical queries, notably those that employ the "near" relationship, the user can be expected to be interested in documents that relate to locations in the vicinity of the specified geographical location as well as those that match exactly with the named place. The query footprint is therefore expanded beyond the boundary of the footprint of the specified geographical location. Along with the other textual query terms specifying the concept of the query $T = \{t_1, t_2, ...t_m\}$, it is then submitted to the search engine. In general a user query Q takes the form $Q = T \cup F^Q$.

The document analysis and metadata extraction component is used to build a database of web documents that are indexed with regard to textual content and to geographic context. The geographic context is encoded in the form of a document footprint F^d derived from footprints of place names in the geographical ontology that have been detected as geographically significant. The individual footprints of a document footprint are equivalent to the place name footprints in the ontology and are used to perform spatial indexing of the document. Typically there will be several individual footprints in a $F^d = \{f_1, f_2, ...f_n\}$ and hence a document may be spatially indexed with respect to multiple locations. The core search engine finds those documents whose footprints intersect the query footprint. The individual documents returned d_i consist of those documents in the document collection D which contain all the non-spatial textual query terms $t_j \in T$ and which have footprints that intersect the query footprint. The set of documents returned is therefore

$$\{d_i \mid d_i \in D, (\ t_j \in d_i\ (\forall j \in 1..m)\ ) \wedge (\ Q \cap f_k), f_k \in F^d i\ (k \in\ 1..n)\ \}$$

where $F^d i$ refers to the document footprint of document d_i.

Relevance ranking determines an overall ranking for a document by combining a score from text ranking, in the form of a BM25 score [16], with a score from spatial ranking. The spatial ranking can be performed in several different ways. It measures the distance between the query footprint and the document footprint(s) primarily as a Euclidean distance but it is also possible to measure angular difference in order to process queries that employ a directional spatial relationship. The textual and spatial scores can be combined using distributed and non-distributed methods [11].

It should be noted that retrieval of the set of document ids whose footprints match the query footprint is not accompanied by any geometric filtering prior to submission to the relevance ranking component. If a spatial indexing method is used in which documents are referenced by spatial cells, all documents referenced by a cell that intersects the query footprint are passed to the relevance ranking component. This is justifiable in that documents that are outside the query footprint will be ranked lowest in the geographical dimension, and will be geographically adjacent to documents within the query footprint.

4 Spatial and Textual Indexing

Here we investigate hybrid indexing schemes that combine inverted files, that list the documents containing indexed document terms, with a spatial access method to maintain the geometric footprints of indexed documents. The spatial indexing methods employed here are all based on a fixed grid scheme. Clearly more sophisticated spatial access methods could be used but a fixed grid lends itself to relatively simple schemes that should be sufficient to demonstrate the relative merits of the approaches presented (note that fixed grids are used successfully in some commercial GIS).

Once a textual index for terms and a spatial index for document footprints are available then either of them can be used first to get a set of results that can be refined by using the other. Thus an important issue is to decide the order of the search on the index types i.e. Text followed by Spatial or Spatial followed by Text. Here we present and implement schemes based on both approaches and compare their performance experimentally with each other and with a pure text indexing scheme. The pure text indexing scheme PT treats geographical terms the same as other text terms and hence relies entirely on exact matching of query terms with document terms. Our first spatio-textual scheme ST uses a spatial index in a first stage and later searches text indexes created separately for each cell of space. Access to the second spatio-textual scheme TS starts with a term index and then exploits spatial indexes associated with each term of the term index. The third scheme T performs textual indexing and spatial indexing of documents independently, before combing the results.

4.1 Pure Text Indexing PT

In the pure text indexing scheme an inverted file scheme is used consisting of a lexicon file, each record of which contains fields for an item of text and a pointer (and associated offset data) to an entry in the "postings" file containing lists of occurrences of those documents from the document set D of size N that contain the text item. There will be L records in the lexicon, where L is the number of indexed terms, and L lists of document ids in the postings file. In a worst case scenario, all documents contain all indexed terms so that the list of document occurrences for a term would be of length N, resulting in $O(L.N)$ storage. (Note that we are including some component factors of some linear complexity functions, such as in the latter expression, in order to help make distinctions between the various indexing schemes). In practice it is generally assumed that following Heap's Law [2] the size of the lexicon is $O(N^{\beta})$, where $0 < \beta < 1$ with typical values between 0.4 and 0.6, with the occurrences storage being $O(N)$. Total storage may therefore be regarded as $O(N)$.

Queries to this index contain all the terms in the user's query, consisting of m non-spatial textual query terms and n geographical query terms. Assuming that, having found a text term, the cost of a read into memory of the corresponding document list is proportional to K_a, the maximum number of documents referenced by a lexicon term, then, if the lexicon is managed with an access structure such as a binary tree or a B-tree, the access time for the PT index is $O((m+n)(\log L + K_a))$.

4.2 Spatial Primary Index ST

In this index, the space corresponding to the geographical coverage of the place names specified in documents is divided into a set of p regular grid cells $C = \{c_1, ... c_p \}$ and for each cell an inverted text index is constructed. Each text index is structured in the same manner as the pure text index PT described above, but the document set S that it refers to consists of those documents d_j whose document footprint F^D_j intersects the corresponding spatial cell. Thus for a particular cell c_i the corresponding documents S = $\{ d_j \mid d_j \in D \wedge F^D_j \cap c_i \}$. Those documents whose document footprints intersect more than one cell will be represented in multiple cell text indexes. The principle of the ST index is illustrated with respect to the set of documents whose footprints are represented as rectangles in Figure 1. Here a collection of 16 documents, $D=\{D_1, D_2, .., D_{16}\}$, is distributed over a document space R divided into 4 cells. Let S_R be the document space associated with the entire set D, where the respective subdivisions for cells $R1$, $R2$, $R3$ and $R4$ are $S_{R1} = \{D_1, D_7, D_{12}, D_{15}\}$, $S_{R2} = \{D_{15}, D_{10}, D_{11}, D_3, D_{13}\}$, S_{R3} = $\{D_2, D_5, D_{14}, D_{12}, D_{15}\}$, $S_{R4} = \{D_{15}, D_{14}, D_9, D_6, D_{11}, D_{16}, D_4, D_8\}$.

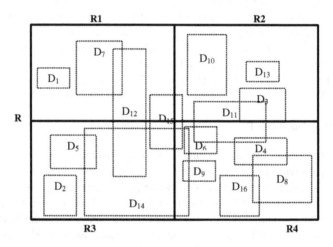

Fig. 1. A spatial index of documents with rectangular footprints

In a worst case scenario the storage cost for this scheme would be p times that of the PT scheme, i.e. $O(p.N)$, corresponding to the event that all document footprints intersected all cells. In practice the process of categorising documents geographically, or geo-tagging, associates the majority of documents with a number of specific areas, reflecting the geographical focus of the documents. Consequently, the individual cell text indexes can be expected to be smaller than the PT index (this is investigated in the experimental results). Query times can be expected to depend on the number of cells r that are intersected by the query footprint and the sizes of the text indexes associated with those cells. Having determined which r cells are intersected by the query footprint, which can be computed relatively trivially in a regular grid spatial index, the subsequent query time would be $O(r((m\log(L_a) + K_b) + \log(p)))$, where m is the number of non-spatial query terms, L_a is the maximum number of terms indexed by the cell-specific indexes and K_b is the maximum number of documents referenced by a term in a cell-specific term index . Note that $\log(p)$ refers to the cost of locating the start of a cell-specific term index, and assumes a sorted list of location codes to identify the spatial cells. If the indexes are stored in separate files then access to the relevant index may be achieved in constant time.

4.3 Text Primary Spatio-textual Index (TS)

In this index a pure text index structure is modified so that the list of documents in the postings file for each term is associated with a spatially-grouped set of documents that contain the term. The spatially organised documents take the form $[Cell_1[DocumentList_1]; Cell_2[DocumentList_2]....Cellp[DocumentList_p]\ ]$, where $Cell_i$ are the cell identifiers of the regular grid cells and $DocumentList_i$ are the lists of documents whose footprints intersect the corresponding cell. For the example given in Figure 1, let us suppose that we have a list of documents associated with the index term "spirit" :

spirit	$\{D_1, D_2, D_3, D_7, D_8, D_9, D_{11}, D_{13}\}$

In the TS index the term "spirit" would be associated with a list of document occurrences grouped as:

spirit	$\{R1(D_1, D_7); R2(D_3, D_{11}, D_{13}); R3(D_2); R4(D_8,D_9,D_{11})\}$

In the worst case, the document footprints of all documents would intersect all cells. This storage may be characterized as $O(p.N)$. As indicated above in the context of the space primary scheme, in practice each document can be expected to be referenced by a subset of the cells, reflecting its geographical focus.

A query to this index consists of the m text terms and the query footprint. Having calculated which r spatial cells are intersected by the query footprint, m queries are required to the main index, corresponding to the individual textual query terms. For each such query, the r cells of the term-specific spatial index are accessed. If, for each term, there are maximum K_c documents referenced per cell then the access time is $O(m\ (\log(L) + r(\log(p) + K_c)\ )\ )$.

4.4 Text Index with Spatial Post-processing (T)

In this scheme we use a pure text index to find those documents D1 that contain the non-geographical query terms. Separately, a spatial index of documents (based on their footprints) is used to find those documents D2 whose document footprint intersects the query footprint. These two sets are then intersected to find those documents that both contain the non-spatial query terms and have a footprint that intersects the query footprint. The storage for this scheme is $O(N)$ for the text index and $O(p.K_d)$ for the spatial index, where K_d is the maximum number of documents referenced by a spatial cell. It may be noted that the storage for ST might also be characterized in the same way, but there is a difference in practice in that the storage for ST is very much bigger, as for each cell a term index is stored, as opposed to the single list of document occurrences per cell in the T scheme.

The query time for accessing the text index is reduced relative to PT in that the geographical terms are not included, giving $O(\ m(\log L + K_a)\ )$. The query time to access the spatial index of documents is $O(\ r(\ \log(p) + K_d\ )\ )$, where r is again the number of cells intersected by the query footprint. Thus each access to a spatial cell will be accompanied by a retrieval of the list of documents referenced by the cell. Having obtained two lists of documents they can be matched to find the common documents, a process that will be enhanced if the documents are stored in both lists in order of their document ids. In this case the match time would be directly proportional to the total numbers of documents.

5 Experimental Comparison of Indexing Schemes

The performance of the four indexing schemes described above has been compared with regard to query time and to the numbers of documents that are returned. The spatio-textual schemes are compared with respect to differing cell size of the spatial index. The number of documents returned is of particular interest as this measures the size of the set passed to the relevance ranking procedure, which is itself a significant cost in the document retrieval process. We do not compare the quality of the results between the schemes from a user's point of view. This would require a geographical test collection of documents that had been ranked manually or semi-manually with regard to their relevance for particular queries. At present no such test collection is freely available though efforts are in place to create one [15]. In the present study all three spatio-textual schemes return exactly the same sets of documents for each spatial cell resolution that is studied. As explained earlier, the pure text scheme will inevitably be inferior with regard to a "recall" measure of quality in that, assuming no query term expansion, it will not find documents that are geographically relevant but which do not include the geographical query term employed in the query.

The document collection consists of 19,956 HTML documents relating to the United Kingdom taken from a terabyte-sized crawl of the Web conducted in 2001. A subset of 19,046 documents was allocated document footprints (geo-tagged) using the GATE (General Architecture for TEXT Engineering) information extraction system [4]. ANNIE, the default Information Extraction system, was used to perform named entity recognition to detect the presence of place names. This uses gazetteer lists (e.g. common names of people and places) and context rules to disambiguate between named entities. These rules assist in distinguishing between place names that are used in a geographical context, and hence are of interest, and those that may be geographically spurious in that they refer for example to people's names and the names of organisations and buildings. The standard GATE gazetteer is enhanced here with the UK Ordnance Survey 1:50,000 gazetteer containing over 250,000 place names of topographic features and settlements, the SABE geo-dataset for the UK, from which more than 10,000 names and footprints were extracted, and the UK Ordnance Survey CodePoint dataset which lists more than a million UK postcodes.

Text indexing facilities are provided by an in-house research IR system called GLASS. All indexing schemes are file-based resulting in much longer access times than for a commercial system, in which most indexes would be maintained in main memory. As the purpose is to compare performance characteristics of spatio-textual indexing methods with pure text indexing, absolute timings are not of particular consequence. For each of the spatio-textual schemes, spatial cell resolutions range from a 2 X 2 subdivision of the geographical region covered by the geo-tagged documents to an 8 X 8 subdivision (and include a 1 X 1, i.e. single cell, subdivision for reference purposes). For each cell resolution we report statistics on the index sizes, the average numbers of documents referenced per cell and the average number of terms indexed within each cell. The purpose of these statistics is to demonstrate the way in which spatial indexing focuses search on geographically-specific documents.

5.1 Implemented Indexing Schemes

5.1.1 PT: Pure Text
The pure text indexing scheme employs the basic GLASS text indexing procedure that is exploited in the SPIRIT search engine. It follows the structure explained above and the file-based lexicon is accessed using a binary search on the sorted index terms. Query expressions include all geographical and non-geographical terms.

5.1.2 ST: Space-Primary Spatio-textual Indexing
This scheme consists of a set of spatial cell-specific text indexes. Each such text index is implemented using the same indexing method as in PT, except that the documents indexed in an individual cell-specific index are those whose footprints intersect the cell. Following calculation of the cells intersected by the query, the files containing each of the relevant text indexes are accessed initially through the unix file system, with the file names being generated from the cell ids.

5.1.3 TS: Text-Primary Spatio-textual Indexing
This indexing scheme is created by modifying the document occurrences lists in the GLASS index. For an individual indexed term, the occurrences list is segmented into cell-specific sub-lists. Each such sub-list contains the identities of documents whose footprint intersects the respective cell. The beginning of the occurrences file contains header data providing the offsets of the start and end of each cell-specific sub-list, supporting direct access reads to the relevant file sections.

5.1.4 T: Separate Text and Spatial Indexes
In the T indexing scheme the pure text index component is identical in structure to that of PT, while the spatial index consists of a table containing records with the structure [cell_id, document_list]. The unix *grep* command is used to the access relevant parts of the file for a given cell id in order to read the respective sub-list into main memory. This may fall short of the theoretical logarithmic access referred to above in this context. The results from the text and spatial index, which are ordered by document id, are intersected using a unix shell script matching procedure.

5.2 Query Schemes

Four query sets were employed , for each of which 100 queries were run. We now describe these query sets:

Random text terms and random place names (Random): Non-geographical concept query terms were selected randomly from the terms in the lexicon and combined with a randomly selected geographical term selected from the SPIRIT list of geographical place names within the UK region. The number of non-geographical query terms was also chosen randomly from the range of 1 to 10.

Selected concept terms and random geography, largest 500 footprints (Top500FP): The non-geographical query terms were selected randomly from 241 concepts (terms or phrases) obtained from the UpMyStreet.com web site, which provides a directory of geographically-specific information. The geographical terms were chosen randomly from the 500 SPIRIT UK place names with the largest footprints. These queries will tend towards larger geographic areas, using "realistic" concept terms.

Selected concept terms and random geography, smallest 500 footprints (Bottom500FP): The same as Top500FP but geographical terms are those in the SPIRIT geo-ontology with the 500 smallest footprints. The geographical search is highly focused and lie often within a single cell of the spatial indexes.

Selected concept terms and random geography from 5 largest footprints (Top5FP): As Top500FP but the query footprints are derived randomly from the 5 SPIRIT UK place names with the largest footprints. It maximises the numbers of spatial cells that need to be accessed to retrieve relevant documents.

5.3 Experimental Results

The experimental results have been used to compare the schemes with regard to the size of the indexes, the time to construct the indexes and the query times for each of the four query sets. Results are presented with respect to the differing spatial index resolutions. We also show how the numbers of documents returned, the numbers of documents that intersect each spatial cell, and the numbers of terms indexed, change with cell size.

The sizes of the indexes for each of the schemes are compared in Figure 2. Here we can see that, for the ST and TS schemes, decreasing cell size, and hence increasing numbers of cells, has a significant negative impact on storage, as predicted in Section 4. For the highest resolution index with $p = 64$ cells, the latter schemes are in fact about 20 times bigger than the PT scheme. This factor demonstrates that there is a definite degree of geographical focus of the documents. This focus is illustrated in Figure 3 which plots the average numbers of documents and of terms per cell against grid resolution. For 8 X 8 grid resolution there are on average about 3000 documents per cell, out of nearly 20,000 documents. This reflects the fact that many documents are represented by multiple individual footprints, averaging 21, with a maximum value of 803 in these experiments. The T scheme shows very little degradation in index size with increasing grid resolution. This is because the spatial index of documents, used here with the PT index, occupies relatively little space compared with the term indexes. Total storage for the 8 X 8 resolution grid index is about 1Mb, whereas the total storage for PT is of the order of 100Mb (see figure 2). Note that in this and subsequent figures "GLASS" in the legend refers to the PT scheme.

The indexing times for the schemes are presented in Figure 4. The ST scheme stands out as having poorer performance with increasing grid resolution. This scheme differs notably from the others in that it is necessary to build separate inverted text indexes for each spatial cell. TS in comparison is more integrated, with a single text index. It is the document occurrences file that is modified in TS relative to PT, with the additional cell-specific document occurrence "sub-lists".

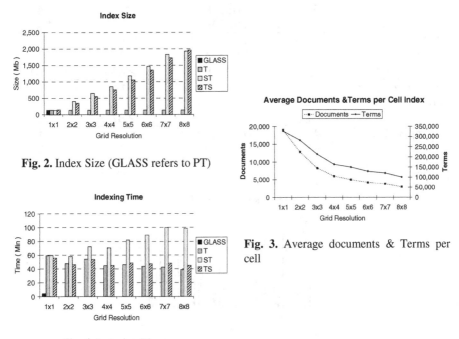

Fig. 2. Index Size (GLASS refers to PT)

Fig. 3. Average documents & Terms per cell

Fig. 4. Indexing Times

Table 2 summarises some statistics of the four query sets that were used to study query timings and numbers of documents retrieved for the different indexing schemes. It presents the minimum, maximum and average size of the query footprint as a percentage of the total area of the indexed region. For the highest resolution spatial indexing scheme, each grid cell would be about 0.016%. Thus in the Top500FP query set the biggest query footprint is similar in size to the smallest cell used in the spatial indexing method. The third query set Bottom500FP has small query footprints relative to spatial index cell size, while the query footprints for the fourth query set Top5FP are extremely large, averaging 0.39 of the entire indexed region. The number of terms in the concept part of the queries is given and it corresponds to the value m in Section 4. The numbers of terms in the place name corresponds to the value n, and only the PT scheme, since these terms are not submitted to the indexes directly in the other schemes (they are converted to a geometric query footprint). All queries use the "near" spatial relation and so increase he size of the target place name footprint in order to generate the query footprint.

Table 2. Query set characteristics

Query Set		Query Footprint (% of Total Space)	Terms in Concept	Terms in Place Name
Random	Min	0.000395	1	1
	Max	0.069283	10	4
	Avg	0.002951	5.94	1.48
Top 500 FP	Min	0.000399	1	1
	Max	0.017344	6	4
	Avg	0.002185	2.65	1.61
Bottom 500 FP	Min	9.47E-08	1	1
	Max	1.67E-06	7	3
	Avg	1.21E-06	2.87	1.55
Top 5 FP	Min	0.061869	1	1
	Max	1	6	2
	Avg	0.391055	2.55	1.2

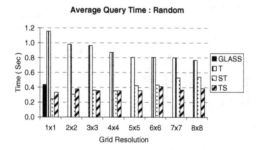

Fig. 5. Average Query Time : Random

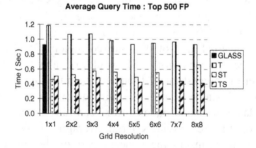

Fig. 6. Average Query Time : Top 500 FP

Figures 5 to 8 illustrate the average query times for each query set respectively, based on 100 queries for each query set. In the first three query sets, the ST and TS schemes are similar or better than PT for all grid resolutions. The T scheme is

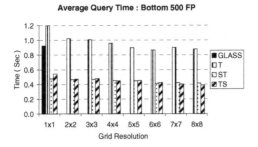

Fig. 7. Average Query Time : Bottom 500 FP

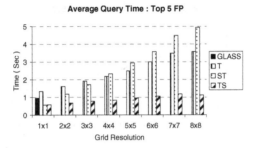

Fig. 8. Average Query Time : Top 5 FP

somewhat inferior to the other schemes, though in the case of the more realistic query sets (Top500FP and Bottom 500FP) it is usually no worse than double the other spatio-textual schemes and not much worse than PT. We regard this as a reflection of an inefficient document merging process (using unix shell scripts) that matches the results of the pure text index with the spatial index of document occurrences.

Figure 8 illustrates the results for the Top5FP query set, which employs very large query footprints. The results here clearly reflect the theoretical analysis whereby timings depend upon the numbers of spatial cells intersected by the query footprint. The average query footprint occupies 0.39 of the entire index and hence intersects a similar proportion of the spatial cells of the respective indexes. This impacts most upon schemes ST and T in which results must be obtained from each intersected cell, prior to merging of the result sets. In both ST and T the merging is performed outside of the main index access programs, using unix scripts. The TS scheme works comparatively well with this query set as all data processing is performed within the shared memory of the modified version of GLASS. In this respect it is the most well integrated spatio-textual scheme. The absolute query times for all schemes here were slow (about a second per query), but this is due to the use of disk-based as opposed to main memory storage methods and the fact that the text indexing methods are not optimised in several respects.

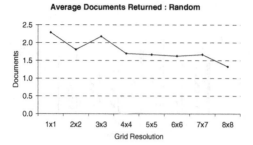

Fig. 9. Average Documents Returned: Random

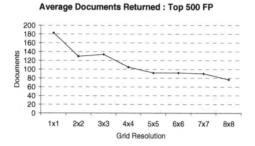

Fig. 10. Average Documents Returned: Top 500 FP

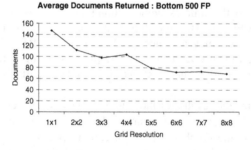

Fig. 11. Average Documents Returned: Bottom 500 FP

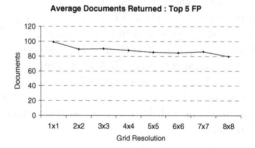

Fig. 12. Average Documents Returned: Top 5 FP

Figures 9 to 12 illustrate the numbers of documents returned for each of the query sets. In the Random query set only about 2 documents are being returned per query, due to the unrealistic random combinations of concept query terms, and no clear pattern emerges. The other three schemes, notably Top500FP and Bottom 500FP, demonstrate a clear trend of reducing numbers of documents returned as grid cell resolution increases. The reason for the decrease in numbers of documents returned is that, as indicated previously, there is no filtering at this stage of the retrieved data against the query footprint. All data in spatial cells that intersect the query footprint are returned. As cell size decreases so there will be a decrease in the numbers of documents that fall outside the query footprint but which lie inside the intersected index cells. The fewer documents that are outside the query footprint the less work is required of the relevance ranking component. In the results here for the Top500FP and Bottom500FP, i.e. the most realistic query sets, the highest resolution spatial indexes result in returning about 50% of those documents returned using a single cell.

6 Conclusions

Spatial indexing of web documents in combination with text indexing of the document content provides a means of managing and retrieving relevant web documents for purposes of geographic search that is superior to conventional text indexing alone. Effective spatio-textual indexing will help to ensure that all relevant documents are retrieved, even when they do not include geographical terms that match exactly with those in a user's query. Spatial indexing also facilitates processing search engine queries that include spatial relationships to a named place, such as *near* and *north of*. Three indexing schemes combining spatial and textual indexing have been presented and compared with each other and with a pure text index (PT), using a web collection of real documents classified with regard to their geographical context. The ST indexing scheme is space primary and creates a set of spatially-specific text indexes. The TS scheme is text primary and associates spatially ordered lists of documents with the indexed terms. The T scheme uses a pure text index to find relevant documents containing the non-geographical query terms and a separate spatial index of documents to find documents whose footprint intersects the query footprint, before intersecting the result sets.

Comparing index sizes, both ST and TS proved expensive relative to PT. The T scheme resulted in very little additional storage cost. The high storage overheads of ST and TS can be explained largely by the fact that the document footprints usually consist of many individual footprints (average 21) reflecting the multiple places referred to in the document. This could be alleviated by more sophisticated geoparsing and geocoding procedures which identified a few dominant individual places to which a document refers as described in [1]. Query times for TS and ST were usually faster than for PT for all spatial index grid resolutions considered (PT has to process all query terms, whereas the other methods convert geographical terms to a query footprint for access to the spatial elements of the indexes and do not use them for access to the text index). An exception to this performance occurred with the

ST index for queries with very large query footprints. The T scheme produced slower query times on the whole than the other spatio-textual schemes but, for the most realistic query sets, this was about double, while in comparison with PT it was only about 25% greater. The slower times reflect the pragmatic but inefficient use of unix script functions such as *grep*. It showed the same degradation with increasing grid resolution for the query set using very large query footprints. For speed, the TS scheme was consistently advantageous. All spatio-textual schemes behaved the same returning fewer documents with increasing spatial grid resolution, reflecting the closer approximation of the grid cells to the query footprint with increasing resolution.

It should be stressed that the objective of this study was to investigate the viability of spatio-textual indexing schemes in comparison with pure text indexing. It is assumed that spatio-textual indexing will retrieve more relevant documents (i.e. improve recall) in comparison with pure text methods, as it will be able to find documents referring to contained and nearby places to the geographical query place, and to places with alternative names to that specified in the query. In summary, the study has demonstrated that scheme T introduced minimal storage overheads while resulting in only a small degradation in query times relative to PT, except for very large query footprints. The TS scheme gave the most consistently good query time performance but was marred by the large storage overheads of multiple footprints.

There is clearly scope for further work to refine the methods described with regard to improved geo-tagging and improved document merging methods. It would also be appropriate to investigate higher spatial grid resolutions and other spatial indexing methods, as well as the use of a much larger web collection. However, spatial index access times were not a significant overhead here. It would be of interest to investigate closer integration of text and spatial indexing, such as the use of spatial cell identifiers (locational keys) as part of the text index.

An issue requiring further attention is that of user evaluation of the results. It has been stated that spatio-textual indexing is assumed to generate superior results relative to pure text indexing. Provided that the geoparsing and geocoding of documents is done effectively, i.e. documents are on the whole correctly categorized with regard to their geographical context, then this appears to be a reasonable assumption. Future studies will conduct such an evaluation to test this assumption when an adequate test collection becomes available.

Acknowledgements

This research was funded by the EC SPIRIT project IST-2001-35047 : Spatially-aware information retrieval on the internet.

References

1. Amitay, E., et al., 2004. Web-a-where: geotagging web content. in *27th ACM SIGIR Conference*: 273-280
2. Baeza-Yates, R. and B. Ribeiro-Neto, 1999. *Modern Information Retrieval*: Addison Wesley.

3. Buyukokkten, O., et al., 1999. Exploiting geographical location information of web pages. in *WebDB'99* (with ACM SIGMOD'99).
4. Cunningham, H., et al., 2002. GATE: A Framework and Graphical Development Environment for Robust NLP Tools and Applications. in *40th Anniversary Meeting of Assoc. for Computational Linguistics (ACL'02)*.
5. Ding, J., L. Gravano, and N. Shivakumar, 2000. Computing Geographical Scopes of Web Resources. in *26th Int. Conf. on Very Large Data Bases (VLDB)*, 545–556.
6. GoogleLocal. http://www.local.google.com
7. Jones, C.B, A.I. Abdelmoty, D. Finch, G. Fu and S. Vaid, 2004. The SPIRIT Spatial Search Engine :Architecture, Ontologies and Spatial Indexing. *in Third Int. Conf on Geographic Information Science GIScience 2004. LNCS* 3234, 125-39.
8. Jones, C.B., A.I. Abdelmoty, and G. Fu., 2003. Maintaining ontologies for geographical information retrieval on the web. in *On The Move to Meaningful Internet Systems 2003, ODBASE'03.LNCS* 2888, 934-51.
9. Jones, C.B., et al., 2002. Spatial information retrieval and geographical ontologies an overview of the SPIRIT project. in Proc ACM *SIGIR 2002*, 387-8.
10. Kornai, A. and B. Sundheim, eds. 2003. HLT-NAACL *Workshop on Analysis of Geographic References*.
11. Kreveld, M.van, I. Reinbacher, A. Arampatzis and R. van Zwol, 2004. Distributed Ranking Methods for Geographic Information Retrieval, in *Developments in Spatial Data Handling*, P.F. Fisher (Ed), Springer, 231-243.
12. McCurley, K.S., 2001. Geospatial mapping and navigation on the web. in *WWW10 Conference*, :http://www10.org/cdrom/papers/278/
13. Mirago. http://www.mirago.com
14. NorthernLight. http://www.northernlight.com
15. Purves, R. and C.B. Jones, 2004. *Workshop on Geographic Information Retrieval, SIGIR* 2004, http://www.sigir.org/forum/2004D/purves_sigirforum_2004d.pdf
16. Robertson, S.E. and S. Walker.1994 Some simple effective approximations to the 2-Poisson model for probabilistic weighted retrieval. *ACM SIGIR 1994*, 232-41.
17. Sagara, T. and M. Kitsuregawa, 2004 Yellow Page driven Methods of Collecting and Scoring Spatial Web Documents. in *SIGIR Workshop on Geographical Information Retrieval*, http://www.geo.unizh.ch/~rsp/gir/
18. Sanderson, M. and J. Kohler, 2004 Analyzing geographic queries. in *SIGIR Workshop on Geographic Information Retrieval*, http://www.geo.unizh.ch/~rsp/gir/
19. Silva, M.J., et al.2004 Adding Geographic Scopes to Web Resources. in *SIGIR Workshop on Geographical Information Retrieval*, http://www.geo.unizh.ch/~rsp/gir/
20. SPIRIT. http://www.geo-spirit.org/

Evaluation of Top-k OLAP Queries
Using Aggregate R–Trees*

Nikos Mamoulis[1], Spiridon Bakiras[2], and Panos Kalnis[3]

[1] Department of Computer Science, University of Hong Kong, Pokfulam Road, Hong Kong
nikos@cs.hku.hk
[2] Department of Computer Science, Hong Kong University of Science and Technology,
Clear Water Bay, Hong Kong
sbakiras@cs.ust.hk
[3] Department of Computer Science, National University of Singapore
kalnis@comp.nus.edu.sg

Abstract. A top-k OLAP query groups measures with respect to some abstraction level of interesting dimensions and selects the k groups with the highest aggregate value. An example of such a query is "find the 10 combinations of product-type and month with the largest sum of sales". Such queries may also be applied in a spatial database context, where objects are augmented with some measures that must be aggregated according to a spatial division. For instance, consider a map of objects (e.g., restaurants), where each object carries some non-spatial measure (e.g., the number of customers served during the last month). Given a partitioning of the space into regions (e.g., by a regular grid), the goal is to find the regions with the highest number of served customers. A straightforward method to evaluate a top-k OLAP query is to compute the aggregate value for each group and then select the groups with the highest aggregates. In this paper, we study the integration of the top-k operator with the aggregate query processing module. For this, we make use of spatial indexes, augmented with aggregate information, like the aggregate R–tree. We device a branch-and-bound algorithm that accesses a minimal number of tree nodes in order to compute the top-k groups. The efficiency of our approach is demonstrated by experimentation.

1 Introduction

Data warehouses integrate and summarize large amounts of historical information, accumulated from operational databases. On-line Analytical Processing (OLAP) refers to the set of operations that are applied on a Data Warehouse to assist analysis and decision support. Data warehouses are usually modeled by the star schema [8], where some measures (e.g., sales) are analyzed with respect to some interesting dimensions (e.g., products, stores, time, etc.), representing business perspectives. A *fact* table stores records corresponding to transactions that have been consolidated in the warehouse. One or more columns in the fact table capture the measures, while each remaining attribute stores values for a dimension at the *most refined* abstraction level. For example, a tuple

---

* Supported by grant HKU 7380/02E from Hong Kong RGC.

C. Bauzer Medeiros et al. (Eds.): SSTD 2005, LNCS 3633, pp. 236–253, 2005.

in the fact table stores a transaction for a particular product-id sold at a particular store-id at some particular time instant. A dimensional table models multi-level hierarchies of a particular dimension. For example, a tuple in the dimensional table `product` stores information about the color, type, manufacturer, etc., for each product-id. Data analysts are interested in summarizing the fact table information with respect to the interesting dimensions at some particular level of their hierarchies, e.g., "retrieve the total sales per month, product color, and store location".

The star schema was extended in [6] to include spatial abstraction levels and dimensions. The location of stores where products are sold is an example of a spatial attribute, with respect to which the sales could be analyzed (possibly together with non-spatial attributes of other dimensions). We can also define hierarchies for spatial attributes. In general, hierarchies of spatial and non-spatial ordinal attributes can be defined either by predefined decompositions of the value ranges (e.g., exact location, city, county, state, country, etc.) or by ad-hoc partitioning techniques (e.g., by a regular spatial grid of arbitrary granularity).

An ideal method to manage a data warehouse, in order to answer OLAP queries efficiently, is to materialize all possible groupings of the measures with respect to every combination of dimensions and hierarchies thereof. In this way, the result of each OLAP query could directly be accessed. Unfortunately, this technique is infeasible, because huge space is required for storing the results for all possible combinations and long time is required to maintain these combinations after updates in the warehouse. In view of this, several partial materialization techniques [7,6] select from the complete hierarchy of possible hyper-cubes those that assist the evaluation of most frequent OLAP queries and at the same time they meet the space and maintenance time constraints. Nevertheless these techniques still cannot deal with ad-hoc groupings of the dimensional ranges, which may still have to be evaluated directly on base tables of the data warehouse. This is particularly the case for spatial attributes, for which the grouping hierarchies are mostly ad-hoc.

Papadias et al. [14] proposed a methodology that remedies the problem of ad-hoc groupings in spatial data warehouses. Their method is based on the construction of an *aggregate* R–tree [10] (simply aR–tree) for the finest granularity of the OLAP dimensions (i.e., for the fact table data). The aR–tree has similar structure and construction/update algorithms as the R*–tree [3]; the difference is that each directory node entry e is augmented with aggregate results on all values indexed in the sub-tree pointed by e. Accordingly, the leaf node entries contain information about measures for some particular combination of dimensional values (i.e., spatial co-ordinates or ordinal values of other dimensions at the finest granularity). This index can be used to efficiently compute the aggregate values of ad-hoc selection ranges on the indexed attributes (e.g., "find the total sales for product-ids 100 to 130 between 10 Jan 2005 and 15 Feb 2005"). In addition, it can be used to answer OLAP group-by queries for ad-hoc groupings of dimensions by spatially *joining* the regions defined by the cube cells with the tree.

An interesting OLAP query generalization is the *iceberg* query [5]; the user is only interested in cells of the cuboid with aggregate values larger than a threshold t (e.g., "find the sum of sales for each combination of product-type and month, only for combinations where the sum of sales is greater than 1000"). In this paper, we study a variant of iceberg queries, which, to our knowledge, has not been addressed in the past. A top-k OLAP query groups measures in a cuboid and returns only the k cells of the cuboid with the largest aggregate value (e.g., "find the 10 combinations of product-type and month with the largest sum of sales"). A naive way to process top-k OLAP queries (and iceberg queries) is to perform the aggregation for each cell and then select the cells with the highest values. Previous work [5] on iceberg queries for ad-hoc groupings employed hashing, in order to early eliminate groups having small aggregates and minimize the number of passes over the base data.

We follow a different approach for top-k OLAP query evaluation, which operates on an aR-tree that indexes the fact table. We traverse the tree in a branch-and-bound manner, following entries that have the highest probability to contribute to cells of large aggregate results. By detecting these dense cells early, we are able to minimize the number of visited tree nodes until the termination of the algorithm. Our method can also be applied for iceberg queries, after replacing the floating bound of the k-th cell by the fixed bound t, expressed in the iceberg query. As we show, our method can evaluate ad-hoc top-k OLAP queries and iceberg queries by *only a part of the base data, only once*. Therefore, it is more efficient than hash-based methods [5] or spatial joins [14], which require *multiple* passes over the *whole* fact table. The efficiency of our approach is demonstrated by extensive experimentation with real datasets.

The remainder of the paper is organized as follows. Section 2 reviews related work. Section 3 formally defines top-k OLAP queries. In Section 4, we describe in detail our proposed solution. Section 5 experimentally demonstrates the efficiency of the proposed algorithm. Finally, Section 6 concludes the paper.

2   Related Work

To date, there is a huge bibliography on data warehousing and OLAP [13], regarding view selection and maintenance [7,12], modeling [8,2], evaluation of OLAP queries [1], indexing [9], etc. In this section, we discuss in more detail past research on indexing spatial data for evaluating aggregate range queries and OLAP queries in the presence of spatial dimensions. In addition, we review past work on iceberg queries and top-k selection queries and discuss their relation to the problem studied in this paper.

2.1   Spatial OLAP

Methods for view selection have been extended for spatial data warehouses [6,15], where the spatial dimension plays an important role, due to the ad-hoc nature of groups there. Papadias et al. [14] proposed a methodology, where a spatial hierarchy is defined by the help of an aggregate R-tree (aR-tree). The aR-tree is structurally similar to the R*-tree [3], however, it is not used to index object-ids, but *measures* at particular loca-tions (which could be mixtures of spatial co-ordinates and ordinal values of non-spatial dimensions at the finest granularity). The main difference to the R*-tree is that each directory node entry e is augmented with aggregate results for all measures indexed in the sub-tree pointed by e. Figure 1 shows an exemplary aR-tree (the ids of entries at the leaf level and the contents of some nodes are omitted). The value shown under each non-leaf entry e_i corresponds to an aggregate value (e.g., sum) for all measures in the subtree pointed by e_i.

The tree can be used to efficiently compute *aggregate range queries*, which summarize the measures contained in a spatial region. These queries are processed similarly to normal range queries on a R–tree. Tree entries (and their corresponding subtrees) are pruned if they do not intersect the query region. If the MBR of an entry e partially overlaps the query, it is followed as usual, however, if e's MBR is totally covered by the query range, the augmented aggregate result $e.agg$ on e is counted and the subtree pointed by e needs not be accessed. For example, consider an aggregate sum query q indicated by the dashed rectangle of Figure 1. From the three root entries, q overlaps only e_2, so the pointer is followed to load the corresponding node and examine entries e_7, e_8, e_9. From these, e_7 is pruned and e_8 partially overlaps q, so it is followed and 10, 5 are added to the partial result. On the other hand, e_9 is totally covered by q, so we can add $e_9.agg = 20$ to the query result, without having to visit the leaf node pointed by e_9. The aR–tree can also be used for *approximate* query processing, if partially overlapped entries are not followed, but their aggregate results are scaled based on the overlapped fraction of e.MBR [10].

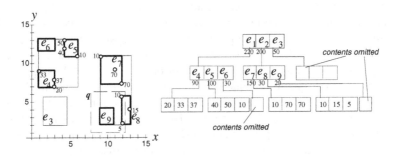

Fig. 1. An aR–tree

[14] showed how the aR–tree can be used to process OLAP group-by queries for groups defined by ad-hoc spatial regions. In this case, a *spatial join* is performed between the tree and the boundaries of the regions for which we want to compute aggregate results. Finally, if there is not enough space to fully materialize the tree, it can be *partially* materialized, by selecting levels that correspond to most significant grouped hierarchies (i.e., the ones that assist most queries).

Another query, related to the top-k OLAP query we study in this paper, is the top-k spatial join [16]. Given two spatial datasets A and B, the top-k spatial join retrieves the k objects from A or B that intersect the largest number of objects from the other dataset. A branch-and-bound algorithm that computes this join is proposed in [16], assuming that both A and B are indexed by R–trees. Our top-k OLAP queries could be considered as a variant of this join query, where one of the joined datasets is the set of regions from which we want to derive the ones with the top-k aggregate result. However, these regions in our problem are ad-hoc and we do not presume an index on them. In addition, the top-k join considers the *count* aggregate function only. Finally, the algorithm proposed in [16] is a join method that essentially accesses both datasets at least once for most typical queries. On the other hand, we do not explicitly access a dataset corresponding to the interesting regions (we compute their results on-the-fly instead) and we access only a part of the base data (using an aR–tree index on them).

2.2   Iceberg Queries

The term *iceberg* query was defined in [5] to characterize a class of OLAP queries that retrieve aggregate values above some specified threshold t (defined by a HAVING clause). An example of an iceberg query in SQL is shown below:

```
SELECT product-type, store-city, sum(quantity)
FROM Sales
GROUP BY product-type, store-city
HAVING sum(quantity) >= 1000 ;
```

In this example, from all groups of product-types and store locations (cities), the user wants only those having aggregate result no smaller than $t = 1000$. The motivation is that the data analyst is often interested in exceptional aggregate values that may be helpful for decision support. A typical query optimizer would first perform the aggregation for each ⟨product-type, store-city⟩ group and then return the ones whose aggregate value exceeds the threshold. In order to avoid useless aggregations for the pairs which disqualify the query, [5] present several hash-based methods with output-sensitive cost. These techniques were later extended for selecting exceptional groups in a whole hierarchy of data cubes [4]. The iceberg query is similar to the top-k OLAP query we study in this paper. In our case, we are interested in the k groups with the largest aggregate values, instead of aggregates above a given threshold. As opposed to the methods in [5], our top-k OLAP algorithm is not based on hashing, but operates on an existing aR–tree that indexes the base data. As we show, our method can also be adapted for iceberg queries.

2.3   Top-k Aggregate Queries

[11] propose methods for processing top-k *range* queries in OLAP cubes. Given an arbitrary query range, the problem is to find the top-k measures in this range. This query is a generalization of *max* range queries (i.e., $k = 1$ for *max* queries). The data cube is partitioned into sufficiently small cells and the top-k aggregate values in each partition are pre-computed. These values can then be used to compute the top-k results in query regions that cover multiple cells. Top-k range queries are essentially different than top-k OLAP queries, since the latter deal with the retrieval of top-k aggregated values of *groups* (as opposed to top-k measures) in the whole space (as opposed to a particular range). To our knowledge, there is no prior work on the problem we study in this paper. In the next section, we formally define top-k OLAP queries and motivate their processing using aR–trees.

3   Problem Formulation

We assume a data warehouse with a star-schema, where the dimensional values recorded in the fact table correspond to either spatial locations (for spatial dimensions) or to ordinal (i.e., numerical) values at finest granularity. We also assume that the user is interested in computing aggregates based on a partitioning of the dimensional domains (e.g., "retrieve the total sales per month, product type, and store location (city)"). This partitioning could be ad-hoc or according to some known hierarchy (e.g., time instants are grouped to hours, days, weeks, months, etc.). We assume that each partition forms a *contiguous* range of values in the domain, and that partitions are disjoint and cover the complete domain of the dimension. Finally, we consider a single aggregate function (sum) on a single measure (e.g., sales quantity). We will later discuss how to extend our methodology for cases where these assumptions do not hold.

Let D be the total number of dimensions. As discussed in [7,14], we typically select a subset of the 2^D dimensional combinations to materialize/index. Consider such a combination of dimensions. We can build an aR–tree index on top of the corresponding cuboid, where we index the summarized information based on the finest granularity values recorded in the fact table. This index can be used to answer OLAP queries (both group-by's and range selections) related to this set of dimensions (or a subset thereof) for any combination of hierarchies (i.e., partitionings) in the individual dimensions. For example, an aR–tree on dimensions ⟨time, product, store-location⟩ could be used to compute the aggregate value for every combination of date/week/month, product-id/type, and street/city/county/state location of stores.

Selecting the combinations of dimensions to materialize can be done with existing techniques (e.g., [7]) and it is out of the scope of this paper. While results for all or some combinations of predefined dimensional partitionings could be pre-computed and materialized, we assume that only the finer granularity summaries for the selected dimensional sets are materialized. The rationale is that (i) it is expensive to store and maintain pre-computed results for all possible combinations of dimensional partitionings, (ii) there could be ad-hoc partitionings, especially in the space dimension (as discussed in [6,15,14]), and (iii) the aR–tree can handle well arbitrary partitionings of the multi-dimensional space [14].

Now, we are ready to formally define the top-k OLAP query:

Definition 1. Let $\mathcal{D} = \{d_1, \ldots, d_m\}$ be a set of m interesting dimensions and assume that the domain of each dimension $d_i \in \mathcal{D}$ is partitioned into a set $R_i = \{r_i^1, \ldots, r_i^{|R_i|}\}$ of $|R_i|$ ad-hoc or predefined ranges based on some hierarchy level. Let k be a positive integer. An OLAP top-k query on $\mathcal{D}$ selects the k groups $g_1, \ldots, g_k$ with the largest aggregate results, such that $g_j = \{r_1, \ldots, r_m\}$ and $r_i \in R_i \forall i \in [1, m]$.

An example top-10 OLAP query could be expressed by the SQL statement that follows. Here, the partition ranges at each dimension are implicitly defined by levels of hierarchy (type for products and city for stores).

```
SELECT product-type, store-city, sum(quantity)
FROM Sales
GROUP BY product-type, store-city
```

```
ORDER BY sum(quantity)
STOP AFTER 10;
```

A naive method to process a top-k OLAP query is to compute the aggregate result for each *cell* (i.e., group of ranges from different dimensions) and while doing so maintain a set of the top-k cells. This method has several shortcomings. First, many cells with small aggregate values will be computed and then filtered out, wasting computations and I/O accesses. Second, since the definition of the dimensional ranges may be ad-hoc, measures within a given cell may be physically located far in the disk. As a result, it may not be possible to compute the aggregates for all cells at a single pass. In order to alleviate the problem, hashing or chunking techniques can be used. Alternatively, a spatial join can be performed if the base data are indexed by an aR–tree, as discussed. Nevertheless it is still desirable to process the top-k OLAP query without having to access all data and without having excessive memory requirements.

Assume that we have only two (spatial) dimensions x and y, with integer values ranging from 0 to 15. In addition, assume that each dimension is partitioned into value ranges $[0, 5)$, $[5, 10)$, $[10, 15]$. Figure 2 shows a set of measures indexed by an aR–tree (the same as in Figure 1) and the 3×3 groups (cells) $c_1, \ldots, c_9$ defined by the combinations of partition ranges. Based on the information we see in the figure (some contents are omitted), we know that $c_1.agg = 120$, since e_6 with $e_6.agg = 30$ is totally contained in c_1. In addition, we know that $c_4.agg \geq 90$ and $c_4.agg \leq 90 + e_3.agg = 140$. Similarly, $c_9.agg \leq 20 + e_9.agg = 40$. Observe that result of a top-1 OLAP query (i.e., the cell with the highest aggregate result, assuming *sum* is the aggregate function) is c_6, with $c_6.agg = 150$, because there is no other cell that can reach this result in any case. Thus, by having browsed the tree partially *we can derive some upper and lower bounds for the aggregate results at each cell*, which can help determining early the top-k OLAP query result. This observation is used by our branch-and-bound algorithm described in the next section.

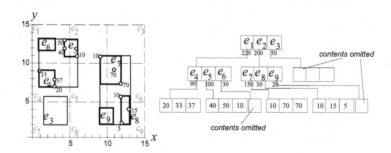

Fig. 2. Top-k grouping example

4   Processing Top-k OLAP Queries Using an aR–Tree

Given an aR–tree that indexes a set of dimensions at a finest granularity and a top-k OLAP query that is based on an ad-hoc partitioning of each dimensional domain, our objective is to evaluate the query by visiting the smallest possible number of nodes in the aR–tree. Assume, for the moment, that we can maintain in memory information about all cells (i.e., the total number of cells is manageable). While traversing the aR–tree, we can compute (partial or total) results for various cells. For example, by visiting the leftmost path of the tree in Figure 2, we know that the result $c_4.agg$ for cell c_4 is between 90 (due to the contents of e_4) and 140 (due to e_3 that overlaps c_4). Thus, we can set *lower* $c_4.lb$ and upper $c_4.ub$ bounds for the aggregate result in c_4, and accordingly for all cells in space. In addition, based on the information derived by traversing the tree, we can maintain a set LB of k cells with the largest lower bounds. The k-th largest lower bound can be used to prune aR–tree entries (and the corresponding sub-trees) as follows:

Lemma 1. Let t be the k-th largest lower bound. Let e_i be an aR–tree entry. If for all cells c that intersect e_i, $c.ub \leq t$, then the subtree pointed by e_i cannot contribute to any top-k result, and thus it can be pruned from search.

The proof of the lemma is straightforward based on the definitions of lower and upper bounds. Intuitively, this lemma can be used to terminate the algorithm after we have computed exactly the contents of some cells and non-visited subtrees overlap only with cells that cannot end up in the top-k result. Now the question is how we can compute and update the lower and upper bounds while traversing the tree. Another question is *how* should we traverse the tree (i.e., in what order should the nodes be visited) if we want to maximize the pruning power of Lemma 1. It turns out that both questions are related; their answers are given by the algorithm described in the next subsection.

4.1   The Basic Algorithm

An entry e_i of a subtree (not visited yet) that intersects a number of cells can contribute at most $e_i.agg$ to the aggregate result of the cell. For example, in Figure 2, even though we do not know the contents of the subtree pointed by e_3, we know that c_4 can contribute at most 50 to this cell. In addition, for an entry e_i which is totally contained in a cell c, we know that it contributes exactly $e_i.agg$ to c, without having to access the subtree pointed by e_i. For example, visiting the leaf node pointed by e_4 is pointless, since the MBR of the entry is totally contained in c_4, thus we know that c_4 gets exactly 90 from this entry. These observations, lead to the design of our top-k OLAP algorithm, which is described in Figure 3.

During the *initialization* phase of the algorithm, we visit the root node of the aR–tree and compute upper and lower bounds for all cells based on their overlap with root entries (lines 1–9). In our running example, we use $e_1.agg, e_2.agg, e_3.agg$ to compute $c_1.ub = 220, c_2.ub = 420, c_3.ub = 200$, etc. In addition, $c_i.lb = 0$ for all cells c_i, since no entry is totally contained in one of them. Assuming that $k = 1$ (in this example) and based on the information so far, the algorithm cannot terminate, since the highest lower bound is smaller than some upper bound.

At this moment, we have to determine which node to visit next. Intuitively, an entry which intersects the cell with the greatest upper bound should be followed first, in order to decrease this upper bound and at the same time increase the lower bounds of other cells, potentially leading to an early termination of the algorithm. In addition, from all entries intersecting the cell with the greatest upper bound the one with the largest $e.agg$ should be visited first, since it is likely to contribute most in the cells it overlaps. Thus, we *prioritize* the entries to be visited according to the above criteria, and follow a *best-first* search order. In other words, all entries (i.e., subtrees) of the aR-tree that have not been followed yet are organized in a heap H (i.e., priority queue). The entry e to be followed next is the one with the greatest $e.agg$ from those intersecting the cell with the greatest upper bound. In our example, after visiting the root, e_1, e_2, e_3 are inserted into H and e_1 becomes the top element, since it intersects c_2 (and c_5) having $c_2.ub = 420$ and $e_1.agg > e_2.agg$ (e_2 also intersects c_2 and c_5). Lines 10–12 of the algorithm compute the heap order key for the root entries and insert them to H.

When de-heaping an entry e from H, we visit the corresponding node n at the aR–tree. Let C be the set of cells intersected by e. The first thing to do is to decrease the upper bounds of cells in C by $e.agg$, since these bounds will be refined by the entries of the new node n. For each entry $e_i \in n$ again we consider two cases; (i) e_i is totally contained in a cell, or (ii) e_i overlaps more than one cells. In the first case, we only update the lower bound of the covering cell. Otherwise, we add $e_i.agg$ to the upper bounds of all cells that intersect e_i. Note that for entries at the leaf level only case (i) applies. After processing all entries, the upper bounds of all cells in C are updated. Based on these new bounds, we compute the heap key of the newly processed entries (only for case (ii) entries) and add them on H. In addition, for entries that are already in H and intersect any cell in C, we change the positions in H, if necessary, considering the new upper bounds of these cells.

The algorithm terminates (line 15) if for the entry e that is de-heaped $e.ub \leq t$, where t is the smallest of the top-k results found so far (stored in LB). Indeed if this condition holds, no cell can potentially have higher aggregate value than the currently k-th result.

Consider again the example of Figure 2 and assume that we want to find the cell with the highest aggregate value (i.e., $k = 1$). We start by examining the three root entries. We add them on H, after having computed $e_1.ub = 420$, $e_2.ub = 420$, and $e_3.ub = 270$. e_1 becomes the top heap element, since $e_1.agg > e_2.agg$.

After de-heaping e_1, we load the aR–tree node pointed by it. First, we reduce the upper bounds of c_1, c_2, c_4, c_5 by $e_1.agg = 220$. Entry e_4 is totally covered by cell c_4, thus we now have $c_4.lb = 90$. Note that we will never have to visit the node pointed by e_4. c_4 now becomes the currently best result and $t = 90$. Entry e_5 overlaps cells c_1 and c_2, increasing their upper bounds by $e_5.agg$. Finally, e_6 is fully contained in c_1 and sets $c_1.lb = 50$. The upper bounds of e_2 and e_3 are updated to 300 (due to c_2) and 140 (due to c_4), respectively. In addition, e_5 has been added to H with $e_5.ub = 300$ (due to c_2). The next entry to be de-heaped is e_2. Since $e_2.ub > t$, the algorithm does not terminate and we load the corresponding node and examine its entries which are all added on H. The top heap entry is now e_7 with $e_7.ub = 250$ (due to c_2). Still $e_7.ub > t$ and we pop the pointed node by it, which is a leaf node. The currently best cell now becomes c_6 with $c_6.lb = 140$. In turn, the top heap entry is e_8, with $e_8.ub = 170$ (due to c_6). After visiting the leaf node pointed by e_8, $c_6.lb$ becomes 150, which is the current t. The algorithm now terminates because the next entry popped from H is e_3 with $e_3.ub = 140 < t$.

Algorithm TopkOLAP(aR–tree T, k)

1. $LB := \varnothing$; $t := 0$; $c.lb := c.ub := 0$, for all cells;
2. $n := \text{root}(R)$;
3. **for** each entry $e_i \in n$ **do**
4.   **if** $e_i.MBR$ is *contained* in a cell c **then**
5. $c.lb := c.lb + e_i.agg$; $c.ub := c.ub + e_i.agg$;
6.   add/update c in LB; /*heap of lower bounds*/
7. $t := k$-th largest value in LB;
8.   **else** /*not contained*/
9.   **for** each cell c intersected by e_i set $c.ub := c.ub + e_i.agg$;
10. **for** each entry $e_i \in n$ **do**
11. $e_i.ub := \max\{c.ub, \forall$ cells c intersected by $e_i\}$;
12.   add e_i on a max-heap H; /*organize H primarily by $e_i.ub$; break ties, using $e_i.agg$*/
13. **while** notempty(H) **do**
14. $e := H$.top;
15.   **if** $e.ub \leq t$ **then break**; /*termination condition*/
16. $n :=$ load aR–tree node pointed by e;
17. $\mathcal{C} :=$ all cells c intersected by e;
18.   **for** each cell $c \in \mathcal{C}$ set $c.ub := c.ub - e_i.agg$;
19.   **for** each entry $e_i \in n$ **do**
20.   **if** $e_i.MBR$ is *contained* in a cell c **then** /* always true if n is a leaf node */
21. $c.lb := c.lb + e_i.agg$; $c.ub := c.ub + e_i.agg$;
22.   add/update c in LB;
23. $t := k$-th largest value in LB;
24.   **else** /*not contained*/
25.   **for** each cell c intersected by e_i set $c.ub := c.ub + e_i.agg$;
26.   **for** each entry $e_i \in n$ not contained in a cell **do**
27. $e_i.ub := \max\{c.ub, \forall$ cells c intersected by $e_i\}$;
28.   add e_i on H;
29.   **for** each entry $e_j \in H$ overlapping some cell in $\mathcal{C}$ **do**
30. $e_j.ub := \max\{c.ub, \forall$ cells c intersected by $e_j\}$;
31.   update e_j's position in H, if necessary;

Fig. 3. The basic algorithm for top-k OLAP queries

4.2 Minimizing the Memory Requirements

The pseudo-code of Figure 3 shows the basic functionality of our top-k OLAP algorithm. Due to the effective branch-and-bound nature of the algorithm, we can avoid accessing a large fraction of the aR–tree nodes, resulting in a sub-linear I/O performance. However, the basic version of algorithm has large space requirements, since for each cell we need to maintain and update a lower and upper aggregate bound. If the total number of cells is larger than the available memory (this can happen when we have very refined partitions at each dimension), then the algorithm is inapplicable. Therefore, it is crucial to minimize the number of cells for which we compute and maintain information. For this purpose, we use the following observations:

- *We need not keep information about cells that are intersected by at most one entry.*
  At the early phases of the algorithm, the MBRs of the high-level entries we have
  seen so far (e.g., the root entries) intersect a large number of cells. However, for a
  cell c that is intersected by only one entry e_i, we know that $c.ub = e_i.agg$, thus
  we do not have to have to explicitly compute and maintain this information. In
  addition, for cells intersected by no entry, we need not keep any information at all.
  Thus, we maintain information only for cells that are intersected by more than one
  entries. This holds only for cells with 0 lower bound; i.e., those for which no partial
  aggregate has been computed. On the other hand, we have to maintain any cell c
  with a partial aggregate (i.e., with $c.lb > 0$), if $c.ub > t$.
- *We can keep a single upper bound for all cells intersected by the same set of entries.*
  Consider two entries that are close in space and jointly affect a contiguous range
  of cells. We need not keep an upper bound for each cell, since we can use a single
  upper bound for the whole range. Later, if one of the two entries is de-heaped and
  the contents of its subtree are loaded, the range of cells is also broken and the
  (different) upper bounds for the individual cells are computed.
- *We need not keep information about cells that may not end up in the top-k result.*
  If, for a cell c, we know that $c.ub < t$, we can prune the cell and never consider it
  in computations of lower and upper bounds, for nodes that are visited next.

We implemented an advanced version of the top-k OLAP algorithm, which has
small memory requirements, based on these observations. Initially, we do not keep explicit upper bounds for the cells, but compute $e_i.ub$ for the examined entries, according
to the common cells they intersect. As soon as lower bounds (i.e., partial result) are
computed for cells, we start maintaining cells with $c.lb > 0$, which may end-up in the
top-k result. In addition, for every entry e_i in H, we keep pointers to all candidate cells
(with $c.lb > 0$) that they overlap, but compute and maintain $e_i.ub$ on-the-fly, considering also cells with $c.lb = 0$, however, without explicitly maintaining those cells.

For example consider again the top-1 OLAP query on the tree of Figure 2 and assume that we are in the phase of examining the root entries (i.e., lines 3–12 of Figure 3).
Instead of explicitly computing $c.ub$ for each cell overlapped by any entry, and then
computing $e_1.ub$, $e_2.ub$, $e_3.ub$ from them, we follow an alternative approach that needs
not materialize $c.ub$. For each entry (e.g., e_1), we compute on-the-fly $c.ub$ for all cells c
it overlaps, by considering the influence of all other entries (e.g., e_2, e_3) in c. Then we
set as $e_i.ub$ the largest $c.ub$. Later, after e_1 is de-heaped and the corresponding node is
loaded, $c_1.lb$ and $c_4.lb$ are computed and stored explicitly, while these cells can end up
in the top-k result.

4.3   Extensions for Related Problems and Generic Problem Settings

So far, we have described our basic algorithm and optimization techniques for it for the
case of OLAP queries, where we look for the top-k cell in a cuboid with the greatest
sum of a single measure. We now discuss variants of this query and how our algorithm
can be adapted for them.

Iceberg queries. Our top-k OLAP algorithm can also be used to process iceberg queries
(described in Section 2.2). We use exactly the same technique for searching the tree.

However, the threshold t used for termination is not floating, based on the current top-k result, but it is a fixed user parameter for the iceberg query. Entries for which $e_i.ub < t$ can be immediately pruned from search. In addition, cells for which the aggregate result has been computed and it is found no smaller than t, are immediately output. We do not need priority queues H and LB, but we can apply a simple depth-first traversal of the tree, to progressively refine the results for cells, until we know that the cell passes the threshold, or can never pass it, based on the potential aggregate value ranges. This method is expected to perform much better than the algorithm of [5] (which operates on raw data) because it utilizes the tree to avoid reading (i.e., prune) cells with aggregate values lower than t. Note that the algorithm of [5] requires reading all base data at least once.

Range-restricted top-k OLAP queries. Our algorithm can be straightforwardly adapted for top-k *range* OLAP queries, where the top-k cells are not searched in the whole space, but only in a sub-space defined by a window. For this case, we combine the window query with the top-k aggregation, by immediately pruning aR–tree entries and cells that do not intersect the window. Apart from that, the algorithm is exactly the same.

Arbitrary partitionings. In spatial OLAP, the regions of interest, for which data are aggregated and the top-k of them are selected, may not be orthocanonical (i.e., defined by some grid), but they could have arbitrary shape (e.g., districts in a city). Our algorithm can be adapted also for arbitrary regions as follows. When the aR–tree root is loaded, we spatially join the MBRs of the root entries to the extents of the regions. Thus, we define a bipartite graph, that connects regions to entries that overlap them. When an entry e is de-heaped, the graph is used to immediately find the regions it affects, in order (i) to compute upper bounds for the regions, based on e's children and (ii) to extend the graph by connecting the newly en-heaped entries (i.e., the children of e) that partially overlap some of these regions.

Query dimensionality. So far, we have assumed that there is an aR–tree for each combination of dimensions that could be in a top-k OLAP query. Nevertheless, as already discussed, it is usually impractical or infeasible to materialize and index even the most refined data level (i.e., the fact table) for *all* combinations of dimensions. Thus, a practicable approach is to index only certain dimensional combinations. Our algorithm can also be applied for top-k OLAP queries, where the set of dimensions is a *subset* of an indexed dimensional set. In this case, instead of cells, the space is divided into *hyper-stripes* defined by the partitionings of only those dimensions of the top-k OLAP query. For the remaining dimensions, the whole dimensional range is considered for each partition. For instance, consider an aR–tree on dimensions ⟨time, product, store-location⟩ and a top-k OLAP query on dimensions ⟨product, store-location⟩. We can use the aR–tree to process the query, however, disregarding the time dimension in the visited entries (and of course in the partitionings).

Non-contiguous ranges. We have assumed each partition of a particular dimension, defines a *contiguous range* on the base data. For example, in an OLAP query about product-types, we assume that the product-ids are ordered or clustered based on product-type. However, this might not be the case, for all hierarchical groupings of the same dimension. For instance, we cannot expect the domain of product-ids to be ordered or clustered by both product-type and product-color. In order to solve this problem, we consider as a different dimension each ordering at the most refined level of an original dimension (according to the hierarchies of the dimension). In other words, we treat as different dimensions two orderings of product-ids; one based on product-type and one based on product-color. Given an arbitrary OLAP query, we use the set of dimensions, where each dimension is ordered (at the finest granularity level) such that the OLAP partitionings are contiguous in the domains of the individual dimensions.

Multiple measures and different aggregate functions. So far, we considered a single measure (e.g., sales quantity) and aggregate function (i.e., sum). As discussed in [10,14], the aR–tree could be augmented with information about multiple measures and more than one aggregate functions (i.e., sum, count, min, max). Our method is straightfor-wardly applied for arbitrary aggregations of the various measures, assuming that the aR–tree on which it operates supports the measure and aggregate function of the query.

5   Experimental Evaluation

We evaluated the efficiency of the proposed top-k OLAP algorithm, using synthetically generated and real spatial data. We compared our algorithm to the naive approach of scanning the data and computing on-the-fly the aggregate results for each cell, while maintaining at the same time the top-k cells with the largest aggregate results. Un-less otherwise stated, we assume that we can allocate a counter (i.e., partial measure) for each cell in memory, a reasonable assumption for most queries. In the case where memory is not enough for these counters, the naive approach first hashes the data into disk-based buckets corresponding to groups of cells and then computes the top-k result at a second pass over these groups. We use I/O cost as a primary comparison factor, as the computational cost is negligible compared to the cost of accessing the data. The naive method and our top-k OLAP algorithm were implemented on a 2.4GHz Pentium 4 PC with 512 Mb of memory.

5.1   Description of Data

Our synthetic data are d-dimensional points generated uniformly in a $[1 : 10000]^d$ map. We use the following approach in order to generate the measure of each point. First, we randomly choose 10 anchor points in the data space. To generate the measure for a point, we fist find its nearest anchor point and its distance to it. All potential distances of points to their nearest anchor were discretized using 1000 bins. The measure assigned to a point follows a Zipfian distribution favoring small distances. The measure value corresponding to the largest distance is 1. The remaining measures were normalized based on this value. The generator simulates an OLAP application, where most of the transactions (i.e., points) have similar and small measures, whereas there are few, large transactions.

We also used a non-uniform dataset; a 2D spatial dataset containing 400K road segments of North America.[1] We assigned a measure at the center of each segment, using the same methodology as for the synthetic data described above. The resulting dataset models a collection of traffic measurements on various roads of a real map.

5.2   Experimental Results

In the first set of experiments, we compare the efficiency and memory requirements of our algorithm for top-k OLAP queries, compared to the naive approach on the synthetic data, for various data generation and query parameters. The default data generation parameter values are N=200K points (i.e., fact table tuples), d=2 dimensions, and θ=1 for the Zipfian distribution of measures. Unless otherwise stated, we set the page (and aR–tree node) size to 1Kb. For top-k queries, the default parameter values are $k = 16$ and $c = 10000$ total group-by cells (e.g., 100×100 cells for a 2D dataset).

The first experiment compares our algorithm to the naive approach for various sizes of the base data, using the default values for the other parameters. Figure 4 shows the results. Our top-k OLAP algorithm incurs an order of magnitude fewer I/O accesses compared to the naive approach, due to its ability to prune early aR–sub-trees that do not contain query results. The performance gap grows with N, because the aR–tree node extents become smaller and there are higher chances for node MBRs to be contained in cells and not accessed.

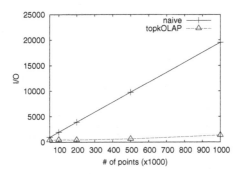

Fig. 4. Performance as a function of database size

Next, we evaluate the efficiency of our approach as a function of the skew on the measures (Figure 5). We used the default data generation parameters and varied the values of θ to $0, 0.5, 1, 1.5, 2$. As expected, the efficiency of our method increases with θ, because the top-k cells become more distinguishable from the majority of cells with low aggregate values. On the other hand, for uniform measures (recall that the points are also uniform), our algorithm becomes worse than the naive approach; it accesses all aR–tree pages (more than the data blocks due to the lower node utilization). In this case, the top-k results are indistinguishable from the remaining cells, since all cells have more or less the same aggregate value.

---

[1] collected and integrated from http://www.maproom.psu.edu/dcw/

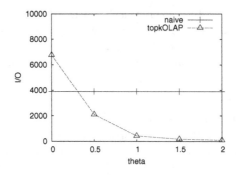

Fig. 5. Performance as a function of skew on the measure values

We also validated efficiency as a function of k; the number of cells with the highest values to be retrieved. As Figure 6 shows, the cost of our method increases with k, although not dramatically. The reason is that for large values of k, the k-th result becomes less indistinguishable from the average cells (or else, the k-th result becomes less different than the $k+1$-th).

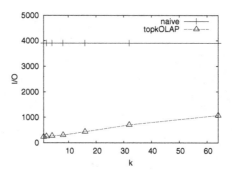

Fig. 6. Performance as a function of k

So far, we have assumed a fixed number of cells (100×100). We now evaluate the performance of our method when this value changes. Figure 7a shows the I/O cost of our method as a function of the number of partitions at each dimension. The dimensionality is fixed to $d = 2$, so the total number of cells is the square of the partitions per dimension. In general, the performance of our algorithm decreases with c, but not dramatically. For as few as $c = 10 \times 10$ cells, our method accesses many nodes because $k = 16$ is relatively high compared to the number of cells and there is no great difference between the k-th and $k+1$-th cell.

We also plot the memory requirements of the two algorithms in Figure 7b. The memory usage is measured in terms of cells for which lower bounds (or partial measures for the naive approach) are explicitly maintained in memory in the worst case. For the naive algorithm this corresponds to the total number of cells, assuming that the memory is large enough to accommodate them. For our algorithm (see Section 4.2), the memory requirements are also dominated by the number of cells, which is expected to be much larger than the number of entries in the heap. The number of cells for which we have to keep information in memory is not constant; initially it increases, it reaches a *peak*, and then decreases. Here, we plot the peak number of cells. Observe that for small c our method has similar memory requirements to the naive approach. However, the memory requirements of our method increase almost linearly with the number of partitions per dimension, as opposed to the naive approach which requires $O(c)$ memory (i.e., quadratic to the number of partitions per dimension). This is a very important advantage of our approach, because memory savings are more important for large values of c, e.g., where c exceeds the available memory.

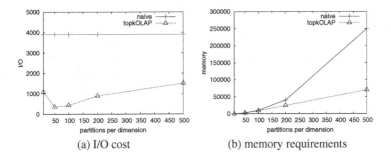

(a) I/O cost   (b) memory requirements

Fig. 7. Performance varying the number of partitions

Figure 8 shows the performance of our algorithm, when varying the problem dimensionality d. The total number of cells is fixed to $c = 10000$, which implies that the partitions per dimension decrease with d. Note that for few dimensions (2 or 3) our method performs well, however, for higher dimensional values, it may access similar or more pages compared to the naive approach. This behavior can be attributed to two facts. First, as the dimensionality increases, the performance of multi-dimensional structures (like the aR–tree) deteriorates. The bounding hyper-rectangles become less tight with more empty space. In addition, their extents are larger and it becomes unlikely that they separate well the top-k cells from cells of small aggregate values. Second, as d increases, the extents of cells at each dimension become larger, thus more aR–tree nodes overlap the cells in the result, as well as other, irrelevant cells. Thus, our method is especially useful for low (2 or 3) dimensional data (like spatial OLAP data), or skewed high-dimensional data.

The next experiment evaluates the efficiency of our approach for top-k spatial OLAP queries. We used the spatial dataset and generated a measure for each point in it, according to their distance to the nearest of 10 random anchor points ($\theta = 1$). Figure 9a shows that our algorithm is very efficient compared to the naive approach for a wide range of spatial partitionings. In Figure 9b, we compare the two methods for 100×100 cells and different values of θ, when generating the measures. Our top-k OLAP algorithm is

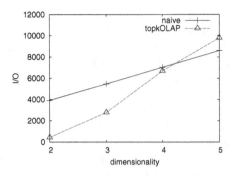

Fig. 8. Performance with respect to dimensionality

more efficient than the naive approach, even for uniform measures. Due to the spatial skew, many cells are empty and there is large difference in the aggregate values of cells, making our method very effective in pruning the search space. In addition, the data are very dense, compared to the uniform points in the synthetically generated datasets, and many nodes of the aR–tree need not be loaded as they are included in cells. Overall, our method is very efficient when either the data points or the measures are skewed, a realistic case especially in spatial data warehouses (since most real spatial data are skewed by nature).

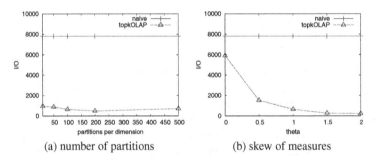

(a) number of partitions   (b) skew of measures

Fig. 9. Performance for spatial OLAP queries

6   Conclusions

In this paper, we studied a new and important query type for on-line analytical processing; the top-k OLAP query. We proposed a branch-and-bound technique that operates on an aR–tree and computes the result of the query, by accessing only a part of the tree. We proposed an effective optimization that greatly reduces the memory requirements of our method, rendering it applicable even to queries with a huge number of candidate results (i.e., cells of the partitioned space). Experiments confirm the efficiency of our approach, compared to a conventional hash-based approach that does not utilize existing indexes.

References

1. S. Agarwal, R. Agrawal, P. Deshpande, A. Gupta, J. F. Naughton, R. Ramakrishnan, and S. Sarawagi. On the computation of multidimensional aggregates. In *Proc. of VLDB*, pages 506–521, 1996.
2. R. Agrawal, A. Gupta, and S. Sarawagi. Modeling multidimensional databases. In *Proc. of ICDE*, pages 232–243, 1997.
3. N. Beckmann, H. Kriegel, R. Schneider, and B. Seeger. The R*–tree: An efficient and robust access method for points and rectangles. In *Proc. of ACM SIGMOD*, pages 220–231, 1990.
4. K. S. Beyer and R. Ramakrishnan. Bottom-up computation of sparse and iceberg CUBEs. In *Proc. of ACM SIGMOD*, 1999.
5. M. Fang, N. Shivakumar, H. Garcia-Molina, R. Motwani, and J. D. Ullman. Computing iceberg queries efficiently. In *Proc. of VLDB*, 1998.
6. J. Han, N. Stefanovic, and K. Koperski. Selective materialization: An efficient method for spatial data cube construction. pages 144–158, 1998.
7. V. Harinarayan, A. Rajaraman, and J. D. Ullman. Implementing data cubes efficiently. In *Proc. of ACM SIGMOD*, pages 205–216, 1996.
8. R. Kimball. *The Data Warehouse Toolkit*. John Wiley, 1996.
9. Y. Kotidis and N. Roussopoulos. An alternative storage organization for ROLAP aggregate views based on cubetrees. In *Proc. of ACM SIGMOD*, pages 249–258, 1998.
10. I. Lazaridis and S. Mehrotra. Progressive approximate aggregate queries with a multi-resolution tree structure. In *Proc. of ACM SIGMOD*, 2001.
11. Z. X. Loh, T. W. Ling, C.-H. Ang, and S. Y. Lee. Analysis of pre-computed partition top method for range top-k queries in OLAP data cubes. In *Proc. of CIKM*, pages 60–67, 2002.
12. I. S. Mumick, D. Quass, and B. S. Mumick. Maintenance of data cubes and summary tables in a warehouse. In *Proc. of ACM SIGMOD*, pages 100–111, 1997.
13. Ondelette.com. Data Warehousing and OLAP: A research-oriented bibliography. In *http://www.ondelette.com/OLAP/dwbib.html*, 2005.
14. D. Papadias, P. Kalnis, J. Zhang, and Y. Tao. Efficient OLAP operations in spatial data warehouses. In *Proc. of SSTD*, 2001.
15. T. B. Pedersen and N. Tryfona. Pre-aggregation in spatial data warehouses. In *Proc. of SSTD*, pages 460–480, 2001.
16. M. Zhu, D. Papadias, J. Zhang, and D. Lee. Top-k spatial joins. *IEEE TKDE*, 17(4):567–579, 2005.

PA-Tree: A Parametric Indexing Scheme for Spatio-temporal Trajectories

Jinfeng Ni and Chinya V. Ravishankar

Department of Computer Science and Engineering,
University of California, Riverside
{jni, ravi}@cs.ucr.edu

Abstract. Many new applications involving moving objects require the collection and querying of trajectory data, so efficient indexing methods are needed to support complex spatio-temporal queries on such data. Current work in this domain has used MBRs to approximate trajectories, which fail to capture some basic properties of trajectories, including smoothness and lack of internal area. This mismatch leads to poor pruning when such indices are used. In this work, we revisit the issue of using parametric space indexing for historical trajectory data. We approximate a sequence of movement functions with single continuous polynomial. Since trajectories tend to be smooth, our approximations work well and yield much finer approximation quality than MBRs. We present the PA-tree, a parametric index that uses this new approximation method. Experiments show that PA-tree construction costs are orders of magnitude lower than that of competing methods. Further, for spatio-temporal range queries, MBR-based methods require 20%–60% more I/O than PA-trees with clustered indicies, and 300%–400% more I/O than PA-trees with non-clustered indicies.

1 Introduction

GPS has been widely used for a number of years in support of a variety of new applications, including tracking of vehicle fleets, navigation of watercraft and aircraft, the emergency E911 service for cellular phones [13]. Such applications would benefit greatly from an ability to make complex spatio-temporal queries on databases containing huge amounts of trajectory data about objects moving in two or higher dimensional space.

Work already exists on developing indices to support spatio-temporal queries. Such work is typically either in support of *predictive* queries, which require the future location of objects based on their current locations and velocities (for example, *"find all objects that will be within Union Square in 10 minutes"*), or in support of *historical* queries, which query the past locations of moving objects (for example, *"find all objects which were at the intersection of Freeway 10 and 15 an hour ago"*). In this paper, we focus on historical queries, intended to search a large set of historical trajectories.

In general, we can classify indexing methods into *Native Space Indexing* methods (NSI), and *Parametric Space Indexing* methods (PSI) [18]. In NSI, motion in a d-dimensional space is represented as a series of line segments (or curves) in $d + 1$ dimensional space, using time as an additional dimension. PSI can be regarded as the dual

C. Bauzer Medeiros et al. (Eds.): SSTD 2005, LNCS 3633, pp. 254–272, 2005.

transformation of NSI, where a parametric space defined by the motion parameters is used. PSI has be shown to be an efficient approach for predictive queries, for example, TPR-tree [19], TPR*-tree [22], STRIPES [16].

PSI has not been advocated in the literature for historical queries. Indeed, Porkaew et al. [18] showed that PSI was actually outperformed by NSI for historical queries. Unlike the predicted trajectory case, which uses only one predicted motion function for each object, each historical trajectory could consist of hundreds or even thousands motion functions. PSI will hence introduce large storage overheads, and significantly degrades query performance. As a result, much previous work on historical queries has attempted to index each trajectory in the native space, using approximations such as Minimum Bounding Rectangles (MBRs) [9,8], Octagons [25], or regular grid cells [4]. However, as shown by Kollios et al [10], MBRs are rather coarse approximation for trajectory. A trajectory typically consists of a series of line segments or curves, and does not have any internal area. Consequently, using MBRs may result in a large amount of dead space, leading to a significant loss in pruning power.

1.1 Our Work

In this paper, we revisit the issue of indexing historical trajectory in parametric space. Unlike previous work in the area [18], we do not represent each line segment or curve with a parametric function. Instead, we try to *approximate* a series of line segments or curves with *a single* continuous polynomial. This approximated trajectory may not perfectly match the original trajectory. However, if we also keep track of the maximum deviation between the approximation and the original movement, we can still ensure that the approximation is *conservative*, and will not generate false negatives. Therefore, as long as the maximum deviation is small, the approximated polynomial function and the maximum deviation together provide a much tighter approximation than the generally used MBRs. We are therefore able to improve query performance significantly.

The fundamental observation behind our scheme is that trajectories, in general, have a certain degree of smoothness, as suggested in [3]. First, object movements are governed by the laws of physics, resulting in smooth motion trajectories. Second, many objects are constrained to move along road networks, which usually have some degree of smoothness. Indeed, for *similarity-based queries*, exploiting the smoothness of trajectories has yielded performance far better than that of previous methods[3].

The work in [3] also uses polynomials to approximate trajectories, but there are major differences between our work and theirs. First, [3] targets similarity-based queries, and defines similarity over entire trajectories of equal length, ignoring the time components. Hence the techniques in [3] are generally not applicable to spatio-temporal databases, where the time component is crucial in answering timestamp or time interval queries [17]. Second, the lower-bound lemma in [3] is only valid for similarity queries, so that other approaches are needed to deal with spatio-temporal queries using polynomial approximations. Further, [3] uses approximations of the same degree for all the trajectories, which can cause serious difficulties when the approximation degree is high. In contrast, we use different polynomials of different degrees for different trajectories, and develop a two-level index structures to avoid this problem.

In this work, we make the following contributions:

- We revisit the issue of indexing historical trajectory in parametric space. Observing the smoothness of object movements, we show that parametric indexing using polynomial approximations can improve query performance significantly over current schemes using native space indexing.
- We develop a cost model to optimize the degree of the polynomial approximation given a trajectory segment. Further, we present the PA-tree, a new index scheme for historical trajectory data, based on polynomial approximations.
- We evaluate the performance of our schemes using synthetic trajectory datasets. Our empirical results indicate that in most cases, the MVR-trees require 20% - 60% more IO than PA-trees with clustered indicies, and 300%–400% more IO than PA-trees with non-clustered indicies. More importantly, the cost of constructing PA-trees is orders of magnitude faster than the construction of MVR-trees, suggesting that PA-trees may be suitable for on-line indexing of trajectories.

2   Related Work

MBRs have been widely used to approximate multi-dimensional data, and consequently R-trees are the most common index structure for multidimensional data. Earlier work using MBRs for trajectories includes the RT-tree [24] and 3D R-tree [23]. However since the RT-tree does not take temporal attributes into account during the insertion/deletion, timestamp or time interval queries are inefficient. 3D R-tree is inefficient for timestamp queries, since the query time depends on the total number of entries in the history [21].

Kollios et al. [11] present methods for indexing linear historical trajectories. They model a long-lived trajectory with multiple MBRs by splitting it into segments to reduce the large dead space resulting from the use of a single MBR, and use partial-persistent R-trees ("PPR-tree") to index the multiple MBRs. This work is extended in [9,8], where the motion function could be arbitrary (In the latest work [8], term "MVR-tree" is used in stead of "PPR-tree"). This method can be more efficient than 3D R-tree, since the total empty volume after splitting would be reduced. However, since this method still uses MBRs for approximating each segment, there remains significant dead space.

Zhu et al [25] used octagonal prisms, which are MBRs whose four corners are cut off to approximate trajectory. However, their experiments demonstrate only small differences between octagonal prisms and MBR when the number of splits increases to a certain point, since little gains will result from cutting off MBR corners when the number of splits becomes large.

Some previous work has been based on a discrete event model, under which an object is assumed to stay at its current position until it issues an update to the server. However, this model can not be used to represent gradual changes in object locations, limiting its applicability[4]. The basic idea is to build a separate R-tree for each timestamp, as in HR-tree [14] and MR-tree [24]. Unchanged nodes are not duplicated in consecutive R-trees to reduce the storage cost. However, these index structures are only efficient for timestamp queries, but are not efficient for time interval queries [4,21]. The MV3R-tree [21] is a hybrid structure that uses a multi-version R-tree for timestamp queries, and a small 3D R-tree for time-interval queries. The two indices share the same leaf pages, in order to reduce the storage cost, resulting in a quite complex algorithm for maintaining the indices [4].

SETI [4] is an indexing method which can support both inserts and searches. SETI uses two-level index structures to decouple the spatial and the temporal dimensions. Space is partitioned into multiple cells, and the temporal attributes of all line segments intersecting a cell will be indexed with a 1-dimensional index structure. However, since multiple line segments of the same trajectory may overlap the query range, SETI must eliminate duplicates, which may be expensive. Also SETI does not have the *trajectory preservation* property [17], since each data page may contain segments of multiple trajectories, with no guarantee all line segments of one trajectory will be stored together. Hence, SETI may not be able to efficiently support trajectory-based queries [17].

Polynomial approximations have been used to approximate *predictive* trajectories by Tao et al. [20], who use *STP-trees* to index the polynomial coefficients. Several major differences exist between their work and ours. First, our query types are different. Second, [20] applies the same degree of approximation for all trajectories, assuming the same motion type for *all* objects. In practice, different objects may have trajectories with different complexities. In contrast, we choose the degree of polynomial approximation based on the complexities of trajectory, a strategy applicable in more general scenarios. Further, in [20], when a k degree polynomial is used for each axis in a d-dimensional space, the STP-tree becomes an index structure in the parametric space of $(k + 1)d$ dimensions, leading to the problem of curse of dimensionality for large k. Unlike [20], we adopt a two-level structure (see Section 6) to address this problem.

3   Problem Definition and Data Model

Data Model. In many location-based services, location data are obtained by periodic sampling. Specifically, the trajectory for an object O_i has the form

$$Trj(O_i) = \{ID_i, t_0, t_1, \cdots, t_n, f_1(t), f_2(t), \cdots, f_n(t)\}$$

Function $f_j(t)$ is a movement function representing movement during time interval $[t_{j-1} : t_j]$, $1 \leq j \leq n$. The interval $[t_0 : t_n]$ is the *lifetime* of the trajectory.

Our approach is applicable to any movement function $f(t)$, as long as we can determine the location of the object at any time instant during its lifetime from $f(t)$. For simplicity of exposition, we adopt a linear mobility model, which is widely used in the literature [4,17]. Each $f_j(t)$ is now a linear function of time, so that a trajectory consists of a series of connected line segments. This representation is refereed to as a *polyline*.

As in previous work [9,8], we assume time is discrete, and the dataset temporal range $[0, T]$ contains the lifetimes of all the trajectories. We assume an object moves in a two-dimensional XY-space. The extension to higher dimensions is straightforward.

Query Types. We focus mainly on historical coordinate-based queries, in particular on spatio-temporal range queries, since spatio-temporal range queries are essential building blocks for all other types of queries. A spatio-temporal range query may be a timestamp query, or a time interval query. A timestamp query $Q(r, t)$ asks for all the objects

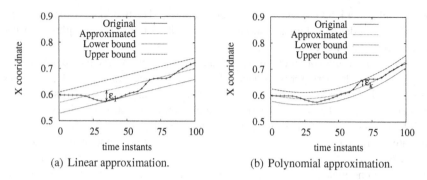

Fig. 1. Approximating a trajectory segment with polynomials

within spatial range r at timestamp t. Similarly, a time interval query $Q(r, t_b, t_e)$ asks for all the objects which were within spatial range r at any timestamp $t \in [t_b : t_e]$.

The PA-tree also supports efficient execution of *trajectory-based queries*, which may take the output of coordinate-based queries as input and retrieve the exact trajectory so that certain properties, such as direction or speed can be derived [17,25]. As we will explain, the PA-tree allows a series of consecutive line segments belonging to the same trajectory to be stored together. This *trajectory preservation* property ensures the trajectory-based queries can be answered efficiently with the PA-tree.

4   Overview of Our Approach

Our approach proceeds in two steps. In the first step, we calculate the parametric representations used to approximate each trajectory. We will approximate a trajectory in the XY-space with two polynomial functions: $\hat{f}^x(t)$ and $\hat{f}^y(t)$ modeling movement in the X direction and in the Y direction, respectively, where t is time. We also determine the maximum deviation of the polynomial approximation from the exact movement in X and Y dimensions. The polynomial coefficients and the maximum deviation suffice for us to make the approximation conservative, guaranteeing no false negatives.

Fig. 1(a) and Fig. 1(b) shows the X-component of a trajectory, and illustrates how we construct a linear and an order-k polynomial approximation to it. Such approximations are not exact, so we create conservative upper and lower bounds for the object's position by offsetting the approximating polynomial upwards and downwards by an amount equal to the maximum deviation between the trajectory and the polynomial. We can now guarantee that the object will be located within these bounds.

In the second step, we build an index structure over the coefficients obtained in the first step. However, not all trajectory are likely to be equally complex, so that we may need polynomials of different degree for different trajectories. This causes problems when building an index structure using the coefficients, since the dimensionalities of the indexed items may be different. Current index structures assume that the dimensionality of all data is the same. Adopting the same polynomial degree for all trajectories is not advisable, since the curse of dimensionality will quickly degrade the performance of any index structure in high dimensional space.

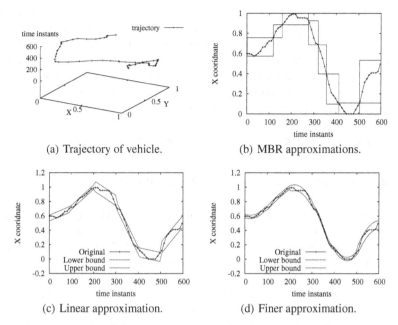

(a) Trajectory of vehicle.   (b) MBR approximations.

(c) Linear approximation.   (d) Finer approximation.

Fig. 2. Comparison of approximations for a moving vehicle trajectory collected in [1]

Two-Level Indexing. We address this problem by using a two-level index structure. The first-level index structure uses only the first two coefficients of each polynomial, so that each data entry is a 6-tuple (two coefficients for each dimension, and the corresponding maximum deviations). This strategy ensures that we are not operating in a high dimensional space, so that an R-tree or its variants can still be efficient for indexing. As we will illustrate in Section 5, by appropriately splitting the temporal domain $[0, T]$ into intervals, we can adopt a piecewise linear approximation in the first level index structure, each linear approximation corresponding to multiple line segments in the trajectory. However, even with this piecewise-linear approximation, we can achieve much smaller dead space than the MBRs with the same size of representation.

The second-level index structure is elaborated within the leaf nodes of the first-level structure. As noted earlier, some trajectories may be complex and require higher-degree polynomial approximations. The higher-degrees coefficients are stored in the second level structure. If we descend to the leaf nodes in the first level structure, and still are unable to determine whether the trajectory satisfies the query predicates, the additional coefficients can be retrieved and used in the filtering step. As our experiments will show, most trajectories can be approximated very well with quadratic or cubic polynomials, so that the second level structure does not introduce significant space overhead.

An Illustrative Example. Figure 2(a) plots the trajectory of a moving vehicle for 10 minutes, collected in the Intellishare project [1] at the University of California–Riverside. Figure 2(b) plots the X-movement against time, and the eight MBRs obtained with the LAGreedy algorithm proposed in [9,8]. We note that the eight MBRs together requires $8 \times 6 = 48$ values. Figure 2(c) plots the result of our method, in which the trajectory is split into 6 segments, each of which is approximated with a linear function. Each segment requires two coefficients, and one maximum deviation each for X-movement or Y-movement, plus the temporal intervals. In all, 48 values are required for the approximations. It is quite clear that our polynomial approximations produce much smaller dead space than the MBR approximations. We should note that since we split all the trajectories at the same split timestamps, there is no need to keep the temporal intervals in the intermediate nodes in the index structure, which would further save storage cost. Figure 2(d) plots the approximation with more coefficients, with significantly reduced dead space. This example motivates our work.

5   Approximating Trajectories with Polynomials

In this paper, we propose an approximation in parametric space by using Chebyshev polynomials. Chebyshev polynomials have been shown to have the near-optimal L_∞ deviation among all approximations with the same degree [12], and perfectly match our requirements. Further, the Chebyshev coefficients are easy to compute [12,3].

We have chosen to split each trajectory into multiple segments by dividing the temporal domain $[0, T]$ into m disjoint time intervals, each of which is approximated with a polynomial. There are two reasons for such splitting. First, approximating the entire trajectory with a single polynomial may require a polynomial of high degree, leading to a high-dimensional indexing problem. Second, the marginal benefit for the first few coefficients will be much larger than that of high-order coefficients. Therefore, it is wise to split the trajectory into multiple segments, and approximate each segment with a lower-degree polynomial.

5.1   Splitting the Time Domain

We split the temporal domain $[0, T]$ into m equal time intervals: $I_1 = [0, T_1), I_2 = [T_1, T_2), \cdots, I_m = [T_{m-1}, T)$, where $T_1, T_2, \cdots, T_{m-1}$ are called splitting timestamps. Each trajectory is split into multiple segments using the same $m - 1$ splitting timestamps. This strategy is different from that in [9], where each trajectory selects different splitting timestamps. We choose our strategy for three reasons. First, since we index in parametric space, a set of segments can not be clustered unless they have the same temporal domain, since it would be meaningless to cluster coefficients corresponding to different temporal domains. Second, even with a equal-sized splitting strategy, we can still use different numbers of coefficients for different trajectories. Indeed, basing the number of coefficients for approximation on the trajectory complexity is equivalent to using different splitting timestamps. Finally, using equal-length splitting intervals obviates the need to maintain time intervals in index nodes. This could significantly reduce the storage cost of the index structure, and eventually lead to a reduction of I/O cost during the filtering step.

One problem in using equal-sized intervals is that some trajectories may begin or end within some time interval. For instance, in Figure 3(a), trajectory Tr_2 begins in the middle of interval I_2. We can simply extend its lifetime to the beginning of I_2, and require that the object remain at its initial location during this extension. This will

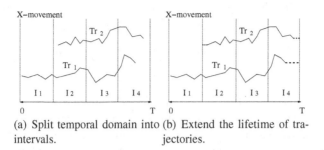

(a) Split temporal domain into intervals.   (b) Extend the lifetime of trajectories.

Fig. 3. Splitting the temporal domain

result in a "faked" trajectory. However, as long as we maintain the actual object lifetime in the second level index structure, we will see no false positives. Figure 3(b) shows the trajectories with extended lifetimes, where the dashed line segments represents the extension.

An important issue is how to choose the number of intervals. This is a complex problem, since we have to minimize the overall query cost. Similar as [8,4] which chose the number of MBRs or grid cells through experiments, we also vary the number of intervals in experimental evaluations, and choose the number of intervals which result in the best query I/O cost.

5.2 Approximating a Trajectory Segment with a Polynomial

We now consider how to obtain polynomial approximation (PA) with Chebyshev polynomials. Consider a trajectory segment in the temporal interval $I_i = [T_{i-1}, T_i)$

$$\{ID_j, t_0, t_1, t_2, \cdots, t_s, f_0^x(t), f_0^y(t), f_1^x(t), f_1^y(t), \cdots, f_{s-1}^x(t), f_{s-1}^y(t)\}$$

where the linear functions f_i^x and f_i^y describe the X-movement and the Y-movement during the interval $[t_i, t_{i+1}]$, and $t_0 = T_{i-1}$, $t_s = T_i$. We illustrate the approximation for the X-movement only, so we will omit the superscript x when no confusion can arise. We can rewrite piecewise linear functions for X-movement in functional form as follows: $f(t) = f_i(t)$, if $t \in [t_i, t_{i+1}]$.

Given function $f(t)$, we can use Chebyshev polynomials as the base functions to get the approximated function $\hat{f}(t)$. We first normalize the temporal domain $[t_0, t_s]$ to the interval [-1,1], by substituting $t' = \frac{2t - t_s - t_0}{t_s - t_0}$. Now, $f(t)$ can be approximated as

$$\hat{f}(t) = c_0 T_0(t) + c_1 T_1(t) + \cdots + c_k T_k(t), \tag{1}$$

where $T_i(t) = \cos(i \arccos(t)), t \in [-1, 1]$ is the Chebyshev polynomial of degree i, and the coefficients $c_0, c_1, \cdots, c_k$ are to be determined. The Gauss-Chebyshev formula leads to the following theorem [12], which gives an explicit way to compute the coefficients:

Theorem 1. *Let $f(t)$ be the function over interval $[-1, 1]$ to be approximated. The polynomial $T_m(t)$ has m roots $\rho_j = \cos\frac{(j-0.5)\pi}{m}$ for $1 \le j \le m$. Now,*

$$c_0 = \frac{1}{m}\sum_{j=1}^{m} f(\rho_j)T_0(\rho_j) = \frac{1}{m}\sum_{j=1}^{m} f(\rho_j)$$
$$c_i = \frac{2}{m}\sum_{j=1}^{m} f(\rho_j)T_i(\rho_j),\ 1 \le i \le k$$

Since $\hat{f}(t)$ only approximates $f(t)$, it may differ from $f(t)$ at time instant $t \in [t_0, t_s]$. To ensure that this approximation leads to no false negatives, we must find a conservative approximation such that the approximation is guaranteed to contain the object's location at all times. This goal can be achieved by computing the maximum deviation $\epsilon_k = \max\left\{|f(t) - \hat{f}(t)|\right\}, t \in [t_0, t_s]$, after obtaining the $k+1$ coefficients. Now the range $[\hat{f}(t) - \epsilon_k, \hat{f}(t) + \epsilon_k]$ is guaranteed to contain $f(t)$ for $t \in [t_0, t_s]$.

The $k + 1$ coefficients can be computed in time $O(mk)$, where m is the highest degree of Chebyshev polynomial used in the approximation, and k is the number of coefficients. The computation of maximum deviation error requires time $O(lk)$, where l is the number of instants between t_0, t_s. Therefore, the total cost is $O(mk + lk)$.

5.3   Clustering Multiple Polynomials

As with any index, each level of the PA-tree must maintain a bound on the key attributes of lower-level nodes. In our case, each non-leaf entry in the index structure for interval $I_i = [T_{i-1}, T_i)$ maintains certain coefficients $\{c_i^\top\}$ and $\{c_i^\perp\}$ that enable us to compute conservative upper and lower bounds for the values of the polynomials $\hat{f}_1(t), \hat{f}_2(t), \cdots, \hat{f}_n(t)$ stored in the child node pointed by the entry. We will now discuss how to compute these bounds if we are using order-k polynomial approximations. In Section 6.2 we discuss the use of these coefficients in query processing. As in Section 5, we first normalize t to $[-1, 1]$.

Let $\hat{f}_j(t) = c_{0,j} + c_{1,j}T_1(t) + \cdots + c_{k,j}T_k(t)$, where $1 \le j \le n$. We store the values $c_i^\top = \max\{c_{i,j}\}$, $c_i^\perp = \min\{c_{i,j}\}$, where $1 \le j \le n$, $0 \le i \le k$ in the index node. For any $t \in [-1, 1]$, the lower- and upper-bounds are computed from the functions

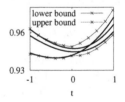

0.96

0.93
-1   0   1
                t

Fig. 4. Bounds

$$\phi^\perp(t) = c_0^\perp + a_1(t)T_1(t) + \cdots + a_k(t)T_k(t), \text{ and}$$
$$\phi^\top(t) = c_0^\top + b_1(t)T_1(t) + \cdots + b_k(t)T_k(t), \text{ where}$$

$$a_i(t) = \begin{cases} c_i^\top, & \text{if } T_i(t) \le 0 \\ c_i^\perp, & \text{otherwise,} \end{cases} \text{ and } b_i(t) = \begin{cases} c_i^\perp, & \text{if } T_i(t) \le 0 \\ c_i^\top, & \text{otherwise,} \end{cases} \text{ for all } i.$$

Theorem 2. *The bounds $\phi^\perp(t)$ and $\phi^\top(t)$ are conservative.*

Proof. Consider any $\hat{f}_j(t) = c_{0,j} + c_{1,j}T_1(t) + \cdots + c_{k,j}T_k(t)$. We will have $\phi^\top(t) - \hat{f}_j(t) = \sum_i (b_i(t) - c_{i,j})T_i(t)$, and $(b_i(t) - c_{i,j}) > 0$ when $T_i(t) > 0$, and $(b_i(t) - c_{i,j}) < 0$ when $T_i(t) < 0$, from $b_i(t)$'s definition. Now, $\phi^\top(t) - \hat{f}_j(t)$ is the sum of positive terms, and is positive. The proof for $\phi^\perp(t)$ is similar.

Figure 4 shows three solid curves representing three 3-order polynomials, while the dotted lines represent the lower- and upper-bounding polynomials computed as above. We note that this bound may not be tight for all t, but it is conservative, guaranteeing that any polynomial $\hat{f}_j(t)$ will be inside the bound. There is an issue with a possibly high computation cost when the query interval $[t_b, t_e]$ is large, since we may have to compute the bound for all $t \in [t_b, t_e]$. Fortunately, in the first level of the PA-tree, only the linear approximations in the form of $c_o + c_1 T_1(t) = c_0 + c_1 t$ are used, which has a much simpler way to compute the bound over $[t_b, t_e]$, due to the monotonicity of $c_0 + c_1 t$ (see Section 6.2).

5.4   Comparing Approximation Quality

To gauge the potential for improvement with our scheme, we compare the dead space obtained using our method with that obtained with the MBR approximation. This metric captures the pruning power of index structures based on the respective approximations. Larger amounts of dead space would suggest smaller pruning power, since it will result in more refinement candidates.

We compare our scheme with the MBR approximations obtained using the LAGreedy algorithm [9]. The volume of each MBR, is simply the product of the edge lengths along the X-dimension, Y-dimension and the temporal dimension. Each entry is a 6-tuple, as discussed in Section 4.

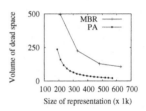

Fig. 5. Dead space

If we use $k + 1$ coefficients each to approximate the X-movements and Y-movements, the volume of dead space can be computed as $4\epsilon_k^x \epsilon_k^y (t_s - t_0)$, where $[t_0, t_s]$ is the temporal domain. The representation size is $2(k + 1) + 5$, since we represent $2(k + 1)$ coefficients in all, the value of k, as well as the maximum deviation and the temporal domain.

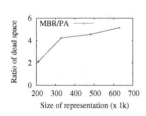

Fig. 6. Ratio

Figure 5 and 6[1] compare the quality of our polynomial approximation with that of MBR approximations for a dataset of 5000 trajectories generated using the network-based generator of [2]. The generator took the road network in San Joaquin County, CA as its input and simulated the movements of objects moving along the road network (see Section 7). Clearly, we see that for a given representation size, the dead space with our our polynomial approximation is much as to 2–5 times smaller than the dead space with MBR approximations. This is expected, since the polynomial approximation captures the inherent smoothness of the movement, and treats the tra-

---

[1] The ratio is computed as the dead space of MBRs over the dead space of PA with the same or smaller size of representation.

jectory as a polyline, rather than a spatial object with extent. Therefore, there is significant potential for improving the overall query performance.

5.5   Choosing the Degree of Approximating Polynomials

In general, the polynomial degree k should be determined based on the characteristics of trajectories. Clearly, there is a trade-off between the approximation quality and the degree k used for approximation. A smaller k value requires less space in the index, as well as less I/O during the filter step. On the other hand, fewer coefficients may result in poorer filtering, causing more trajectories to be examined during the refinement step, increasing its I/O cost.

Another consideration is the complexity of the trajectory segment. Obviously, if the trajectory segment has a relatively simple form, a few coefficients will suffice to get small deviation error. However, since we are not aware of any well-defined notion of complexity for this context, it is not easy to estimate the optimal degree.

We present a heuristic method to estimate the degree k, aimed at minimizing the expected size of representations to be retrieved during query evaluation. We will make the following reasonable assumptions. First, we assume the spatial range r of the query is an $l_x \times l_y$ rectangle, with l_x and l_y uniformly distributed between 0 and some maximum value L. Second, the spatial range r itself is uniformly distributed in the region normalized to a unit square.

Let S_k be the size of representation of k-degree polynomials approximation. Let S be the size of the exact representation for the trajectory segment. Next, we derive the expected size of representations that have to be retrieved for a random query.

If the approximation does not intersect the query range, we can safely prune it out during the filter step. If a segment's approximation lies *completely* inside r, we can safely say it is a true hit during the filter step, and no further checking is needed. We call this category of true hit a *filtering true hit (FT)*. Otherwise, the segment becomes a *candidate (CD)* for the refinement step, in which its exact representation must be retrieved. If a segment lies outside r, we have a *false hit (FH)*. If a segment truly lies inside r, we will record a true hit during refinement, and refer to it as a *refinement true hit (RT)*.

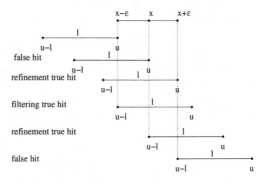

Fig. 7. True and false hits

To estimate the expected I/O cost, we must estimate the probability that the trajectory segment is a candidate for refinement. A candidate can be either a false hit or a refinement true hit. In the following, we consider the X and Y dimensions separately, and omit the x and y superscripts and the subscript k when no confusion is likely.

Let the query range r's projection on X-dimension be $[u - l, u]$. Let the exact location of an object on a trajectory segment at query timestamp be x. As shown in Figure 7, if $u \in [x - \epsilon, x) \cup (x + l, x + l + \epsilon]$, we have a false hit, since r will overlap with our trajectory approximation, but x does not belong to $[u - l, u]$. Therefore, the probability a false hit on the X-dimension is $\Pr[FH_X] = 2\epsilon$.

Further, as shown in Figure 7, if $l \leq 2\epsilon$, there can be no filtering true hit, since the segment's approximation along the X-dimension can not be completely inside r's project on X-dimension. However, when $l \geq 2\epsilon$, and $u \in [x + \epsilon, x + l - \epsilon]$, we have a filtering true hit on the X-dimension. When $u \in [x, x + \epsilon] \cup [x + l - \epsilon, x + l]$, we have a refinement true hit. Therefore, the probability of a refinement true hit on X-dimension is $\Pr[RT_X] = \frac{1}{L} \left(\int_0^{2\epsilon} l\,dl + \int_{2\epsilon}^{L} 2\epsilon\,dl \right) = 2\epsilon - 2\epsilon^2/L$. Now, the probability the trajectory segment is a candidate for refinement on X-dimension is $\Pr[CD_X] = \Pr[FH_X] + \Pr[RT_X] = 4\epsilon - 2\epsilon^2/L$.

Now, since a candidate occurs only when it is a candidate on the X- or Y-dimensions, the probability of a candidate is $\Pr[CD] = 1 - (1 - \Pr[CD_X])(1 - \Pr[CD_Y]) = 1 - (1 - 4\epsilon_k^x + 2(\epsilon_k^x)^2/L)(1 - 4\epsilon_k^y + 2(\epsilon_k^y)^2/L)$.

Now, the expected I/O cost for the trajectory segment when using degree k for approximation is $\overline{IO}_k = S_k + \Pr[CD]S$. This metric provides us a heuristics for estimating the degree k required to be used in the polynomial approximation. More specifically, we would like to find k such that $\overline{IO}_k$ is minimized.

6   PA-Trees and Query Processing

We now present the PA-tree, a new method for indexing polynomial approximations of 2-D trajectories. PA-trees resemble R*-trees, but each entry consists of polynomial coefficients, rather than MBRs. We recall that the temporal domain $[0, T]$ is split into m intervals. In a gross sense, the root node of a PA-tree has m index trees as children, each responsible for indexing trajectory segments within one of these intervals.

Figure 8 shows a PA-tree. Indexing in PA-trees actually occurs at two levels. The first level of indexing is an R*-tree like structure, and is used to index the two leading coefficients of the polynomial describing movement along each dimension. It is reasonable to see this as a 4-dimensional indexing problem, with each dimension corresponding to one coefficient. Each entry in the index structure also holds the maximum deviation errors ϵ_1^x and ϵ_1^y.

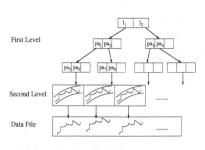

First Level

Second Level

Data File

Fig. 8. An example of PA-tree

As in R*-trees, an entry in a leaf node has the form (ptr, pa), where ptr is the pointer to the exact representation of the trajectory segment, and pa is a tuple of 6 values: $\langle c_0^x, c_1^x, c_0^y, c_1^y, \epsilon_1^x, \epsilon_1^y \rangle$. Entries in non-leaf nodes are of the form (ptr, pa), where ptr is the pointer to a child node, and pa has the form $\langle c_0^{x\perp}, c_0^{y\perp}, c_1^{x\perp}, c_1^{y\perp}, c_0^{x\top}, c_0^{y\top}, c_1^{x\top}, c_1^{y\top}, \epsilon_1^x, \epsilon_1^y \rangle$, representing the lower (upper) bounds of the coefficients for the entries stored in the child node pointed by ptr. Also, pa maintains the maximum ϵ_1^x and ϵ_1^y for all the entries in the subtree.

In the second level, we store more coefficients as well as the corresponding maximum deviation for each trajectory segment, if the estimated degree is larger than 1 (See subsection 5). This information provides more pruning power than the linear approximation used in the first level structure.

Insertions and deletions are similar to the corresponding operations for R*-tree. The primary difference is that we need to ensure that the $\epsilon_1^x, \epsilon_1^y$ values in the non-leaf nodes are the maximum $\epsilon_1^x, \epsilon_1^y$ for all the segments in its subtree.

6.1 Improving Query Performance with Clustered Indices

As suggested in [9], clustered indices can significantly reduce the I/O cost for the refinement step. This optimization can also be applied to the PA-tree, so that all data associated with a leaf node entry is stored sequentially on the disk next to the leaf node itself, resulting in sequential retrieval of data. Clustered indices can be created in two steps. In the first step, a non-clustered index is created. In the second step, we can reorganize the disk pages to store data pages sequentially next to the leaf pages.

We note that clustered indices may not be an appropriate choice in some applications. For example, some applications may need indices clustered on other attributes, say object ID. Also, some applications that may already have collected large amounts of trajectory data, may not allow data reorganization due to its high cost. Consequently, we consider both clustered and non-clustered indices in our experimental evaluation. For both cases, PA-tree shows significant improvements over current methods.

6.2 Query Processing

Given a query $Q(r, t_b, t_e)$, we start with PA-tree root which contains the pointers to the segment index roots and the corresponding temporal intervals. We check whether the temporal interval intersects $[t_b, t_e]$. If they do not, the subtree rooted at that root node is discarded. Otherwise, we search the corresponding subtree.

Let $I_i = [T_{i-1}, T_i]$ be the temporal interval corresponding to an entry in a non-leaf node in the PA-tree. Given $Q(r, t_b, t_e)$, we must check whether there is a trajectory segment inside r at any time $t \in [\max\{t_b, T_{i-1}\}, \min\{t_e, T_i\}]$. Let the index entry be $\langle c_0^{x\perp}, c_0^{y\perp}, c_1^{x\perp}, c_1^{y\perp}, c_0^{x\top}, c_0^{y\top}, c_1^{x\top}, \epsilon_1^x, \epsilon_1^y \rangle$. In the following discussion, we will omit the superscripts x and y for the sake of clarity.

As in Section 5, t is first normalized to $[-1, 1]$. Let t_1 and t_2 be the normalized values of $\max\{t_b, T_{i-1}\}$ and $\min\{t_e, T_i\}$, respectively. Now, the non-leaf entry represents all movement in the approximated linear form $c_0 + c_1 T_1(t) = c_0 + c_1 t$, where $c_0 \in [c_0^\perp, c_0^\top]$, and $c_1 \in [c_1^\perp, c_1^\top]$. In principle, we can apply the dual transformation technique of [10] to check whether there are linear trajectories intersecting r during $[t_1, t_2]$. However, the slope c_1 and the temporal attribute t could be either positive or negative, making it hard to apply duality transformations. Instead, we determine the upper and lower bounding polynomials for the motion segment in the form $c_0 + c_1 t$, where $c_0 \in [c_0^\perp, c_0^\top]$, $c_1 \in [c_1^\perp, c_1^\top]$, and $t \in [t_1, t_2]$. If ϵ_1 is the maximum deviation error, we use the monotonicity of $c_0 + c_1 t$ to compute the lower bound as: $x^\perp = c_0^\perp + \min\{c_1^\perp t_1, c_1^\perp t_2, c_1^\top t_1, c_1^\top t_2\} - \epsilon_1$, and the upper bound as:

Table 1. Characteristics of the datasets used in the experiments

Dataset	Description	Total objects	Average movement functions per object	Total num. of line segments	Dataset size (MB)
CA5k	San Joaquin, CA	5,000	300	1,500,000	48
OD5k	Oldenburg	5,000	258	1,290,000	41

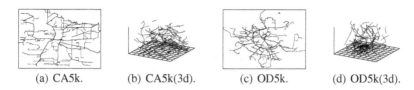

(a) CA5k.   (b) CA5k(3d).   (c) OD5k.   (d) OD5k(3d).

Fig. 9. A snapshot of datasets

$x^\top = c_0^\top + \max\{c_1^\perp t_1, c_1^\perp t_2, c_1^\top t_1, c_1^\top t_2\} + \epsilon_1$. If the computed range intersects with the query range r, we know there may be candidates satisfying the query predicates. We now descend the tree and repeat the process for the subtree rooted at this entry, down to the leaf nodes.

At the leaf node, we will first retrieve the $k + 1$ coefficients in the second level structure, stored sequentially in the leaf nodes. The approximate location $\hat{f}(t)$ at any normalized time instant $t \in [t_1, t_2]$ can be computed using Equation 1, as well as the spatial range $[\hat{f}(t) - \epsilon_k, \hat{f}(t) + \epsilon_k]$. If there is a time $t \in [t_1, t_2]$ such that the spatial computed compass is completely inside r, the trajectory segment is a filtering true hit, its ID will be reported. If this range does not intersect query r for any $t \in [t_1, t_2]$, the trajectory segment is pruned out. Otherwise, refinement is required for determining whether this trajectory segment is a true hit or false hit.

7   Experimental Evaluation

Since no real trajectory data sets are currently publicly available, we generated synthetic data sets using Brinkhof's network-based generator [2]. We used the TIGER data files for the road network in San Joaquin County, CA, and the road network in the city of Oldenburg, German. Our datasets were obtained by running the simulation for a total of 1000 timestamps. We focus mainly on the results of the datasets generated by the network-based generator, since it is has been extensively used in the previous work in this area [4,8,25]. Further, as indicated by some recent work [15,6], movement along roads has practical significance in real-world applications.

Datasets CA5k and OD5k have 5,000 trajectories in all, and were generated with 6 object classes, 3 external object classes, 3,000 initial objects, and 2 new objects per time-instant. We note that each object reports its position and movement function at every time instant during its lifetime, so the number of movement functions for each object will be the same as the duration of its lifetime. Table 1 shows the characteristics of our datasets.

We implemented the PA-tree with the Spatial Index Library of [7]. Our method is compared with the MVR-tree approach [9,8], which uses the LAGreedy algorithm

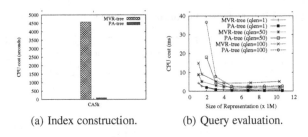

(a) Index construction.   (b) Query evaluation.

Fig. 10. CPU cost

to model each trajectory with multiple MBRs. In the following figures, the legend
`PA-tree` represents our method, while `MVR-tree` represents the method of [9,8].

Our experiments were run on an Intel Pentium IV 1.7Ghz processor, with 512
Mbytes of main memory. The page size is 4Kbyte in all experiments. We use a buffer
with size being about 10% of the original dataset. Unlike [8], we do not reset the buffer
before executing every query, since reseting the buffer will render the buffer useless
when evaluating a workload of multiple queries. Further, we assume the ratio of cost of
sequential I/O to that of random I/O is 1 : 20 [5].

We use three types of query workloads, each containing 1000 queries with varying $qlen$, the length of temporal interval. The three workloads consist of queries with
$qlen = 1$ for timestamp queries, $qlen = 50$ and $qlen = 100$ for medium and large time
interval queries, respectively. Each query range is a rectangle uniformly distributed in
the unit square, with the edge length being uniformly distributed in $[0, 0.1]$. In the following figures, the average query performances per query are reported.

We evaluated performance with respect to the size of index structures, by varying
the number of MBRs for the MVR-tree or the number of interals for the PA-tree. For the
MVR-tree, let $S\%$ represents $(1 + S\%)N$ MBRs are used for a dataset of N trajectories.
S is varied from 10 to 1000. For the PA-tree, we varied m, the number of intervals that
the temporal domain is split into, from 5 to 50.

Clustered Index vs. Non-clustered Index. We tested the query performance for both
clustered index and non-clustered index. For a clustered index, all the trajectory segments associated with the entries in a leaf node will be stored sequentially to that leaf
node. For non-clustered index, same as the TB-tree [17], each data page consists of line
segments belonging to the same trajectory. All the data page will be stored sequentially,
according to the order of the start-time of the line segments (in case of a tier, trajectory
id will be used), while each entry of leaf nodes will have a pointer to its data page.

Further, for clustered index, we notice that assigning all the available buffer to the
index structure can reduce the overall I/O cost. This is because the data pages are sequentially retrieved, while the index pages are retrieved via random I/O. In contrast, for
non-clustered index, both index pages and data pages could be random I/O. Therefore,
we assign 50% buffer to the index structure, while 50% buffer to the data file.

7.1   Performance of Index Construction

For the MVR-tree, building the index structures involved assigning MBRs to each trajectory, creating MBRs for each trajectory, and loading the MBRs into MVR-trees. As pointed in [8], the first two steps are extremely expensive, since it requires one full database scan in order to compute the best approximation per trajectory. In contrast, building PA-trees is much more efficient, since each trajectory can be processed individually. We split each trajectory into segments according to the temporal domain splits, estimate the degree of polynomial approximation and insert the polynomial approximations into the PA-tree. Therefore, as Figure 10(a) shows, the cost of building the MVR-tree is about 50 times higher than that of building the PA-tree for the CA5k dataset. This clearly demonstrates the PA-tree will be a more appropriate choice when the dataset is extremely large, or when the trajectory data is collected at high rate, requiring on-line processing.

7.2   Query Performance

Executing query over the PA-tree requires us to compute the polynomials during the filtering step, so that the CPU cost will be higher than that of the MVR-tree. However, since the PA-tree has higher pruning power, we have a much smaller candidate set for the refinement step, so the CPU cost during the refinement step will be much smaller than the MVR-tree. As a result, in Figure 10(b), we can see that in most cases the overall CPU cost for the PA-tree is actually better than that of the MVR-tree. At any rate, the bottleneck is typically I/O, since CPU speeds tend to improve much faster than I/O speeds. We will therefore focus on the I/O cost, due to space limitations.

Figure 11 and 12 plots the IO performance for the dataset CA5k and OD5K, respectively. We only discuss CA5k in detail, since results for OD5k are quite similar. From Figure 11(a), we observe that both PA-tree and MVR-tree reduce the size of candidate set for the refinement step more effectively with larger index structures. This is expected, since increasing index size implies smaller dead space, and higher approximation quality results in fewer candidates. However, we can clearly see the PA-tree has significantly smaller candidate set than that of the MVR-tree, which is consistent with the comparison shown in Figure 5. Further, this disparity increases with $qlen$, since longer query period implies higher chance of false hits with MBR approximations.

Performance with Clustered Indices. Figure 11(b) shows the total I/O cost including both filtering and refinement steps, in terms of the numbers of equivalent random I/O operations. For all types of workloads, the PA-tree incurs lower overall I/O cost than the MVR-tree. The improved approximation quality in the PA-tree requires checking of fewer index nodes and fewer candidates.

Figure 11(b) captures some interesting trade-offs. A larger index allows better pruning, lowering the number of candidates and I/O cost for the refinement step. However, since the buffer size is fixed, increasing the index size beyond a certain point causes the filtering-step I/O cost to overwhelm the benefits of better pruning. After that point, increasing index size yields no benefit. This results in an upward trend in the I/O cost, which is quite noticeable for the MVR-tree. This effect is stronger with clustered indices, for which a larger fraction of I/O costs are incurred in the filter step.

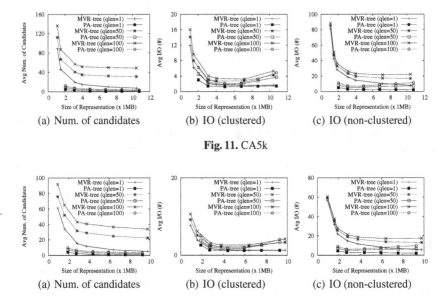

(a) Num. of candidates   (b) IO (clustered)   (c) IO (non-clustered)

Fig. 11. CA5k

(a) Num. of candidates   (b) IO (clustered)   (c) IO (non-clustered)

Fig. 12. OD5k

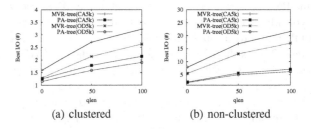

(a) clustered   (b) non-clustered

Fig. 13. Best IO

For a given index size, the MVR-tree is able to split a trajectory into more segments than the PA-tree, since the PA-tree must hold more coefficients per segment. When a clustered index is used, only the line segments of the candidate segment, stored sequentially adjacent to the leaf nodes, will be retrieved. Therefore, the MVR-tree incurs lower I/O cost per candidate trajectory segment. (This advantage disappears for non-clustered indices, as we see shortly.) However, the PA-tree still requires lower refinement step costs than the MVR-tree due to its superior ability to reduce candidate set size.

Figure 13(a) plots the best I/O performance for MVR-tree and PR-tree over all possible index sizes. For clustered index, the MVR-tree is 20%–60% more expensive than the PA-tree.

Performance with Non-clustered Indices. Figure 11(c) plots the overall IO cost using non-clustered indices. PA-tree performance shows even greater improvements over the MVR-tree, and mirrors the improvements in candidate set size. For non-clustered indices, a candidate requires at least one disk I/O, except for a buffer hit. Overall, as Figure 13(b) shows, the best I/O cost achieved with PR-tree is about 3–4 times lower than that for MVR-tree.

8 Conclusions and Future Work

In this paper, we have presented a new parametric indexing method suitable for large trajectory datasets, and for answering historical spatio-temporal queries efficiently. Our polynomial approximations method achieves much better performance than the general used MBR approximation. We present the PA-tree, a two-level structure for indexing trajectories using polynomial approximations. Our comprehensive experimental evaluations demonstrate that the PA-tree significantly outperforms current methods which uses MBR approximation, such as the MVR-tree. Consequently, the PA-tree is an extremely efficient and practical indexing structure for evaluating historical queries over trajectory data. As a future work, we are investigating the applicability of our methods to domains other than trajectory data, such as complex spatial objects.

Acknowledgments

This work was supported in part by grants from Tata Consultancy Services, Inc., the Digital Media Innovations Program of the University of California, and by award FTN F30602-01-2-0536 from the Defense Advanced Research Projects Agency.

References

1. M. Barth.   UCR IntelliShare Project.   http://evwebsvr.cert.ucr.edu/ intellishare/.
2. T. Brinkhoff. Generating Network-Based Moving Objects. In *SSDBM'00*, page 253. IEEE Computer Society, 2000.
3. Y. Cai and R. Ng. Indexing Spatio-temporal Trajectories With Chebyshev Polynomials. In *SIGMOD Conference*, pages 599–610. ACM Press, 2004.
4. V. P. Chakka, A. Everspaugh, and J. M. Patel. Indexing Large Trajectory Data Sets With SETI. In *CIDR*, 2003.
5. L. Chung, B. Worthington, R. Horst, and J. Gray. Windows 200 Disk IO Performance. Microsoft technical report, MS-TR-2000-55, June 2000.
6. S. Gupta, S. Kopparty, and C. V. Ravishankar. Roads, Codes and Spatiotemporal Queries. In *PODS*, pages 115–124, 2004.
7. M. Hadjieleftheriou. Spatial Index Library. http://www.cs.ucr.edu/ marioh/ spatialindex/index.html.
8. M. Hadjieleftheriou, G. Kollios, D. Gunopulos, and V. J. Tsotras. Indexing Spatio-temporal Archives. *The VLDB Journal*. to appear.
9. M. Hadjieleftheriou, G. Kollios, V. J. Tsotras, and D. Gunopulos. Efficient Indexing of Spatiotemporal Objects. In *EDBT '02*, pages 251–268. Springer-Verlag, 2002.
10. G. Kollios, D. Gunopulos, and V. J. Tsotras. On Indexing Mobile Objects. In *PODS '99*, pages 261–272. ACM Press, 1999.

11. G. Kollios, V. J. Tsotras, D. Gunopulos, A. Delis, and M. Hadjieleftheriou. Indexing Animated Objects Using Spatiotemporal Access Methods. *TKDE*, 13(5):758–777, 2001.
12. J. C. Mason and D. Handscomb. *Chebyshev Polynomials*. Chapman and Hall, 2003.
13. Federal Communications Commision. Enhanced 911. http://www.fcc.gov/911/enhanced/.
14. M. A. Nascimento and J. R. O. Silva. Towards Historical R-trees. In *Proceedings of the 1998 ACM symposium on Applied Computing*, pages 235–240. ACM Press, 1998.
15. D. Papadias, J. Zhang, N. Mamoulis, and Y. Tao. Query Processing in Spatial Network Databases. In *VLDB*, pages 802–813, 2003.
16. J. M. Patel, Y. Chen, and V. P. Chakka. STRIPES: An Efficient Index for Predicted Trajectories. In *SIGMOD Conference*, pages 637–646, 2004.
17. D. Pfoser, C. S. Jensen, and Y. Theodoridis. Novel Approaches in Query Processing for Moving Object Trajectories. In *VLDB'00*, pages 395–406. Morgan Kaufmann Publishers Inc., 2000.
18. K. Porkaew, I. Lazaridis, and S. Mehrotra. Querying Mobile Objects in Spatio-Temporal Databases. In *SSTD*, pages 59–78, 2001.
19. S. Saltenis, C. S. Jensen, S. T. Leutenegger, and M. A. Lopez. Indexing the Positions of Continuously Moving Objects. In *SIGMOD Conference*, pages 331–342, 2000.
20. Y. Tao, C. Faloutsos, D. Papadias, and B. Liu. Prediction and Indexing of Moving Objects with Unknown Motion Patterns. In *SIGMOD Conference*, pages 611–622, 2004.
21. Y. Tao and D. Papadias. MV3R-Tree: A Spatio-Temporal Access Method for Timestamp and Interval Queries. In *VLDB '01*, pages 431–440. Morgan Kaufmann Publishers Inc., 2001.
22. Y. Tao, D. Papadias, and J. Sun. The TPR*-Tree: An Optimized Spatio-Temporal Access Method for Predictive Queries. In *VLDB*, pages 790–801, 2003.
23. Y. Theodoridis, M. Vazirgiannis, and T. Sellis. Spatio-Temporal Indexing For Large Multimedia Applications. In *ICMCS '96*. IEEE Computer Society, 1996.
24. X. Xu, J. Han, and W. Lu. RT-tree: An Improved R-tree Index Structure For Spatiotemporal Databases. In *Proc. of the 4th Intl. Symposium on Spatial Data Handling*, 1990.
25. H. Zhu, J. Su, and O. H. Ibarra. Trajectory Queries and Octagons in Moving Object Databases. In *CIKM*, pages 413–421, 2002.

On Trip Planning Queries in Spatial Databases*

Feifei Li, Dihan Cheng, Marios Hadjieleftheriou,
George Kollios, and Shang-Hua Teng

Computer Science Department, Boston University
{lifeifei, dcheng, marioh, gkollios, steng}@cs.bu.edu

Abstract. In this paper we discuss a new type of query in Spatial Databases, called the Trip Planning Query (TPQ). Given a set of points of interest P in space, where each point belongs to a specific category, a starting point S and a destination E, TPQ retrieves the *best* trip that starts at S, passes through at least one point from each category, and ends at E. For example, a driver traveling from Boston to Providence might want to stop to a gas station, a bank and a post office on his way, and the goal is to provide him with the best possible route (in terms of distance, traffic, road conditions, etc.). The difficulty of this query lies in the existence of multiple choices per category. In this paper, we study fast approximation algorithms for TPQ in a metric space. We provide a number of approximation algorithms with approximation ratios that depend on either the number of categories, the maximum number of points per category or both. Therefore, for different instances of the problem, we can choose the algorithm with the best approximation ratio, since they all run in polynomial time. Furthermore, we use some of the proposed algorithms to derive efficient heuristics for large datasets stored in external memory. Finally, we give an experimental evaluation of the proposed algorithms using both synthetic and real datasets.

1 Introduction

Spatial databases has been an active area of research in the last two decades and many important results in data modeling, spatial indexing, and query processing techniques have been reported [29,17,40,37,42,26,36,4,18,27]. Despite these efforts, the queries that have been considered so far concentrate on simple range and nearest neighbor queries and their variants. However, with the increasing interest in intelligent transportation and modern spatial database systems, more complex and advanced query types need to be supported.

In this paper we discuss a novel query in spatial databases, the Trip Planning Query (TPQ). Assume that a database stores the locations of spatial objects that belong to one or more categories from a fixed set of categories $\mathcal{C}$. The user specifies two points in space, a starting point S and a destination point E, and a subset of categories $\mathcal{R}$, ($\mathcal{R} \subseteq \mathcal{C}$), and the goal is to find the *best* trip (route) that starts at S, passes through at least one point from each category in $\mathcal{R}$ and ends at E. An example of a TPQ is the following: A user plans to travel from Boston to Providence and wants to stop

---

* This work was partially supported by NSF grants IIS-0133825, IIS-0308213, CCR-0311430, and ITR CCR-0325630.

C. Bauzer Medeiros et al. (Eds.): SSTD 2005, LNCS 3633, pp. 273–290, 2005.

at a supermarket, a bank, and a post office. Given this query, a database that stores the locations of objects from the categories above (as well as other categories) should compute efficiently a feasible trip that minimizes the total traveling distance. Another possibility is to provide a trip that minimizes the total traveling time.

Efficient TPQ evaluation could become an important new feature of advanced navigation systems and can prove useful for other geographic applications as has been advocated in previous work [12]. For instance, state of the art mapping services like MapQuest, Google Maps, and Microsoft Streets & Trips, currently support queries that specify a starting point and only one destination, or a number of user specified destinations. The functionality and usefulness of such systems can be greatly improved by supporting more advanced query types, like TPQ. An example from Streets & Trips is shown in Figure 1, where the user has explicitly chosen a route that includes an ATM, a gas station and a Greek restaurant. Clearly, the system could not only optimize this route by re-arranging the order in which these stops should be made, but it could also suggest alternatives, based on other options available (e.g., from a large number of ATMs that are shown on the map), that the user might not be aware of.

Fig. 1. A route from Boston University (1) to Boston downtown (5) that passes by a gas station (2), an ATM (3), and a Greek restaurant (4) that have been explicitly specified by the user in that order. Existing applications do not support route optimization, nor do they give suggestions of more suitable routes, like the one presented to the right

TPQ can be considered as a generalization of the Traveling Salesman problem (TSP) [2,1,10] which is NP-hard. The reduction of TSP to TPQ is straightforward. By assuming that every point belongs to its own distinct category, any instance of TSP can be reduced to an instance of TPQ. TPQ is also closely related to the group minimum spanning/steiner tree problems [24,20,16], as we discuss later. From the current spatial database queries, TPQ is mostly related to *time parameterized* and *continuous* NN queries [5,41,36,37], where we assume that the query point is moving with a constant velocity and the goal is to incrementally report the nearest neighbors over time as the query moves from an initial to a final location. However, none of the methods developed to answer the above queries can be used to find a good solution for TPQ.

Contributions. This paper proposes a novel type of query in spatial databases and studies methods for answering this query efficiently. Approximation algorithms that achieve various approximation ratios are presented, based on two important parameters: The total number of categories m and the maximum category cardinality ρ. In particular:

- We introduce four algorithms for answering TPQ queries, with various approximation ratios in terms of m and ρ. We give two practical, easy to implement solutions

better suited for external memory datasets, and two more theoretical in nature algorithms that give tighter answers, better suited for main memory evaluation.
- We present various adaptations of these algorithms for practical scenarios, where we exploit existing spatial index structures and transportation graphs to answer TPQs.
- We perform an extensive experimental evaluation of the proposed techniques on real transportation networks and points of interest, as well as on synthetic datasets for completeness.

In parallel and independently with our work, Sharifzadeh et al. [31], addressed a similar query called the Optimal Sequenced Route (OSR) Query. The main difference between the TPQ and the OSR query is that in the latter, the user has to specify the order of the groups that must be visited.

2   Preliminaries

This section defines formally the general TPQ problem and introduces the basic notation that will be used in the rest of the paper. Furthermore, a concise overview of related work is presented.

2.1   Problem Formulation

We consider solutions for the TPQ problem on *metric graphs*. Given a connected graph $G(\mathcal{V}, \mathcal{E})$ with n vertices $\mathcal{V} = \{v_1, \ldots, v_n\}$ and s edges $\mathcal{E} = \{e_1, \ldots, e_s\}$, we denote the cost of traversing a path $v_i, \ldots, v_j$ with $c(v_i, \ldots, v_j) \geq 0$.

Definition 1. *G is a metric graph if it satisfies the following conditions:*

1. $c(v_i, v_j) = 0$ *iff* $v_i = v_j$
2. $c(v_i, v_j) = c(v_j, v_i)$
3. *The triangle inequality* $c(v_i, v_k) + c(v_k, v_j) \geq c(v_i, v_j)$

Given a set of m categories $\mathcal{C} = \{C_1, \ldots, C_m\}$ (where $m \leq n$) and a mapping function $\pi : v_i \longrightarrow C_j$ that maps each vertex $v_i \in \mathcal{V}$ to a category $C_j \in \mathcal{C}$, the TPQ problem can be defined as follows:

Definition 2. *Given a set $\mathcal{R} \subseteq \mathcal{C}$ ($\mathcal{R} = \{R_1, R_2, \ldots, R_k\}$), a starting vertex S and an ending vertex E, identify the vertex traversal $\mathcal{T} = \{S, v_{t_1}, \ldots, v_{t_k}, E\}$ (also called a trip) from S to E that visits at least one vertex from each category in $\mathcal{R}$ (i.e., $\cup_{i=1}^{k} \pi(v_{t_i})$ $= \mathcal{R}$) and has the minimum possible cost $c(\mathcal{T})$ (i.e., for any other feasible trip $\mathcal{T}'$ satisfying the condition above, $c(\mathcal{T}) \leq c(\mathcal{T}')$).*

In the rest, the total number of vertices is denoted by n, the total number of categories by m, and the maximum cardinality of any category by ρ. For ease of exposition, it will be assumed that $\mathcal{R} = \mathcal{C}$, thus $k = m$. Generalizations for $\mathcal{R} \subset \mathcal{C}$ are straightforward (as will be discussed shortly).

2.2   Related Work

In the context of spatial databases, the TPQ problem has not been addressed before. Most research has concentrated on traditional spatial queries and their variants, namely range queries [18], nearest neighbors [15,19,29], continuous nearest neighbors [5,37,41], group nearest neighbors [26], reverse nearest neighbors [22], etc. All these queries are fundamentally different from TPQ since they do not consider the computation of optimal paths connecting a starting and an ending point, given a graph and intermediate points.

Research in spatial databases also addresses applications in spatial networks represented by graphs, instead of the traditional Euclidean space. Recent papers that extend various types of queries to spatial networks are [27,21,30]. Most of the solutions therein are based on traditional graph algorithms [10,23]. Clustering in a road network database has been studied in [43], where a very efficient data structure was proposed based on the ideas of [32]. Likewise, here we study the TPQ problem on road networks, as well.

The Traveling Salesman Problem (TSP) has received a lot of attention in the last thirty years. A simple polynomial time 2-approximation algorithm for TSP on a metric graph can be obtained using the Minimum Spanning Tree (MST) [10]. The best constant approximation ratio for metric TSP is the $\frac{3}{2}$-approximation that can be derived by the Christofides algorithm [9]. Recently, a polynomial time approximation scheme (PTAS) for Euclidean TSP has been proposed by Arora [1]. For any fixed $\varepsilon > 0$ and any n nodes in $\mathbb{R}^2$ the randomized version of the scheme can achieve a $(1 + \varepsilon)$-approximation in $O(n \log^{O(\frac{1}{\varepsilon})} n)$ running time. Unfortunately, it seems that the TPQ does not admit a PTAS. Furthermore, there are many approximation algorithms for variations of the TSP problem, e.g., TSP with neighborhoods [11]. Nevertheless, the solutions to these problems cannot be applied directly to TPQ, since the problems are fundamentally different. For more approximation algorithms for different versions of TSP, we refer to [2] and the references therein. Finally, there are many practical heuristics for TSP [33], e.g., genetic and greedy algorithms, that work well for some practical instances of the problem, but no approximation bounds are known about them.

TPQ is also closely related to the Generalized Minimum Spanning Tree (GMST) problem. The GMST is a generalized version of the MST problem where the vertices in a graph G belong to m different categories. A tree T is a GMST of G if T contains at least one vertex from each category and T has the minimum possible cost (total weight or total length). Even though the MST problem is in P, it is known that the GMST is in NP. There are a few methods from the operational research and economics community that propose heuristics for solving this problem [24] without providing a detailed analysis on the approximation bounds. The GMST problem is a special instance of an even harder problem, the Group Steiner Tree (GST) problem [16,20]. For example, polylogarithmic approximation algorithms have been proposed recently [14,13]. Since the GMST problem is a special instance of the GST problem, such bounds apply to GMST as well.

3 Fast Approximation Algorithms

In this section we examine several approximation algorithms for answering the trip planning query in main memory. For each solution we provide the approximation ratios in terms of m and ρ. For simplicity, consider that we are given a complete graph G^c, containing one edge per vertex pair v_i, v_j ($1 \leq i, j \leq n$) representing the cost of the shortest path from v_i to v_j in the original graph G. Let $T_k = \{v_{t_0}, v_{t_1}, \ldots, v_{t_k}\}$ denote the partial trip that has visited k vertices, excluding S (where $S = v_{t_0}$). Trivially, it can be shown that a trip T_k constructed on the induced graph G^c, has exactly the same cost as in graph G, with the only difference being that a number of vertices visited on the path from a given vertex to another are hidden. Hiding irrelevant vertices by using the induced graph G^c guarantees that any trip T produced by a given algorithm will be represented by exactly m significant vertices, which will simplify exposition substantially in what follows. In addition, by removing from graph G^c all vertices that do not belong to any of the m categories in R, we can reduce the size of the graph and simplify the construction of the algorithms. Given a solution obtained using the reduced graph and the complete shortest path information for graph G^c, the original trip on graph G can always be acquired. In the following discussion, T_a^P denotes an approximation trip for problem P, while T_o^P denotes the optimal trip. When P is clear from context the superscript is dropped. Furthermore, due to lack of space the proofs for all theorems appear in the full version of this paper.

3.1 Approximation in Terms of m

In this section we provide two greedy algorithms with tight approximation ratios with respect to m.

Nearest Neighbor Algorithm. The most intuitive algorithm for solving TPQ is to form a trip by iteratively visiting the nearest neighbor of the last vertex added to the trip from all vertices in the categories that have not been visited yet, starting from S. Formally, given a partial trip T_k with $k < m$, T_{k+1} is obtained by inserting the vertex $v_{t_{k+1}}$ which is the nearest neighbor of v_{t_k} from the set of vertices in R belonging to categories that have not been covered yet. In the end, the final trip is produced by connecting v_{t_m} to E. We call this algorithm A_{NN}, which is shown in Algorithm 1..

Algorithm 1. $A_{NN}(G^c, R, S, E)$

1: $v = S, I = \{1, \ldots, m\}, T_a = \{S\}$
2: **for** $k = 1$ to m **do**
3:     $v = $ the nearest $NN(v, R_i)$ for all $i \in I$
4:     $T_a \leftarrow \{v\}$
5:     $I \leftarrow I - \{i\}$
6: **end for**
7: $T_a \leftarrow \{E\}$

Theorem 1. A_{NN} *gives a* $(2^{m+1} - 1)$-*approximation (with respect to the optimal solution). In addition, this approximation bound is tight.*

Minimum Distance Algorithm. This section introduces a novel greedy algorithm, called A_{MD}, that achieves a much better approximation bound, in comparison with the previous algorithm. The algorithm chooses a set of vertices $\{v_1, \ldots, v_m\}$, one vertex per category in $\mathcal{R}$, such that the sum of costs $c(S, v_i) + c(v_i, E)$ per v_i is the minimum cost among all vertices belonging to the respective category R_i (i.e., this is the vertex from category R_i with the minimum traveling distance from S to E). After the set of vertices has been discovered, the algorithm creates a trip from S to E by traversing these vertices in nearest neighbor order, i.e., by visiting the nearest neighbor of the last vertex added to the trip, starting with S. The algorithm is shown in Algorithm 2.

Algorithm 2. $\mathcal{A}_{MD}(G^c, \mathcal{R}, S, E)$

1: $U = \emptyset$
2: **for** $i = 1$ to m **do**
3:     $U \leftarrow \pi(v) = R_i : c(S, v) + c(v, E)$ is minimized
4: $v = S, T_a \leftarrow \{S\}$
5: **while** $U \neq \emptyset$ **do**
6:     $v = NN(v, U)$
7:     $T_a \leftarrow \{v\}$
8:     Remove v from U
9: **end while**
10: $T_a \leftarrow \{E\}$

Theorem 2. *If m is odd (even) then $\mathcal{A}_{MD}$ gives an m-approximate ($m + 1$-approximate) solution. In addition this approximation bound is tight.*

3.2   Approximation in Terms of ρ

In this section we consider an Integer Linear Programming approach for the TPQ problem which achieves a linear approximation bound w.r.t. ρ, i.e., the maximum category cardinality. Consider an alternative formulation of the TPQ problem with the constraint that $S = E$ and denote this problem as Loop Trip Planning Query(LTPQ) problem. Next we show how to obtain a $\frac{3}{2}\rho$-approximation for LTPQ using Integer Linear Programming.

   Let $A = (a_{ji})$ be the $m \times (n + 1)$ incidence matrix of G, where rows correspond to the m categories, and columns represent the $n + 1$ vertices (including $v_0 = S = E$). A's elements are arranged such that $a_{ji} = 1$ if $\pi(v_i) = R_j$, $a_{ji} = 0$ otherwise. Clearly, $\rho = max_j \sum_i a_{ji}$, i.e., each category contains at most ρ distinct vertices. Let indicator variable $y(v) = 1$ if vertex v is in a given trip and 0 otherwise. Similarly, let $x(e) = 1$ if the edge e is in a given trip and 0 otherwise. For any $\mathcal{S} \subset \mathcal{V}$, let $\delta(\mathcal{S})$ be the edges contained in the cut $(\mathcal{S}, \mathcal{V} \setminus \mathcal{S})$. The integer programming formulation for the LTPQ problem is the following:

   Problem $IP_{LTPQ} = $ minimize $\sum_{e \in \mathcal{E}} c(e)x(e)$, subject to:

1. $\sum_{e \in \delta(\{v\})} x(e) = 2y(v)$, for all $v \in \mathcal{V}$,
2. $\sum_{e \in \delta(\mathcal{S})} x(e) \geq 2y(v)$, for all $\mathcal{S} \subset \mathcal{V}, v_0 \notin \mathcal{S}$, and all $v \in \mathcal{S}$,
3. $\sum_{i=1}^{n} a_{ji}y(v_i) \geq 1$, for all $j = 1, \ldots, m$,
4. $y(v_0) = 1$,
5. $y(v_i) \in \{0, 1\}, x(e_i) \in \{0, 1\}$

Condition 1 guarantees that for every vertex in the trip there are exactly two edges incident on it. Condition 2 prevents subtrips, that is the trip cannot consist of two disjoint subtrips. Condition 3 guarantees that the chosen vertices cover all categories in $\mathcal{R}$. Condition 4 guarantees that v_0 is in the trip. In order to simplify the problem we can relax the above Integer Programming into LP_{LTPQ} by relaxing Conditions 5 to: $0 \leq y(v), x(e) \leq 1$. Any efficient algorithm for solving Linear Programming could now be applied to solve LP_{LTPQ} [34]. In order to get a feasible solution for IP_{LTPQ}, we apply the randomized rounding scheme stated below:

Randomized Rounding: For solutions obtained by LP_{LTPQ}, set $y(v_i) = 1$ if $y(v_i) \geq \frac{1}{\rho}$. If the trip visits vertices from the same category more than once, randomly select one to keep in the trip and set $y(v_j) = 0$ for the rest.

Theorem 3. LP_{LTPQ} *together with the randomized rounding scheme above finds a* $\frac{3}{2}\rho$-*approximation for* IP_{LTPQ}, *i.e., the integer programming approach is able to find a* $\frac{3}{2}\rho$-*approximation for the LTPQ problem.*

We denote any algorithm for LTPQ as $\mathcal{A}_{LTPQ}$. A TPQ problem can be converted into an LTPQ problem by creating a special category $C_{m+1} = E$. The solution from this converted LTPQ problem is guaranteed to pass through E. Using the result returned by $\mathcal{A}_{LTPQ}$, a trip with constant distortion could be obtained for TPQ:

Lemma 1. *A β-approximation algorithm for LTPQ implies a 3β-approximation algorithm for TPQ.*

Therefore, by combining Theorem 3 and Lemma 1:

Lemma 2. *There is a polynomial time algorithm based on Integer Linear Programming for the TPQ problem with a $\frac{9}{2}\rho$-approximation.*

3.3   Approximation in Terms of m and ρ

In Section 2 we discussed the Generalized Minimum Spanning Tree (GMST) problem which is closely related to the TPQ problem. Recall that the TSP problem is closely related to the Minimum Spanning Tree (MST) problem, where a 2-approximation algorithm can be obtained for TSP based on MST. In similar fashion, it is expected that one can obtain an approximate algorithm for TPQ problem, based on an approximation algorithm for GMST problem.

Unlike the MST problem which is in P, GMST problem is in NP. Suppose we are given an approximation algorithm for GMST problem, denoted $\mathcal{A}_{GMST}$. We can construct an approximation algorithm for TPQ problem as shown in Algorithm 3.

Lemma 3. *If we use a β-approximation algorithm for GMST problem, then Algorithm 3. for TPQ problem is a 2β-approximation algorithm.*

We can get a solution for TPQ by using Lemma 3 and any known approximation algorithm for GST, as GMST is a special instance of GST. For example, the $O(\log^2 \rho \log m)$ algorithm proposed in [14], which yields a solution to TPQ with the same complexity.

Algorithm 3. APPROXIMATION ALGORITHM FOR TPQ BASED ON GMST

1: Compute a β-approximation $Tree_a^{GMST}$ for G rooted at S using $\mathcal{A}_{GMST}$.
2: Let LT be the list of vertices visited in a pre-order tree walk of $Tree_a^{GMST}$.
3: Move E to the end of LT.
4: Return $\mathcal{T}_a^{TPQ}$ as the ordered list of vertices in LT.

---

4   Algorithm Implementations in Spatial Databases

In this section we discuss implementation issues of the proposed TPQ algorithms from a practical perspective, given disk resident datasets and appropriate index structures. We show how the index structures can be utilized to our benefit, for evaluating TPQs efficiently. We opt at providing design details only for the greedy algorithms, $\mathcal{A}_{NN}$ and $\mathcal{A}_{MD}$ since they are simpler to implement in external memory, while the Integer Linear Programming and GMST approaches are more appropriate for main memory and are not easily applicable to external memory datasets.

4.1   Applications in Euclidean Space

First, we consider TPQs in a Euclidean space where a spatial dataset is indexed using an R-tree [18]. We show how to adapt $\mathcal{A}_{NN}$ and $\mathcal{A}_{MD}$ in this scenario. For simplicity, we analyze the case where a single R-tree stores spatial data from all categories.

Implementation of $\mathcal{A}_{NN}$. The implementation of $\mathcal{A}_{NN}$ using an R-tree is straightforward. Suppose a partial trip $\mathcal{T}_k = \{S, p_1, \ldots, p_k\}$ has already been constructed and let $\mathcal{C}(\mathcal{T}_k) = \cup_{i=1}^k \pi(p_i)$, denote the categories visited by $\mathcal{T}_k$. By performing a nearest neighbor query with origin p_k, using any well known NN algorithm, until a new point p_{k+1} is found, such that $\pi(p_{k+1}) \notin \mathcal{C}(\mathcal{T}_k)$, we iteratively extend the trip one vertex at a time. After all categories in $\mathcal{R}$ have been covered, we connect the last vertex to E and the complete trip is returned. The main advantage of $\mathcal{A}_{NN}$ is its efficiency. Nearest neighbor query in R-tree has been well studied. One could expect very fast query performance for $\mathcal{A}_{NN}$. However, the main disadvantage of $\mathcal{A}_{NN}$ is the problem of "searching without directions". Consider the example shown in Figure 2. $\mathcal{A}_{NN}$ will find the trip $T1 = \{S \rightarrow A1 \rightarrow B1 \rightarrow C1 \rightarrow E\}$ instead of the optimal trip $T2 = \{S \rightarrow C2 \rightarrow A2 \rightarrow B2 \rightarrow E\}$. In $\mathcal{A}_{NN}$, the search in every step greedily expands the point that is closest to the last point in the partial trip without considering the end destination, i.e., without considering the direction. The more intuitive approach is to limit the search within a vicinity area defined by S and E. The next algorithm addresses this problem.

Implementation of $\mathcal{A}_{MD}$. Next, we show how to implement $\mathcal{A}_{MD}$ using an R-tree. The main idea is to locate the m points, one from each category in $\mathcal{R}$, that minimize the Euclidean distance $\mathcal{D}(S, E, p) = c(S, p) + c(p, E)$ from S to E through p. We call this the minimum distance query. This query meets our intuition that the trip planning query should be limited within the vicinity area of the line segment defined by S, E (as in the example in Figure 2). The minimum distance query can be answered by modifying the NN search algorithm for R-trees [29], where instead of using the traditional $MinDist$

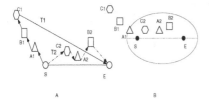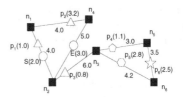

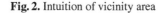

Fig. 2. Intuition of vicinity area   **Fig. 3.** A simple road network

measure for sorting candidate distances, we use $\mathcal{D}$. In that case, the vicinity area is an ellipse and not a circle (Figure 2). Given S and E we run the modified NN search once for locating all m points incrementally, and report the final trip.

All NN algorithms based on R-trees compute the nearest neighbors incrementally using the tree structure to guide the search. An interesting problem that arises in this case is how to geometrically compute the minimum possible distance $\mathcal{D}(S, E, p)$ between points S, E and any point p inside a given MBR M (similar to the $MinDist$ heuristic of the traditional search). This problem can be reduced to that of finding the point p on line segment AB (where AB is a boundary of M) that minimizes $\mathcal{D}(S, E, p)$, which can then be used to find the minimum distance from M, by applying it on the MBR boundaries lying closer to line segment SE. Point p can be computed by projecting the mirror image E' of E, given AB. It can be proved that:

Lemma 4. *Given line segments AB and SE, the point p that minimizes $\mathcal{D}(S, E, p)$ is:*
Case A: If EE' intersects AB, then p is the intersection of AB and SE'.
Case B: If EE' and SE do not intersect AB, then p is either A or B.
Case C: If SE intersects AB, then p is the intersection of SE and AB.

Using the lemma, we can easily compute the minimum distances $\mathcal{D}(S, E, M)$ for appropriately sorting the R-tree MBRs during the NN search. The details of the minimum distance query algorithm is shown in Algorithm 4.. For simplicity, here we show the algorithm that searches for a point from one particular category only, which can easily be extended for multiple categories. In line 8 of the algorithm, if c is a node then $\mathcal{D}(S, E, c)$ is calculated by applying Lemma 4 with line segments from the borders of the MBR of c; if c is a point then $\mathcal{D}(S, E, c)$ is the length $|Sc| + |cE|$. Straightforwardly, the algorithm can also be modified for returning the top k points.

4.2 Applications in Road Networks

An interesting application of TPQs is on road network databases. Given a graph $\mathcal{N}$ representing a road network and a separate set $\mathcal{P}$ representing points of interest (gas stations, hotels, restaurants, etc.) located at fixed coordinates on the edges of the graph, we would like to develop appropriate index structures in order to answer efficiently trip planning queries for visiting points of interest in $\mathcal{P}$ using the underlying network $\mathcal{N}$. Figure 3 shows an example road network, along with various points of interest belonging to four different categories.

Algorithm 4. ALGORITHM MINIMUM DISTANCE QUERY FOR R-TREES

Require: Points S, E, Category R_i, R-tree rtree
 1: PriorityQueue $QR = \emptyset$, $QS = \{(rtree.root, 0)\}$; $B = \infty$
 2: **while** QS not empty **do**
 3:      $n = QS.top$;
 4:      **if** $n.dist \geq B$ **then**
 5:          return $QR.top$
 6:      **for** all children c of n **do**
 7:          $dist = \mathcal{D}(S, E, c)$
 8:          **if** n is an index node **then**
 9:              $QS \leftarrow (c, dist)$
10:          **else if** $\pi(M) = R_i$ **then**   ▷ (c is a point)
11:              $QR \leftarrow (c, dist)$
12:              **if** $dist \leq B$ **then** $B = dist$

For our purposes we represent the road network using techniques from [32,43,27]. In summary, the adjacency list of $\mathcal{N}$ and set $\mathcal{P}$ are stored as two separate flat files indexed by B^+-trees. For that purpose, the location of any point $p \in \mathcal{P}$ is represented as an offset from the road network node with the smallest identifier that is incident on the edge containing p. For example, point p_4 is 1.1 units away from node n_3.

Implementation of $\mathcal{A}_{NN}$. Nearest neighbor queries on road networks have been studied in [27], where a simple extension of the well known Dijkstra algorithm [10] for the single-source shortest-path problem on weighted graphs is utilized to locate the nearest point of interest to a given query point. As with the R-tree case, straightforwardly, we can utilize the algorithm of [27] to incrementally locate the nearest neighbor of the last stop added to the trip, that belongs to a category that has not been visited yet. The algorithm starts from point S and when at least one stop from each category has been added to the trip, the shortest path from the last discovered stop to E is computed.

Implementation of $\mathcal{A}_{MD}$. Similarly to the R-tree approach, the idea is to first locate the m points from categories in $\mathcal{R}$ that minimize the network distance $c(S, p_i, E)$ using the underlying graph $\mathcal{N}$, and then create a trip that traverses all p_i in a nearest neighbor order, from S to E. It is easy to show with a counter example that simply finding a point p that first minimizes cost $c(S, p)$ and then traverses the shortest path from p to E, does not necessarily minimize cost $c(S, p, E)$. Thus, Dijkstra's algorithm cannot be directly applied to solve this problem. Alternatively, we propose an algorithm for identifying such points of interest. The procedure is shown in Algorithm 5.

The algorithm locates a point of interest $p : \pi(p) \in R_i$ (given R_i) such that the distance $c(S, p, E)$ is minimized. The search begins from S and incrementally expands all possible paths from S to E through all points p. Whenever such a path is computed and all other partial trips have cost smaller than the tentative best cost, the search stops. The key idea of the algorithm is to separate partial trips into two categories: one that contains only paths that have not discovered a point of interest yet, and one that contains paths that have. Paths in the first category compete to find the shortest possible route from S to any p. Paths in the second category compete to find the shortest path from their respective p to E. The overall best path is the one that minimizes the sum of both costs.

Algorithm 5. ALGORITHM Minimum Distance Query FOR ROAD NETWORKS

Require: Graph $\mathcal{N}$, Points of interest $\mathcal{P}$, Points S, E, Category R_i
1: For each $n_i \in \mathcal{N} : n_i.c_p = n_i.c_{\neg p} = \infty$
2: PriorityQueue $PQ = \{S\}$, $B = \infty$, $T_B = \emptyset$
3: **while** PQ not empty **do**
4:      $T = PQ.top$
5:      **if** $T.c \geq B$ **then** return T_B
6:      **for** each node n adjacent to $T.last$ **do**
7:          $T' = T$ ▷ (create a copy)
8:          **if** T' does not contain a p **then**
9:              **if** $\exists p : p \in \mathcal{P}, \pi(p) = R_i$ on edge $(T'.last, n)$ **then**
10:                  $T'.c+ = c(T'.last, p)$
11:                  $T' \leftarrow p, PQ \leftarrow T'$
12:              **else**
13:                  $T'.c+ = c(T'.last, n), T' \leftarrow n$
14:                  **if** $n.c_{\neg p} > T'.c$ **then**
15:                      $n.c_{\neg p} = T'.c, PQ \leftarrow T'$
16:          **else**
17:              **if** edge (T', n) contains E **then**
18:                  $T'.c+ = c(T'.last, E), T' \leftarrow E$
19:                  Update B and T_B accordingly
20:              **else**
21:                  $T'.c+ = c(T'.last, n), T' \leftarrow n$
22:                  **if** $n.c_p > T'.c$ **then**
23:                      $n.c_p = T'.c, PQ \leftarrow T'$
24:              **endif**
25:      **endfor**
26: **endwhile**

The algorithm proceeds greedily by expanding at every step the trip with the smallest current cost. Furthermore, in order to be able to prune trips that are not promising, based on already discovered trips, the algorithm maintains two partial best costs per node $n \in \mathcal{N}$. Cost $n.c_p$ ($n.c_{\neg p}$) represents the partial cost of the best trip that passes through this node and that has (has not) discovered an interesting point yet. After all k points(one from each category $R_i \in \mathcal{R}$) have been discovered by iteratively calling this algorithm, an approximate trip for TPQ can be produced. It is also possible to design an incremental algorithm that discovers all points from categories in $\mathcal{R}$ concurrently.

5   Extensions

5.1   I/O Analysis for the Minimum Distance Query

In this section we study the I/O bounds for the minimum distance query in Euclidean space, i.e., the expected number of I/Os when we try to find the point p that minimizes $\mathcal{D}(S, E, p)$ from a point set indexed with an R-tree. By carefully examining Algorithm 4. and Lemma 4, we can claim the following:

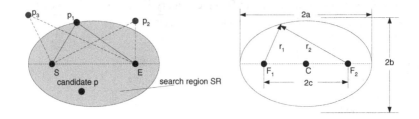

Fig. 4. The search region of a minimum distance query

Claim. The lower bound of I/Os for minimum distance queries is the number of MBRs that intersect with line segment SE.

For the average case, the classical cost models for nearest neighbor queries can be used [39,7,6,28,38]. On average the I/O for any type of queries on R-trees is given by the expected node access: $NA = \sum_{i=0}^{h-1} n_i P_{NA_i}$ where h is the height of the tree, n_i is the number of nodes in level i and P_{NA_i} is the probability that a node at level i is accessed. The only peculiarity of minimum distance queries is that their search region *SR*, i.e., the area of the data space that may contain candidate results, forms an ellipse with focii the points S, E. It follows immediately that, on average, in order to answer a minimum distance query we have to visit all MBRs that intersect with its respective *SR*. Thus, if we quantify the size of *SR* we can estimate P_{NA_i}.

Consider the example in Figure 4, and suppose p_1 is currently the point that minimizes $\mathcal{D}(S, E, p_1)$. Then the ellipse defined by S, E, p_1 will be the region that contains possible better candidates, e.g., p in this example. This is true due to the property of the ellipse that $r_1 + r_2 = 2a$, i.e., any point p' on the border of the ellipse satisfies $\mathcal{D}(S, E, p') = 2a$. Therefore, to estimate the I/O cost of the query all we need to do is estimate quantity a. Assuming uniformity and a unit square universe, we have $Area_{SR} = k/|P|$. We also know that $Area_{SR} = Area_{ellipse} = 2\pi/\sqrt{4ac - b^2} = 2\pi/\sqrt{4ac - (a^2 - c^2)}$. Hence, $a = 2c + \sqrt{5c^2 - (\frac{2\pi|P|}{k})^2}$

With $S, E, c = |SE|/2$, and a, we could determine the search region for a k minimum distance query. With the search region being identified, one could derive the probability of any node of the R-tree being accessed. Then, the standard cost model analysis in [7,6,28,38] can be straightforwardly be applied, hence the details are omitted. Generalizations for non-uniform distributions can also be addressed similarly to the analysis presented in [38], where few modifications are required given the ellipsoidal shape of the search regions. The I/O estimation for queries on road networks is much harder to analyze and heavily depends on the particular data structures used, therefore it is left as future work.

5.2 Hybrid Approach

We also consider a hybrid approach to the trip planning query for disk based datasets (in both Euclidean space and road networks). Instead of evaluating the queries using

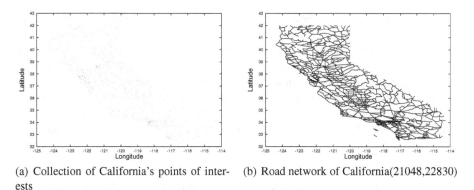

(a) Collection of California's points of inter-   (b) Road network of California(21048,22830)
ests

Fig. 5. Real dataset from California

the proposed algorithms, the basic idea is to first select a sufficient number of good candidates from disk, and then process those in main memory. We apply the minimum distance query to locate the top k points from each respective category and then, assuming that the query visits a total of m categories, the $k \times m$ points are processed in main memory using any of the strategies discussed in Section 3. In addition, an exhaustive search is also possible. In this case, there are m^k number of instances to be checked. If m^k is large, a subset can be randomly selected for further processing, or the value of k is reduced. Clearly, the hybrid approach will find a solution at least as good as algorithm $\mathcal{A}_{MD}$. In particular, since the larger the value of k the closer the solution will be to the optimal answer, with a hybrid approach the user can tune the accuracy of the results, according to the cost she is willing to pay.

6   Experimental Evaluation

This section presents a comprehensive performance evaluation of the proposed techniques for TPQ in spatial databases. We used both synthetic datasets generated on real road networks and real datasets from the state of California. All experiments were run on a Linux machine with an Intel Pentium 4 2.0GHz CPU.

Experimental Setup. To generate synthetic datasets we obtained two real road networks, the city of Oldenburg(OL) with 6105 nodes and 7035 edges and San Joaquin county(TG) with 18263 nodes and 23874 edges, from [8]. For each dataset, we generated uniformly at random a number of points of interest on the edges of the network. Datasets with varying number of categories, as well as varying densities of points per category were generated. The total number of categories is in the range $m \in [5, 30]$, while the category density is in the range of $\rho \in [0.01N, 0.25N]$, where N is the total number of edges in the network. For Euclidean datasets, points of interest are generated using the road networks, but the distances are computed as direct Euclidean distances between points, without the network constraints. Our synthetic dataset has the flexibility of controlling different densities and number of categories, however it is based on uniform distribution on road network (not necessarily uniform in the Euclidean space).

To study the general distribution of different categories, we also obtain a real dataset for our experiments. First we get a collection of points of interests that fall into different categories for the state of California from [35] as shown in Figure 5(a), then we obtain the road network for the same state from [25] as shown in Figure 5(b). Both of them represent the locations in a longitude/latitude space, which makes the merging step straightforward. The California dataset has 63 different categories, including airports, hospitals, bars, etc., and altogether more than $100,000$ points. Different categories exhibit very different densities and distributions. The road network in California has $21,048$ nodes and $22,830$ edges. For all experiments, we generate 100 queries with randomly chosen S and E.

Road Network Datasets. In this part we study the performance of the two algorithms for road networks. First, we study the effects of m and ρ. Due to lack of space we present the results for the OL based datasets only. The results for the TG datasets were similar. Figure 6(a) plots the results for the average trip length as a function of m, for $\rho = 0.01N$. Figure 6(b) plots the average trip length as a function of ρ, for $m = 30$. In both cases, clearly $\mathcal{A}_{MD}$ outperforms $\mathcal{A}_{NN}$. In general, $\mathcal{A}_{MD}$ gives a trip that is 20%-40% better (in terms of trip length) than the one obtained from $\mathcal{A}_{NN}$. It is interesting to note that with the increase of m and the decrease of ρ the performance gap between the two algorithms increases. $\mathcal{A}_{NN}$ is greatly affected by the relative locations of points as it greedily follows the nearest point from the remaining categories irrespective of its direction with respect to the destination E. With the increase of m, the probability that $\mathcal{A}_{NN}$ wanders off the correct direction increases. With the decrease of ρ, the probability that the next nearest neighbor is close enough decreases, which in turn increases the chance that the algorithm will move far away from E. However, for both cases $\mathcal{A}_{MD}$ is not affected.

We also study the query cost of the two algorithms measured by the average running time of one query. Figure 7(a) plots the results as a function of density, and $m = 15$. In general, $\mathcal{A}_{NN}$ has smaller runtime. The reason is that the $\mathcal{A}_{MD}$ query in the road network is much more complex and needs to visit an increased number of nodes multiple times.

Euclidean Datasets. Due to lack of space we omit the plots for Euclidean datasets. In general, the results and conclusions were the same as for the road network datasets. A small difference is that the performance of the two algorithms is measured with respect to the total number of R-tree I/Os. In this case, $\mathcal{A}_{NN}$ was more efficient than $\mathcal{A}_{MD}$, especially for higher densities as shown in Figure 7(b).

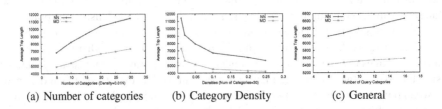

(a) Number of categories   (b) Category Density   (c) General

Fig. 6. Average trip length of $\mathcal{A}_{NN}$ and $\mathcal{A}_{MD}$

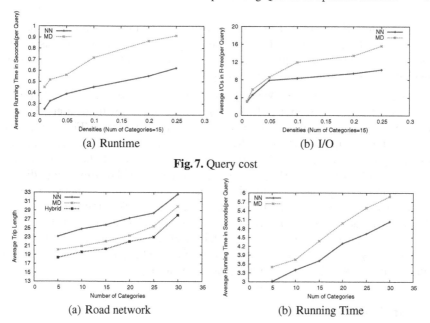

(a) Runtime

(b) I/O

Fig. 7. Query cost

(a) Road network

(b) Running Time

Fig. 8. Experiments with real dataset

General Datasets and Query Workloads. In the previous experiments datasets had a fixed density for all categories. Furthermore, queries had to visit all categories. Here, we examine a more general setting where the density for different categories is not fixed and queries need to visit a subset $\mathcal{R}$ of all categories. Figure 6(c) summarizes the results. We set $m = 20$ and ρ uniformly distributed in $[0.01N, 0.20N]$. We experiment with subsets of varying cardinalities per query and measure the average trip length returned by both algorithms. $\mathcal{A}_{MD}$ outperforms $\mathcal{A}_{NN}$ by 15% in the worst case. With the increase of the cardinality of $\mathcal{R}$, the performance gain on $\mathcal{A}_{MD}$ increases.

Real Datasets. So far we have tested our algorithm on synthetic datasets To compare the performance of the algorithms in a real setting, we apply $\mathcal{A}_{NN}$ and $\mathcal{A}_{MD}$ on the real dataset from California. There are 63 different categories in this dataset, hence we show the query workload that requires visits to a subset of categories (up to 30 randomly selected categories). Figure 8(a) compares the average trip length obtained by $\mathcal{A}_{NN}$ and $\mathcal{A}_{MD}$ in the road network case. In this case, we simply use longitude and latitude as the point coordinates and calculate the distance based on that. So the absolute value for the distance is small. As we have noticed, $\mathcal{A}_{MD}$ still outperforms $\mathcal{A}_{NN}$ in terms of trip length, however, with the price of a higher query cost as indicated in Figure 8(b). Notice that the running time in this experiment is much higher than the one in Figure 7(a) as we are dealing with a much larger network as well as more data points. Similar results have been observed for the same dataset in Euclidean space (where the cost is measured in I/Os) and they are omitted. It is interesting to note that the trip length is increasing w.r.t. the number of categories in a non-linear fashion (e.g., from 25 categories to 30 categories), as compared to the same experiment on the synthetic dataset shown in Figure 6(a). This could be explained by the non-uniformity property and skewness of the real dataset. For example, there are more than 900 airports and only about 50 harbors. So when a query category for harbors is included, one expect to see a steep increase in the trip length.

Study of the Hybrid Approach. We also investigate the effectiveness of the hybrid approach as suggested in Section 5.2. Our experiments on synthetic datasets show that the hybrid approach improves results over $\mathcal{A}_{MD}$ by a small margin (Figure 8(a)). This is expected due to the uniformity of the underlying datasets. With the real dataset, as we can see in Figure 8(a), there is a noticeable improvement with the hybrid approach over $\mathcal{A}_{MD}$ (we set $m = 5$). This is mainly due to the skewed distribution in different categories in the real dataset. The hybrid approach incurs additional computational cost in main memory (i.e., cpu time) but identifies better trips. We omit the running time of hybrid approach from Figure 8(b) as it exhibits exponential increase($O(m^k)$) with the number of categories. However, when the number of categories is small, the running time of hybrid approach is comparable to $\mathcal{A}_{NN}$ and $\mathcal{A}_{MD}$, e.g., when $m = 5$ its running time is about 3.8 seconds for one query, on average.

7   Conclusions and Future Work

We introduced a novel query for spatial databases, namely the Trip Planning Query. First, we argued that this problem is NP-Hard, and then we developed four polynomial time approximation algorithms, with efficient running time and varying worst case guarantees. We also showed how to apply these algorithms in practical scenarios, both for Euclidean spaces and Road Networks. Finally, we presented a comprehensive experimental evaluation. For future work we plan to extend our algorithms to support trips with user defined constraints. Examples include visiting a certain category during a specified time period [3], visiting categories in a given order, and more.

References

1. S. Arora. Polynomial time approximation schemes for euclidean tsp and other geometric problems. In *FOCS*, page 2, 1996.
2. S. Arora. Approximation schemes for NP-hard geometric optimization problems: A survey. *Mathematical Programming*, 2003.
3. N. Bansal, A. Blum, S. Chawla, and A. Meyerson. Approximation algorithms for deadline-tsp and vehicle routing with time-windows. In *STOC*, pages 166–174, 2004.
4. N. Beckmann, H. Kriegel, R. Schneider, and B. Seeger. The R*-tree: An efficient and robust access method for points and rectangles. In *SIGMOD*, pages 220–231, 1990.
5. R. Benetis, C. S. Jensen, G. Karciauskas, and S. Saltenis. Nearest neighbor and reverse nearest neighbor queries for moving objects. In *IDEAS*, pages 44–53, 2002.
6. S. Berchtold, C. Böhm, D. A. Keim, and H.-P. Kriegel. A cost model for nearest neighbor search in high-dimensional data space. In *PODS*, pages 78–86, 1997.
7. C. Böhm. A cost model for query processing in high dimensional data spaces. *TODS*, 25(2):129–178, 2000.
8. T. Brinkhoff. A framework for generating network-based moving objects. *GeoInformatica*, 6(2):153–180, 2002.

9. N. Christofides. Worst-case analysis of a new heuristic for the travelling salesman problem. Technical report, Computer Science Department, Carnegie Mellon University, 1976.

10. T. Cormen, C. Leiserson, R. Rivest, and C. Stein. *Introduction to Algorithms*. The MIT Press, 1997.

11. A. Dumitrescu and J. S. B. Mitchell. Approximation algorithms for tsp with neighborhoods in the plane. In *SODA*, pages 38–46, 2001.

12. Max J. Egenhofer. What's special about spatial?: database requirements for vehicle navigation in geographic space. In *SIGMOD*, pages 398–402, 1993.

13. G. Even and G. Kortsarz. An approximation algorithm for the group steiner problem. In *SODA*, pages 49–58, 2002.

14. J. Fakcharoenphol, S. Rao, and K. Talwar. A tight bound on approximating arbitrary metrics by tree metrics. *Journal of Computer and System Sciences*, 69(3):485–497, 2004.

15. H. Ferhatosmanoglu, I. Stanoi, D. Agrawal, and A. E. Abbadi. Constrained nearest neighbor queries. In *SSTD*, pages 257–278, 2001.

16. N. Garg, G. Konjevod, and R. Ravi. A polylogarithmic approximation algorithm for the group steiner tree problem. *Journal of Algorithms*, 37(1):66–84, 2000.

17. R. Hartmut Guting, M. H. Bohlen, M. Erwig, C. S. Jensen, N. A. Lorentzos, M. Schneider, and M. Vazirgiannis. A foundation for representing and querying moving objects. *TODS*, 25(1):1–42, 2000.

18. A. Guttman. R-trees: A dynamic index structure for spatial searching. In *SIGMOD*, pages 47–57, 1984.

19. G. Hjaltason and H. Samet. Distance Browsing in Spatial Databases. *TODS*, 24(2):265–318, 1999.

20. E. Ihler. Bounds on the Quality of Approximate Solutions to the Group Steiner Problem. Technical report, Institut fur Informatik, Uiversity Freiburg, 1990.

21. M. R. Kolahdouzan and C. Shahabi. Voronoi-based k nearest neighbor search for spatial network databases. In *VLDB*, pages 840–851, 2004.

22. F. Korn and S. Muthukrishnan. Influence sets based on reverse nearest neighbor queries. In *SIGMOD*, pages 201–212, 2000.

23. R. Motwani and P. Raghavan. *Randomized Algorithms*. Cambridge University Press, 1995.

24. Y. S. Myung, C. H. Lee, and D. W. Tcha. On the Generalized Minimum Spanning Tree Problem. *Networks*, 26:231–241, 1995.

25. Digital Chart of the World Server. http://www.maproom.psu.edu/dcw/.

26. D. Papadias, Q. Shen, Y. Tao, and K. Mouratidis. Group nearest neighbor queries. In *ICDE*, pages 301–312, 2004.

27. D. Papadias, J. Zhang, N. Mamoulis, and Y. Tao. Query processing in spatial network databases. In *VLDB*, pages 802–813, 2003.

28. A. Papadopoulos and Y. Manolopoulos. Performance of nearest neighbor queries in r-trees. In *ICDT*, pages 394–408, 1997.

29. N. Roussopoulos, S. Kelley, and F. Vincent. Nearest neighbor queries. In *SIGMOD*, pages 71–79, 1995.

30. C. Shahabi, M. R. Kolahdouzan, and M. Sharifzadeh. A road network embedding technique for k-nearest neighbor search in moving object databases. In *GIS*, pages 94–100, 2002.

31. M. Sharifzadeh, M. Kolahdouzan, and C. Shahabi. The Optimal Sequenced Route Query. Technical report, Computer Science Department, University of Southern California, 2005.

32. S. Shekhar and D.-R. Liu. Ccam: A connectivity-clustered access method for networks and network computations. *TKDE*, 9(1):102–119, 1997.

33. TSP Home Web Site. http://www.tsp.gatech.edu/.

34. D. A. Spielman and S.-H. Teng. Smoothed analysis of algorithms: why the simplex algorithm usually takes polynomial time. In *STOC*, pages 296–305, 2001.

35. U.S. Geological Survey. http://www.usgs.gov/.

36. Y. Tao and D. Papadias. Time-parameterized queries in spatio-temporal databases. In *SIG-MOD*, pages 334–345, 2002.

37. Y. Tao, D. Papadias, and Q. Shen. Continuous nearest neighbor search. In *VLDB*, pages 287–298, 2002.

38. Y. Tao, J. Zhang, D. Papadias, and N. Mamoulis. An Efficient Cost Model for Optimization of Nearest Neighbor Search in Low and Medium Dimensional Spaces. *TKDE*, 16(10):1169–1184, 2004.

39. Y. Theodoridis, E. Stefanakis, and T. Sellis. Efficient cost models for spatial queries using r-trees. *TKDE*, 12(1):19–32, 2000.

40. M. Vazirgiannis and O. Wolfson. A spatiotemporal model and language for moving objects on road networks. In *SSTD*, pages 20–35, 2001.

41. X. Xiong, M. F. Mokbel, and W. G. Aref. Sea-cnn: Scalable processing of continuous k-nearest neighbor queries in spatio-temporal databases. In *ICDE*, pages 643–654, 2005.

42. X. Xiong, M. F. Mokbel, W. G. Aref, S. E. Hambrusch, and S. Prabhakar. Scalable spatio-temporal continuous query processing for location-aware services. In *SSDBM*, pages 317–327, 2004.

43. M. L. Yiu and N. Mamoulis. Clustering objects on a spatial network. In *SIGMOD*, pages 443–454, 2004.

Capacity Constrained Routing Algorithms for Evacuation Planning: A Summary of Results*

Qingsong Lu**, Betsy George, and Shashi Shekhar

Department of Computer Science and Engineering,
University of Minnesota, 200 Union St SE,
Minneapolis, MN 55455, USA
{lqingson, bgeorge, shekhar}@cs.umn.edu
http://www.cs.umn.edu/research/shashi-group/

Abstract. Evacuation planning is critical for numerous important applications, e.g. disaster emergency management and homeland defense preparation. Efficient tools are needed to produce evacuation plans that identify routes and schedules to evacuate affected populations to safety in the event of natural disasters or terrorist attacks. The existing linear programming approach uses time-expanded networks to compute the optimal evacuation plan and requires a user-provided upper bound on evacuation time. It suffers from high computational cost and may not scale up to large transportation networks in urban scenarios. In this paper we present a heuristic algorithm, namely Capacity Constrained Route Planner(CCRP), which produces sub-optimal solution for the evacuation planning problem. CCRP models capacity as a time series and uses a capacity constrained routing approach to incorporate route capacity constraints. It addresses the limitations of linear programming approach by using only the original evacuation network and it does not require prior knowledge of evacuation time. Performance evaluation on various network configurations shows that the CCRP algorithm produces high quality solutions, and significantly reduces the computational cost compared to linear programming approach that produces optimal solutions. CCRP is also scalable to the number of evacuees and the size of the network.

Keywords: evacuation planning, routing and scheduling, transportation network.

---

* This work was supported by Army High Performance Computing Research Center contract number DAAD19-01-2-0014 and the Minnesota Department of Transportation contract number 81655. The content of this work does not necessarily reflect the position or policy of the government and no official endorsement should be inferred. Access to computing facilities was provided by the AHPCRC and the Minnesota Supercomputing Institute.
** Corresponding author.

C. Bauzer Medeiros et al. (Eds.): SSTD 2005, LNCS 3633, pp. 291–307, 2005.

1 Introduction

Evacuation planning is critical for numerous important applications, e.g. disaster emergency management and homeland defense preparation. Traditional evacuation warning systems simply convey the threat descriptions and the need for evacuation to the affected population via mass media communication. Such systems do not consider capacity constraints of the transportation network and thus may lead to unanticipated effects on the evacuation process. For example, when Hurricane Andrew was approaching Florida in 1992, the lack of effective planning caused tremendous traffic congestions, general confusion and chaos [1]. Therefore, efficient tools are needed to produce evacuation plans that identify routes and schedules to evacuate affected populations to safety in the event of natural disasters or terrorist attacks [12,14,7,8].

The current methods of evacuation planning can be divided into two categories, namely traffic assignment-simulation approach and route-schedule planning approach. The traffic assignment-simulation approach uses traffic simulation tools, such as DYNASMART [27] and DynaMIT [5], to conduct stochastic simulation of traffic movements based on origin-destination traffic demands and uses queuing methods to account for road capacity constraints. However, it may take a long time to complete the simulation process for a large transportation network. The route-schedule planning approaches use network flow and routing algorithms to produce origin-destination routes and schedules of evacuees on each route. Many research works have been done to model the evacuation problem as a network flow problem [15,4] and to find the optimal solution using linear programming methods. Hamacher and Tjandra [17] gave an extensive literature review of the models and algorithms used in these linear programming methods. Based on the triple-optimization results by Jarvis and Ratliff [20], linear programming method for evacuation route planning works as follows. First, it models the evacuation network into a network graph, as shown by network G in Figure 1, and it requires the user to provide an estimated upper bound T of the evacuation egress time. Second, it converts evacuation network G to a time-expanded network, as shown by G_T in Figure 2, by duplicating the original evacuation network G for each discrete time unit $t = 0, 1, \ldots, T$. Then, it defines the evacuation problem as a minimum cost network flow problem [15,4] on the time-expanded network G_T. Finally, it feeds the expanded network G_T to minimum cost network flow solvers, such as NETFLO [21], to find the optimal solution. For example, EVACNET [9,16,22,23] is a computer program based on this approach which computes egress time for building evacuations. It uses NETFLO code to obtain the optimal solution. Hoppe and Tardos [18,19] gave a polynomial time bounded algorithm by using ellipsoid method of linear programming to find the optimal solution for the minimum cost flow problem. Theoretically, ellipsoid method has a polynomial bounded running time. However, it performs poorly in practice and has little value for real application [6].

Linear programming approach can produce optimal solutions for evacuation planning. It is useful for evacuation scenarios with moderate size networks, such as building evacuation. However, this approach has the following limita-

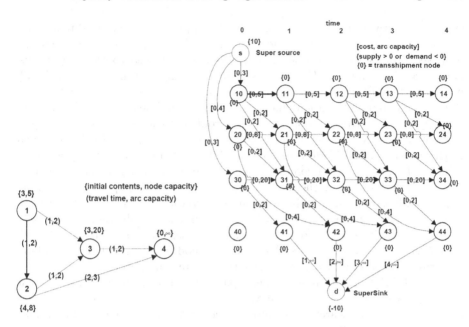

Fig. 1. Evacuation Network G, **Fig. 2.** Time-expanded Network G_T, with $T=4$,
(source: [17])   (source: [17])

tions. First, it significantly increases the problem size because it requires time-expanded network G_T to produce a solution. As can been seen in Figures 1 and 2, if the original evacuation network G has n nodes and the time upper bound is T, the time-expanded network G_T will have at least $(T + 1)n$ nodes. This approach may not be able to scale up to large size transportation networks in urban evacuation scenarios due to high computational run-time caused by the tremendously increased size of the time-expanded network. Second, linear programming approach requires the user to provide an upper bound T of the evacuation time in order to generate the time-expanded network. It is almost impossible to precisely estimate the evacuation time for an urban scenario where the number of evacuees is large and the transportation network is complex. An under-estimated time bound T will result in failure of finding a solution. In this case, the user will have to increase the value of T and re-run the algorithm until a solution can be reached. On the other hand, an over-estimated T will result in an over-expanded network G_T and hence lead to unnecessary storage and run-time.

Heuristic routing and scheduling algorithms can be used to find sub-optimal evacuation plan with reduced computational cost. It is useful for evacuation scenarios with large size networks and scenarios that do not require an optimal plan, but need to produce an efficient plan within a limited amount of time. However, old heuristic approaches only compute the shortest distance route from a source to the nearest destination without considering route capacity constraints. It cannot produce efficient plans when the number of evacuees is large and the evacuation network is complex. New heuristic approaches are needed to account for capacity constraints of the evacuation network. Lu, Huang and Shekhar [26] proposed prototypes of two heuristic capacity constrained routing algorithms, namely SRCCP and MRCCP, and tested its performance using small size building networks. SRCCP assigns only one route to each source node. It has very fast run-time but the solution quality is very poor and hence has little value for real application. MRCCP assigns multiple routes to each source node and produces high quality solution with much less run-time compared to that of linear programming approach. However, its scalability to large size networks is unsatisfactory because it has a computational cost of $O(p \cdot n^2 log n)$ (where n the is number of nodes and p is the number of evacuees). In this paper, we present an improved algorithm called Capacity Constrained Route Planner (CCRP). CCRP can reduce the run-time to $O(p \cdot n log n)$ by conducting only one shortest path search in each iteration instead of the multiple searches used in MRCCP. We also present the analysis of its algebraic cost model and provide the results of performance evaluation using large size transportation networks.

In the CCRP algorithm, we model capacity as a time series because available capacity of each node and edge may vary during the evacuation. We use a generalized shortest path search algorithm to account for route capacity constraints. This algorithm can divide evacuees from each source into multiple groups and assign a route and time schedule to each group of evacuees based on an order that is prioritized by each group's destination arrival time. It then reserves route capacities for each group subject to the route capacity constraints. The quickest route available for one group is re-calculated in each iteration based on the available capacity of the network. Performance evaluation on various network configurations shows that the CCRP algorithm produces high quality solutions, and significantly reduces the computational cost compared to linear programming approach. CCRP is also scalable to the number of evacuees and the size of the network. A case study using a nuclear power plant evacuation scenario shows that this algorithm can be used to improve existing evacuation plans by reducing evacuation time.

We also explored the possibility of formulation of a new optimal algorithm using A* search[28,29]. It addresses the limitations of linear programming approach by using only the original evacuation network to find the optimal solution and it does not require the user to provide an upper bound of the evacuation time. Details of the A* search formulation and the proof of monotonicity and admissibility of this A* search algorithm are available in [25]. It is not included in this paper due to space constraints.

Outline: The rest of the paper is organized as follows. In Section 2, the problem formulation is provided and related concepts are illustrated by an example evacuation network. Section 3 describes the Capacity Constrained Route Planner (CCRP) algorithm and the algebraic cost model. In Section 4, we present the experimental design and performance evaluation. We summarize our work and discuss future directions in Section 5.

2   Problem Formulation

We formulate the evacuation planning problem as follows:

Given: A transportation network with non-negative integer capacity constraints on nodes and edges, non-negative integer travel time on edges, the total number of evacuees and their initial locations, and locations of evacuation destinations.

Output: An evacuation plan consisting of a set of origin-destination routes and a scheduling of evacuees on each route. The scheduling of evacuees on each route should observe the capacity constraints of the nodes and edges on this route.

Objective: (1) Minimize the evacuation egress time, which is the time elapsed from the start of the evacuation until the last evacuee reaches the evacuation destination. (2) Minimize the computational cost of producing the evacuation plan.

Constraint: (1) Edge travel time preserves FIFO (First-In First-Out) property. (2) Edge travel time reflects delays at intersections. (3) Limited amount of computer memory.

We illustrate the problem formulation and a solution with an example evacuation network, as shown in Figure 3. In this evacuation network, each node is shown by an ellipsis. Each node has two attributes: maximum node capacity and initial node occupancy. For example, at node N1, the maximum capacity is 50, which means this node can hold at most 50 evacuees at each time point, while the initial occupancy is 10, which means there are initially 10 evacuees at this node. In Figure 3, each edge, shown as an arrow, represents a link between two nodes. Each edge also has two attributes: maximum edge capacity and travel time. For example, at edge N4-N6, the maximum edge capacity is 5, which means at each time point, at most 5 evacuees can start to travel from node N4 to N6 through this link. The travel time of this edge is 4, which means it takes 4 time units to travel from node N4 to N6. This approach of modelling a evacuation scenario to a capacitated node-edge graph is similar to those presented in Hamacher [17], Kisko [23] and Chalmet [9].

As shown in Figure 3, suppose we initially have 10 evacuees at node N1, 5 at node N2, and 15 at node N8. The task is to compute an evacuation plan that evacuates the 30 evacuees to the two destinations (node N13 and N14) using the least amount of time.

Example 1 (An Evacuation Plan). Table 1 shows an example evacuation plan for the evacuation network in Figure 3. In this table, each row shows one group of evacuees moving together during the evacuation with a group ID, source node, number of evacuees in this group, the evacuation route with time schedule, and the destination time. The route is shown by a series of node number and the time schedule is shown by a start time associated with each node on the route. Take source node N8 for example; initially there are 15 evacuees at N8. They are divided into 3 groups: Group A with 6 people, Group B with 6 people and

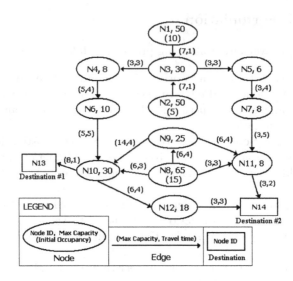

Fig. 3. Node-Edge Graph Model of Example Evacuation Network

Group C with 3 people. Group A starts from node N8 at time 0 to node N10, then starts from node N10 at time 3 to node N13, and reaches destination N13 at time 4. Group B follows the same route of group A, but has a different schedule due to capacity constraints of this route. This group starts from N8 at time 1 to N10, then starts from N10 at time 4 to N13, and reaches destination N13 at time 5. Group C takes a different route. It starts from N8 at time 0 to N11, then starts from N11 at time 3 to N14, and reaches destination N14 at time 5. The procedure is similar for other groups of evacuees from source node N1 and N2. The whole evacuation egress time is 16 time units since the last groups of people (Group H and I) reach destination at time 16. This evacuation plan is an optimal plan for the evacuation scenario shown in Figure 3.

In our problem formulation, we allow time dependent node capacity and edge capacity, but we assume that edge capacity does not depend on the actual flow amount in the edge. We also allow time dependent edge travel time, but we require that the network preserve the FIFO (First-In First-Out) property.

Alternate problem formulations of the evacuation problem are available by changing the objective of the problem. The main objective of our problem formulation is to minimize the evacuation egress time. Two alternate objectives are: (1) Maximize the number of evacuees that reach destination for each time unit; (2) Minimize the average evacuation time for all evacuees. Jarvis and Ratliff presented and proved the *triple optimization theorem* [20], which illustrated the properties of the solutions that optimize the above objectives of the evacuation problem. A review of linear programming approaches to solve these problem formulations was given by Hamacher and Tjandra [17].

Table 1. Example Evacuation Plan

Group of Evacuees			Route with Schedule	Dest.Time
ID	Source	Number		
A	N8	6	N8(T0)-N10(T3)-N13	4
B	N8	6	N8(T1)-N10(T4)-N13	5
C	N8	3	N8(T0)-N11(T3)-N14	5
D	N1	3	N1(T0)-N3(T1)-N4(T4)-N6(T8)-N10(T13)-N13	14
E	N1	3	N1(T0)-N3(T2)-N4(T5)-N6(T9)-N10(T14)-N13	15
F	N1	1	N1(T0)-N3(T1)-N5(T4)-N7(T8)-N11(T13)-N14	15
G	N2	2	N2(T0)-N3(T1)-N5(T4)-N7(T8)-N11(T13)-N14	15
H	N2	3	N2(T0)-N3(T3)-N4(T6)-N6(T10)-N10(T15)-N13	16
I	N1	3	N1(T1)-N3(T2)-N5(T5)-N7(T9)-N11(T14)-N14	16

3 Proposed Approach

Linear programming approach can produce optimal solutions for evacuation planning. It is useful for evacuation scenarios with moderate size networks, such as building evacuation. However, it may not be able to scale up to large size transportation networks in urban evacuation scenarios due to high computational cost caused by the tremendously increased size of the time-expanded network. Heuristic routing and scheduling algorithms can be used to find sub-optimal evacuation plan with reduced computational cost. It is useful for evacuation scenarios with large size networks and scenarios that do not require an optimal plan, but need to produce an efficient plan within a limited amount of time.

In this section, we present a heuristic algorithm, namely Capacity Constrained Route Planner (CCRP), that produces sub-optimal solutions for evacuation planning. We model edge capacity and node capacity as a time series instead of fixed numbers. A time series represents the available capacity at each time instant for a given edge or node. We propose a heuristic approach based on an extension of shortest path algorithms [13,11] to account for capacity constraints of the network.

3.1 Capacity Constrained Route Planner (CCRP)

The Capacity Constrained Route Planner (CCRP) uses an iterative approach. In each iteration, the algorithm first searches for route R with the earliest destination arrival time from any source node to any destination node, taking previous reservations and possible waiting time into consideration. Next, it computes the actual amount of evacuees that will travel through route R. This amount is affected by the available capacity of route R and the remaining number of evacuees. Then, it reserves the node and edge capacity on route R for those evacuees. The algorithm continues to iterate until all evacuees reach destination. The detailed pseudo-code and algorithm description are shown in Algorithm 1..

The CCRP algorithm keeps iterating as long as there are still evacuees left at any source node (line 1). Each iteration starts with finding the route R with the

---

Algorithm 1. Capacity Constrained Route Planner (CCRP)

```
Input:
    1) G(N, E): a graph G with a set of nodes N and a set of edges E;
       Each node n ∈ N has two properties:
```
$Maximum\_Node\_Capacity(n)$: non-negative integer
$Initial\_Node\_Occupancy(n)$: non-negative integer
```
       Each edge e ∈ E has two properties:
```
$Maximum\_Edge\_Capacity(e)$: non-negative integer
$Travel\_time(e)$: non-negative integer
```
    2) S: set of source nodes, S ⊆ N;
    3) D: set of destination nodes, D ⊆ N;
Output: Evacuation plan:Routes with schedules of evacuees on each route
Method:
Pre-process network: add super source node s₀ to network,
    link s₀ to each source nodes with an edge which
```
$Maximum\_Edge\_Capacity() = \infty$ and $Travel\_time() = 0$;   (0)
```
while any source node s ∈ S has evacuee do {                                    (1)
    find route R < n₀, n₁, ..., nₖ > with time schedule < t₀, t₁, ..., tₖ₋₁ >
        using one generalized shortest path search from super source s₀
            to all destinations, (where s ∈ S, d ∈ D, n₀ = s, nₖ = d)
        such that R has the earliest destination arrival time among
                                        routes between all (s,d) pairs,
```
and $Available\_Edge\_Capacity(e_{n_i n_{i+1}}, t_i) > 0$, $\forall i \in \{0, 1, \ldots, k-1\}$,
and $Available\_Node\_Capacity(n_{i+1}, t_i + Travel\_time(e_{n_i n_{i+1}})) > 0$,
$$\forall i \in \{0, 1, \ldots, k-1\}; \quad (2)$$
$flow = \min($ number of evacuees still at source node s,
$Available\_Edge\_Capacity(e_{n_i n_{i+1}}, t_i)$, $\forall i \in \{0, 1, \ldots, k-1\}$,
$Available\_Node\_Capacity(n_{i+1}, t_i + Travel\_time(e_{n_i n_{i+1}}))$,
$$\forall i \in \{0, 1, \ldots, k-1\};$$
```
        );                                                                     (3)
    for i = 0 to k − 1 do {                                                     (4)
```
$Available\_Edge\_Capacity(e_{n_i n_{i+1}}, t_i)$ reduced by $flow$;   (5)
$Available\_Node\_Capacity(n_{i+1}, t_i + Travel\_time(e_{n_i n_{i+1}}))$ reduced by $flow$;
```
                                                                               (6)
    }                                                                          (7)
}                                                                              (8)
Output evacuation plan;                                                        (9)
```

---

earliest destination arrival time from any sources node to any destination node based on the current available capacities (line 2). This is done by generalizing Dijkstra's shortest path algorithm [13,11] to work with the time series node and edge capacities and edge travel time. Route R is the route that starts from a source node and gets to a destination node in the least amount of time and available capacity of the route allows at least one person to travel through route R to a destination node.

Compared with the earlier MRCCP algorithm [26], major improvements in CCRP lie in line 0 and line 2. In MRCCP, finding route R (line 2) is done by running generalized shortest path searches from each source node. Each search is terminated when any destination node is reached. In CCRP, this step is improved by adding a super source node s_0 to the network and connecting s_0 to all source nodes(line 0). This allows us to complete the search for route R by using only one single generalized shortest path search, which takes the super source s_0 as the start node. This search terminates when any destination node is reached. Since the super source s_0 is connected to each source nodes by an edge with infinite capacity and zero travel time, it can be easily proved that the shortest route found by this search is the route R we need in line 2. This improvement significantly reduces the computational cost of the algorithm by one degree of magnitude compared with MRCCP. We give a detailed analysis of the cost model of CCRP algorithm in the next section.

3.2   Algebraic Cost Model of CCRP

We now provide the algebraic cost model for the computational cost of the proposed CCRP algorithm. We assume that n is the number of nodes in the evacuation network, m is the number of edges, and p is the number of evacuees.

The CCRP algorithm is an iterative approach. In each iteration, the route for one group of people is chosen and the capacities along the route are reserved. The total number of iterations equals the number of groups generated. In the worst case, each individual evacuee forms one group. Therefore, the upper bound of the number of groups is p, i.e. the number of iterations is $O(p)$. In each iteration, the computation of the route R with earliest destination arrival time is done by running one generalized Dijkstra's shortest path search. The worst case computational complexity of Dijkstra's algorithm is $O(n^2)$ for dense graphs [11]. Various implementations of Dijkstra's algorithm have been developed and evaluated extensively [4,10,32]. Many of these implementations can reduce the computational cost by taking advantage of the sparsity of the graph. Transportation road networks are very sparse graphs with a typical edge/node ratio around 3. In CCRP, we implement Dijkstra's algorithm using heap structures, which runs in $O(m + n log n)$ time [4,10]. For sparse graphs, $n log n$ is the dominant term. The generalization of Dijkstra's algorithm to account for capacity constraints affects only how the shortest distance to each node is defined. It does not affect the computational complexity of the algorithm. Therefore, we can complete the search for route R with $O(n log n)$ run-time. The reservation step is done by updating the node and edge capacities along route R, which has a cost of $O(n)$. Therefore, each iteration of the CCRP algorithm is done in $O(n log n)$ time. As we have seen, it takes $O(p)$ iterations to complete the algorithm. The cost model of the CCRP algorithm is $O(p \cdot n log n)$. CCRP is an improved algorithm based on the same heuristic method of MRCCP [26] which has a run-time of $O(p \cdot n^2 log n)$. CCRP reduces the computational cost of MRCCP by one degree of magnitude.

The computational cost of linear programming approach depends on the method used to solve the minimum cost flow problem. Hoppe and Tardos [18] showed that this problem can be solved using ellipsoid method which is theoretically polynomial time bounded. However, the computational complexity of

Table 2. Comparison of Computational Costs (n: number of nodes, p: number of evacuees, T: user-provided upper-bound on evacuation time)

Algorithm	Computational Cost	Solution Quality
CCRP	$O(p \cdot n log n)$	Sub-optimal
MRCCP	$O(p \cdot n^2 log n)$	Sub-optimal
Linear Programming Approach	at least $O((T \cdot n)^6)$	Optimal

ellipsoid method is at least $O(N^6)$[6](where N is the number of nodes in the network). Since linear programming approach requires a time-expanded network, N equals to $(T+1)n$ (where n is the number of nodes in the original evacuation network, T is the user-provided evacuation time upper bound).

Table 2 provides a comparison of CCRP, MRCCP, and the linear programming approach. As can be seen, linear programming approach produces optimal solutions but suffers from high computational cost. Both CCRP and MRCCP reduce the computation cost by producing sub-optimal solution, while CCRP gives better computational cost than MRCCP.

Lemma 1: CCRP is strictly faster than MRCCP.

The computational costs of CCRP and MRCCP are $O(p \cdot n log n)$ and $O(p \cdot n^2 log n)$ respectively, as shown in Table 2.

4   Experiment Design and Performance Evaluation

Performance evaluation of the CCRP algorithm was done by conducting experiments using various evacuation network configurations. In this section, we present the experiment design and an analysis of the experiment results.

4.1   Experiment Design

Figure 4 describes the experiment design to evaluate the performance of the CCRP algorithm. The purpose is to compare the algorithm run-time and solution quality of the proposed CCRP algorithms with that of MRCCP [26] and NETFLO [21] which is a popular linear programming package used to solve minimum cost flow problems.

First, we used NETGEN [24] to generate evacuation networks with evacuees. NETGEN is a program that generates transportation networks with capacity constraints and initial supplies based on input parameters. In our experiments, the following three were selected as independent parameters to test their impacts on the the performance of the algorithms: number of evacuees initially in the network, number of source nodes, and network size represented by number of nodes. Number of edges is treated as a dependent parameter as we set the number of edges to be equal to 3 times the number of nodes because 3 is the typical edge/node ratio for real transportation road networks. Next, the same evacuation network generated by NETGEN was fed to the CCRP and MRCCP algorithms. Before feeding the network to NETFLO, we used a network transformation tool to transform the evacuation network into a time-expanded network, which is required by minimum cost flow solvers as NETFLO to solve evacuation problems [17,9]. This process requires an input parameter T which is the estimated upper-bound on evacuation egress time. If the evacuation cannot be completed by time T, NETFLO will return no solution. In this case, we must increase T to create a new time-expanded network and try to run NETFLO again until a solution can be reached. Finally, after CCRP, MRCCP and NET-FLO produced a solution for each test case, the evacuation egress time, which represents the solution quality, and the algorithm run-time were collected and analyzed in the data analysis module.

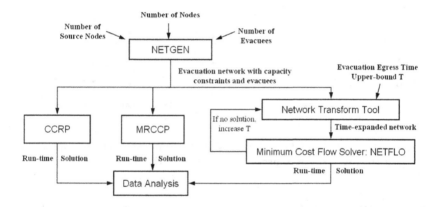

Fig. 4. Experiment Design

The experiments were conducted on a workstation with Intel Pentium IV 2GHz CPU, 2GB RAM and Debian Linux operating system.

4.2   Experiment Results and Analysis

We want to answer three questions: (1) How does the number of evacuees affect the performance of the algorithms? (2) How does the number of source nodes affect the performance of the algorithms? (3) Are the algorithms scalable to the size of the network, particularly will they handle large size transportation networks as in urban evacuation scenarios?

Experiment 1: How does the number of evacuees affect the performance of the algorithms?
   The purpose of the first experiment is to evaluate how the number of evacuees affects the performance of the algorithms. We fixed the number of nodes and the number of source nodes of the network, and varied the number of evacuees

to observe the quality of the solution and the run-time of CCRP, MRCCP and NETFLO algorithms.

The experiment was done with four test groups. Each group had a fixed network size of 5000 nodes and fixed number of source nodes at 1000, 2000, 3000, and 4000 respectively. We varied the number of evacuees from 5000 to 50000. Here we present the experiment results of the test group with number of source nodes fixed at 2000. We omit the results from the other three groups since this group shows a typical result of all test groups. Figure 5 shows the solution quality represented by evacuation egress time and Figure 6 shows the run-times of the three algorithms.

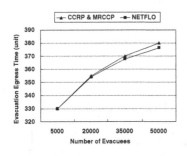
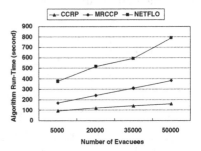

Fig. 5. Quality of Solution With Respect to Number of Evacuees

Fig. 6. Run-time With Respect to Number of Evacuees

Since CCRP and MRCCP use the same heuristic method to find solution, it is expected that CCRP and MRCCP produced solutions with the same evacuation egress time for each test case. As seen in Figure 5, CCRP and MRCCP produced very high quality solution compared with the optimal solution produced by NETFLO. The solution quality of CCRP and MRCCP drops slightly as the the number of evacuees grows. In Figure 6, we can see that, in each case, the run-time of CCRP remains half that of MRCCP and less than 1/3 that of NETFLO. In addition, the CCRP run-time is scalable to the number of evacuees while the run-time of NETFLO grows much faster.

This experiment shows: (1) CCRP produces high quality solutions with much less run-time than that of NETFLO. (2) The run-time of CCRP is scalable to the number of evacuees.

Experiment 2: How does the number of source nodes affect the performance of the algorithms?

In the second experiment, we evaluate how the number of source nodes affects the performance of the algorithms. We fixed the number of nodes and the number of evacuees in the network, and varied the number of source nodes to observe the quality of the solution and the run-time. In this experiment, by varying the number of source nodes, we actually create different evacuee distributions in the network. A higher number of source nodes means that the evacuees are more scattered in the network.

Again, the experiment was done with four test groups. Each group had a fixed network size of 5000 nodes and fixed number of evacuees at 5000, 20000, 35000, and 50000 respectively. We varied the number of source nodes from 1000 to 4000. Here we present the experiment results of the test group with number of evacuees fixed at 5000. It shows a typical result of all test groups. Figure 7 shows the solution quality represented by evacuation egress time and Figure 8 shows the run-times of the three algorithms.

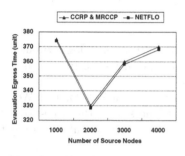

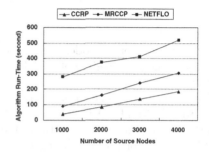

Fig. 7. Quality of Solution With Respect to Number of Source Nodes

Fig. 8. Run-time With Respect to Number of Source Nodes

As seen in Figure 7, in each test case, CCRP and MRCCP produced high quality solution (within 5 percent longer evacuation time) and the number of source nodes has little effect on the solution quality. It is also noted that the evacuation time is non-monotonic with respect to the number of source nodes and we plan to explore the potential reasons in future works.

Figure 8 shows that the run-time of all three algorithms are scalable to the number of source nodes. However, the run-time of CCRP remains less than half that of NETFLO.

This experiment shows: (1)The solution quality of CCRP is not affected by the number of source nodes. (2) The run-time of CCRP is scalable to the number of source nodes.

Experiment 3: Are the algorithms scalable to the size of the network?

In the third experiment, we evaluate how the network size affects the performance of the algorithms. We fixed the number of evacuees and the number of source nodes in the network, and varied the network size to observe the quality of solution and the run-time of the algorithms.

The experiment was done with a fixed number of evacuees at 5000 and the number of source nodes at 10. We varied the number of nodes from 50 to 50000. Figure 9 shows the solution quality represented by evacuation egress time and Figure 10 shows the run-times.

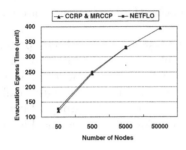

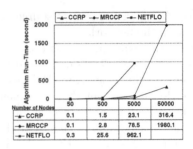

Fig. 9. Quality of Solution With Respect to Network Size

Fig. 10. Run-time With Respect to Network Size

Note: x-axis(number of nodes) in Figure 9 and 10 is on a logarithmic scale rather than linear. Run-time of CCRP and MRCCP in Figure 10 grow in small polynomial.

There is no data point for NETFLO at network size of 50000 nodes. We were unable to run NETFLO for this setup because the size of the time-expanded network became too large (more than 20 million nodes and 80 million edges)that NETFLO could not produce solution.

As seen in Figure 9, in each of the first three test case, CCRP and MRCCP produced high quality solution (within 5 percent longer evacuation time) and the solution quality becomes closer to optimal solution as the network size increases. Figure 10 is shown with a data table of each run-time. The x-axis(number of nodes) of Figure 10 is on a logarithmic scale rather than linear and the run-time of CCRP and MRCCP grow in small polynomial. It can be seen that the run-time of CCRP is scalable to the network size while the NETFLO run-time grows exponentially.

This experiment shows: (1) Given a fixed number of evacuees and source nodes, the solution quality of CCRP increases as the network size increases. (2) The run-time of CCRP is scalable to the size of the network.

We also conducted experiments using a real evacuation scenario. The Monticello nuclear power plant is about 40 miles to the northwest of the Twin Cities. Evacuation plans need to be in place in case of accidents or terrorist attacks. The evacuation zone is a 10-mile radius around the nuclear power plant as defined by Minnesota Homeland Security and Emergency Management [3].

The experiment was done using the road network around the evacuation zone provided by the Minnesota Department of Transportation [2], and the Census 2000 population data for each affected city. The total number of evacuees is about 42,000. The old hand-crafted evacuation plan has an evacuation egress time of 268 minutes. CCRP algorithm produced a much better plan with evacuation time of only 162 minutes. This experiment shows that our algorithm is effective in real evacuation scenarios to reduce evacuation time and improve existing plans.

Our approach was presented in the UCGIS Congressional Breakfast Program on homeland security[30], and the Minnesota Homeland Security and Emergency Management newsletter[31]. It was also selected by the Minnesota Department of Transportation to be used in the evacuation planning project for the Twin Cities Metro Area, which involves a road network of about 250,000 nodes and a population of over 2 million people.

5   Conclusions and Discussions

In this paper, we proposed a new capacity constrained routing algorithm for evacuation planning problem. Existing linear programming approach uses time-expanded network and requires user provided upper bound on evacuation time. To address these limitations, we presented a heuristic algorithm, namely Capacity Constrained Route Planner(CCRP), which produces sub-optimal solution for evacuation planning problem without using time-expanded networks. We provided the algebraic cost model and the performance evaluations using various network configurations. Experiments show that CCRP algorithm produces high quality solution and significantly reduces the computational cost compared to linear programming approach which produces optimal solution. It is also shown that the CCRP algorithm is scalable to the number of evacuees and the size of the transportation network. A case study using real evacuation scenario shows that CCRP algorithm can be used to improve existing evacuation plans by reducing total evacuation time.

The limitation of CCRP algorithm remains the follows. First, we assume that maximum capacity of an edge does not depend on traffic flow amount on the edge. We understand that it is a challenging task to accurately model the capacity of each road segment in a real evacuation scenario as the actual traffic flow rate may depend on vehicle speed as well as road occupancy. Second, the generalized shortest path algorithm we used in CCRP requires that the edge travel time reflects traffic delays at intersections. For future work, we plan to incorporate existing research results, such as Ziliaskopoulos and Mahmassani [33], to better address this problem.

To address the sub-optimality issue of the CCRP algorithm, we also explored the possibility of formulating the evacuation problem as a search problem using A* algorithm. Our A* search formulation addresses the limitations of linear programming approach by only using the original evacuation network to find optimal solution. Thus, it does not require prior knowledge of evacuation time. We proved that the heuristic function used in our A* formulation is monotone and admissible thus guaranteeing the optimality of the solution. Details of the A* search formulation can be found in [25]. It is not included in this paper due to space constraints.

Acknowledgment

We are particularly grateful to members of the Spatial Database Research Group at the University of Minnesota for their helpful comments and valuable discussions. We would also like to express our thanks to Kim Koffolt for improving the readability of this paper.

This work is supported by the Army High Performance Computing Research Center (AHPCRC) under the auspices of the Department of the Army, Army Research Laboratory under contract number DAAD19-01-2-0014 and the Minnesota Department of Transportation under contract number 81655. The content does not necessarily reflect the position or policy of the government and no official endorsement should be inferred. AHPCRC and the Minnesota Supercomputer Institute provided access to computing facilities.

References

1. *Hurricane Evacuation web page.* http://i49south.com/hurricane.htm, 2002.
2. *Minnesota basemap web site.* http://www.dot.state.mn.us/tda/basemap/, Minnesota Department of Transportation, 2004.
3. *Monticello evacuation planning web site.* http://www.hsem.state.mn.us/, Minnesota Homeland Security and Emergency Management, 2004.
4. R.K. Ahuja, T.L. Magnanti, and J.B. Orlin. *Network Flows: Theory, Algorithms, and Applications.* Prentice Hall, 1993.
5. M. Ben-Akiva et al. *Development of a Deployable Real-Time Dynamic Traffic Assignment System: DynaMIT and DynaMIT-P User's Guide.* Intelligent Transportation Systems Program, Massachusetts Institute of Technology, 2002.
6. D. Bertsimas and J.N. Tsitsiklis. *Introduction to Linear Optimization* . Athena Scientific, 1997.
7. S. Brown. Building America's Anti-Terror Machine: How Infotech Can Combat Homeland Insecurity. *Fortune*, pages 99–104, July 2002.
8. The Volpe National Transportation Systems Center. Improving Regional Transportation Planning for Catastrophic Events(FHWA). *Volpe Center Highlights*, pages 1–3, July/August 2002.
9. L. Chalmet, R. Francis, and P. Saunders. Network Model for Building Evacuation. *Management Science*, 28:86–105, 1982.
10. B.V. Cherkassky, A.V. Goldberg, and T. Radzik. Shortest Paths Algorithms: Theory and Experimental Evaluation . *Mathematical Programming*, 73:129–174, 1996.
11. T. Cormen, C. Leiserson, R. Rivest, and C. Stein. *Introduction to Algorithms.* MIT Press, 2nd edition, 2001.
12. The Homeland Security Council. Planning Scenarios, Executive Summaries, Created for Use in National, Federal, State, and Local Homeland Security Preparedness Activities. July 2004.
13. E.W. Dijkstra. A Note on Two Problems in Connexion with Graphs. *Numerische Mathematik*, 1:269–271, 1959.
14. ESRI. GIS for Homeland Security, An ESRI white paper, November 2001.
15. L.R. Ford and D.R. Fulkerson. *Flows in Network.* Princeton University Press, 1962.
16. R. Francis and L. Chalmet. A Negative Exponential Solution To An Evacuation Problem. *Research Report No.84-86, National Bureau of Standards, Center for Fire Research*, October 1984.
17. H.W. Hamacher and S.A. Tjandra. Mathematical Modeling of Evacuation Problems: A state of the art. *Pedestrian and Evacuation Dynamics*, pages 227–266, 2002.
18. B. Hoppe and E. Tardos. Polynomial Time Algorithms For Some Evacuation Problems. *Proceedings of the 5th Annual ACM-SIAM Symposium on Discrete Algorithms*, pages 433–441, 1994.
19. B. Hoppe and E. Tardos. The Quickest Transshipment Problem. *Proceedings of the 6th annual ACM-SIAM Symposium on Discrete Algorithms*, pages 512–521, January 1995.

20. J.J. Jarvis and H.D. Ratliff. Some Equivalent Objectives for Dynamic Network Flow Problems. *Management Science*, 28:106–108, 1982.
21. J.L. Kennington and R.V. Helgason. *Algorithm for Network Programming*. Wiley and Sons, 1980.
22. T. Kisko and R. Francis. Evacnet+: A Computer Program to Determine Optimal Building Evacuation Plans. *Fire Safety Journal*, 9:211–222, 1985.
23. T. Kisko, R. Francis, and C. Nobel. *EVACNET4 User's Guide*. University of Florida, 1998.
24. D. Klingman, A. Napier, and J. Stutz. NETGEN: A Program for Generating Large Scale Capacitated Assignment, Transportation, and Minimum Cost Flow Network Problems. *Management Science*, 20:814–821, 1974.
25. Q. Lu, B. George, and S. Shekhar. Capacity Constrained Routing Algotithm for Evacuation Planning: A Summary of Results. *Technical Report, Department of Computer Science and Engineering, University of Minnesota*, (05-023), 2005.
26. Q. Lu, Y. Huang, and S. Shekhar. Evacuation Planning: A Capacity Constrained Routing Approach. *Proceedings of the First NSF/NIJ Symposium on Intelligence and Security Informatics*, pages 111–125, June 2003.
27. H.S. Mahmassani, H. Sbayti, and X. Zhou. *DYNASMART-P Version 1.0 User's Guide*. Maryland Transportation Initiative, University of Maryland, September 2004.
28. N.J. Nilsson. *Principles of Artificial Intelligence*. Tioga Publishing Co., 1980.
29. J. Pearl. *Heuristics: Intelligent Search Strategies for Computer Problem Solving*. Addison Wesley, 1984.
30. S. Shekhar. Evacuation Planning: Presentation at UCGIS Congressional Breakfast Program on Homeland Security. *http://www.ucgis.org/winter2004/program.htm*, February 2004.
31. S. Shekhar and Q. Lu. Evacuation Planning for Homeland Security. *Minnesota Homeland Security and Emergency Management newsletter*, October 2004.
32. F.B. Zhan and C.E. Noon. Shortest Paths Algorithms: An Evaluation Using Real Road Networks. *Transportation Science*, 32:65–73, 1998.
33. A.K. Ziliaskopoulos and H.S. Mahmassani. A Note on Least Time Path Computation Considering Delays and Prohibitations for Intersection Movements. *Transportation Research B*, 30(5):359–367, 1996.

High Performance Multimodal Networks

Erik G. Hoel, Wee-Liang Heng, and Dale Honeycutt

Environmental Systems Research Institute,
380 New York Street, Redlands, CA 92373
{ehoel, wheng, dhoneycutt}@esri.com

Abstract. Networks often form the core of many users' spatial databases. Networks are used to support the rapid navigation and analysis of linearly connected data such as that found in transportation networks. Common types of analysis performed on such networks include shortest path, traveling salesman, allocation, and distance matrix computation.

Network data models are usually represented as a small collection of tables: a junction table and an edge table. In the context of networks used to model transportation infrastructure, it is also necessary to model turn restrictions and impedances (delays). Network data is frequently persisted in normalized relational tables that are accessible via standard SQL-based queries. We propose a different approach where the network connectivity information is persisted using a compressed binary storage representation in a relational database. The connectivity information is accessible via standard Java, .NET, and COM APIs that are tailored to common access patterns used in the support of high performance network engines. These network engines run on the client or application server tier rather than as extensions on the relational server.

In this paper, we discuss the problem of building a robust and scalable implementation of a network data model. The fundamental and central requirements are enumerated. These requirements include support for hierarchical networks, turn restrictions, and logical z elevations. We propose a different approach to representing network topology that addresses many of the high-end modeling requirements of network systems. Our approach supports all of the listed requirements in addition to multimodal modeling (e.g., coexistent road, bus, and rail networks) within the context of multi-user, long transaction databases.

1 Introduction

Network data models have been used to represent geographic information for well over thirty years [15], [18], [19]. These models have been incorporated into a number of operational systems (see, for example TransCAD [3] or ARC/INFO [22]). Despite the relative maturity of such technology, most systems have fallen short of meeting the most sophisticated requirements of transportation network modeling. Such requirements include the ability to model multimodal (or intermodal) transportation systems (transportation networks where two or more different transportation modes are linked – e.g., roads and rail) and the ability to handle coincident features participating in different modes of the model (e.g, subways underneath streets, or bus

C. Bauzer Medeiros et al. (Eds.): SSTD 2005, LNCS 3633, pp. 308–327, 2005.

routes along city streets) when geometric analysis of the participant features is used to derive network connectivity [29]. In addition, some systems fail to address the requirement to model turns and maneuvers without applying complex graph transformations to represent permissible turns as explicit edges [30].

In this paper, we describe a design for modeling multimodal networks that are persisted in relational databases. This design is the basis for our implementation of networks in the ArcGIS geographic information system. The design satisfies the fundamental goal of supporting sophisticated network models that are consumed by high performance network engines, and is tailored for fast retrieval of connectivity information within network analysis algorithms. Network engines provide fine grained (i.e., forward star [6]) access to very large external networks persisted in an RDBMS or the file system, and are intended to reside on the client in the case of traditional two-tier systems, or on the application server in n-tier architectures. The network engine supports a rich set of network analysis algorithms, such as shortest path finding, traveling salesman problems, and network resource allocation operations, that also execute in the desktop or application server tier and are used in a variety of desktop and server-based network analysis applications.

In the first section of this paper, we review the logical model of network topology. The major requirements of a high-end network data model are discussed and a connectivity model that supports multimodal networks is presented. We then consider the issue of representing turn restrictions and maneuvers (multi-part turns) – a critical component in transportation networks [9]. Existing approaches to representing turns are reviewed and our modeling approach is presented. The access model of consuming networks is reviewed in the context of common workflows as well as a query model that is tailored to the support of high performance network engines. We then address the issue of the physical storage representation of a network. The conventional physical database implementation is briefly detailed. We then present our alternative physical representation and highlight the reasons and motivation behind its departure from the conventional implementation. We conclude with a brief discussion of our implementation experience and outline our ongoing future research in this domain.

2 Logical Model

The movement of people, the transportation of goods and services, as well as the distribution of resources, energy, and communication are commonly modeled with network systems. Network data structures for representing geographic information are a standard topic in geographic information science [18], [27].

In this paper, we use the term *network* to refer to a connectivity graph of junctions and their connecting edges, where each junction and edge is associated with a feature with point or line geometry respectively. The term *network element* is used to refer to the collection of junctions and edges comprising the network. All network elements have a set of numeric properties, called *network attributes*. Attributes capture information about network elements, such as the travel time across an element, and are used to define the navigation context during an analysis.

Junction attribute values provide a high-level view of traversing intersections. For example, the travel time attribute value on a junction element describes how long it takes to cross the element, ignoring the edge elements used to enter and exit the junction element. For more detailed modeling of traversing intersections, we use turns. In the simplest case, a turn element models entering a junction from a particular edge element and exiting to another. A multipart turn element, also known as a maneuver, enters the junction element from a path of two or more connected edge elements.

Turn elements are not strictly part of the graph model. They represent a relationship rather than being an abstraction of a real-world entity. Turns do not modify the junction-edge connectivity of the network; instead they affect traversability of the network elements. Turns are not considered an attribute of a network junction, though they occur at every junction. This is because they are intrinsically dependent upon the properties of the associated network edges.

The connectivity graph of a network is derived from the source data during a process called *network building*. During a build, junction, edge and turn elements are generated from point, line and turn features, and connectivity relationships are established. The connectivity graph is typically stored separately from the source data, with network analysis algorithms (including the build process) consuming it.

2.1 Requirements

The primary requirements for any robust implementation of network data models are:

- **Multimodal models.** In the context of transportation networks, a multimodal network is one in which two or more types of transportation modes (such as walking, riding a train, or driving a car) are modeled. Alternatively, with utility networks, a multimodal network may consist of the differing transmission and distribution systems.

- **Hierarchical models.** Hierarchy is used within network models to further control the flow within the network [19]. Differing elements may be assigned to different levels of hierarchy, with flow through the higher levels of the hierarchy taking precedence over the lower levels when performing path or route finding operations. Within transportation networks, interstate highways are commonly associated with the highest level of the hierarchy, state highways and major feeders the next lower level, and city streets the lowest level of the hierarchy.

- **Turns and maneuvers.** Support for turning movements, both two-part turns and multi-part turns (known as maneuvers), is necessary in order to more accurately model transportation networks. The definition of a turn should be separated from its attribution. A turn is not simply one restriction or penalty; instead it should be regarded as a first-class entity with attribution.

- **Fast network navigation.** The persisted representation must support fast retrieval of connectivity information for use within network analysis algorithms, and should be structured according to the most common access patterns.

- **Z elevations.** In order to refine network connectivity with planar network datasets (e.g., modeling freeway over and underpasses), logical z elevation values are supplied by commercial data vendors on the ends of each line feature. These elevation values must be respected when establishing network connectivity.

- **Rich attribution of network elements.** To capture real-world constraints, such as one-way travel restrictions, height/weight limits, and time-of-day travel times, we need a rich attribute model that supports multiple attributes on a network element.

- **Uniform attribute access model.** Clients of the model should be insulated from the details of where attribute values originate. For example, the travel time attribute for an edge element may be derived from the properties of the associated street feature, or it may be a real-time value. In each case, client applications should be able to retrieve attribute values without knowledge of the underlying storage.

In addition to these network specific requirements, other standard system requirements such as performance, editability, persistence in a relational database, support for long transactions, and scalability (e.g, a continental dataset of 50+ million edges) also apply.

2.2 Connectivity Model

Connectivity in a network is generally based upon spatial coincidence of the endpoints of line (real-world) features and other point features. This leads to a 1:1 mapping between features participating in a network and the network elements used to represent the network connectivity. This approach works reasonably well for simpler planar network datasets (e.g., TIGER/Line [20], or others commonly available from commercial data vendors such as Tele Atlas or NAVTEQ). However, with non-planar datasets (e.g., long linear features such as highways in transportation networks), it is useful to allow network connectivity partway along a linear feature (we term this *mid-span connectivity*). The familiar one-to-one mapping between linear features and edge elements must be generalized into a one-to-many mapping. Mid-span connectivity is supported in some network models such as the ArcGIS Geometric Network [31]. The example shown in Fig. 1 depicts the one-to-many mapping between line features and edge elements when mid-span connectivity is supported.

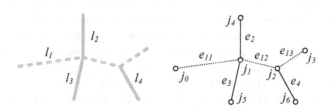

Fig. 1. Mid-span connectivity example; on the left, the long linear feature l_1 (dashed line) will correspond to edge elements e_{11}, e_{12}, and e_{13} if mid-span connectivity is supported

Multimodal Models. As discussed previously, multimodal network models are particularly important in the context of transportation modeling. We employ the concept of connectivity groups within the connectivity model to allow users to group together line classes that should be connected when geometric coincidence is present. A line class may participate in only one group. The number of groups is not constrained. All connectivity is local to a group; line features are not connected to other line features that are found in different connectivity groups. In order to establish connectivity between two groups (e.g., road network in one group, subway network in another group), point feature classes are allowed to participate in one or more groups. Thus, a point feature that is coincident with a road feature in one group and a subway feature in a second group will connect the two groups together in its role as a junction element. Connectivity groups may be employed to model networks containing multiple overlapping subnetworks – e.g., street networks, subway networks, and bus route networks.

An example highlighting connectivity groups is shown in Fig. 2. In this example, the line features participate in two different groups. The first line class contains line l_1 which is depicted by the dashed line. The second line class contains two line features, l_2 and l_3, depicted with solid lines. A point feature class, containing point feature p_1, participates in both connectivity groups. On the right side of Fig. 2, the resulting connectivity is shown. Note that l_1 (edges e_{11} and e_{12}) and l_2 (edges e_{21} and e_{22}) are connected at point p_1 (junction j_3). There is no connectivity between line l_1 and line l_3 as they are in different groups and there is no point feature where they intersect.

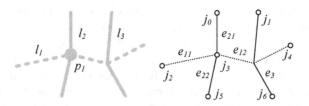

Fig. 2. Example of connectivity groups. Linear feature l_1 (dashed line) is in one group, and features l_2 and l_3 are in a second group. Point feature p_1 is in both groups

Z Elevations. Z elevations (sometimes termed 'z-levs') are a critical component for modeling overpasses and underpasses, tunnels, and highway interchanges with planar datasets (most commercial and governmental datasets are planar). At each endpoint of a line feature, there may be associated z elevation information that is used to refine network connectivity. This elevation information is typically logical – it does not correspond to actual geographical elevations, but rather a logical (ordinal) elevation value. For example, the endpoints of line features representing roads that comprise an underpass may have a z elevation value of 0, while the lines representing the overpass roads may have a value of 1. This logical vertical ordering can extend to support very complex highway interchanges.

Fig. 3 contains an example of four lines meeting at a location that corresponds to an overpass. In the example, lines l_1 and l_3 pass beneath lines l_2 and l_4 (note that all four lines $l_1 - l_4$ share a coincident endpoint; if one did not consider the z elevations when determining network connectivity, all four lines would be connected together). The z elevations are shown (0, 0, 1, and 1 respectively). The resulting connectivity is shown on the right side of the figure. Edges e_1 and e_3 are connected at junction j_{13}; edges e_2 and e_4 are connected at junction j_{24}. Junctions j_{13} and j_{24} appear coincident in the figure.

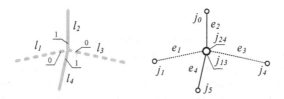

Fig. 3. Example of z elevations and their impact upon connectivity. Z elevations are shown in the left. On the right, the two junctions j_{13} and j_{24} are coincident

The extensions to the simple endpoint connectivity model are reflected in the network build algorithm. During the geometric analysis and connectivity discovery phase of the build process, the connectivity model and the z elevations are used to refine the connectivity between spatially coincident geometries.

2.3 Turns and Maneuvers

Turn restrictions and impedances (or delays) present a problem to most network models. The presence of turns can greatly impact the movement through a network [7], [21]. A common way to model turns within a network is with a turn table [30]. A turn table represents each explicitly specified turn restriction (or penalty) as a row with references to the associated two edges. Turn tables may be augmented with an impedance attribute if the turns may also represent delays or impedances. When traversing the network, the turn table is queried as necessary. An alternative approach is to employ a transition matrix that represents possible transitions at an intersection [10]. The matrix can be encoded into a bitmap for a smaller physical representation.

In order to overcome the performance problems (as perceived by some) of representing turns in an extra table that is disjoint from the network connectivity tables, graph modification techniques have been employed. The goal behind these techniques is to allow the turns to be more directly imbedded within the network connectivity information in order to achieve better performance during network traversals.

Graph Modification – Node Expansion. Node expansion is one technique to imbed turns within a graph by expanding each junction in the graph to a subgraph where permissible turns are explicitly represented as edges [1], [15], [24], [28]. The primary advantage of this approach is that the turns are represented within the connectivity graph of the network (thereby possibly improving traversal performance).

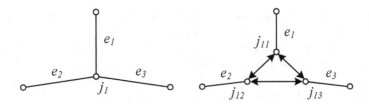

Fig. 4. Example highlighting node expansion where a junction connected to three edges is expanded to a set of three junctions with nine edges representing possible turns (u-turns omitted for clarity)

However, for an intersection of n edges, there are n^2 possible turns (including u-turns). This highlights the fundamental problem with this approach, namely, the significant bloating of the network storage requirements. This adversely impacts both storage costs and traversal performance [21]. In Fig. 4, the intersection junction j_1 is expanded and replaced with three junctions (labeled j_{11} through j_{13}), and edges are used to explicitly indicate permissible turns (the bidirectional edges on the right side of Fig. 4).

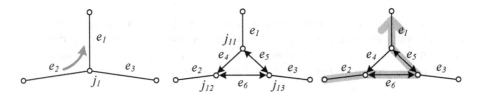

Fig. 5. Example of a turn restriction, the equivalent expanded graph, and an incorrect traversal in the expanded graph

Node expansion also introduces an algorithmic issue caused by traversing the edges in the expanded subgraph in sequence. Such a traversal corresponds to making multiple turns at the same junction in the original graph, and is meaningless. In Fig. 5, the turn from e_2 to e_1 is restricted. The restricted turn is reflected in the expanded graph with the directed edge e_4. However, we can still incorrectly go from e_2 to e_1 via the edges e_6 and e_5. Analysis algorithms that operate on the expanded graph have to avoid such traversals in order to generate correct results.

Graph Modification – Line Graphs. Line graphs (sometimes inappropriately termed dual graphs) are also used to explicitly model turns within a network [2], [30]. Line graphs are a transformation of the original (or primal) graph where edges in the primal are replaced with junctions in the line graph, and edges in the line graph represent turns in the primal. An example is shown in Fig. 6 where a simple (primal) graph consisting of three edges is transformed into the line graph on the right side of the figure. Edge e_1 in the primal is transformed into junction j_{11} in the line graph, edge e_2 into junction j_{12}, and edge e_3 into junction j_{13} respectively. Presuming that all turn movements are allowed, bidirectional edges in the line graph will be created between the three junctions in the line graph (edges e_{12}, e_{13}, and e_{23}).

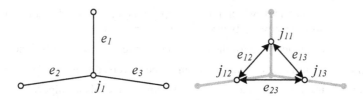

Fig. 6. Example line graph representation. On the right, the primal graph is represented in gray and the line graph is in black. In this example, all turns are possible (u-turns omitted)

Similar to the node expansion technique, the advantage of this approach is that the turns are explicitly represented in the graph. In addition, it results in a smaller graph than with the node expansion technique. However, line graphs require that the primal graph be retained in order to complete certain types of operations such as route drawing [30].

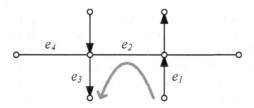

Fig. 7. Example of a three part maneuver e_1-e_2-e_3 at an intersection with a dual carriageway

Maneuvers. A maneuver is a turn that spans three or more edges. Maneuvers are used to model turning movements at complex street intersections within transportation networks. Consider the following intersection formed by a dual carriageway (i.e., a street where each travel direction is represented as a separate line feature) and a two-way street in Fig. 7. To restrict the u-turn from edge e_1 to edge e_3, we need a maneuver composed of the edges e_1, e_2 and e_3 in sequence. The maneuver cannot be synthesized from the two overlapping turns e_1-e_2 and e_2-e_3, since restricting the e_1-e_2 turn also incorrectly restricts the left turn specified by the sequence e_1-e_2-e_4.

Maneuvers can get arbitrarily complicated. We have observed instances of maneuvers with high part counts in transportation networks, such as a nine-part maneuver in the street network for Osaka, Japan. It is awkward to adapt graph modification techniques to model maneuvers.

2.4 Network Attributes

Network attributes are numeric properties of network elements that are used to define the navigation context during an analysis [21]. Examples of common attributes found on network elements include travel time, one-way restrictions, speed along an edge, and hierarchy value. The various types of attributes can be classified as:

- *cost*   impedances, which may be apportioned if the line feature is associated with multiple network edges (e.g., travel time),
- *descriptor* a characteristic of the entire element (e.g., speed limits, lane count),
- *restriction* identify which elements cannot be traversed (e.g., one-way), and
- *hierarchy*   used in conjunction with hierarchical analysis algorithms (e.g., an order or grade – highways, arteries, and city streets).

Network attributes are usually persisted along with the network elements (e.g., the attribute columns in the standard relational model depicted in Fig. 9). The network attributes often are mapped to attributes found in the associated feature; during the process of building the network and establishing network connectivity, attribute values are read from the features and persisted into the network. The reason for doing so is to minimize the number of tables that must be queried during network analysis, in order to achieve better performance.

However, with very dynamic environments such as are found in location-based service applications, it is sometimes advantageous not to have to persist the value of a network attribute along with the network connectivity. This is particularly the case if the attributes on the feature that are mapped to a network attribute are subject to frequent change. Evaluator components serve to abstract away the underlying storage of the network attributes. Client applications (including the build algorithm and the forward star cursors, as well as analysis algorithms) instead query the evaluators that are associated with attributes and feature classes. The evaluators may return values that are persisted directly in the connectivity network, or they may derive an attribute value on the fly (or even query a web service). In this manner, client applications are presented a uniform view of accessing attribute values.

3 Access Model

There are various approaches to effectively building, maintaining, and navigating the elements contained within a network. Some systems (e.g., [23]) have placed the onus upon the client application for the discovery and maintenance of network elements; client applications are responsible for determining the connectivity and appropriately setting the foreign keys that are used to specify the connectivity in the persisted representation (e.g., setting the from and to junctions on the edge elements). Other previous systems (e.g., [22], [31]) have instead provided mechanisms that perform geometric analysis in order to automatically determine connectivity and persist the information. The choice of when to establish or update the persisted connectivity information is based in part upon the user workflows that the network solution is trying to address.

In addition to building the networks, various approaches have been taken to how the network should be queried. Some systems have relied upon low level querying of the persisted network representation (e.g., which two junctions are connected to the specified edge), while other have provided alternative query mechanisms.

3.1 Workflows

There are two common usage classes among users of network data (from a maintenance standpoint); one class of user purchases their network data (or obtains it from external sources) and infrequently edits or modifies the data. They are instead focused on performing analysis upon the obtained network data. The second class of user is actively engaged in editing and maintaining their data. Most often, the second class of users are large organizations such as the government, utilities, or data providers. We have observed that the first class of user is far more common – most people do not actively edit the features participating in their networks.

For the first class of user that infrequently edits their network data, it is sufficient to support a build process that can complete all geometric and connectivity analysis and network element persistence across the dataset in its entirety. For such users, the network is built immediately following network definition and creation. If the user chooses to edit the features in the network, the entire network will have to be rebuilt in order to guarantee correctness of the connectivity used during network analysis.

With the second class of user that is actively editing the features participating in their network, it may still prove viable to only support a global network build process if the organization can tolerate a build occurring on a periodic basis (e.g., over the weekend; see Section 5 for details concerning building the entire US road network in less than two days). If the organization is editing smaller datasets, the build operation can be staged on a more frequent basis (e.g,, overnight).

There is however a subset of this second class of user (the frequent editors) that needs to have correct network connectivity during the course of editing. For such users, it becomes necessary to support a user-initiated incremental build process where only those portions of the network that correspond to edited features are rebuilt. Techniques may be employed (such as dirty area management with ArcGIS Topology [11]) that will assist in the incremental build of the network.

The need for incremental builds during frequent editing can be obviated if network connectivity is "live", i.e., network connectivity is automatically re-generated after individual edits to the source data. This alternative approach is used in the ArcGIS Geometric Network [31]; however, it is not as viable here given our rich connectivity model and the complexities introduced by turns and maneuvers.

3.2 Connectivity Queries

For the conventional normalized relational representation, standard SQL queries may be employed. Navigation at this level can prove cumbersome and slow. In some instances, in order to overcome this problem, middleware libraries have been developed [16] that provide analysis functions (e.g., shortest path between two junctions, or the traveling salesman problem [4], [21]). This is useful; however, the navigation is at a high level, precluding clients from developing their own analysis functionality that may require low-level navigation.

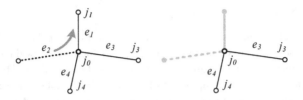

Fig. 8. Example highlighting difference between connectivity and traversability when turn restrictions are present. Connectivity is depicted on the left, with traversability on the right

Alternatively, low-level network navigation may be performed using a forward star adjacency query [6]. A forward star query returns the elements in a network that are immediately reachable from another element. The query is constrained by a set of restrictions (e.g., one-way streets, prohibited turns) that controls which elements are traversable. For example, consider the situation depicted in Fig. 8. In this example, there is a turn restriction at junction j_0 when moving from edge e_2 to edge e_1. This is shown on the left side of the figure. A forward star query at junction j_0 from edge e_2 will result in two edge-junction pairs being returned; namely (e_3, j_3), and (e_4, j_4). The edge-junction pair (e_1, j_1) is not returned as it is not traversable from edge e_2 at junction j_0 because of the turn restriction. From a performance standpoint, forward star queries (and storage representations – see Section 4.2) are the preferred method for querying network connectivity during network analysis operations [26].

4 Physical Model

4.1 Standard Physical Implementation

Network topology can be implemented for relational databases in a straightforward fashion as a normalized relational model with explicit representation of network primitives and connectivity using primary and foreign keys (see Fig. 9). This model has been employed in both research and commercial systems [5], [14], [21], [23]. We term this the standard relational network model. A fundamental implementation choice is whether or not the tables representing the network elements (junctions and edges) contain any associated geometry (in Fig. 9, we depict an implementation where geometry is persisted in the network element tables). If geometry is absent from the network tables, they are sometimes referred to as a logical network. If geometry is present, they may be termed spatial networks [16].

The network connectivity is represented by the *from* and *to* junction id foreign keys in the edge table. This representation is definition-based and follows naturally from the mathematical definition of the edges as being a binary relation on the junctions. Attributes may be added to both the junction and edge tables as necessary. It is common to associate impedances or hierarchy values with network elements in this manner.

The normalized relational model is suited to a class of SQL-based connectivity queries. For a given junction (presuming the junction id is known), the connected edges may be obtained via a selection query of the edge table where either the from or to junction id foreign keys match the specified junction's id value. When traversing a network (e.g, a shortest path computation), each junction that is explored will require a separate SQL query. This can be quite expensive in terms of server loading and suffers from a performance standpoint.

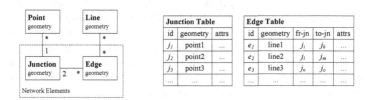

Fig. 9. Standard relational network model with geometry represented in the network tables

In order to address this problem, middleware based solutions have been proposed that cache network connectivity information on the client (or application server) and provide access to the information through a conventional API (e.g., Java) on a collection of higher level components [23]. Data management is usually performed via low-level SQL, while navigation and connectivity analysis is via the higher-level API.

A modified adjacency structure is presented in [10] which stores for each edge in the network, a list of possible outgoing edges from its ending junction, taking into account permissible turns between edges. The modified structure does not satisfy our modeling requirements because it only considers turn prohibitions, which are always enforced to constrain the outgoing edges for each incoming edge. In contrast, we regard a turn as a first-class entity with attribution, e.g., one left turn can be used to specify turning restrictions and penalties for different vehicle types using multiple network attributes. Turns do not modify network connectivity, but affect traversability and costs based on the attributes applied during a network analysis. Furthermore, the modified adjacency structure is limited to two-part turns and cannot represent multi-part turns.

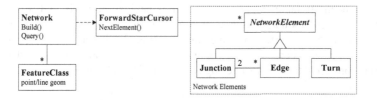

Fig. 10. Basic components in the network engine object model

4.2 Alternative Object Model and Physical Implementation

In order to address some of the problems inherent in the standard network physical implementation and better support the aforementioned requirements (e.g., high performance network analysis functions residing on the client or application server tier)

and workflows, we describe a new object model that is currently hosted within the ArcGIS 9.1 Geodatabase [31]. The basic architecture is a small collection of components that are exposed through industry standard Java, .NET, and COM APIs.

The principal components in the network engine implementation are shown in Fig. 10. The Network is the central component to the system. Chief among its functionality is that which allows the client to build the connectivity of the persisted network representation through geometric analysis of the line and point Features found in the associated FeatureClasses. Each Feature will ultimately correspond to one (or more in the case of line features when the connectivity model is configured to support mid-span connectivity) NetworkElement. The NetworkElement (an abstract class, with three concrete subclasses – Junction, Edge, and Turn) provides an API that allows the direct navigation to the other immediately traversable NetworkElements. It additionally provides a general method for accessing the values of the associated network attributes.

The Network component supports a query model where clients (such as high-performance network analysis algorithms) issue forward star queries [6]. When such a query is issued, a ForwardStarCursor component is returned. This cursor allows the client to index or iterate through the returned traversable NetworkElements (i.e., connected NetworkElements that satisfy the traversability requirements such as respecting turn restrictions, etc.). If necessary, a client can also use lower-level query models supported by the Network component, such as 'give me all NetworkElements that are associated with the specified Feature.'

Storage Representation. The network consists of a collection of tables within a geographical database. The network contains metadata (network definition and connectivity configuration information), junction, edge, and turn elements, the connectivity relationships between them, and the attributes necessary for traversing the network and performing analysis.

Junction Table. The physical storage representation of the network differs, however, from the conventional relational implementation discussed previously (r.e., Fig. 9). While the connectivity information in the conventional implementation is represented as foreign keys within the edge table, we instead represent the connectivity as a set of (edge id, junction id) foreign key tuples that are associated with a junction record in the junction connectivity table. This representation is navigation-based, and is designed to answer the most common adjacency query during network analysis, which is to find the edges and junctions connected to a given junction (r.e., the forward star query in Section 3.2). Each junction record can have four such tuples; if more are needed (e.g., the junction is connected to five or more edges), an overflow table is used.

Junction Table							Edge Table	
id	edge1	junc1	...	edge4	junc4		id	from-jn
j_1	e_i	j_l	...	e_k	j_k		e_1	j_i
j_2	e_l	j_l	...	e_m	j_m		e_2	j_l
j_3	e_n	j_n	...	e_o	j_o		e_3	j_n
...	...	...	...	...	...		...	...

Fig. 11. Network storage representation optimized for forward star queries

It is important to note that this storage representation utilizes fixed-length records (r.e., the need for an overflow table). Fixed length records allow us to have direct access into the connectivity data. The number of adjacency tuples in the fixed-length record was determined empirically. For transportation networks, almost all junctions have degree four or less (i.e., the number of edges connected to each junction is typically four or less). This is evident in Table 1, which shows the frequency distribution of junction degrees for a network on Southern California (715,286 junctions and 978,965 edges).

For this network, with four tuples per fixed-length record, the space utilization is 68% with 10 bytes per junction unoccupied or being used for overflow record information. This compares favorably with variable-length records, which would have similar overhead. Only the junctions whose degrees are five or higher (which is less than 0.5% in this network) require two or more records to hold adjacency information; almost all the junctions require only one record for adjacency information. Note that the three-tuple per record representation has a slightly higher space utilization of 76%, but 21% of the junctions would require two or more records.

Table 1. Frequency distribution of junction degrees for a Southern California road network

Degree	1	2	3	4	5	6	7	8
Count	147,86	43,737	375,34	145,40	2,689	234	11	1
Percentage	20.7%	6.1%	52.5%	20.3%	0.4%	0.03	<0.01	<0.01%

Each fixed-length record shown in Fig. 11 is not stored as a row in a relational table; instead, we chose to serialize and compress the rows into larger collections of data (pages) and persist the pages in BLOB tables (an RDBMS column/data type capable of storing binary large objects [13]) within the relational database. The relational database in effect is being used as a paged file system. The network engine components (that reside at either the client or application tiers as described in Section 1 and shown in Fig. 10) provide caching mechanisms and APIs that support both data management and analysis functionality.

Edge Table. The edge table in our storage representation contains the foreign key of the from-junction associated with the edge. If the to-junction is needed, the junction table is queried using the from-junction and edge identifiers. We have observed that finding the from- and to-junctions associated with an edge is actually a fairly uncommon operation during efficient network analysis operations. Thus, we have optimized our storage representation to more effectively support the most common connectivity access pattern - the forward star adjacency query (see Section 3.2). In its simplest form, the forward star adjacency query takes as input a junction element, and returns the set of connected edges and the junctions at the other end of those edges.

Turn Tables. We have chosen not to employ a graph modification technique (e.g., node expansion or line graphs) to represent turning movements as edge elements within the network. As noted earlier, such techniques are awkward for representing complex turns (maneuvers), and the modified graphs are also difficult to maintain in a dynamic editing environment. Instead, we store turn elements in a turn table, with a representation that is optimized for the most common client access patterns.

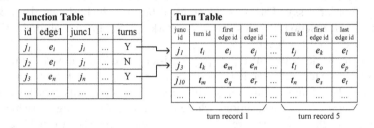

Fig. 12. Turn table representation

The turn table concept that we employ is generalized to effectively support maneuvers as well as the forward star adjacency queries. For each junction in the network, we indicate if there are any associated turns anchored (i.e., the last junction participating in the turn) at the junction. If there are any associated turns anchored at the junction, the turn table contains up to five (turn id, first edge id, last edge id) triplets. An example of a turn table is shown in Fig. 12. If more than five turns are anchored at the junction, an overflow mechanism is supported (similar to that employed with the junction table). As is the case with the junction and edge connectivity tables, we utilize a fixed length record format to facilitate the compression and serialization of the turn table into pages persisted within a BLOB column in the relational database.

During a forward star adjacency query, for a given junction and inbound edge, the turn table may be queried with the specified junction and (first) edge. If an entry matches the (junction, first edge) pair, then the last edge information in the turn entry allows the pairing of the turn with the correct outgoing edge in the forward star result.

4.3 Network Building

As noted earlier, our implementation supports a network building process where the connectivity graph of a network is derived from its source data via geometric analysis. The steps during building are:

1. Extract the geometries of the features in the source data. The extracted coordinates and their feature parentage are stored in a vertex information table.

2. Sort the vertex information table by coordinate values, so that coincident vertices are grouped together.

3. Analyze each group of coincident vertices according to the connectivity model, and generate the appropriate junction elements. During this analysis, vertices that do not connect to other vertices are discarded, while the remaining vertices may be further partitioned into disjoint subsets.

4. Re-sort the vertex information table by vertex, so that vertices from each line feature are re-grouped together.
5. Scan the vertex information table, and generate edge elements connecting adjacent vertices on each line.
6. Analyze turn features and generate associated turn elements.
7. Populate the attribute values of the generated network elements.

Spatial Clustering. When a network analysis algorithm is executing (e.g., a shortest path search between two locations), it typically does not examine the network in a haphazard manner. Instead, there is spatial locality of reference [24]. Areas of the network that are queried next are usually near areas that have already been explored. We exploit this locality by spatially clustering the network elements during the network build process using a space-filling curve (we employed a Peano curve [25]), and persist the network elements in the clustering order within the BLOB pages of the network tables. Other spatial clustering techniques of network elements have been tested and found to be superior to both non-clustered and topologically clustered elements [12].

5   Implementation Experience

This new network model has been implemented and is currently shipping with ESRI's ArcGIS 9.1 product. It addresses each of the requirements enumerated in Section 2.1. It has been used to build very large continent-wide transportation networks, including a dataset derived from the features contained within the entire continental United States (35.9 million line features).

Performance statistics on several different size network build operations are shown in Table 2 (number of linear source features, number of vertices in their geometries, number of network elements created, and the wallclock build time). A reasonable PC (2.4GHz, 2GB RAM) running ArcGIS 9.1 was utilized on the client side and a commercial relational database was employed on the server side. Reported build times include the geometric analysis of the feature geometry in order to establish connectivity, as well as the population of attributes within the persisted network representation (e.g., travel time along an edge). We observed that in the *typical* case, geometry and connectivity analysis consumed 45% of the build process time, while creation of the persistent network elements took 30% of the build time, and population of network attributes the remaining 25% of the processing time.

Table 2. Summary statistics of large networks built

Dataset	Features	Vertices	Net Elements	Build Time
U.S. National	35.9 million	128.3 million	65.1 million	43 hours
Northeast U.S.	5.3 million	27.0 million	9.6 million	1.8 hours
Major U.S. Streets	1.8 million	20.4 million	3.1 million	0.5 hours
Paris Metro	0.4 million	0.8 million	0.7 million	< 3 minutes

6 Future Work

There are several areas of ongoing research and development with our network model that will be incorporated into the ArcGIS product following the 9.1 release. These include the direct support for the class of user that is a heavy editor of the features participating in the network (as described in the Section 3.1). In order to support this group, it will be necessary to support the incremental build of the network in the versioned environment. An analogous capability was developed and provided with ArcGIS Topology [11]; this entailed dirty area management, the development of an incremental topology validation process, and incorporation of topology into the ArcGIS Version Management infrastructure. Analogous development tasks will occur with this new network model.

Dirty Areas. A network can have an associated *dirty area* – a dirty area corresponds to the regions within the network extent where features participating in the network have been modified (added, deleted, or updated) but whose connectivity has not been re-established. When the geometry of a feature that participates in a network is modified, the extent of the dirty area is enlarged to encompass the extent of the bounding rectangle of the modified geometry (note that other simplified geometry representations may also be employed - e.g., convex hulls). The dirty area is persisted with the network. In order to ensure that the network is correct, the portion of the network encompassed in the dirty areas will need to be rebuilt.

It is not necessary to build the entire space spanned by the dirty area at one time; instead, a subset of the dirty area can be built. If the dirty area is partially built, the original dirty area will be clipped by the extent of the region that is built.

Allowing users the ability to build a portion of the dirty area is a pragmatic requirement of supporting extremely large seamless network. For example, when a network is first defined, or when the network metadata (e.g., connectivity model, etc.) is modified, the entire extent of the network is dirty. If users were not provided with the capability to build a portion of the dirty area, the user would be required to build the entire network which could prove to be a very lengthy process (e.g., a couple days of processing time for large continent-wide network datasets). As was discussed in [11], the dirty area model effectively supports partial processing in computationally intensive areas of GIS such as topology.

Incremental Build. In order to minimize the amount of processing necessary to maintain a consistent connectivity network, the dirty area mechanism may be exploited in conjunction with an incremental build algorithm. In an incremental build, the connectivity information associated with features in a dirty area is deleted and rebuilt (recreated) in order to achieve a consistent state of the network. The high level algorithm is as follows:

1. Delete the network elements associated with the line features intersecting the dirty area (or portion thereof) being built.
2. Load the geometries of all the line features intersecting the area being built along with the associated network metadata (connectivity model, ternary mapping of evaluators to network attributes and feature classes).

3. Load the geometries of all point features that are connected to the line features intersecting the area being built (these point features may lie outside the area being built if the line feature extends outside the area).

4. Sort the vertices associated with the features, retaining the parentage information.

5. Discard all isolated line vertices (i.e., the vertices on the line features that are not coincident with other vertices from differing line or point features).

6. Discard interior line vertices if the connectivity model does not support mid-span connectivity on the associated line feature class.

7. Create network junctions as necessary for the remaining line vertices and other isolated vertices associated with point features.

8. Create network edges between the junctions as appropriate based upon the connectivity model.

9. Update the dirty areas associated with the network.

It is important to note that the network build process does not need to span all features within the network. A build can be performed on a subset of the space spanned by the dataset. This is a complex task since the re-built portion of the network has to be properly stitched together with the rest of the network.

7 Conclusion

In this paper we described the logical model of GIS network topology and several extensions to the standard network model that directly facilitate the modeling of multimodal systems, supporting mid-span connectivity on line features, as well as supporting endpoint elevation data that often accompanies large planar datasets from commercial data vendors. We reviewed a common physical database implementation that uses the conventional notions for mapping entities and relationships to tables and the standard primary key / foreign key referential integrity model. Problems with this approach were discussed. We then presented an alternative implementation of the network model which used a different physical approach to persisting network connectivity. This new model additionally supports turn restrictions and impedances, both two-part turns as well as multi-part turns (maneuvers). Efficient mechanisms for navigating the network connectivity were discussed (the forward star adjacency query), as well as a more flexible mechanism (network evaluators) for maintaining and querying attributes on the network elements. This design serves as the basis for our implementation of transportation networks in the ArcGIS geographic information system; this new model has been implemented and is currently shipping with the ArcGIS 9.1 product.

Our future work will focus on extending the network with support for dirty area management policies and an incremental build algorithm that is more useful for organizations that frequently edit their network data (e.g., governmental organizations and commercial data providers). In addition, we will be supporting this network model in the distributed database environment, incorporating aspatial features (features without geometry) into the network model, as well as other performance enhancements.

Acknowledgements

Numerous other individuals within ESRI Development were involved in the design and implementation of this network model in the ArcGIS 9.1 product. Key people included Frederic Albert, Gillian Allen, Hal Bowman, Jai Chakrapani, Matt Crowder, Craig Gillgrass, Alan Hatakeyama, Rich Krabill, Jim McKinney, Sudhakar Menon, Scott Morehouse, Jay Sandhu, Frederic Schettini, Doug Sterling, and Jeff Wickstrom.

References

1. J. Añez, T. de la Barra, and B. Pérez. Dual Graph Representation of Transport Networks. *Transportation Research* 30(3), June 1996
2. T. Caldwell. On Finding Minimum Routes in a Network with Turn Penalties. *Communications of the ACM* 4(2), February 1961
3. Caliper. *TransCAD: Transportation GIS Software Ref. Man.*, Newton, MA, 1996
4. T. Cormen, C. Leiserson, R. Rivest, and C. Stein. Introduction to Algorithms, 2nd Edition. MIT Press, Cambridge, Massachusetts, 2001
5. K. Dueker and J. Butler. *GIS-T Enterprise Data Model with Suggested Implementation Choices.* PR101, Center for Urban Studies, Portland State, 1997
6. J. Evans and E. Minieka. *Optimization Algorithms for Networks and Graphs.* M. Dekker (editor), New York, 1992
7. ESRI. *Network Analysis; Workspation ARC/INFO version 8.1.* Prepared by Environmental Systems Research Institute, Redlands, California, 2001
8. T. Foresman (editor). *The History of Geographical Information Systems.* Prentice Hall PTR, Upper Saddle River, New Jersey, 1998
9. M. Goodchild. Geographic Information Systems and Disaggregate Transportation Planning. *Geographical Systems* 5, 1998
10. R. Güting, V. de Almeida, and Z. Ding. *Modeling and Querying Moving Objects in Networks.* FernUniversität in Hagen, Informatik-Report 308, 2004
11. E. Hoel, S. Menon, and S. Morehouse. Building a Robust Relational Implementation of Topology. In *Proc. of the 8th Intl. Symp. on Spatial and Temporal Databases (SSTD 2003).* Santorini Island, Greece, July 2003
12. Y.-W. Huang, N. Jing, and E. Rundensteiner. Optimizing Path Query Performance: Graph Clustering Strategies. *Transportation Research Part C 8*, 2000
13. International Organization for Standardization (ISO). *ISO International Standard: Database Language SQL – Part 2: Foundation (SQL/Foundation)*, ANSI/ISO.IEC 9075-2:99, September 1999
14. C. Jensen, T. Pedersen, L. Speičys, and I. Timko. Data Modeling for Mobile Services in the Real World. In *Proc. of the 8th Intl. Symp. on Spatial and Temporal Databases (SSTD 2003).* Santorini Island, Greece, July 2003
15. R. Kirby and R. Potts. The Minimum Route Problem for Networks with Turn Penalties and Prohibitions. *Transportation Research* 3, 1969
16. R. Kothuri, A. Godfrind, E. Beinat. *Pro Oracle Spatial.* Apress, Berkeley, 2004
17. L. Lang. *Transportation GIS.* ESRI Press, Redlands, California, 1999
18. P. Longley, M. Goodchild, D. Maguire, and D. Rhind (editors). *Geographical Information Systems, Volume 2.* John Wiley & Sons, New York, 1999

19. M. Mainguenaud. Modeling the Network Component of Geographical Information Systems. *International Journal of Geographic Information Systems* 9(6), 1995

20. R. Marx. The TIGER System: Automating the Geographic Structure of the United States Census. *Government Publications Review* 13, 1986

21. H. Miller and S.-L. Shaw. *Geographic Information Systems for Transportation.* Oxford University Press, Oxford, England, 2001

22. S. Morehouse. ARC/INFO: A Geo-Relational Model for Spatial Information. In *Proc. of the 7th Intl. Symp. on Computer Assisted Cartography (Auto-Carto 7)*, Washington, DC, March 1985

23. Oracle. *Oracle Database 10g Network Data Model.* Prepared by Oracle Corporation, Redwood Shores, California, 2003

24. D. Papadias, J. Zhang, N. Mamoulis, and Y. Tao. Query Processing in Spatial Network Databases. In *Proc. of the 29th VLDB Conf. (VLDB 2003).* Berlin, 2003

25. G. Peano. Sur une Courbe Qui Remplit Toute une Aire Plaine. *Mathematische Annalen*, 36: 157-160, 1890

26. B. Ralston. GIS and ITS Traffic Assignment: Issues in Dynamic User-Optimal Assignments. *GeoInformatica* 4(2), June 2000

27. P. Rigaux, M. Scholl, and A. Voisard. *Spatial Databases with Application to GIS.* Morgan Kaufmann, San Francisco, 2002

28. L. Speičys, C. Jensen, and A. Kligys. Computational Data Modeling for Network-Constrained Moving Objects. In *Proc. of the 11th ACM Intl. Workshop on Advances in Geographic Information Systems (ACM-GIS'03)*, New Orleans, Nov. 2003

29. F. Southworth and B. Peterson. Intermodal and International Freight Network Modeling. *Transportation Research Part C 8*, 2000

30. S. Winter. Modeling Costs of Turns in Route Planning. *GeoInformatica* 6(4), 2002

31. M. Zeiler. *Modeling Our World: The ESRI Guide to Geodatabase Design.* ESRI Press, Redlands, California, 1999

Nearest Neighbor Search on Moving Object Trajectories

Elias Frentzos[1], Kostas Gratsias[1,2], Nikos Pelekis[1], and Yannis Theodoridis[1,2]

[1] Department of Informatics, University of Piraeus, 80 Karaoli-Dimitriou St,
GR-18534 Piraeus, Greece
{efrentzo, gratsias, npelekis, ytheod}@unipi.gr
http://isl.cs.unipi.gr/db
[2] Research and Academic Computer Technology Institute,
11 Aktaiou St & Poulopoulou St, GR-11851 Athens, Greece
{gratsias, ytheod}@cti.gr

Abstract. With the increasing number of Mobile Location Services (MLS), the need for effective k-NN query processing over historical trajectory data has become the vehicle for data analysis, thus improving existing or even proposing new services. In this paper, we investigate mechanisms to perform NN search on R-tree-like structures storing historical information about moving object trajectories. The proposed branch-and-bound algorithms vary with respect to the type of the query object (stationary or moving point) as well as the type of the query result (continuous or not). We also propose novel metrics to support our search ordering and pruning strategies. Using the implementation of the proposed algorithms on a member of the R-tree family for trajectory data (the TB-tree), we demonstrate their scalability and efficiency through an extensive experimental study using synthetic and real datasets.

1 Introduction

With the integration of wireless communications and positioning technologies, the concept of Moving Object Databases (MOD) has become increasingly important, and has posed a great challenge to the database community. In such implicitly formulated location-aware environments, moving objects are continuously changing locations; nevertheless existing DBMSs are not well equipped to handle continuously changing data. Emerging location-dependent services (including nearby information accessing and enhanced 911 services) call for new query processing algorithms and techniques to deal with both the spatial and temporal domains.

Unlike traditional databases, MODs have some distinctive characteristics: First of all, spatio-temporal queries are continuous in nature. In contrast to snapshot queries, which are invoked only once, continuous queries require continuous evaluation as the query result becomes invalid after a short period of time. Secondly, we typically have to deal with vast volumes of historical data which correspond to a large number of mobile and stationary objects. As a consequence, querying functionality embedded in an extensible DBMS that supports moving objects has to present robust behavior in the above mentioned issues.

C. Bauzer Medeiros et al. (Eds.): SSTD 2005, LNCS 3633, pp. 328–345, 2005.

An important class of queries that definitely turns out to be useful for MOD processing is the so-called k nearest neighbor (k-NN) queries, where one is interested in finding the k closest trajectories to a predefined query object Q. To our knowledge, in the literature such queries primarily deal with either static ([8], [2], [4]) or continuously moving query points ([11], [13]) over stationary datasets, or queries about the future positions of a set of continuously moving points ([1], [12], [5]). Apparently, these types of queries do not cover NN search on historical trajectories.

The challenge accepted in this paper is to describe diverse mechanisms to perform k-NN search on R-tree-like structures [6] storing historical information. To illustrate the problem, consider an application tracking the positions of rare species of wild animals. Such an application is composed of a MOD storing the location dependent data, together with a spatial index for searching and answering k-NN queries in an efficient manner. Experts in the field would be advantaged if they could pose queries about the nearest trajectories of animals to a stationary point (lab, source of food or other non-emigrational species) or an animal moving from location P_1 to P_2 during a period of time. By these types of queries an expert may figure out motion habits and patterns of wild species or deviations from natural emigration, which could be interrelated with environmental and/or ecological changes or destructions. Having in mind that users of MODs are usually interested in continuous types of queries, the above queries can be extended to their continuous counterparts, where the result is a time-varying number (the nearest distance depends on time) along with a collection of trajectory ids and the appropriate time intervals for which each moving object is valid.

To make the previous example more intelligible, Fig. 1 illustrates the trajectories of six moving animals $\{O_1, O_2, O_3, O_4, O_5, O_6\}$ along with two stationary points (Q_1 and Q_2) representing two sources of food. Now, consider the following queries demonstrated in Fig. 1 (Queries 2 and 4 are the continuous counterparts of Queries 1 and 3, respectively):

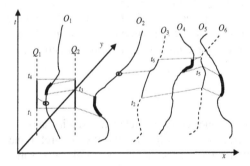

Fig. 1. Continuous and non-continuous point and trajectory NN queries over moving objects trajectories

- Query 1. *"Find which animal was nearest to the stationary food source Q_1 during the time period $[t_1, t_4]$"*, resulting to animal O_1.
- Query 2. *"Find which animal was nearest to the stationary food source Q_2 at any time instance of the time period $[t_1, t_4]$"*, resulting to a list of objects: O_2 for the interval $[t_1, t_3)$; O_1 for the interval $[t_3, t_4]$.

- Query 3. *"Find which animal was nearest to animal O_3 during the time period $[t_2,t_6]$"*, resulting to O_2.
- Query 4. *"Find which animal was nearest to animal O_6 at any time instance of the time period $[t_2,t_6]$"*, resulting to a list of objects: O_5 for the interval $[t_2,t_5)$; O_4 for the interval $[t_5,t_6]$.

To the best of our knowledge, this is the first work on continuous k-NN query processing over historical trajectories of moving objects. Outlining the major issues that will be addressed in this paper, our main contributions are as follows:

- We propose a set of four query processing algorithms to perform NN search on R-tree-like structures storing historical information about moving objects. The description of our branch-and-bound traversal algorithms for different queries depends on the type of the query object as well as on whether the query itself is continuous or not. The algorithms are generalized to find the k nearest neighbors.
- We propose novel metrics to support our search ordering and pruning strategies. More specifically, the definition of the minimum distance metric MINDIST between points and rectangles, initially proposed in [8] and extended in [13], is further extended in order for our algorithms to calculate the minimum distance between trajectories and rectangles.
- We conduct a comprehensive set of experiments over synthetic and real datasets demonstrating that the algorithms are highly scalable and efficient in terms of node accesses and pruned space.

The rest of the paper is structured as follows. Related work is discussed in Section 2, while Section 3 introduces, at an abstract level, the set of k-NN algorithms over moving object trajectories, as well as the metrics that support our search ordering and pruning strategies. Sections 4 and 5 constitute the core of the paper describing in detail the query processing algorithms to perform NN search over historical trajectory information (Section 4) together with their continuous counterparts (Section 5). Section 6 presents the results of our experimental study and Section 7 provides the conclusions of the paper and some interesting research directions.

2 Related Work

In the last decade, NN queries have fueled the spatial and spatiotemporal database community with a series of interesting noteworthy research issues.

The first algorithm for k nearest neighbor search over a moving query point was proposed in [11]. The algorithm assumes that sites (landmark points) are static and their locations (known in advance) are stored in an R-tree-like structure. A discrete time dimension is assumed, thus a periodical sampling technique is applied on the trace of the moving query point. The location of the query point that lies between two consecutive sampled locations is estimated using linear or polynomial splines.

Using the TPR-tree (Time Parameterized Tree) structure [9], Benetis et al. [1] presented efficient solutions for NN and RNN (Reverse Nearest Neighbor) queries for moving objects. (An RNN query returns all the objects that the query object is the nearest neighbor of.) The proposed algorithm was the first to address continuous RNN queries, since previous existing RNN algorithms were developed under the assumption that the query point is stationary. The algorithms for both NN and RNN queries in [1] refer to future (estimated) locations of the query and data points, which are assumed to be continuously moving on the plane. In the same paper, an algorithm for answering CNN queries is also proposed.

Tao et al. [13] also studied CNN queries and proposed an R-tree based algorithm (for moving query points and static data points) that avoids the pitfalls of previous ones (false misses and high processing cost). The proposed tree pruning heuristics exploit the MINDIST metric presented in [8]. At each leaf entry, the algorithm focuses on the accurate calculation of the split points (the points of the query segment that demonstrate a change of neighborhood). A theoretical analysis of the optimal performance for CNN algorithms was presented and cost models for node accesses were proposed. Finally, the CNN algorithm was extended for the case of k neighbors and trajectory inputs.

Shahabi et al. [10] presented the first algorithm for processing the k-NN queries for moving objects in road networks. Their proposed algorithm, which utilizes the network distance between two locations instead of the Euclidean, is based on transforming the road network into a higher dimensional space, in which simpler distance functions can be applied. Using this embedding space, efficient techniques are proposed for finding the shortest path between two points in the road network. The above procedure, which is utilized in the case of static query points, is slightly modified in order to support the case of moving query points.

Acknowledging the advantages of the above fundamental techniques, in this paper we present the first complete treatment of historical NN queries over moving object trajectories, handling both stationary and moving query objects.

3   Problem Statements and Metrics

We first define the NN queries that are considered in this paper. Subsequently, we present the heuristics utilized by our algorithms to implement the metrics needed to formulate our ordering and pruning strategy.

3.1   Problem Statement

Let D be a database of N moving objects with objects ids $\{O_1, O_2, ..., O_N\}$. The trajectory T_i of a moving object O_i consists of M_i 3D-line segments $\{L_{i1}, L_{i2}, ..., L_{iM_i}\}$. Each 3D line segment L_j is of the form $((x_{j\text{-}start}, y_{j\text{-}start}, t_{j\text{-}start}), (x_{j\text{-}end}, y_{j\text{-}end}, t_{j\text{-}end}))$, where $t_0 \leq t_{j\text{-}start} < t_{j\text{-}end} \leq now$. Obviously, as we treat only historical moving object trajectories, each partial linear movement is temporally restricted between t_0, the beginning of the calendar, and *now*, the current time point.

We have already stated that NN queries search for the closest trajectories to a query object Q. In our case, we distinguish two types of query objects: Q_p, a point (x,y) that remains stationary during the time period of the query $Q_{per}[t_{start}, t_{end}]$, and Q, a moving object with trajectory T. Furthermore, the MOD is indexed by an R-tree like structure such as the 3D R-tree [16], the STR-tree or the TB-tree [7]. Having in mind the previous discussion, we define the following two types of NN queries:

- NN_Q_p (D, Q_p, Q_{per}) query searches database D for the NN over a point Q_p that remains stationary during a time period Q_{per}, and returns the closest to Q_p point p_c from which a moving object O_i passed during the time period Q_{per}, as well as the implied minimum distance.
- NN_Q_T (D, Q_T, Q_{per}) query is similar to the previous with the difference being upon the query object Q which in the current case is a moving object with trajectory T.

The extensions of the above queries to their continuous counterparts vary in the output of the algorithms. In the continuous case, each query returns a time-varying real number, as the nearest distance depends on time. We introduce the following two types of CNN queries:

- CNN_Q_p (D, Q_p, Q_{per}) query over a point Q_p that remains stationary during a time period Q_{per} returns a list of triplets consisting of the time-varying real value R_i along with a moving object O_i (belonging in database D) and the corresponding time period $[t_{i\text{-}start}, t_{i\text{-}end})$ for which the nearest distance between Q_p and O_i stands. These time-varying real values R_i are, in any time instance of their lifetime, smaller or equal to the distance between any moving object O_j in D and the query point Q_p. The time periods $[t_{i\text{-}start}, t_{i\text{-}end})$ are mutually disjoint and their union forms Q_{per}.
- Similarly, CNN_Q_T (D, Q_T, Q_{per}) differs, compared to the previous, upon the query object Q which in the current case is a moving object with trajectory T. These time-varying real values R_i are, in any time instance of their lifetime, smaller or equal to the distance between any moving object O_j and the query trajectory Q_T. The time periods $[t_{i\text{-}start}, t_{i\text{-}end})$ are mutually disjoint and their union forms Q_{per}.

The above four queries are generalized to produce the corresponding k-NN queries. The generalization of the first two queries is straightforward by simply requesting the 1-st, 2-nd, ..., k-th nearest point – with respect to a query point or a query trajectory – from which a moving object O_i passed during the time period Q_{per}, excluding at the same time points belonging to a moving object already marked as the j-th nearest ($1 \le j < k$). The continuous queries are generalized to produce k-CNN requesting to provide with k lists of $\{R_i, [t_{i\text{-}start}, t_{i\text{-}end}), O_i\}$ triplets. Then, for any time during the time period Q_{per}, the i-th list ($1 \le i \le k$) will contain the i-order NN moving object (with respect to the query point or the query trajectory) at this time instance.

To exemplify the proposed k-NN extensions, let us recall Fig. 1. Searching for the 2-NN versions of the four queries (Query 1, 2, 3 and 4) presented in Section 1, we will have the following results:

- Query 1 (non-continuous): O_1 (1st NN) and O_2 (2nd NN)
- Query 2 (continuous): 1-NN list includes O_2 for the interval $[t_1, t_3)$ and O_1 for the interval $[t_3, t_4]$; 2-NN list includes O_1 for the interval $[t_1, t_3)$ and O_2 for the interval $[t_3, t_4]$
- Query 3 (non-continuous): O_2 (1st NN) and O_4 (2nd NN)
- Query 4 (continuous): 1-NN list includes O_5 for the interval $[t_2, t_5)$ and O_4 for the interval $[t_5, t_6]$; 2-NN list includes O_4 for the interval $[t_2, t_5)$ and O_5 for the interval $[t_5, t_6]$.

3.2 Metrics

We exploit on the definition of the minimum distance metric (MINDIST) presented in [8] between points and rectangles, in order to calculate, on the one hand, the minimum distance between line segments and rectangles and, on the other hand, the minimum distance between trajectories and rectangles that are needed to implement the above discussed algorithms.

Initially, in [8], Roussopoulos et al. defined the Minimum Distance (MINDIST) between a point P in the n-dimensional space and a rectangle R in the same space as the square of the Euclidean distance between P and the nearest edge of R, if P is outside R (or zero, if P is inside R).

In the sequel, Tao et al. [13] proposed a method to calculate the MINDIST between a 2D line segment L and a rectangle M. They initially determine whether L intersects M; if so, MINDIST is set to zero. Otherwise, they choose the shortest among six distances, namely the four distances between each corner point of M and L and the two minimum distances from the start and end point of L to M. Therefore, the calculation of MINDIST between a line segment and a rectangle involves an intersection check, four segment-to-point MINDIST calculations and two point-to-rectangle MINDIST calculations.

In this paper, we propose a more efficient method to calculate MINDIST between a line segment L and a rectangle M (Fig. 2). As before, if L intersects M, then MINDIST is obviously zero. Otherwise, we decompose the space in four quadrants using the two axes passing through the center of M and we determine the quadrants Q_s and Q_e in which the start ($L.start$) and the end ($L.end$) point of L lie in, respectively.

Then, MINDIST is the minimum among:

- Case 1 ($L.start$ and $L.end$ belong to the same quadrant ($Q_s = Q_e$)): (i) MINDIST between the corner of M in Q_s and L, (ii) MINDIST between $L.start$ and M or (iii) MINDIST between $L.end$ and M.
- Case 2 ($L.start$ and $L.end$ belong to adjacent quadrants Q_s and Q_e, respectively): (i) MINDIST between the corner of M in Q_s and L, (ii) MINDIST between the corner of M in Q_e and L, (iii) MINDIST between $L.start$ and M or (iv) MINDIST between $L.end$ and M.
- Case 3 ($L.start$ and $L.end$ belong to non adjacent quadrants Q_s and Q_e, respectively): two MINDIST between the two corners of M, that do not belong in either Q_s or Q_e, and L.

This method utilizes a smaller number of (point-to-segment and point-to-rectangle) distance calculations compared to the corresponding algorithm in [13]. Finally, we extend the above method in order to calculate the MINDIST metric between the projection of a trajectory T on the plane (usually called route) and a rectangle M. Since a route can be viewed as a collection of 2D line segments, the MINDIST between a route of a trajectory and a rectangle can be computed as the minimum of all MINDIST between the rectangle and each line segment composing the route. The efficiency of this calculation can be enhanced by simply not computing twice, with respect to the query rectangle, the quadrant and the MINDIST of the end and the start of adjacent line segments.

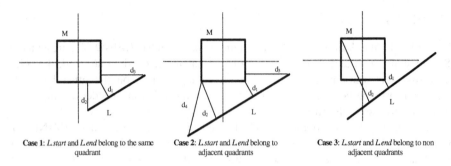

Case 1: *L.start* and *L.end* belong to the same quadrant

Case 2: *L.start* and *L.end* belong to adjacent quadrants

Case 3: *L.start* and *L.end* belong to non adjacent quadrants

Fig. 2. The proposed calculation method of MINDIST between a line segment and a rectangle

4 NN Algorithms over Trajectories

In this section we describe in details the algorithms answering the four types of NN queries presented in Section 3.1 and, then, we generalize them in order to support the respective k-NN queries.

4.1 NN Algorithm for Stationary Query Objects (Points)

The NN algorithm for stationary query objects (PointNNSearch algorithm, illustrated in Fig. 3, provides the ability to answer NN queries for a static query object Q_p, during a certain query time period $Q_{per}[t_{start}, t_{end}]$. The algorithm uses the same heuristics as in [8] and [2], pruning the search space according to Q_{per}.

The algorithm accesses the tree structure (which indexes the trajectories of the moving objects) in a depth-first way pruning the tree nodes according to Q_{per} rejecting those being fully outside it. At leaf level, the algorithm iterates through the leaf entries checking whether the lifetime of an entry overlaps Q_{per} (Line 4); if the temporal component of the entry is fully inside Q_{per}, the algorithm calculates the actual Euclidean distance between Q and the (spatial component of the) entry; otherwise, if the temporal component of the entry is only partially inside Q_{per}, a linear interpolation is applied so as to compute the entry's portion being inside Q_{per} (Line 5) and calculate the Euclidean distance between Q and the portion of that entry. When a candidate nearest is selected, the algorithm, backtracking to the upper level, prunes the nodes in the active branch list (Line 20) applying the MINDIST heuristic [8] [2].

4.2 NN Algorithm for Moving Query Objects (Trajectories)

PointNNSearch algorithm can be modified in order to support the second type of NN query where the query object is a trajectory of a moving point (TrajectoryNN-Search algorithm, illustrated in Fig. 5). At the leaf level, the algorithm calculates the minimum horizontal Euclidean Distance between each leaf entry and each query trajectory segment using the Min_Horizontal_Dist function (Line 10) which computes the minimum horizontal Euclidean Distance between two 3D line segments. In addition, for each segment of trajectory Q and before calculating its distance from the current entry we first check whether its temporal extent overlaps the temporal extent of the bounding rectangle of node N.

```
Algorithm PointNNSearch(node N, 2D point Q, time period Qper, struct
Nearest)
 1.   IF N Is Leaf
// Iterate by computing actual Euclidean distance from point Q
 2.      FOR i = 1 to N.EntriesCount
 3.        E = N.Entry(i)
// If entry is (fully or partially) inside the period
 4.          IF Qper Overlaps (E.Ts, E.TE)
// Compute entry's spatial extent inside the period
 5.            nE = Interpolate(E, Max(Qper.Ts, E.Ts), Min(Qper.TE, E.TE))
// Compute actual distance from Q. Update Nearest if necessary
 6.            Dist = Euclidean_Dist_2D(Q, nE)
 7.            IF Dist < Nearest.Dist
 8.              Nearest.Entry = nE
 9.              Nearest.Dist = Dist
10.           END IF
11.         END IF
12.      NEXT
13.   ELSE
// Generate branch list with entries overlapping the query period
14.      BranchList = GenBranchList(Q, N, Qper)
// Sort active branch List by MinDist
15.      SortBranchList(BranchList)
// Iterate through active branch List
16.      FOR i = 1 TO BranchList.Count
17.        E = N.Entry(i)
// Visit Child Nodes
18.        NN = E.ChildNode
19.        PointNNSearch(NN, Q, Qper, Nearest)
// Apply MinDist heuristic to do pruning
20.        PruneBranchList(BranchList)
21.      NEXT
22.   END IF
```

Fig. 3. Historical NN search algorithm for stationary query points (PointNNSearch)

```
Algorithm genTrajectoryBranchList(node N, trajectory Q)
 1.   FOR i = 1 TO N.EntriesCount
 2.     E = N.Entry(i)
// If entry is (fully or partially) inside the trajectory lifetime
 3.     IF (Q.Ts, Q.TE) Overlaps (E.Ts, E.TE)
// Compute trajectory's spatial extent inside E's lifetime
 4.       nQ = Interpolate(Q, Max(Q.Ts, E.Ts), Min(Q.TE, E.TE))
// Compute MinDist between the resulted trajectory and the rectangle
 5.       Dist=MinDist_Trajectory_Rectangle(nQ, E)
// Add the rectangle along with its calculated distance in the list
 6.       List.Add(nQ, Dist)
 7.     END IF
 8.   NEXT
 9.   RETURN List
```

Fig. 4. Generating Branch List of Node N against Trajectory Q

```
Algorithm TrajectoryNNSearch(node N, trajectory Q, time period Q_per,
struct Nearest)
  1.   Q = Interpolate(Q, Max(Q.T_s, Q_per.T_s), Min(Q.T_E, Q_per.T_E))
  2.   IF N Is Leaf
  3.     FOR j = 1 to Q.Entries
  4.       QE=Q.Entry(j)
  5.       IF (QE.T_s, QE.T_e) Overlaps (N.T_s, N.T_E)
  6.         FOR i = 1 to N.EntriesCount
  7.           E = N.Entry(i)
  8.           IF (QE.T_s, QE.T_e) Overlaps (E.T_s, E.T_E)
  9.             nE = Interpolate(E, Max(QE.T_s, E.T_s), Min(QE.T_E, E.T_E))
 10.             Dist = Min_Horizontal_Dist(QE, nE)
 11.             IF Dist < Nearest.Dist
 12.               Nearest.Entry = nE
 13.               Nearest.Dist = Dist
 14.             END IF
 15.           END IF
 16.         NEXT
 17.       END IF
 18.     NEXT
 19.   ELSE
 20.     BranchList = GenTrajectoryBranchList(Q, N)
 21.     SortBranchList(BranchList)
 22.     FOR i = 1 TO BranchList.Count
 23.       E = N.Entry(i)
 24.       NN = E.ChildNode
 25.       nQ = Interpolate(Q, Max(Q.T_s NN.T_s), Min(Q.T_E NN.T_E))
 26.       TrajectoryNNSearch(NN, nQ, Nearest)
 27.       PruneBranchList(BranchList)
 28.     NEXT
 29.   END IF
```

Fig. 5. Historical NN search algorithm for moving query points (TrajectoryNNSearch)

At the non-leaf levels, the algorithm utilizes GenTrajectoryBranchList function (pseudo-code in Fig. 4) instead of GenBranchList. GenTrajectory-BranchList(*node N, Trajectory Q*) utilizes the MinDist_Trajectory_ Rectangle metric introduced in Section 3.2 in order to calculate the MINDIST between the query trajectory and the rectangle of each entry of the node. Here, we have to point out that we do not calculate MinDist_Trajectory_Rectangle against the original query trajectory Q, but against the part of Q being inside the temporal extent of the bounding rectangle of N, and therefore (if necessary) we have to interpolate to produce the new query trajectory nQ.

4.3 Extending to *k*-NN Algorithms

In the same fashion as in [8], we generalize the above two algorithms to searching the k-nearest neighbors by considering the following:

- Using a buffer of at most k (current) nearest objects sorted by their actual distance from the query object (point or trajectory).
- Pruning according to the distance of the (currently) furthest object in the buffer.
- Updating the distance of each moving object inside the buffer when visiting a node that contains an entry of the same object closer to the query object.

5 CNN Algorithms over Trajectories

The continuous counterparts of the previously described algorithms are also of branch-and-bound type.

5.1 CNN Algorithm for Stationary Query Objects (Points)

We first discuss the query that searches for the nearest moving objects to a stationary query point at any time during a given time period. ContPointNNSearch algorithm used to process this type of query is illustrated in Fig. 6.

All the continuous algorithms use a MovingDist structure (Fig. 6, Line 6), storing the parameters of the distance function, along with the entry's temporal extent and the associated minimum and maximum (D_{min} and D_{max} respectively) of the function during its lifetime. We also store the actual entry inside the structure in order to be able to return it as the query result. ConstructMovingDistance simply calculates this structure.

In Line 8, the Nearests structure is introduced. Nearests is a list of adjacent *"Moving Distances"* temporally covering the period Q_{Per}. Roof is the maximum of all moving distances stored inside the Nearests list and is used to quickly reject those entries (and prune those branches at the non-leaf level) having their minimum distance greater than Roof (consequently, greater than all moving distances stored inside the *Nearests* list). More details on the maintenance of the Nearests structure can be found in [3].

```
Algorithm ContPointNNSearch(node N, 2D point Q, Period Qper, List
Nearests, Roof)
 1.   IF N Is Leaf
 2.     FOR i = 1 to N.EntriesCount
 3.       E = N.Entry(i)
 4.       IF Qper Overlaps (E.Ts, E.TE)
 5.         nE = Interpolate(E, Max(Qper.Ts, E.Ts), Min(Qper.TE, E.TE))
 6.         MovingDist = ConstructMovingDistance(nE, Q)
 7.         IF MovingDist.Dmin < Roof
 8.           UpdateNearests(Nearests, MovingDist, Roof)
 9.         END IF
10.       END IF
11.     NEXT
12.   ELSE
13.     BranchList = GenBranchList(Q, N, Qper)
14.     SortBranchList(BranchList)
15.     PruneContBranchList(BranchList, Nearests, Roof)
16.     FOR i = 1 TO BranchList.Count
17.       E = N.Entry(i)
18.       NN = E.ChildNode
19.       ContPointNNSearch(NN, Q, Qper, Nearests, Roof)
20.       PruneContBranchList(BranchList, Nearests, Roof)
21.     NEXT
22.   END IF
```

Fig. 6. Historical CNN search algorithm for stationary query points (ContPointNNSearch)

When backtracking at non-leaf levels, `ContPointNNSearch` applies `Prune-eContBranchList`, which prunes the branch list using the MINDIST heuristic: First, it compares the MINDIST of each entry with `Roof`, then it calculates the maximum distance inside the *Nearests* list during the entry's lifetime and prunes all entries having MINDIST greater than the calculated one.

5.2 CNN Algorithm for Moving Query Objects (Trajectories)

The fourth type of NN query is the continuous version of the NN query where the query object is the trajectory of a moving point. The algorithm `ContTrajectoryNNSearch`, used to process this type of query is illustrated in Fig. 7.

```
Algorithm ContTrajectoryNNSearch (node N, Trajectory Q, time period
Q_per, List Nearests, Roof)
  1.  Q = Interpolate(Q, Max(Q.T_S, Q_per.T_S), Min(Q.T_E, Q_per.T_E))
  2.  IF N Is Leaf
  3.    FOR j = 1 to Q.Entries
  4.      QE=Q.Entry(j)
  5.      IF (QE.T_s, QE.T_e) Overlaps (N.T_S, N.T_E)
  6.        FOR i = 1 to N.EntriesCount
  7.          E = N.Entry(i)
  8.          IF (QE.T_s, QE.T_e) Overlaps (E.T_S, E.T_E)
  9.            nE = Interpolate(E, Max(QE.T_S, E.T_S), Min(QE.T_E,E.T_E))
 10.            MovingDist = ConstructMovingDistance(nE, QE)
 11.            IF MovingDist.D_min < Roof
 12.              UpdateNearests(Nearests, MovingDist, Roof)
 13.            END IF
 14.          END IF
 15.        NEXT
 16.      END IF
 17.    NEXT
 18.  ELSE
 19.    BranchList = GenTrajectoryBranchList(Q, N)
 20.    SortBranchList(BranchList)
 21.    PruneContBranchList(BranchList, Nearests, Roof)
 22.    FOR i = 1 TO BranchList.Count
 23.      E = N.Entry(i)
 24.      NN = E.ChildNode
 25.      nQ = Interpolate(Q, Max(Q.T_S, NN.T_S), Min(Q.T_E, NN.T_E))
 26.      ContTrajectoryNNSearch(NN, nQ, Nearests, Roof)
 27.      PruneContBranchList(BranchList, Nearests, Roof)
 28.    NEXT
 29.  END IF
```

Fig. 7. Historical CNN search algorithm for moving query points (ContTrajectoryNNSearch algorithm)

`ContTrajectoryNNSearch` differs from `ContPointNNSearch` at two points only: Firstly, at leaf level, `ConstructMovingDistance` calculates the "Moving distance" between two moving points, instead of one moving and one stationary in the non-continuous case (Line 10). As in `TrajectoryNNSearch`, we perform a loop through all the 3D line segments of the query trajectory Q and, for each segment of Q and before processing the leaf entries, we first check whether the lifetime of Q overlaps the temporal extent of the bounding rectangle of N (Line 8). Secondly, at the non-leaf level, GenBranchList is replaced by GenTrajectoryBranchList introduced in the description of TrajectoryNNSearch algorithm (Line 19).

5.3 Extending to k-CNN Algorithms

The two continuous algorithms can be also generalized to searching the k- nearest neighbors by considering the following:

- Using a buffer of at most k current Nearests Lists
- Pruning according to the distance of the furthest Nearests Lists in the buffer – therefore Roof is calculated as the maximum distance of the furthest Nearests List
- Processing each entry against the i-th list (with i increasing, from 1 to k) checking whether it qualifies to be in a list
- Testing each moving distance, replaced by a new entry in the i-th list, against the $(i+1)$-th list to find whether it qualifies to be in a list.

6 Performance Study

The above illustrated algorithms can be implemented in any R-tree-like structure storing historical moving object information such as the 3D R-tree [16], the STR-tree [7] and the TB-tree [7]. Among them, we have chosen to implement the algorithms using the TB-tree due to its proven efficiency regarding historical trajectory information, as demonstrated in [7]. In our implementation, we set a page size of 4096 bytes and a (variable size) buffer fitting the 10% of the index size, thus leading to a maximum of 1000 pages. The experiments were performed in a PC running Microsoft Windows XP with AMD Athlon 64 3GHz processor, 512 MB RAM and several GB of disk size.

6.1 Datasets

While several real spatial datasets are around for experimental purposes, this is not true for the moving object domain. Nevertheless, in this paper, we have exploited on two real-world datasets: a fleet of trucks and a fleet of school buses illustrated in Fig. 8(a) and (b), respectively, and consisting of 276 (112203) and 145 (66096) trajectories (entries in the index), respectively. We have also used synthetic datasets generated by the GSTD data generator [14] in order to achieve a scalability in the volumes of the datasets. A snapshot of the generated data using GSTD is illustrated in Fig. 8(c). The synthetic trajectories generated by GSTD correspond to 20, 50, 100, 250, 500 and 1000 moving objects with the position of each object sampled approximately 1500 times.

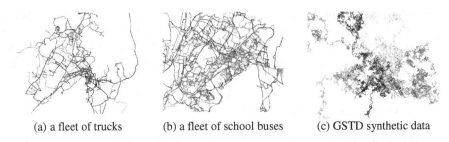

(a) a fleet of trucks   (b) a fleet of school buses   (c) GSTD synthetic data

Fig. 8. Snapshots of real and synthetic spatiotemporal data

Table 1 illustrates summary information about the datasets used. The number of pages occupied by the index for each dataset will be used for calculating the pruning acheived in the search space.

Table 1. Summary Dataset Information

	Real Data		GSTD					
	Trucks	*Buses*	*20*	*50*	*100*	*250*	*500*	*1000*
# trajectories	276	145	20	50	100	250	500	1000
# entries	112203	66096	30277	75717	151482	378803	757360	1514844
index size in pages (of 4kb)	835	466	205	507	1010	2521	5040	10073

6.2 Results on the Search Cost of the Non-continuous Algorithms

The performance of the proposed algorithms was measured in terms of node accesses. Several queries were used in order to evaluate the performance of the proposed algorithms over the synthetic and real data. In particular, we have used the following query sets:

- Q1, Q2: `PointNNSearch` was evaluated with two sets of 500 NN queries increasing the number of moving objects over the GSTD datasets. The queries used a random point in the 2D space and a time period of 1% (5%) of the temporal dimension for Q1 (Q2).
- Q3, Q4: `TrajectoryNNSearch` was evaluated with two sets of 500 NN queries increasing the number of moving objects over the GSTD datasets. The 500 query objects (trajectories) were produced using GSTD also employing a Gaussian initial distribution and a random movement distribution. Then, in Q3 (Q4) we used a random 1% (5%) part of each trajectory as the query trajectory.
- Q5, Q6: two sets of 500 k-NN queries over the real Trucks dataset increasing the number of k with fixed time and increasing the size of the time interval (with fixed k=1) respectively. For `PointNNSearch` we used a random point in the 2D space with a 5% of time as query period, while for `TrajectoryNNSearch` we used a random part of a random trajectory belonging to Buses dataset, temporally covering 1% of time.

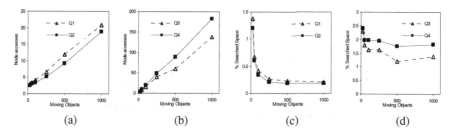

Fig. 9. Node Accesses and searched space in queries Q1-Q4 with the number of moving objects

Fig. 9 illustrates the average number of node accesses per query for the query sets Q1-Q4 evaluating `PointNNSearch` and `TrajectoryNNSearch`. In particular, Fig. 9(a) shows the average number of node accesses per query using the point query sets Q1 and Q2, while Fig. 9(b) shows the average number of node accesses per query using the trajectory query sets Q3 and Q4. As it is clearly illustrated, the performance of the algorithm depends linearly on the dataset cardinality and degrades (more pages are accessed) as the cardinality grows. It is worth to point out that comparing query sets Q1 and Q2, the algorithm accesses more pages in query set Q1, although the lifetime of Q2 is longer than that of Q1 (5% against 1% of the total time). This observation can be explained bearing in mind that decreasing the query temporal extent, the expected nearest distance increases, resulting in fewer pruned nodes in the backtracking procedure of the algorithm. As expected, `TrajectoryNNSearch` tends to be much more expensive than `PointNNSearch`.

The results in Fig. 9(c) and (d) demonstrate the percentage of the indexed space actually used for searching. As illustrated, in all cases, increasing the index size, the percentage of the space to be searched decreases, resulting (for over 1000 moving objects) in a 0.20% of the whole index space for point NN queries and in a 1.2% - 2% for trajectory NN queries. So as to make the results more readable, we have to point out that a range search over the index with zero spatial and 1% temporal extent would lead to a searching among the 10% of the whole indexed space – showing that the pruning performed by our algorithms is much more efficient than a sequential search. The conclusion gathered from the previous observations is that the algorithms presented show high pruning ability, well bounding the space to be searched in order to answer NN queries.

The performance of the two non-continuous NN algorithms increasing the number of k is shown in Fig. 10(a) against Buses dataset.

Clearly, the number of node accesses needed for the processing of a k-NN query increases linearly with k. Fig. 10(b) illustrates the average number of node accesses per non-continuous point and trajectory query increasing the temporal extent against the real "trucks" dataset. It is clear that the cost of `TrajectoryNNSearch` tends to increase with greater rate than the increase of `PointNNSearch`. This observation can be easily explained since when increasing the temporal interval, the spatial extent of the query trajectory also increases leading to a greater spatial space to be searched.

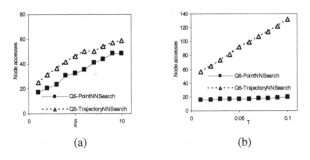

Fig. 10. Node Accesses in queries (a) Q5 increasing the number of k and (b) Q6 increasing the query temporal extent

6.3   Results on the Search Cost of the Continuous Algorithms

In coincidence with the experiments conducted for the non-continuous algorithms, the continuous NN search algorithms were evaluated with the following query sets:

- Q7, Q8: `ContPointNNSearch` was evaluated with two sets of 500 NN queries increasing the number of moving objects over the GSTD datasets like what was done for query sets Q1 and Q2.
- Q9, Q10: `ContTrajectoryNNSearch` was evaluated with two sets of 500 NN queries increasing the number of moving objects over the GSTD datasets like what was done for query sets Q3 and Q4.
- Q11, Q12: two sets of 500 k-CNN queries over the real dataset of buses increasing the number of k with fixed time and increasing the size of the time interval (with fixed $k=1$) respectively. For `ContPointNNSearch` we used a random point in 2D space with a 5% of time as query period, while for `ContTrajectoryNN-Search` we used a random part of a random trajectory belonging to the buses dataset, temporally covering 1% of time.

Fig. 11 illustrates similar results as in Fig. 9, regarding the continuous counterpart of the NN algorithms, thus, illustrating the average number of node accesses per query for the queries sets Q7- Q10. In particular, Fig. 11(a) presents the average number of node accesses per query using `ContPointNNSearch` against query sets Q7 and Q8 while Fig. 11(b) presents the average number of node accesses per query using `ContTrajectoryNNSearch` against query sets Q9 and Q10.

Again, the performance of the algorithms linearly depends on the dataset cardinality and degrades (more pages are accessed) as the cardinality grows. Fig. 11(c) and (d) show the accessed index part as a percentage of the indexed space, illustrating that in all cases, increasing the index size the percentage of the space to be searched decreases, resulting (for over 1000 moving objects) in a 0.50% of the whole index space for point CNN search and in a 2.5% - 3 % for trajectory CNN search.

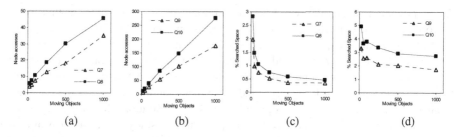

Fig. 11. Node Accesses and searched space in queries Q7-Q10 increasing the number of moving objects

A comparison between the non-continuous NN algorithms with their continuous counterparts (e.g. Fig. 9 vs. Fig. 11), shows that the continuous algorithms are much more expensive than the non-continuous ones, which is expected since the continuous algorithms prune the search space by using a list of moving distances instead of a single distance.

The performance of the continuous NN algorithms increasing the number of k is illustrated in Figure 12(a) for the real Buses dataset. The number of node accesses required for the processing of a k-NN query increases linearly with k. Figure 12(b) illustrates the average number of node accesses per continuous point and trajectory query increasing the temporal extent for Trucks dataset. Presenting the same behavior as with the non-continuous queries, the performance of ContTrajectoryNN-Search tends to degrade with greater rate than that of ContPointNNSearch, having the same explanation (by increasing the temporal interval, the spatial extent of the query trajectory also increases leading to a greater spatial space to be searched).

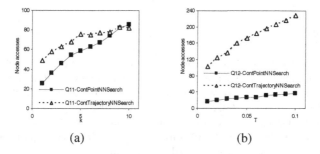

Fig. 12. Node Accesses in queries (a) Q11 increasing the number of k and (b) Q12 increasing the query temporal extent

7 Conclusions and Future Work

NN queries have been in the core of the spatial and spatiotemporal database research during the last decade. The majority of the algorithms processing such queries so far mainly deals with either stationary or moving query points over static datasets or future (predicted) locations over a set of continuously moving points. In this work, acknowledging the contribution of related work, we presented the first complete treatment of historical NN queries over moving object trajectories stored on R-tree like structures. Based on our proposed novel metrics, which support our searching and pruning strategies, we presented algorithms answering the NN and CNN queries for stationary query points or trajectories and generalized them to search for the k nearest neighbors. The algorithms are applicable to R-tree variations for trajectory data, among which, we used the TB-tree for our performance study due to its proven efficiency regarding historical trajectory information. Under various synthetic datasets (generated by GSTD) and two real trajectory datasets, we illustrated that our algorithms show high pruning ability, well bounding the space to be searched in order to answer NN and CNN queries. The pruning power of our algorithms is also verified in the case of the k-NN and k-CNN queries (for various values of k).

As such, future work includes the development of algorithms to support distance join queries ("*find pairs of objects passed nearest to each other (or within distance d from each other) during a certain time interval and/or under a certain space constraint*"). A second research direction includes the development of selectivity estimation formulae for query optimization purposes investing on the work presented in [15] for predictive spatiotemporal queries.

Acknowledgements

Research partially supported by the Archimedes and Pythagoras EPEAEK II Programmes of the Greek Ministry of National Education and Religious Affairs, co-funded by the European Union. We are also grateful to Emphasis Telematics S.A. for providing the real Buses and Trucks datasets used for our experimentation purposes.

References

1. Benetis, R., Jensen, C., Karciauskas, G., and Saltenis, S., Nearest Neighbor and Reverse Nearest Neighbor Queries for Moving Objects. *Proceedings of IDEAS*, 2002
2. Cheung, K.L., and Fu, A.,W., Enhanced Nearest Neighbour Search on the R-tree. *SIGMOD* Record, vol. 27(3), pp. 16-21, September 1998
3. Frentzos, E., Gratsias, K., Pelekis, N., and Theodoridis, Y., Nearest Neighbor Search on Moving Object Trajectories. UNIPI-ISL-TR-2005-02, Technical Report Series, University of Piraeus, 2005. Available at: http://isl.cs.unipi.gr/db.
4. Hjaltason, G., and Samet, H., Distance Browsing in Spatial Databases, *ACM Transactions in Database Systems*, vol. 24(2), pp. 265-318, 1999
5. Iwerks, G.S., Samet, H., and Smith, K., Continuous K-Nearest Neighbor Queries for Continuously Moving Points with Updates, *Proceedings of VLDB*, 2003
6. Manolopoulos, Y., Nanopoulos, A., Papadopoulos, A. N., and Theodoridis, Y., *R-trees: Theory and Applications*, Springer-Verlag, 2005
7. Pfoser D., Jensen C. S., and Theodoridis, Y., Novel Approaches to the Indexing of Moving Object Trajectories, *Proceedings of VLDB*, 2000

8. Roussopoulos, N., Kelley, S., and Vincent, F., Nearest Neighbor Queries, *Proceedings of ACM SIGMOD*, 1995
9. Saltenis, S., Jensen, C. S., Leutenegger, S. and Lopez, M., Indexing the Positions of Continuously Moving Objects, *Proceedings of ACM SIGMOD*, 2000
10. Shahabi, C., Kolahdouzan, M., and Sharifzadeh, M., A Road Network Embedding Technique for K-Nearest Neighbor Search in Moving Object Databases, *GeoInformatica*, vol. 7(3), pp. 255-273, 2003
11. Song, Z., and Roussopoulos, N., K-Nearest Neighbor Search for Moving Query Point, *Proceedings of SSTD*, 2001
12. Tao, Y., and Papadias, D., Time Parameterized Queries in Spatio-Temporal Databases, *Proceedings of ACM SIGMOD*, 2001
13. Tao, Y., Papadias, D., and Shen, Q., Continuous Nearest Neighbor Search, *Proceedings of VLDB*, 2002
14. Theodoridis, Y., Silva, J. R. O., and Nascimento, M. A., On the Generation of Spatio-temporal Datasets, *Proceedings of SSD*, 1999
15. Tao, Y., Sun, J., and Papadias, D., Analysis of predictive spatio-temporal queries, *ACM Transactions on Database Systems* vol. 28(4), pp. 295-336, December 2003
16. Theodoridis, Y., Vazirgiannis, M., and Sellis, T., Spatio-temporal Indexing for Large Multimedia Applications. *Proceedings of ICMCS*, 1996

Opportunistic Data Dissemination in Mobile Peer-to-Peer Networks*

A. Prasad Sistla[1], Ouri Wolfson[1], and Bo Xu[1]

[1] Department of Computer Science, University of Illinois at Chicago,
University of Illinois at Chicago, USA
{sistla, wolfson, boxu}@cs.uic.edu

Abstract. In this paper we examine the dissemination of availability reports about resources in mobile peer-to-peer networks, where moving objects communicate with each other via short-range wireless transmission. Each disseminated report represents an observed spatial-temporal event, and the relevance of the report to a moving object decays as the age of the reported resource and the distance from its location increase. We propose an opportunistic approach, in which an object propagates the reports it carries (namely the information that it has about these resources) to encountered objects and obtains new reports in exchange. Least relevant reports are discarded after each exchange so as to limit the communication data volume of future exchanges. Our theoretical and experimental analysis indicates that the opportunistic dissemination algorithm automatically limits the global distribution of a report to a bounded spatial area and to the duration for which it is of interest. We propose two variants of the opportunistic dissemination algorithm and compare them with the traditional client-server architecture in terms of data accuracy. The proposed system has the potential to create a completely new information marketplace.

1 Introduction

A mobile peer-to-peer network is a set of moving objects that communicate via short-range wireless technologies such as IEEE 802.11 [1], Bluetooth [2], or Ultra Wide Band (UWB) [3]. With such communication mechanisms, a moving object receives information from its neighbors, or from remote objects by multi-hop transmission relayed by intermediate moving objects. A killer application of mobile peer-to-peer networks is resource discovery in transportation. For example, the mobile peer-to-peer approach can be used to disseminate the information of available parking slots, which enables a vehicle to continuously display on a map to the driver, at any time, the available parking spaces around the current location of the vehicle. Or, the driver may use this approach to get the traffic conditions (e.g. average speed) one mile ahead. Similarly, a cab driver may use this approach to find a cab customer, or vice versa.

A mobile peer-to-peer network can also be used in matching resource producers and consumers among pedestrians. For example, an individual wishing to sell a pair

---

* Research supported by NSF Grants 0326284, 0330342, ITR-0086144, and 0209190.

C. Bauzer Medeiros et al. (Eds.): SSTD 2005, LNCS 3633, pp. 346–363, 2005.

of tickets for an event (e.g. ball game, concert), may use this approach right before the event, at the event site, to propagate the resource information. For another example, a passenger who arrives at an airport may use this approach to find another passenger for cab-sharing from the airport to downtown, so as to split the cost of the cab. Furthermore, the approach can be used in social networks; when two singles whose profiles match are in close geographic proximity, then one can call the other's cell phone and suggest a short face-to-face meeting.

The approach can also be used for emergency response and disaster recovery, in order to match specific needs with expertise (e.g. burn victim and dermatologist) or to locate victims. For example, scientists are developing cockroach-sized robots or sensors that are carried by real cockroaches, which are able to search victims in exploded or earthquake-damaged buildings [4]. These robots or sensors are equipped with radio transmitters. When a robot discovers a victim, it can use the data dissemination among mobile sensors to propagate the information to human rescuers. Sensors can also be installed on wild animals for endangered species animal assistance. A sensor monitors its carrier's health condition, and it disseminates a report when an emergency symptom is detected. Thus we use the term moving objects to refer to all, vehicles, pedestrians, robots, and animals.

In this paper we propose to examine an *opportunistic* approach to dissemination of reports regarding availability of resources (parking slot, taxi-cab customer, dermatologist, etc.). In this approach, a moving object propagates the reports it carries to encountered objects, and obtains new reports in exchange. For example, a vehicle finds out about available parking spaces from other vehicles. These spaces may either have been vacated by these encountered vehicles or these vehicles have obtained this information from other previously encountered ones. Thus the parking space information transitively spreads out across vehicles. Similarly, information about an accident or a taxi cab customer is propagated transitively. In this paper we explore this information propagation paradigm, which we call *opportunistic peer-to-peer* (or OP2P).

With OP2P, a moving object constantly receives availability reports from the peers it encounters. If not controlled, the number of reports saved and communicated by a peer may continuously increase. In order to limit the data exchange volume, we employ a relevance function that prioritizes the availability reports. The relevance of a report to a moving object o is clearly spatio-temporal, namely the relevance decreases the older the report gets, and the farther the reported resource is from m. In this paper, we introduce a simple spatio-temporal relevance function, and assume that each moving object saves only the M most relevant reports. We call this method *Opportunistic Report Dissemination* (ORD). In this paper we study ORD from three aspects.

First, we examine the pattern of report propagation with ORD. Mathematical modeling and analysis of the report-distribution is intractable in general. However, it can be solved for the case in which the relevance is purely temporal, i.e., no spatial component. This is the case for hotspots that broadcast to neighboring moving objects, for example, the current stock-market average. Thus we first devise differential equations that model object distribution in case relevance of a report to a moving object is temporal. We show that a report R generated at a certain time disappears from the system after a limited period of time, t. Using the maximum speed of a moving object, t can be easily translated to a limited geographic area where the information about R spreads. This area is a circle around the point in space where the report was generated, namely its home location.

Then, using simulations we analyze the more general case, where relevance to a moving object is spatio-temporal. We show that again, a report only spreads within a limited geographic area around its home location (e.g. the location of a parking space, or the location of a cab customer). Second, within this limited area, the replication-density of a report varies with time, in a way which will be explained. Finally, the report starts disappearing from the system, until a time threshold beyond which there is no copy of the report in the system.

Then we compare ORD with the client/server model. In the client/server model, a sensor senses the availability of the resource, and sends a report to a central database when the resource becomes available. The moving objects access the server through a cellular network. There are several drawbacks of the client/server model. First, it is difficult for the model to scale to a large number of moving objects. One possible solution to increase the scalability is to divide a geographic area into service regions (similar to cells in a cellular infrastructure). There is a server in each service region that handles resources and moving objects within that region. However, this solution introduces the complexity of hand-over, which occurs when a moving object crosses the border between two service regions. Second, the client/server model is vulnerable to the failure of the central server. Finally, in the client/server mode, a moving object user has to pay for the cellular communication and the information service. In the peer-to-peer model a user only needs to pay for the initial installation of the communication module. The operation of the communication module is virtually free. A back of the envelope calculation reveals that the cost (in terms of fuel) of communicating with encountered vehicles is less than a cent per day, even if the communication is continuous throughout the day.

With all the above extra cost, what does client/server buy us? We compare the quality of data received with ORD and client/server. Observe that at any time instance an availability report in the local database of a moving object o may be incorrect in the sense that a resource that shows as available, actually becomes unavailable before o reaches it. So we propose a set of data quality measures that mimic the precision/recall measures used in information retrieval. Our simulations show that with very reasonable object density and wireless transmission range, the quality of the ORD method reaches that of client/server. This indicates that OP2P could serve well as an alternative to the client/server model but with much less operational cost. We also study the performance of ORD when there are failures in peer-to-peer interactions. In an OP2P environment, interaction failures may be caused by packet loss, communication module sleeping for power reservation, or the limited connection time between highly mobile objects.

Finally, we study a variant of ORD, in which invalidation messages are propagated when a resource becomes unavailable. This algorithm is called Opportunistic Report Dissemination with Invalidation (ORDI). We compare ORDI with ORD and client/server with invalidation. It would appear that clearly the invalidation algorithm will have a higher data quality, but this is misleading since we compare the algorithms on the same size of local database; therefore, the invalidation messages may occupy space of other correct reports. However, our experimental analysis shows that ORDI is indeed superior to ORD.

In summary, this paper makes the following contributions. First, we devise a mathematical model for dissemination of information about purely temporal resources and experimentally analyze the dissemination of information about spatio-temporal resources. Then we propose an algorithm for opportunistic dissemination of information about spatio-temporal resources, and we compare it with the tradition client/server model. Finally we analyze the performance of the algorithm when using invalidation messages.

Although some concepts employed in this paper including gossiping and invalidation have been analyzed in the past, this paper applies these concepts to a mobile peer-to-peer environment. Furthermore, it combines these concepts with a novel aspect, which is filtering and ranking of spatio-temporal information based on a relevance function.

The rest of the paper is organized as follows. Section 2 introduces the model and the ORD algorithm. Section 3 analyzes the report propagation pattern. Section 4 compares ORD with the client/server model. Section 5 describes ORDI and compares it with ORD and the client/server model. Section 6 discusses relevant work and section 7 concludes the paper.

2   The Model

2.1   Resource Model

In our system, resources may be spatial, temporal, or spatio-temporal. Information about the location of a gas station is a spatial resource. Information about the price of a stock on 11/12/03 at 2pm is temporal. There are various types of spatio-temporal resources, including parking slots, car accidents (reports about such resources provide traffic-jam information), taxi-cab requests, ride-sharing invitations, demands of expertise in disaster situations, and so on. These resources are spatial in the sense that they are tied to a location, and are temporal in the sense that they are valid or available only for a limited time-duration.

Formally, in our model there are N *resource types* $T_1, T_2, ..., T_N$. At any point in time there are M *resources* $R_1, R_2, ..., R_M$, where each resource belongs to a resource type. We assume that resources are located at points in two-dimensional geospace. The location of the resource is referred to as the *home* of the resource. This is the spatial aspect of resources. For example, the home of an available parking space is the location of the space, and the home of a cab request or a cab-sharing invitation is the location of the customer. The state of each resource alternates between *valid* (i.e. available) and *invalid*. The period of time during which the resource is valid is called the *valid duration*. This is the temporal aspect of resources. For example, the valid duration of the cab request resource is the time period since the request is issued, until the request is satisfied or canceled.

2.2 Peers and Validity Reports

The system consists of two types of peers, namely fixed hotspots and moving objects. Each peer o that senses the validity of resources produces *validity reports*. Denote by $a(R)$ a report for a resource R. For each resource R there is a single peer o that produces validity reports, called the *report producer* for R. A peer may be the report producer for multiple resources. Each report $a(R)$ contains four attributes, namely *resource-type*, *resource-id*, *timestamp*, and *home-location*. Attribute resource-type indicates the resource type of R. Resource-id is the identification of R that is unique among all the resources in the system. In our model time is a sequence of discrete atomic time units, 1, 2, 3, ..., and timestamp is a natural number indicating the time at which $a(R)$ is transmitted to a peer by its producer.

For each resource type T, a peer o has a *validity reports database*, or *reports database*. Denote by $DB_o(T)$ the reports database of o for the resource type T.

2.3 Relevance Model

In order to limit the data exchange volume, for each resource type T, a moving object keeps in the reports database the top M relevant reports of type T that the object knows at that time. M is referred to as the *interest threshold*. For example, a user who is looking for a resource type T has the reports database that keeps the top 10 relevant validity reports of T. In other words, the user wants only the 10 most relevant reports to be saved and displayed. In this paper we use the following relevance function:

$$\text{Rel}(a(R)) = e^{-(\alpha \cdot t + \beta \cdot d)} \quad (\alpha, \beta \geq 0) \tag{1}$$

where t is the age of $a(R)$, namely the number of time units since $a(R)$ is transmitted by its producer, and d is the travel distance from the home-location of R to the moving object. α and β are constants that represent the decay of relevance with respect to time and distance respectively. α and β may vary per resource type. Observe that this function is always positive, indicating that each report always has some relevance, and it decreases as t and d increase. We assume that each moving object is equipped with a GPS system so that (i) the object knows its location at any point in time and (ii) the clock is synchronized among all the objects. Thus both the age t and the distance d can be computed by the moving object.

Let us consider the resources that require a moving object to physically reach them ahead of other objects in order to occupy or possess them (e.g. parking slots, cab requests, or highway assistance requests). In our prior work ([22]) we have shown that for such a resource R, under some conditions the relevance of a report $a(R)$ equals to the probability that R is valid when the moving object reaches it.

Theorem 1. Assume that the length of the valid duration (see subsection 2.1) of R is a random variable with an exponential distribution having mean u. Let the speed of the moving object be v. If $\alpha = 1/u$ and $\beta = 1/(u \cdot v)$, then the relevance of a report $a(R)$ is the probability (at report acquisition time) that the resource R is valid when the moving object reaches R.

For the proof of Theorem 1 a reader is referred to [22]. The theorem motivates our definition of the relevance function (at least for resources with exponentially distributed valid-duration).

2.4   The Opportunistic Report Dissemination (ORD) Algorithm

We assume that each peer is capable of communicating with the neighboring peers within a maximum of a few hundred meters. One example is an 802.11 hotspot or a PDA with Bluetooth support. The underlying communication module provides a mechanism to resolve interference and collisions. Each peer is also capable of discovering peers that enter into or leave out of its transmission range (see e.g. [13]).

Recall that each moving object keeps a database of size M for each resource type. When two moving objects A and B encounter each other (i.e. they come within transmission range)[1], for each resource type T, A and B exchange their local databases, i.e. $DB_A(T)$ and $DB_B(T)$, and each one keeps the M most relevant reports. When A encounters a hotspot C, C transmits to A the reports it produces. Again A keeps the M most relevance reports.

In the rest of this paper we will assume that there is a single resource type in the system. However, most of our results also apply when there are multiple resource types, and we will specify when they do not.

3   Pattern of Report Propagation

In subsection 3.1 we study the propagation pattern of reports for temporal resources and in 3.2 we study the propagation pattern of reports for spatio-temporal resources.

3.1   Propagation Pattern of Reports for Temporal Resources

In this subsection we theoretically analyze how a report is propagated per time and per distance with the ORD algorithm. We consider a special case for the relevance function, where the decay factor of distance β is zero. Thus the relevance is purely a function of the age. This relevance function models the decay of the reports that are only specific to time, e.g. the Dow Jones Industrial Average at a particular time. In this section we first introduce the parameters and our assumptions, and then we develop a mathematical model that describes the propagation of a report. Finally we use the mathematical model to analyze the propagation of reports.

3.1.1   Parameters and Assumptions
Let N be the total number of peers in the system. We assume that the value of M is 1 for all the peers although our result can be extended to the general case where M is more than 1. Each peer interacts with other peers by a Poisson process with intensity

---

[1] If A uses broadcast, then the broadcast is used to select a peer to interact with. After the peer is selected, the interaction follows the one-to-one exchange procedure. Extension of report exchange to take the advantage of broadcast is a subject of our future work.

λ. We assume that the length of the valid duration of a resource is 0, and the wireless transmission range is small enough such that at most one peer can receive the report when it is produced. Observe that with this assumption, the age of a report is always 0 when it is acquired. Further observe that the report may still be relevant even after it is invalid. For example, although a Dow Jones Industrial Average report is invalid right after it is generated, it is still of interest for a period of time. For another example, even after occupied, a parking slot report is relevant because peers do not always know whether it is occupied. We consider only the reports that are received by a peer when they are produced. Such reports are generated within the system by a Poisson process with intensity μ. Peers are randomly distributed in the space at any point in time, and therefore each peer is equally probable to receive each produced report. A newly generated report is sent to exactly one peer. Thus each peer receives newly generated reports according a Poisson process with the rate μ/N. Finally we assume that each report exchange is finished instantaneously.

3.1.2 A Mathematical Model for Report Propagation

Let us define two variables: $q(t)$: The conditional probability that a peer has a report for a resource R at time t ($t>0$) given that R is created at time 0. $g(t)$: The probability that at time t ($t>0$) a report that is created after 0 is in the reports database of a peer. Now consider $q(t+\Delta t)$ which is the probability that at time $t+\Delta t$ a report R that is created at time 0 is in the reports database of a peer o. Let Δt be small enough such that at most one report is generated in the system between t and $t+\Delta t$ and o can interact with at most one peer during the same time interval. $q(t+\Delta t)$ is the probability that one of the following mutually exclusive events happens:

1. o has R at time t, and it does not acquire any new report between t and $t+\Delta t$, and it does not interact with any peer between t and $t+\Delta t$. The probability for this to happen is $q(t) \cdot (1 - \lambda \cdot \Delta t) \cdot (1 - \frac{\mu}{N} \cdot \Delta t)$.

2. o has R at time t, and it does not acquire any new report between t and $t+\Delta t$, and it interacts with one peer o' between t and $t+\Delta t$, and o' does not have a report that is created after 0. The probability for this to happen is $q(t) \cdot (1 - \frac{\mu}{N} \cdot \Delta t) \cdot \lambda \cdot \Delta t \cdot (1 - g(t))$.

3. At time t o has no reports, or has a report that is created before 0, and it does not acquire any new report between t and $t+\Delta t$, and it interacts with one peer m' between t and $t+\Delta t$, and m' has R. The probability for this to happen is $(1 - q(t) - g(t)) \cdot (1 - \frac{\mu}{N} \cdot \Delta t) \cdot \lambda \cdot \Delta t \cdot q(t)$.

Thus

$$q(t + \Delta t) = q(t) \cdot (1 - \lambda \cdot \Delta t) \cdot (1 - \frac{\mu}{N} \cdot \Delta t) + q(t) \cdot (1 - \frac{\mu}{N} \cdot \Delta t) \cdot \lambda \cdot \Delta t \cdot (1 - g(t)) +$$

$$(1 - q(t) - g(t)) \cdot (1 - \frac{\mu}{N} \cdot \Delta t) \cdot \lambda \cdot \Delta t \cdot q(t)$$

By similar analysis, we obtain the following equation:

$$g(t+\Delta t) = \frac{\mu}{N}\cdot\Delta t + (1-\frac{\mu}{N}\cdot\Delta t)\cdot g(t) + (1-\frac{\mu}{N}\cdot\Delta t)\cdot(1-g(t))\cdot\lambda\cdot\Delta t\cdot g(t)$$

After simplification of the above difference equations, we get the following differential equations:

$$\begin{cases} \dfrac{dq(t)}{dt} = -(\lambda+\dfrac{\mu}{N})\cdot q(t) + 2\cdot\lambda\cdot q(t)\cdot(1-g(t)) - \lambda\cdot q(t)^2 \\ \dfrac{dg(t)}{dt} = \dfrac{\mu}{N} - \dfrac{\mu}{N}\cdot g(t) + \lambda\cdot g(t) - \lambda\cdot g(t)^2 \end{cases} \tag{2}$$

Since each peer is equally probable to acquire the report, $q(0)=1/N$. Finally, $g(0)=0$. Let $C(t)$ be the number of copies of a report t time units after its creation. We have the following theorem.

Theorem 2: $C(t)$ is a random variable with expected value $q(t)\cdot N$ where $q(t)$ is given by the equation group (2).

We used Theorem 2 to compute the expected number of copies as a function of time. We used the following set of parameter values: $N=2500$, $\lambda=0.12$, $\mu=10$. The solid line in Figure 1 shows the result. Observe that the number first increases until a maximum value is reached. And then it decreases until disappearing from the system. From this figure we can estimate how far away a report can be propagated. Take the cut-off age beyond which the expected number of copies is below 1, which is about 60 seconds. Assume that the wireless transmission range is zero. Multiplying this cut-off age by the maximum speed of moving objects gives the maximum distance the report can be propagated to. For example, if the maximum speed is 60 miles/hour, then the maximum distance is 1 mile.

3.1.3 Validation of the Mathematical Model

We conducted a simulation to validate the analytical model. In this simulation, 2500 objects are initially uniformly distributed within a 5mile×5mile square area and they randomly move with a constant speed 40 miles/hour. The transmission range is 50

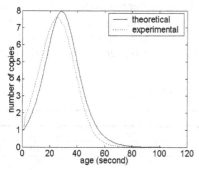
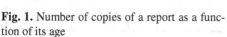
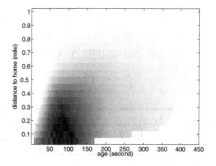

Fig. 1. Number of copies of a report as a function of its age

Fig. 2. Propagation pattern of reports

meters. This setup gives on average 0.12 interactions per each object per each second. Reports are generated with intensity 10 and each report is randomly assigned to an object. Figure 1 shows the results. The dashed line represents the experimental result. It can be seen that Theorem 1 accurately describes the behavior of the system.

3.2 Propagation Pattern of Reports for Spatio-temporal Resources

In this subsection we study by simulations the propagation pattern of reports for spatio-temporal resources. First we describe the simulation method and then we present the simulation results.

3.2.1 Simulation Method

We synthetically generated and moved objects within a 10mile×10mile square area. The objects move in a random-walk model. Specifically, for each object i, we randomly chose two points within the square area, and assigned them as the start point and the first stop of i respectively. i moves along line segment between the two points at a constant speed. When the first stop is reached, another random point is chosen as the second stop of i, and i moves from the first stop to the second stop at the same constant speed. And so on. The motion speed of i is randomly picked up from the interval $[v-5, v+5]$ where v is a parameter.

Hotspots are randomly distributed in the square area with density 500 hotspots per square mile. Resources are generated only at hotspots. At each hotspot, the length of the valid duration of a resource follows an exponential distribution with mean u seconds, and the time length of the invalid duration follows an exponential distribution with mean 360 seconds. The home of all the reports announced by a hotspot is the location of the hotspot. All the hotspots and the moving objects have the same wireless transmission range. All the moving objects have the same interest threshold. The value of the time decay factor α is $1/u$ and the distance decay factor β is $1/(u \cdot s)$ where s is the motion speed of a moving object.

There are five parameters for each simulation run, namely the interest threshold M, the wireless transmission range r, the constant speed v, the objects density g (i.e. the number of objects per square mile), and the mean of valid duration u. M is fixed to be 10, v is fixed to be 40 miles/hour, and u is fixed to be 120 seconds. The time unit is second. All the parameters and their values are listed in Table 1.

Each simulation run is executed as follows. At the beginning of the simulation run, 10×10×g objects are generated and they start to move at the same time (time 0). Resources are generated and the status of each resource alternates between valid and invalid as described earlier. When the distance between two peers is smaller than r in

Table 1. Simulation parameters and their values

Parameter	Symbol	Unit	Value
Mean of valid duration	u	second	120
Interest threshold	M		10
Transmission range	r	meter	50
Motion speed	v	miles/hour	40
Object density	g	objects/mile^2	100

a time unit, they exchange their reports, re-evaluate the relevance, and purge the least relevant reports if needed. Each exchange is finished instantaneously. The length of each run is 10 simulated hours.

During a simulation run, we trace the distribution of each report $a(R)$ at each time unit. For this purpose, we generate 50 rings centered at the home of R, each with the width of 0.05 mile. At each time unit we calculate the density of the copies of $a(R)$ at each ring, and average among all the time units of a simulation run.

3.2.2 Simulation Results

Figure 2 shows the average density as a function of the age and the distance to home for ORD. The density is coded by the gray-level, such that a deeper gray-level represents a higher density. The lowest gray-level (white) represents zero density. We make the following observations.

Observation 1: At any point in time, there is a spatial boundary for the distribution of the report, beyond which the density is zero. This boundary first expands, until a maximum value (about 0.9 mile) is reached. Then the boundary starts to shrink until finally the report disappears from the system (at the age of 450 seconds or so). The boundary expands at beginning because of the propagation of the report caused by opportunistic exchanges. However, as time passes, the relevance decreases, causing two effects: (i) more objects purge the report out; and (ii) less objects save it. These two effects make the number of copies start to decrease, and thus the boundary starts to shrink. After some time, the relevance becomes so low that all the objects that have carried it have purged it out, and no objects save it upon exchange. The report thus disappears from the system.

Observation 2: The gray-level tends to be deep when the distance to the home is small and it fades as the distance to home increases. In other words, the copies are more densely distributed in the areas close to the home than in the areas farther away. This is a useful behavior, because it means that the report has a higher availability in the area to which it is of interest.

The above propagation pattern shows that, by very simple local decisions made at each moving object, the opportunistic dissemination algorithm automatically limits the global distribution of a report to a bounded spatial area, which is a circle around the home location of the resource. The algorithm also limits the distribution to the time-duration for which the report is of interest. We conducted experiments with more parameter configurations. These experiments show that the spatial and temporal boundaries automatically adapt depending on the number of resources in the system, the traffic density and speed, and other parameters that dictate the amount of storage, processing power, and bandwidth that should be allocated to each resource. For example, if resources are generated less frequently, then each report will stay in the system longer, and spread farther.

4 Comparison with Client/Server Model

With the ORD algorithm, a moving object o may have validity reports that are incorrect, i.e. the resources these reports refer to become invalid before o reaches them.

When the validity reports in an object are used for decision making, an important measure is how many out of them are correct and how many are incorrect. In this section we compare ORD with the client/server model with this regard. In the client/server model, the validity reports are stored in a central database, and transmitted to the moving objects as query answers.

In subsection 4.1 we describe the client/server model. In 4.2 we define the performance measure. In 4.3 we describe the simulation setup, and in 4.4 we present the simulation results.

4.1 Client/Server Access (CS)

In the client/server model, there is a centralized database that stores validity reports. The database is updated by report producers and is queried by moving objects. A report producer inserts a validity report $a(R)$ to the database when a resource R is sensed. Each moving object o issues a continuous query to the centralized database. For example, when approaching the destination, o issues query "acquire the top 10 relevant validity reports about parking slots". At each time unit the centralized database evaluates each query and transmits o the top M reports (top with regard to o) where M is the interest threshold. The timestamp of each transmitted report is set to the transmission time. o replaces the current reports in its local reports database with the received reports. The centralized database knows the location of o at any point in time.

4.2 Performance Measure

Definition 1. A resource R is *correct* for a moving object o at time t if R remains valid when o reaches it under the condition that o goes to R at t. Otherwise R is *incorrect* for o at t.

According to the above definition, if R is invalid at t, then it is definitely incorrect for o at t. However, R may be incorrect for o at t even if R is valid at t. What matters is whether R is still valid when o reaches it.

Definition 2. A validity report $a(R)$ is *correct* for a moving object o at time t if R is correct for o at t. Otherwise $a(R)$ is *incorrect* for o at t.

Notice that in reality, at the time when o receives $a(R)$, it usually does not know whether $a(R)$ is correct or not, because it does not know whether R will remain valid when o reaches R. In this paper "correctness" and "incorrectness" of reports are defined solely for the purpose of performance evaluation.

Definition 3. Let K be the sum of the relevance values of the validity reports in o's reports database at time t. Let K' be the sum of the relevance values of validity reports in o's reports database that are correct for o at t. The *precision* of o's reports database at time t is K'/K.

We call the above measure "precision" because it indicates how many of the reports that o know are correct. This mimics the precision measure that is used in the information retrieval area. However, in our definition, precision is not the fraction of the correct reports known by o out of all the reports known by o. Instead, it is the ratio between the total relevance of the two sets. The reason for this is that the reports are unequal in relevance and therefore are different in importance for decision making. Thus when each report is counted for precision, it should be weighted by the relevance. With the same fraction of the correct reports, the higher relevance the correct reports occupy, the better data quality of the database. Next we define the notion of recall.

Definition 4. Let R be a resource that is correct for o at time t. The *relevance* of R to o at t is the relevance of the report $a(R)$ assuming that its timestamp is t.

Notice that above we define the relevance of a resource, as opposed to the relevance of a report defined in section 2.3. Intuitively the relevance of R to o at t is the relevance of the report $a(R)$ that has age 0. In other words, the relevance of R to o depends only on the distance of o from the resource.

Definition 5. Let C be the sum of the relevance values of the correct validity reports in o's reports database at time t. Let C' be the sum of the relevance values of the M correct resources in the system that are most relevant to o (i.e. the M closest correct resources). The *recall* of m's reports database at time t is C/C'.

We call the above measure "recall" because it indicates how many of the correct resources that o is interested in knowing (i.e. the correct top M in the system), are actually known by it. However, notice that in our definition, recall is not the fraction of resources correctly known by o out of the top M resources in the system. Instead, it is the ratio between the total relevance of the two sets.

Definition 6. Let P be the precision of o's reports database at time t, and Q the recall of the database at t. The *precision-recall product* of the database at t is $P \cdot Q$.

In this paper the precision-recall product is used as the performance measure.

4.3 Simulation Method

The simulation setup is similar to the one used in 3.2, except for the following. First, the CS algorithm is implemented. With the CS algorithm, at each simulated time unit the system sorts all the resources in the order of their relevance to o and puts the top M to o's database. Second, in ORD an exchange between two moving objects succeeds with probability p. p is called the *successful interaction probability*.

Let us emphasize that in the simulations for the CS model we assume that the bandwidth available for the communication between the server and the clients is infinite, and we ignore the contention/collisions that may cause transmission failures. In the simulations for the ORD algorithm, on the other hand, we take into considerations the bandwidth constraint and contention/collisions, in the following way. First, we calculate the bandwidth required by the ORD algorithm with the most bandwidth-consuming parameter configuration in Table 1. This configuration is object density $g = 2500$ per square mile and transmission range $r = 200$ meters. With this configuration, the bandwidth consumption of the ORD algorithm is 40KBytes per second per object[2]. On the other hand, we estimated, for the above network configuration, with contention and collisions, the effective bandwidth available for each object is 56KBytes per second when 802.11g ([6]) is used. The estimation is extrapolated from the empirical results of [7][3]. This suggests that even with the most bandwidth-consuming parameter configuration, the bandwidth consumption of the ORD algorithm does not exceeds the network capacity. Therefore in all the experiments with ORD, the successful interaction probability p is set to be 1, except for the group that study the impact of p.

The above justification assumes a single resource type. If there are multiple resource types, then with the above network configuration, the bandwidth consumption of ORD may exceeds the network capacity. In that case the successful interaction probability p is lower than 1, which we will also study.

Table 2. Variable simulation parameters

Parameter	Symbol	Unit	Value
Transmission range	r	meter	50, 100, 150, 200
Object density	g	objects/mile^2	100, 500, 1000, 1500, 2000, 2500
Successful interaction probability	p		0.05, 0.1, 0.15, 0.2, 0.25, 0.5, 0.75, 1

Finally, let us mention that we repeated the experiments for motion in a grid network rather than Euclidean space. The reason is that in our traffic applications vehicles move along a road network. We determined that the network results are very similar to those reported here for Euclidean space. Due to space limitations these network results are omitted. The parameters are listed in Table 2.

4.4 Simulation Results

Impact of object density. Figures 3, 4, and 5 show the precision-recall product of ORD as functions of the object density and compare these measures with those of CS. For ORD, the performance increases as the object density increases. Intuitively, as the object density increases, the interactions among objects become more frequent, and thus the newly generated reports get propagated more quickly. These reports purge the old reports out of the databases of moving objects. Since the new reports are more likely to be correct than the old ones, both the precision and the recall increase. Notice that the precision-recall product of CS is not 1. This is because the central database does not know when a resource R will become invalid, and therefore the report $a(R)$ returned by CS to a moving object o may be incorrect (i.e. R is invalid when o reaches R).

The figures show that the performance of ORD approaches to CS when the transmission range is 50 meters and the object density is 2500 objects per square mile. In

---

[2] Notice that the moving object does not transmit any report to a hotspot.

[3] The reference analyzes the effective bandwidth available per object for a short-range wireless technology, with contention and collisions taken into considerations.

this situation, the average distance between two neighboring moving objects is about 32 meters, smaller than the transmission range. In other words, most of the time the network formed by moving objects is connected. In this case, with ORD the propagation of a report reaches its maximum spatial boundary almost instantaneously. However, this is different than a simple flooding in a connected network. In our model each moving object only transmits top M reports and the spreading of each report is automatically restricted within a small portion of the whole network (see the propagation analysis in section 3).

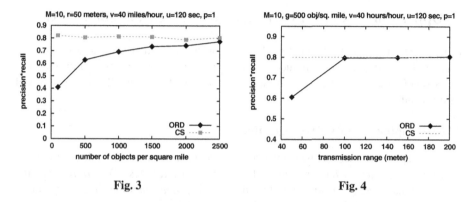

Fig. 3   Fig. 4

Impact of transmission range. Figure 4 shows the precision-recall of ORD as functions of the transmission range and compare with those of CS. Intuitively, as the object density increases, the interactions among objects become more frequent, which generates similar effects as when the object density increases.

Impact of successful interaction probability. Figure 5 shows the performance measures of ORD as functions of the successful interaction probability. Intuitively, as the successful interaction probability increases, the effective interactions among objects become more frequent, which generates similar effects as when the object density increases. In fact, comparing Figure 5 and Figure 3, we notice that reducing the successful interaction probability is the same as simply reducing the object density.

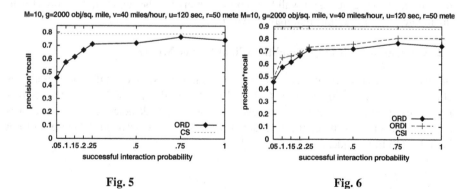

Fig. 5   Fig. 6

For example, the effect of a successful interaction probability of 0.25 is the same as reducing the object density from 2000 to 1000.

Observe that we tested with a long range of probability values, from 0.05 to 1. This is because the successful interaction probability is used to model not only communication reliability, but also the ORD implementations in which not every encounter generates an interaction. For example, if a moving object broadcasts for every 20 time units, or the communication module is awake for 5% of time, then the successful interaction probability is 0.05.

5   Opportunistic Report Dissemination with Invalidation (ORDI)

In this section we first present the ORDI algorithm in which invalidity reports are generated and propagated to reduce the fraction of incorrect reports in a moving object's reports database. Then we compare ORD, ORDI, and the client/server model with invalidation.

5.1   Description of the ORDI Algorithm

At each report producer o, whenever a resource R is detected invalid, o creates an *invalidity report* $i(R)$. $i(R)$ contains the following four attributes: (i) *resource-type* the resource type of R; (ii) *resource-id* the id of R; (iii) *timestamp* the time when report $i(R)$ is created; and (iv) *home-location* the home of R.

After $i(R)$ is produced, the validity report of R is removed from the reports database $DB_o(resource\text{-}type)$, and $i(R)$ is inserted into $DB_o(resource\text{-}type)$. In order to distinguish between validity reports and invalidity reports, each report is given an extra attribute *report-type* when it is inserted into a reports database. *report-type* indicates whether the report is a validity report or an invalidity report.

The invalidity report uses the same relevance function as a validity report, and is exchanged similarly to a validity report. The only difference is as follows. When an invalidity report $i(R)$ is received by an object o, o uses the resource-id attribute to search $a(R)$ in o's reports database. If $a(R)$ is found and its timestamp is smaller than that of $i(R)$, then the validity report is replaced by $i(R)$. If the validity report is not found, then $i(R)$ is either discarded or saved into $DB_m(resource\text{-}type)$ based on its relevance, in the same way a validity report is treated. The reason $i(R)$ is saved is to invalidate a validity report that may arrive later.

5.2   Comparison with ORD and the Client/Server Model

In this subsection we compare ORDI with two algorithms. One is ORD and another is client/server with invalidation, or *CSI*. CSI works similarly as CS except the following. At each report producer o, whenever a resource R is detected invalid, o removes the report $a(R)$ from the centralized database. Thus the centralized database will not include $a(R)$ in any query answer since then.

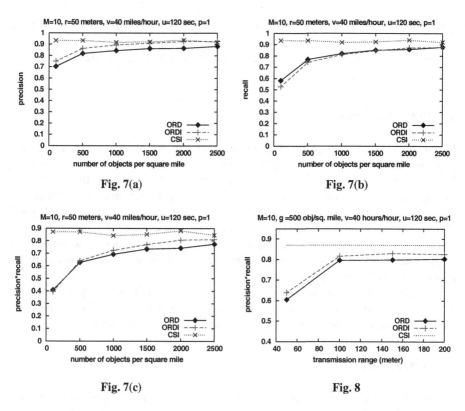

Fig. 7(a)

Fig. 7(b)

Fig. 7(c)

Fig. 8

The simulation setup is the same as described in subsection 4.3. The results are shown in Figures 6, 7, 8. First let us compare ORDI and ORD. The precision of ORDI is higher than that of ORD (Figure 7(a)). This is because invalidation reports push incorrect reports out of the reports database and therefore the fraction of correct reports increases. However, there is little difference between ORDI and ORD in recall (Figure 7(b)). This is somehow surprising because invalidity reports share the same database with validity reports and therefore they may occupy spaces that could have been used to accommodate correct reports. The fact that ORDI is as good as ORD in recall suggests that in most cases invalidity reports occupy the spaces of incorrect reports, which is a desirable behavior. As a result, the precision-recall product of ORDI is better than that of ORD (Figure 7(c)).

Now let us compare ORDI and ORD with CSI. The precision of ORDI approaches to CSI beyond certain object density (Figure 7(a)) or transmission range (figure omitted due to space limitations). This is similar to the observation we have in subsection 4.4 when comparing ORD and CS, and the explanation is also similar. However, neither ORDI nor ORD ever reaches CSI in recall (Figure 7(b)). ORDI never reaches CSI in recall, because with ORDI, invalidity reports share the same database with validity reports, whereas with CSI there are no invalidity reports in a moving object's reports database. In other words, the number of validity reports in a reports database with CSI is higher than that with ORDI. ORD never reaches CSI in recall, because with ORD there can be reports that refer to the resources that have become invalid;

with CSI there are no such reports. As the result, ORDI and ORD never reach CSI in precision-recall product (Figures 7, 8).

The effect of the successful interaction probability to ORDI is similar to that to ORD (see Figure 6).

6   Relevant Work

Resource discovery and publish/subscribe in mobile ad hoc networks are usually implemented by building a routing structure for resource information (see e.g. [8, 9, 11, 15]). Most of these works rely on routing structures. However, the constructed routing structure may easily become obsolete in a highly dynamic and partitionable network environment. Work has also been done on data dissemination in mobile peer-to-peer networks [10, 12, 14, 16, 18, 20, 21]. These methods use the gossiping/epidemic communication paradigm. However, they consider dissemination of regular data objects rather than spatio-temporal resources, and they do not rank the resources for determining what to broadcast.

This paper differs from our prior work (e.g. [19, 22]) on the same topic in multiple aspects. The theoretical analysis of the propagation pattern is new. The comparison with the client/server model is new, and so is the invalidation algorithm.

7   Conclusion

In this paper we devised an algorithm, ORD, for dissemination of spatial and temporal resource-information in a mobile peer-to-peer environment, in which the resource-information database is distributed among the hotspots and moving objects. We analyzed ORD theoretically, using differential equations, and experimentally, using simulations. We compared ORD with the client/server model by simulations. The performance measures are relevance-weighted precision and recall. We determined that ORD performs better when the object density and the wireless transmission range increase. ORD reaches the client/server model in performance when the object density or the transmission range is high enough. We also studied the impact of successful interaction probability, and determined that reducing the successful interaction probability is the same as simply reducing the object density. Thus ORD can perform as well as the client/server model in low success probability environment by increasing the object density. Finally, we studied a variant of ORD, ORDI, which uses invalidity reports to increase precision. The experimental results show that ORDI is better than ORD.

References

1. IEEE Computer Society. Wireless LAN Medium Access Control (MAC) and Physical Layer (PHY) Specifications.1997.
2. J. Haartsen, et al. Bluetooth: Vision, Goals, and Architecture. ACM Mobile Computing and Communications Review, 2(4):38-45, October 1998.
3. http://www.ubisense.net/technology/uwb.html

4. http://firechief.com/ar/firefighting_roborescuers_increase_disaster/
5. Markowetz, et al. Exploiting the Internet As a Geospatial Database, International Workshop on Next Generation Geospatial Information, 2003.
6. IEEE 802.11g Standard. http://grouper.ieee.org/groups/802/11/index.html
7. J. Li, et al. Capacity of Ad Hoc Wireless Networks. MobiCom 2001.
8. Saumitra M. Das, Himabindu Pucha, and Y. Charlie Hu. Ekta: An efficient dht substrate for distributed applications in mobile ad hoc networks. WMCSA 2004.
9. Y. Hu, et al. Exploiting the synergy between peer-to-peer and mobile ad hoc networks. In HotOS-IX, 2003.
10. W. Zhao, M. Ammar, E. Zegura. A Message Ferrying Approach for Data Delivery in Sparse Mobile Ad Hoc Networks. Mobihoc, Tokyo Japan, May 2004.
11. C. Frank, et al. Consistency challenges of service discovery in mobile ad hoc networks. MSWiM 2004.
12. F. Perich et al. On Data Management in Pervasive Computing Environments. IEEE Trans. Knowledge and Data Engineering, 16(5), May 2004.
13. M. J. McGlynn and S. A. Borbash. Birthday protocols for Low Energy Deployment and Flexible Neighbor Discovery in Ad Hoc Wireless Networks. Proc. of MobiHoc 2001.
14. K. Rothermel, et al. Consistent Update Diffusion in Mobile Ad Hoc Networks. Technical Report 2002/04, CS Department, University of Stuttgart, 2002.
15. Y. Huang and H. Garcia-Molina. Publish/Subscribe Tree Construction in Wireless Ad-Hoc Networks. MDM 2003, pages 122-140.
16. M. Papadopouli and H. Schulzrinne. Effects of Power Conservation, Wireless Coverage and Cooperation on Data Dissemination Among Mobile Devices. MobiHoc 2001.
17. Datta, et al. Updates in Highly Unreliable, Replicated Peer-to-Peer Systems. ICDCS 2003.
18. Vahdat and D. Becker. Epidemic Routing for Partially Connected Ad Hoc Networks, Technical Report CS-200006, Duke University, April 2000.
19. O. Wolfson, B. Xu. Opportunistic dissemination of spatio-temporal resource information in mobile peer-to-peer networks. In PDMST'04, 2004.
20. U. Centintemel, et al. Power-efficient Data Dissemination in Wireless Sensor Networks. MobiDE 2003.
21. S. Goel, et al, Grassroots: A Scalable and Robust Information Architecture. Technical Report DCS-TR-523, CS Department, Rutgers University, June 2003.
22. O. Wolfson, B. Xu, Y. Yin. Dissemination of Spatial-Temporal Information in Mobile Networks with Hotspots. *DBISP2P*, 2004.

On Discovering Moving Clusters in Spatio-temporal Data

Panos Kalnis[1], Nikos Mamoulis[2], and Spiridon Bakiras[3]

[1] Department of Computer Science, National University of Singapore
kalnis@comp.nus.edu.sg
[2] Department of Computer Science, University of Hong Kong, Pokfulam Road, Hong Kong
nikos@cs.hku.hk
[3] Department of Computer Science, Hong Kong University of Science and Technology,
Clear Water Bay, Hong Kong
sbakiras@cs.ust.hk

Abstract. A moving cluster is defined by a set of objects that move close to each other for a long time interval. Real-life examples are a group of migrating animals, a convoy of cars moving in a city, etc. We study the discovery of moving clusters in a database of object trajectories. The difference of this problem compared to clustering trajectories and mining movement patterns is that the identity of a moving cluster remains unchanged while its location and content may change over time. For example, while a group of animals are migrating, some animals may leave the group or new animals may enter it. We provide a formal definition for moving clusters and describe three algorithms for their automatic discovery: (i) a straight-forward method based on the definition, (ii) a more efficient method which avoids redundant checks and (iii) an approximate algorithm which trades accuracy for speed by borrowing ideas from the MPEG-2 video encoding. The experimental results demonstrate the efficiency of our techniques and their applicability to large spatio-temporal datasets.

1 Introduction

With the advances of telecommunication technologies we are able to record the movements of objects over a long history. Data analysts are often interested in the automatic discovery of trends or patterns from large amounts of recorded movements. An interesting problem is to find dense clusters of objects which move similarly for a long period. For example, migrating animals usually move in groups (clusters). Another example could be a convoy of cars that follow the same route in a city.

We observe that in many cases such moving clusters do not retain the same set of objects in their lifetime, but objects may enter or leave, while the clusters are moving. In the migrating animals example, during the movement of the cluster, some new animals may enter the group (e.g., those passing nearby the cluster's trajectory), while some animals may leave the group (e.g., those attacked and eaten by lions). Nevertheless the cluster itself retains its density during its whole lifetime, no matter whether it ends up with a totally different set of objects compared to its initial formation.

The automatic discovery of such *moving* clusters is arguably an important problem with several applications. For example, ecologists may want to study the evolution of

C. Bauzer Medeiros et al. (Eds.): SSTD 2005, LNCS 3633, pp. 364–381, 2005.

moving groups of animals. Military applications may monitor troops that move in parallel and merge/evolve over time. The identification of moving dense areas of traffic is useful to traffic surveillance systems. Finally, intelligence and counterterrorism services may want to identify suspicious activity of individuals moving similarly.

The contribution of this paper is the formal definition of moving clusters and the proposal of methods that automatically discover them from a long history of recorded trajectories. Intuitively, a moving cluster is a sequence of spatial clusters that appear in consecutive snapshots of the object movements, such that two consecutive spatial clusters share a large number of common objects. Here, we propose three methods to identify moving clusters in spatio-temporal datasets. Based on the problem definition, our first algorithm, MC1, performs spatial clustering at each snapshot and combines the results into a set of moving clusters. We prove that we can speed-up this process by pruning object combinations that cannot belong to the same cluster, without affecting the correctness of the solution; the resulting algorithm is called MC2. Then, we observe that many clusters remain relatively stable in consecutive snapshots. The challenge is to identify them without having to perform clustering for the entire set of objects. We propose an approximate algorithm, called MC3, which uses information from the past to predict the set of clusters at the current snapshot. In order to minimize the approximation error, we borrow from MPEG-2 video encoding the idea of interleaving approximate with exact cluster sets. We minimize the number of the expensive computations of exact cluster sets by employing a method inspired by the TCP/IP protocol. Our experiments show that MC3 reduces considerably the execution time and produces high quality results.

Previous work has focused mainly on the identification of static dense areas over time [1], or on the clustering of object trajectories for sets that contain the same objects during a time interval [2]. Our problem is different, since both the location and the set of objects of a moving cluster change over time. Related to our work is the incremental maintenance of clusters [3]. Such methods are efficient only if a small percentage of objects is updated. This is not true in our case since potentially all object may be updated (i.e., move) in consecutive snapshots. Recent methods which use approximation to improve the efficiency of incremental clustering [4] are also not applicable in our problem since they do not maintain the continuity of clusters in the time dimension. To the best of our knowledge, this is the first work which deals with the identification of moving clusters.

The rest of the paper is organized as follows: First, we present a formal definition of our problem in Section 2, followed by a survey of the related work in Section 3. Our methods are presented in Section 4, while Section 5 contains the results of our experiments. Finally, Section 6 summarizes the paper and presents some directions for future work.

2   Problem Formulation

Let $H = \{t_1, t_2, \ldots, t_n\}$ be a long, timestamped history. Let $\mathcal{S} = \{o_1, o_2, \ldots, o_m\}$ be a collection of objects that have moved during H. An object o_i not necessarily existed throughout the whole history, but during a contiguous subsequence $o_i.T$ of H. Without

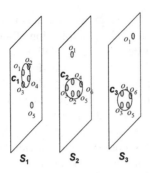

Fig. 1. Example of a moving cluster

loss of generality, we assume that the locations of each object were sampled at every timestamp during $o_i.T$. We refer to $o_i.T$, as the *lifetime* of o_i.

A *snapshot* S_i of H is the set of objects and their locations at time t_i. S_i is a subset of $\mathcal{S}$, since not all objects in $\mathcal{S}$ necessarily existed at t_i. Formally, $S_i = \{o_j \in \mathcal{S} : t_i \in o_j.T\}$. Given a snapshot S_i, we can employ a standard spatial clustering algorithm, like DBSCAN [5] to identify dense groups of objects in S_i which are close to each other and the density of the group meets the density constraints ($MinPts$ and ϵ) of the clustering algorithm.

Let c_i and c_{i+1} be two such *snapshot clusters* for S_i and S_{i+1}, respectively. We say that $c_i c_{i+1}$ is a *moving* cluster if $\dfrac{|c_i \cap c_{i+1}|}{|c_i \cup c_{i+1}|} \geq \theta$, where θ ($0 < \theta \leq 1$) is an integrity threshold for the contents of the two clusters. Intuitively, if two spatial clusters at two consecutive snapshots have a large percentage of common objects then we consider them as a *moving* cluster that moved between these two timestamps. The definition of a spatio-temporal cluster can be generalized as follows:

Definition 1. *Let* $g = c_1, c_2, \ldots, c_k$ *be a sequence of snapshot clusters such that for each* $i(1 \leq i < k)$, *the timestamp of* c_i *is exactly before the timestamp of* c_{i+1}. *Then* g *is a moving cluster, with respect to an integrity threshold* θ ($0 < \theta \leq 1$), *if* $\dfrac{|c_i \cap c_{i+1}|}{|c_i \cup c_{i+1}|} \geq \theta, \forall i : 1 \leq i < k$.

Figure 1 shows an example of a moving cluster. S_1, S_2, and S_2 are three snapshots. In each of them there is a timeslice cluster (c_1, c_2, and c_3). Let $\theta = 0.5$. $c_1 c_2 c_3$ is a moving cluster, since $\dfrac{|c_1 \cap c_2|}{|c_1 \cup c_2|} = \dfrac{3}{6}$ and $\dfrac{|c_2 \cap c_3|}{|c_2 \cup c_3|} = \dfrac{4}{5}$ are both at least θ. Note that objects may enter or leave the moving cluster during its lifetime.

3   Related Work

3.1   Clustering Static Spatial Data

Clustering static spatial data (i.e., static points) is a well-studied subject. Different clustering paradigms have been proposed with different definitions and evaluation criteria, based on the clustering objective. *Partitioning* methods, like k-medoids [6,7], divide the objects into k groups and iteratively exchange objects between them until the quality of the clusters does not further improve. First, k medoids are chosen randomly from the dataset. Each object is assigned to the cluster corresponding to their nearest medoid and the quality of the clusters is defined by summing the distances of all points to their nearest medoid. Then, a medoid is replaced by a random object and the change is committed only if it results to clusters of better quality. A local optimum is reached after a large sequence of unsuccessful replacements. This process is repeated for a number of initial random medoid-sets and the clusters are finalized according to the best local optimum found.

Another class of (agglomerative) *hierarchical* clustering techniques define the clusters in a bottom-up fashion, by first assuming that all objects are individual clusters and gradually merging the closest pair of clusters until a desired number of clusters remain. Algorithms like BIRCH [8] and CURE [9] were proposed to improve the scalability of agglomerative clustering and the quality of the discovered partitions. C2P [10] is another hierarchical algorithm similar to CURE, which employs closest pairs algorithms and uses a spatial index to improve scalability.

Density-based methods discover dense regions in space, where objects are close to each other and separate them from regions of low density. DBSCAN [5] is the most representative method in this class. First, DBSCAN selects a point p from the dataset. A range query, with center p and radius ϵ is applied to verify if the neighborhood of p contains at least a number $MinPts$ of points (i.e., it is dense). If so, these points are put in the same cluster as p and this process is iteratively applied again for the new points of the cluster. DBSCAN continues until the cluster cannot be further expanded; the whole dense region where p falls is discovered. The process is repeated for unvisited points until all clusters and outlier points have been discovered. OPTICS [11] is another density based method. It works similarly to DBSCAN but it does not compute the set of clusters. Instead, it outputs an ordering of the points in the dataset which is used in a second step to identify the clusters for various values of ϵ.

Although these methods can be used to discover snapshot clusters at a given timeslice of the history, they cannot be applied directly for the identification of moving clusters.

3.2   Clustering Spatio-temporal Data

Previous methods on clustering spatio-temporal data have focused on grouping trajectories of similar shape. The one-dimensional version of this problem is equivalent to clustering time-series that exhibit similar movements. Ref. [12] formalized a LCSS (Least Common Subsequence) distance, which assists the application of traditional clustering algorithms (e.g., partitioning, hierarchical, etc.) on object trajectories. In Ref. [2], regression models are used for clustering similar trajectories. Finally, Ref. [13,14] use traditional clustering algorithms on features of segmented time series. The problem of clustering similar trajectories or time-series is essentially different to that of finding moving clusters. The key difference is that a trajectory cluster has a constant set of objects throughout its lifetime, while the contents of a moving cluster may change over time. Another difference is that the input to a moving cluster discovery problem does not necessarily include trajectories that span the same lifetime. Finally, we require the segments of trajectories that participate in a moving cluster to move similarly *and* to be close to each other in space.

A similar problem to the discovery of moving clusters is the identification of areas that remain dense in a long period of time. Ref. [1] proposed methods for discovering such regions in the future, given the locations and velocities of currently moving objects. This problem is different to moving clusters discovery in several aspects. First, it deals with the identification of static, as opposed to moving, dense regions. Second, a sequence of such static dense regions at consecutive timestamps does not necessarily correspond to a moving cluster, since there is no guarantee that there are common objects between regions in the sequence. Third, the problem refers to predicting dense regions in the future, as opposed to discovering them in a history of trajectories.

Our work is also related to the incremental maintenance of clusters in data warehouses. Many researchers have studied the incremental updating of association rules for data mining. Closer to our problem are the incremental implementations of DBSCAN [3] and OPTICS [15]. The intuition of both methods is that, due to the density-based nature of the clustering, the insertion or deletion of a new object o_j affects only the objects in the neighborhood of o_j. Updates are applied in batch and it is assumed that the updated objects are only a small percentage of the dataset. This is not true for our problem due to the movement of objects. Potentially, the entire dataset can be updated at each timestamp rendering the incremental maintenance of clusters prohibitively expensive. Another method, proposed by Nassar et. al. [4], minimizes the updating cost by employing *data bubbles* [16] which are approximate representations of clusters. The method attempts to redistribute the updated objects inside the existing data bubbles. It is not suitable for our problem, since it does not maintain the continuity of clusters in the time dimension.

4   Retrieval of Moving Clusters

In this section we describe three algorithms for the retrieval of moving clusters. The first one, *MC1*, is a straight forward implementation of the problem definition. The next algorithm, *MC2*, improves the efficiency by avoiding redundant checks. Finally, *MC3* is an approximate algorithm which trades accuracy for speed.

4.1   MC1: The Straight-Forward Approach

A direct method for retrieving moving clusters is to follow the problem definition. Starting from S_1, density-based clustering is applied at each timeslice and consecutive timeslice clusters are merged to moving clusters. A pseudocode for the algorithm is given in Figure 2.

The algorithm scans through the timeslices, maintaining at each step a set $\mathcal{G}$ of *current* moving clusters. When examining timeslice S_i, $\mathcal{G}$ includes the moving clusters containing a timeslice cluster in S_{i-1}. After discovering the timeslice clusters in S_i, every pair (g, c), $g \in \mathcal{G}$, $c \in S_i$ is checked to verify whether $g \circ c$ (i.e., g extended by c in S_i) forms a valid moving cluster. Clusters in $\mathcal{G}$ that were not extended at S_i,

---

Algorithm. MC1(Spatio-temporal history H, real ϵ, int $MinPts$, real θ)

1. $\mathcal{G} := \varnothing$; // set of current clusters
2.   **for** $i:=1$ to n // for each timestamp
3.   **for each** current moving cluster $g \in \mathcal{G}$
4. g.extended := $false$
5. $\mathcal{G}_{next} := \varnothing$; // next set of current clusters
6.   // retrieve timeslice clusters at S_i
7. $L :=$ DBSCAN(S_i, e, $MinPts$);
8.   **for each** timeslice cluster $c \in L$
9. $assigned := false$;
10.   **for each** current moving cluster $g \in \mathcal{G}$
11.   **if** $g \circ c$ is a valid moving cluster **then**
12. g.extended := $true$;
13. $\mathcal{G}_{next} := \mathcal{G}_{next} \cup g \circ c$;
14. $assigned := true$;
15.   **if** (**not** $assigned$) **then**
16. $\mathcal{G}_{next} := \mathcal{G}_{next} \cup c$;
17.   **for each** current moving cluster $g \in \mathcal{G}$
18.   **if** (**not** g.extended) **then**
19.   **output** g;
20. $\mathcal{G} := \mathcal{G}_{next}$;

Fig. 2. The MC1 algorithm for computing moving clusters

are output. The $\mathcal{G}_{next}$ set of moving clusters to be used at the next iteration (i.e., for S_{i+1}) consists of (i) clusters in $\mathcal{G}$ that were extended in S_i and (ii) timeslice clusters $c \in S_i$ that were not used as extensions to some $g \in \mathcal{G}$. In this way, MC1 does not miss any cluster and does not output any redundant clusters, whose extensions are also valid clusters.

We assume that all points of the *current* timeslice S_i fit in memory. In practice this means that a relatively low-end computer with 1GB of RAM supports more than 10M points per timeslice. Notice that there is no restriction on the number of timeslices. Having the entire S_i in the main memory eliminates the need to build a spatial index in order to speed-up the clustering algorithm. Instead, we developed a main-memory version of DBSCAN. We divide the 2-D space into a grid where the cell size is $\frac{\epsilon}{\sqrt{2}} \times \frac{\epsilon}{\sqrt{2}}$ and hash each point of S_i to its corresponding cell. Observe that the distance between any two points in the same cell is at most ϵ. Therefore, any cell containing more than $MinPts$ points is part of a cluster; such cells are called *dense*. The main-memory DBSCAN proceeds by merging neighboring dense cells and finally it handles the remaining points which belong to sparse cells.

For each iteration of Line 3, we only need to keep one cluster g of $\mathcal{G}$ in the memory. Therefore, the memory required by MC1 is $O(|S_i| + |g| + \frac{\epsilon^2}{2})$. The if-statement in Line 11 is executed $|G| \cdot |L|$ times and calculates the similarity criterion of Definition 1 which involves an intersection and a union operation. We employ a hash table to implement these operations efficiently. The cost is $O(|g| + |c|)$, where g and c are clusters belonging to $\mathcal{G}$ and L, respectively.

4.2 MC2: Minimizing Redundant Checks

MC1 contains two time-consuming operations: the call to DBSCAN (line 7) and the computation of intersection/union of clusters (line 11). Especially the later exhibits significant redundancy since we expect each cluster $g \in \mathcal{G}$ to share objects only with a few clusters $c \in L$. In this section we present an improved algorithm, called *MC2* which minimizes the redundant combinations of (g, c).

The idea is simple: We select a random object $o_j \in g_i$ and search for it in all clusters of L. Let $c_i \in L$ be the cluster which contains o_j. Then we calculate the intersection and union only for the pair (g_i, c_i). If they satisfy the similarity criterion, we insert $g_i \circ c_i$ in the result. Else we must select another object $o_k \in g_i$ and repeat the process. Notice that some objects are pruned: o_k is selected from $g_i - c_i$ since the common objects of (g_i, c_i) cannot belong to any other cluster of L. The interesting observation is that we never need to test more than $(1 - \theta)|g_i|$ points. The following lemma has the details:

Lemma 1. *Let c_1 and c_2 be clusters. $c_1 c_2$ is not a moving cluster if $|c_1 - c_2| > (1 - \theta)|c_1|$.*

Proof. For any set c_1, c_2, it holds that $|c_1 \cap c_2| \leq |c_1|$. We know that there are more than $(1-\theta)|c_1|$ points in c_1 which do not exist in c_2. Therefore, $|c_1 \cap c_2| < |c_1| - (1-\theta)|c_1|$. By using this value in the formula of Definition 1, we have:

$$\frac{|c_1 \cap c_2|}{|c_1 \cup c_2|} < \frac{|c_1| - (1-\theta)|c_1|}{|c_1 \cup c_2|} \leq \frac{|c_1| - (1-\theta)|c_1|}{|c_1|} = \theta$$

Since $\dfrac{|c_1 \cap c_2|}{|c_1 \cup c_2|} < \theta$, $c_1 c_2$ is not a moving cluster.

Figure 3 presents the pseudocode of MC2. The algorithm is similar to MC1, except that the expensive union/intersection between clusters (Line 15) is executed at most $(1 - \theta)|g_i|$ times for every $g_i \in \mathcal{G}$. Notice that another potentially expensive operation is the search for object o_j in the clusters of L (Line 12). To implement this efficiently, at each timeslice S_i we generate a hash table which contains all objects of S_i. The cost is $O(|S_i|)$ on average. Then we can find an object in constant time, on average. The tradeoff is that we need $O(|S_i|)$ additional memory to store the hash table. If memory size is a concern, we can use an in-place algorithm (e.g., quicksort) to sort the objects of S_i and locate objects by binary search.

Note that if $\theta \leq \dfrac{1}{2}$ ties may appear during the generation of moving clusters; in this case, MC2 does not necessarily produce the same output as MC1. This is illustrated in the example of Figure 4, where $\theta = \dfrac{1}{3}$. The original cluster c_0 splits into two smaller clusters c_1 and c_2 at timeslice S_{i+1}. Both of them satisfy the criterion of Definition 1 since $\dfrac{|c_0 \cap c_1|}{|c_0 \cup c_1|} = \dfrac{|c_0 \cap c_2|}{|c_0 \cup c_2|} = \dfrac{1}{3}$. Therefore, both $\{c_0 c_1, c_2\}$ and $\{c_0 c_2, c_1\}$ are legal sets of moving clusters. The same behavior is also observed in the symmetric case, where two small clusters are merged into a larger one. Since MC1 and MC2 break such ties arbitrarily, the outputs may differ; nevertheless, the result of MC2 is *not* approximate.

Algorithm. MC2(Spatio-temporal history H, real ϵ, int $MinPts$, real θ)

1. $\mathcal{G} := \varnothing$; // set of current clusters
2.   **for** $i:=1$ to n // for each timestamp
3. $\mathcal{G}_{next} := \varnothing$; // next set of current clusters
4. $L := \text{DBSCAN}(S_i, e, MinPts)$; // retrieve timeslice clusters at S_i
5.   **for each** timeslice cluster $c \in L$
6. $c.assigned := false$;
7.   **for each** current moving cluster $g \in \mathcal{G}$
8.   g.extended := false;
9. $k := (1 - \theta)|g|$;
10.   **while** $(k > 0)$
11. o_j is a random object of g;
12. $c := \textbf{find}(o_j$ inside $L)$; // $c \in L$ contains o_j
13.   **if** $(o_j$ not found) **then** $k := k - 1$;
14.   **else**
15.   **if** $g \circ c$ is a valid moving cluster **then**
16.   g.extended := $true$;
17. $\mathcal{G}_{next} := \mathcal{G}_{next} \cup g \circ c$;
18. $c.assigned := true$;
19. $k = k - |g - c|$;
20.   **if** (**not** g.extended) **then** output g;
21.   **for each** cluster $c \in L$
22.   **if** (**not** $c.assigned$) **then** $\mathcal{G}_{next} := \mathcal{G}_{next} \cup c$;
23. $\mathcal{G} := \mathcal{G}_{next}$;

Fig. 3. The MC2 algorithm for computing moving clusters

4.3 MC3: Approximate Moving Clusters

Although MC2 avoids checking redundant combinations of clusters, it still needs to perform an expensive call to DBSCAN for each timeslice. In this section we present an alternative algorithm, called *MC3*, which decreases the execution time by minimizing the set of objects processed by DBSCAN. MC3 is an approximate algorithm which trades speed for accuracy. A cluster c may be approximate because: (i) there exists an object $o_j \in c$ which is a core object but has less than $MinPts$ objects in its ϵ-neighborhood (see [5] for details), or (ii) the distance of o_j from any other object in c is larger than ϵ.

MC3 works as follows: given a current moving cluster set $\mathcal{G}$ and a timeslice S_i, it maps all objects $o_j \in S_i$ to a set of clusters $\mathcal{G}'$. This mapping is done by assuming that all objects which are common in S_{i-1} and S_i remain in the same clusters, irrespectively of their new position in S_i. Any new objects appearing in S_i are assigned to the closest existing cluster within distance $\sqrt{2}\epsilon$, or they are considered as noise if no such cluster exists. Let L_1 be the subset of $\mathcal{G}'$ containing the clusters that do not overlap with each other, and $S_i' \subseteq S_i$ be a set containing these objects that do not belong to any cluster of L_1. The algorithm assumes that L_1 contains valid clusters and does not process them further. On the other hand, the objects in S_i' probably define some new clusters. Therefore, MC3 applies DBSCAN only on S_i'. The output of DBSCAN together with

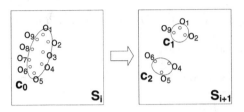

Fig. 4. Cluster split ($\theta = \frac{1}{3}$)

L_1 is the set L of clusters in timeslice S_i. Finally, the next set of moving clusters $\mathcal{G}_{next}$ is computed from $\mathcal{G}$ and L by employing the fast intersection technique of MC2. The details of MC3 are presented in Figure 5.

Algorithm. MC3(Spatio-temporal history H, real ϵ, int $MinPts$, real θ)

1. $\mathcal{G} := \varnothing$; // set of current clusters
2. $timer := 0$;
3. $period := 1$;
4.   **for** $i:=1$ to n // for each timestamp
5. $\mathcal{G}_{next} := \varnothing$; // next set of current clusters
6.   **if** ($timer < period$) **then** // Approximate clustering
7. $\mathcal{G}' :=$ Use $\mathcal{G}$ to assign all objects of S_i to clusters;
8. $L_1 := \{g | g \in \mathcal{G}' \land g$ is disjoint$\}$;
9. $S_i' := S_i - \{$all objects belonging to a cluster of $L_1\}$;
10. $L_2 :=$ DBSCAN(S_i', e, $MinPts$); // retrieve timeslice clusters at S_i'
11. $L := L_1 \cup L_2$;
12. $timer := timer + 1$;
13.   **else**
14. $L :=$ DBSCAN(S_i, e, $MinPts$); // retrieve timeslice clusters at S_i
15.   **if** (($deleted + inserted$ clusters in MC2) $> \alpha |G|$) **then**
16. $period := min(1, period/2)$;
17.   **else** $period := period + 1$;
18. $timer := 0$;
19.   Use the fast intersection method of MC2 to compute $\mathcal{G}_{next}$;
20. $\mathcal{G} := \mathcal{G}_{next}$;

Fig. 5. The MC3 algorithm for computing moving clusters

The intuition behind the algorithm is that several clusters will continue to exist in consecutive timeslices. Therefore, by employing the incremental approach of MC3, $|S_i'|$ (i.e., the input to the expensive DBSCAN procedure) is expected to be much smaller than $|S_i|$. Notice that S_i' can be computed with linear time complexity. First we create a hash table which maps objects to moving clusters of $\mathcal{G}$. We use this hash table to generate $\mathcal{G}'$; the entire process costs $O(|S_{i-1}| + |S_i|)$ on average. Next, we divide the space into a regular grid with cell size equal to $\epsilon \times \epsilon$ and we assign each object to its corresponding cell; the cost is $O(|S_i|)$. During this step we can identify if two clusters *intersect* with each other. Then we check that (i) every cluster c is connected (i.e, all cells belonging to c *meet* each other) and (ii) no pair of clusters *meet* each other. The complexity of this step is $O(9\epsilon^2)$, where ϵ is constant. Figure 6 presents an example with 3 clusters $c_{1...3}$. c_1 is connected and does not meet or intersect with any other cluster; therefore it is placed in L_1. On the other hand, the objects of c_2 and c_3 must be added to S_i' because the two clusters meet.

Observe that in order to minimize the computation cost, we determine the relationships among clusters by considering only the corresponding grid cells. This introduces inaccuracies to the cluster identification process. For instance, the distance of an object $o_j \in c_1$ from any other object in c_1 may be up to $2\sqrt{2}\epsilon$, therefore o_j may need to be removed from c_1. Also, there may be some noise object o_k in the space between c_1 and c_2. By considering these two clusters together with o_k, c_1 and c_2 may need to be merged. When dealing with moving clusters which span several timeslices, an error at the initial assignment propagates to the following timeslice. Therefore, errors tend to accumulate and the quality of the result degrades fast.

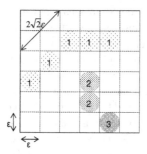

Fig. 6. Checking cluster intersection on an $\epsilon \times \epsilon$ grid

In order to minimize the approximation error, we use a method from video compression. The MPEG-2 encoder achieves high compression rates at high quality by using two types of video frames[1]: I-frames which are static JPEG images and P-frames which represent the difference between $frame_t$ and $frame_{t-1}$; in general I-frames are larger that P-frames. First an I-frame is sent followed by a stream of P-frames (Figure 7.a). When the encoding error exceeds some threshold (e.g., a different scene in the movie), a new I-frame is sent to increase the quality (i.e., eliminate the error). We employ a similar technique in our algorithm. Observe that the set L of clusters generated for each timeslice by MC3, is analogous to P-frames. To decrease the approximation error, we need to introduce periodically new reference cluster sets (i.e., similar to I-frames). We achieve this by interleaving the approximate clustering algorithm with exact clustering. This is shown in Line 14 of MC3, where DBSCAN is executed on the entire S_i. Notice that, contrary to MPEG-2, even after performing exact clustering the error is not necessarily eliminated. This is due to the fact that moving clusters span several timeslices, while exact clustering assigns correctly only the *static* clusters of the current timeslice.

---

[1] There is also a B-frame, which is not relevant to our problem.

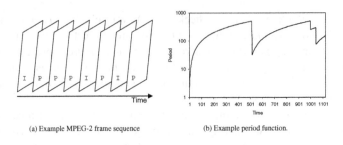

(a) Example MPEG-2 frame sequence   (b) Example period function.

Fig. 7. Example of MPEG-2 and TCP/IP-based period adjustment

Therefore, in order to recover from errors, we may need to execute exact clustering several times within a short interval. On the other hand, if exact clustering is performed too frequently, MC3 degenerates to MC2.

Obviously, the *period* between calls to exact clustering cannot be constant, since the data distribution (and consequently the error), vary among timeslices. In order to adjust the *period* adaptively, we borrowed an idea from the networks area and specifically from the TCP/IP protocol. When a node starts transmitting, TCP/IP assigns a medium transmission rate. If the percentage of packet loss is low, it means that there is spare capacity in the network; therefore the transmission rate is slowly increased. However, if the percentage of packet loss increases considerably, the transmission rate decreases quickly in order to resolve fast the congestion problem in the network. We employ a similar method: initially the *period* for executing exact clustering is set to 1. If the error is low, the *period* is increased linearly at each timestamp. In this way, as long as the error does not increase too much, the expensive clustering is performed infrequently. On the other hand, if the error exceeds some threshold, the *period* is decreased exponentially (Figure 7.b). Therefore, the incorrect cluster assignments are not allowed to propagate through many timeslices.

The reader should note that we cannot perform exact clustering only when there is an error, because we cannot estimate the error unless we compute the exact clustering. The above mentioned method minimizes the unnecessary computations, but some redundancy is unavoidable. We estimate the error in the following way: when an exact clustering is performed, we count the number of moving clusters which are created or removed at the current timeslice. If this number is greater that $\alpha|\mathcal{G}|$, $0 < \alpha \leq 1$, we assume that the error is high. The intuition is that if many moving clusters are changing this may be due to the fact that there were many incorrect assignments at the previous timeslices. Other methods for error estimation are also possible. For instance, we could execute approximate and exact clustering at the same timeslice and compare the results. However, this would increase the execution time, while our heuristic poses minimal overhead and works well in practice.

5   Experimental Evaluation

In this section we present the experimental evaluation of our methods. Due to the unavailability of real datasets, we developed a generator which generates synthetic

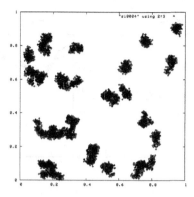

Fig. 8. Output from the generator: a sample timeslice S_i with 9000 objects

datasets with various distributions. The generator accepts several parameters including the number of clusters per timeslice, the average number of objects per cluster, the neighborhood radius ϵ and the density $MinPts$. The average velocity of the clusters and the change probability P_c are also given as input. The output of the generator is a series of timeslices. At each timeslice each cluster may move from the previous position; the velocity vector changes with probability P_c. With the same probability, a cluster may rotate around its center. Also, objects are inserted or deleted with probability P_c. Figure 8 shows a sample of the distribution of objects at a given timeslice. We generated several datasets by varying the number of objects per timeslice from 10K to 50K and the number of timeslices from 50 to 100. Therefore, the size of each dataset was between 500K and 5M objects. We implemented our algorithms in C++ and run all experiments on a relatively low-end Linux system with 1.2GB RAM and 1.3GHz CPU.

In order to evaluate the accuracy of MC3's approximation, we compared at each timeslice the current set of moving clusters produced by MC3 against the set generated by MC2. The quality of the solution is defined as:

$$F = \frac{2 \cdot precision \cdot recall}{precision + recall}$$

This metric is commonly used in data mining [17]. Obviously, F is always 1 for MC1 and MC2, since their solution is exact.

In the first set of experiments we test the scalability of our methods to the size of the dataset. We generated datasets with 100 timeslices. Each timeslice contained 10K to 50K objects and 800 clusters on average, resulting to a database size of 1M to 5M objects. We set $\theta = 0.9$ and $\alpha = 0.1$ (recall that α is used in Line 15 of MC3). The results are shown in Figure 9. As expected, the execution time of all algorithms increases with the dataset size. MC2 is much faster than MC1 and the difference increases for larger datasets. This demonstrates clearly that MC2 manages to prune a large number of redundant cluster combinations. MC3 is faster than MC2 but there is some error introduced to the solution. In Figure 9.b we draw the quality F of MC3's solution for each timeslice. Notice that the quality is reduced at first, because the error is not high enough to trigger the execution of exact clustering. However, when the error exceeds the threshold value, the algorithm adjusts fast to high quality solutions.

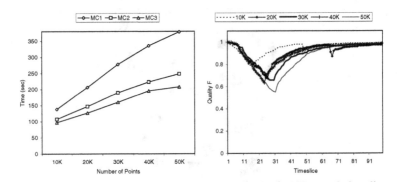

(a) Execution time vs. database size   (b) Quality F of MC3 at each timeslice

Fig. 9. Varying the number of objects. 800 clusters, $\theta = 0.9$, $\alpha = 0.1$

Table 1. Average quality $\overline{F}$ of MC3. 800 clusters, $\theta = 0.9$, $\alpha = 0.1$

	10K objects	20K objects	30K objects	40K objects	50K objects
$\overline{F}$	94.7%	91.1%	90.0%	90.8%	87.0%

The average quality of MC3 for the entire database is shown in Table 1. Given that the number of clusters remains the same for all datasets, when the database size increases, so does the average number of objects per clusters. When this happens, the extends of clusters tend to grow and therefore more errors are introduced due to incorrect assignment of cluster splits or noise objects (see Figure 6 for details). Nevertheless, the average quality remained at acceptable levels (i.e, at least 87%) in all cases.

The second set of experiments test the scalability of the algorithms to the number of clusters. All datasets contain 100 timeslices with 50K objects each (i.e., database size is 5M objects). The average number of clusters per timeslice is varied from 100 to 800. Again, we set $\theta = 0.9$ and $\alpha = 0.1$. The results are presented in Figure 10. The trend is similar to the previous experiment with the exception that the relative performance of MC3 has improved. To explain this, observe the graph of the quality F. For the case of 100 clusters, the quality remains high during the entire lifespan of the dataset. This happens because there are fewer clusters in the space, therefore there is a smaller probability of interaction among them which decreases the probability of errors in MC3. Consequently, the expensive exact clustering in Line 14 of MC3 is called very infrequently and the total execution time is improved.

Table 2 shows the average quality of MC3 for the entire dataset. Notice the strange trend: the quality first decreases and then increases again when there are more clusters. To understand this behavior, refer again to Figure 10.b. We already explained why quality is high for the 100 clusters dataset. On the other hand, when there are 800 clusters, there are a lot of interactions among them which introduce a large margin for error. Therefore, MC3 reaches the error threshold fast and starts performing exact clustering in order to recover. Now look at the 200 and 400 clusters datasets. There is a large number of clusters, therefore quality drops. Nevertheless, the error does not exceed

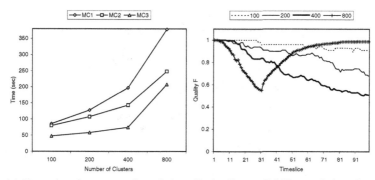

(a) Execution time vs. number of clus- (b) Quality F of MC3 at each timeslice
ters

Fig. 10. Varying the number of clusters. 50K objects, $\theta = 0.9$, $\alpha = 0.1$

Table 2. Average quality $\overline{F}$ of MC3. 50K objects, $\theta = 0.9$, $\alpha = 0.1$

	100 clusters	200 clusters	400 clusters	800 clusters
F	95.7%	86.9%	72.7%	87.0%

the threshold and MC3 does not attempt to recover. The cause of this problem is the inappropriate value for parameter α.

In order to investigate further how parameter α affects the result, we used the 800 clusters, 50K objects dataset from the previous experiment and varied α from 0.05 to 0.2. The quality of MC3 for each timeslice, is shown in Figure 11. When α is low, MC3 reaches the error threshold fast. When this happens, it starts using the exact clustering function in order to recover fast. When α is larger, MC3 needs more time before initiating the recovery process. A point to note here is that for large values of α, the algorithm does not recover completely. For instance, if $\alpha = 0.2$, the algorithm starts oscillating around $F = 0.4$ after some time.

The execution time of MC3 also depends on α. This is expected, because the algorithm trades accuracy for speed. To test this, we generated two more datasets with the same characteristics as before, but with different object agility. Assuming that the previous dataset has *Medium* agility, one of the new datasets contains objects with *High* agility and the other contains objects with *Low* agility. The results are shown in Figure 12. As expected, MC3 is faster for larger values of α. This is due to the fact that when we accept higher error, MC3 uses most of the time the fast approximate function for clustering. Observe that MC3 is faster for low agility datasets. Such datasets contain fewer interactions among clusters; therefore the approximation error is low resulting to very few calls to the expensive exact clustering function.

Table 3 compares the execution time and the solution quality of MC3 against MC2 (recall that $F = 1$ always for MC2). In contrast to MC3, MC2 does not exhibit sensitivity to the agility of the dataset. Therefore, for the low agility dataset, the speedup of MC3 is very high. However, the average quality drops. To compare the various cases,

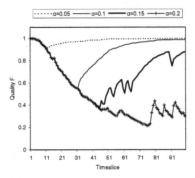

Fig. 11. Quality F of MC3 at each timeslice. 800 clusters, 50K objects, $\theta = 0.9$

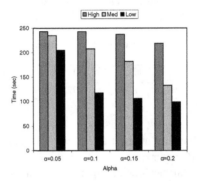

Fig. 12. Execution time of MC3 for varying α. 800 clusters, 50K objects, $\theta = 0.9$

we use the relative quality per time unit, defined as:

$$F(MC3|MC2) = \frac{\frac{F_{MC3}}{t_{MC3}} - \frac{1}{t_{MC2}}}{\frac{1}{t_{MC2}}}$$

Observe that $F(MC3|MC2)$ is much higher for the low agility dataset, meaning that MC3 traded a small percentage of accuracy in order to achieve a much lower execution time.

The last set of experiments investigates the effect of θ, which is the integrity threshold of Definition 1. We used the 800 clusters, 50K objects, medium agility dataset from the previous experiment and set $\alpha = 0.1$. We varied θ from 0.7 to 0.95. Figure 13 shows the execution time for the three algorithms. There is a very small increase of the execution time when θ increases. This is more obvious for MC1. The reason is that when θ is large, there is a smaller probability to find matching moving clusters in consecutive timeslices. To conclude that there is no match, MC1 must check all cluster pairs, whereas, the search would stop as soon as one match was found.

Figure 14 and Table 4 demonstrate the effect of theta on the quality of MC3. Interestingly, the quality improves when θ increases. This happens because, even if the

Table 3. Average quality of MC3 for datasets with varying agility

	MC2: Time (sec)	MC3: Time (sec)	MC3: Quality (F)	$F(MC3\|MC2)$
High	252	246	98.6%	1.2%
Med	248	208	87.0%	3.9%
Low	256	118	73.0%	59.0%

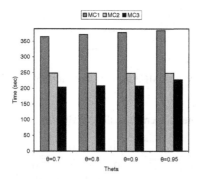

Fig. 13. Execution time vs. θ. 800 clusters, 50K objects, $\alpha = 0.1$

approximate clustering function generates some incorrect clusters, there is a high possibility that the corresponding moving clusters will not satisfy the θ criterion and will be eliminated. Therefore, there is a smaller probability of errors.

6   Conclusions

In this paper we investigated the problem of identifying moving clusters in large spatio-temporal datasets. The availability of accurate location information from embedded GPS devices, will enable in the near future numerous novel applications requiring the extraction of moving clusters. Consider, for example, a traffic control data mining system which diagnoses the causes of traffic congestion by identifying convoys of similarly moving vehicles. Military applications can also benefit by monitoring, for instance, the movement of groups of soldiers.

We defined formally the problem and proposed exact and approximate algorithms to identify moving clusters. MC2 is an efficient algorithm which can be used if 100% accuracy is essential. MC3, on the other hand, generates faster an approximate solution. In order to minimize the approximation error without sacrificing the efficiency, we borrowed methods from MPEG-2 and TCP/IP. Our experimental results demonstrate the applicability of our methods to large datasets with varying object distribution and agility. To the best of our knowledge, this is the first work to focus on the automatic extraction of moving clusters.

The efficiency and accuracy of MC3 depends on the appropriate selection of parameter α and the accurate estimation of error. Currently we are working on a self-tuning

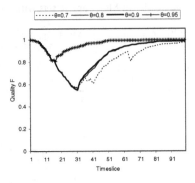

Fig. 14. Quality F of MC3 vs. θ. 800 clusters, 50K objects, $\alpha = 0.1$

Table 4. Average quality of MC3 for varying θ

	$\theta = 0.7$	$\theta = 0.8$	$\theta = 0.9$	$\theta = 0.95$
MC3: Quality $(\overline{F})$	83.0%	86.8%	87.0%	96.5%
$F(MC3 \mid MC2)$	1.1%	3.4%	3.9%	5.3%

method for parameter selection. We also plan to explore sophisticated methods for error estimation.

References

1. Hadjieleftheriou, M., Kollios, G., Gunopulos, D., Tsotras, V.J.: On-line discovery of dense areas in spatio-temporal databases. In: Proc. of SSTD. (2003)
2. Gaffney, S., Smyth, P.: Trajectory clustering with mixtures of regression models. In: Proc. of ICDM. (1999) 63–72
3. Ester, M., Kriegel, H.P., Sander, J., Wimmer, M., Xu, X.: Incremental clustering for mining in a data warehousing environment. In: Proc. of VLDB. (1998) 323–333
4. Nassar, S., Sander, J., Cheng, C.: Incremental and effective data summarization for dynamic hierarchical clustering. In: Proc. of ACM SIGMOD. (2004) 467–478
5. Martin, E., Kriegel, H.P., Sander, J., Xu, X.: A density-based algorithm for discovering clusters in large spatial databases with noise. In: Proc. of KDD. (1996)
6. Kaufman, L., Rousueeuw, P.: Finding Groups in Data: an Introduction to Cluster Analysis. John Wiley and Sons (1990)
7. Ng, R.T., Han, J.: Efficient and effective clustering methods for spatial data mining. In: Proc. of VLDB. (1994)
8. Zhang, T., Ramakrishnan, R., Livny, M.: BIRCH: An efficient data clustering method for very large databases. In: Proc. of ACM SIGMOD. (1996)
9. Guha, S., Rastogi, R., Shim, K.: CURE: An efficient clustering algorithm for large databases. In: Proc. of ACM SIGMOD. (1998)
10. Nanopoulos, A., Theodoridis, Y., Manolopoulos, Y.: C2P: Clustering based on closest pairs. In: Proc. of VLDB. (2001)

11. Ankerst, M., Breunig, M., Kriegel, H.P., Sander, J.: OPTICS: Ordering points to identify the clustering structure. In: Proc. of ACM SIGMOD. (1999) 49–60

12. Vlachos, M., Kollios, G., Gunopulos, D.: Discovering similar multidimensional trajectories. In: Proc. of ICDE. (2002) 673–684

13. Das, G., Lin, K.I., Mannila, H., Renganathan, G., Smyth, P.: Rule discovery from time series. In: Proc. of KDD. (1998) 16–22

14. Li, C.S., Yu, P.S., Castelli, V.: Malm: A framework for mining sequence database at multiple abstraction levels. In: Proc. of CIKM. (1998) 267–272

15. Kriegel, H.P., Kröoger, P., Gotlibovich, I.: Incremental OPTICS: Efficient computation of updates in a hierarchical cluster ordering. In: Proc. of DaWaK. (2003) 224–233

16. Breunig, M.M., Kriegel, H.P., Kröger, P., Sander, J.: Data bubbles: Quality preserving performance boosting for hierarchical clustering. In: Proc. of ACM SIGMOD. (2001)

17. Larsen, B., Aone, C.: Fast and effective text mining using linear-time document clustering. In: Proc. of KDD. (1999) 16–22

Semantic Caching for Multiresolution Spatial Query Processing in Mobile Environments

Sai Sun, Xiaofang Zhou, and Heng Tao Shen

School of Information Technology and Electrical Engineering,
The University of Queensland, Australia
{sunsai, zxf, shenht}@itee.uq.edu.au

Abstract. Spatial data are particularly useful in mobile environments. However, due to the low bandwidth of most wireless networks, developing large spatial database applications becomes a challenging process. In this paper, we provide the first attempt to combine two important techniques, multiresolution spatial data structure and semantic caching, towards efficient spatial query processing in mobile environments. Based on the study of the characteristics of multiresolution spatial data (MSD) and multiresolution spatial query, we propose a new semantic caching model called Multiresolution Semantic Caching (MSC) for caching MSD in mobile environments. MSC enriches the traditional three-category query processing in semantic cache to five categories, thus improving the performance in three ways: 1) a reduction in the amount and complexity of the remainder queries; 2) the redundant transmission of spatial data already residing in a cache is avoided; 3) a provision for satisfactory answers before 100% query results have been transmitted to the client side. Our extensive experiments on a very large and complex real spatial database show that MSC outperforms the traditional semantic caching models significantly.

1 Introduction

Spatial data have been increasingly used in mobile environments such as electronic navigation, dynamic map generalization, and location-dependent applications. This raises many new research challenges and opportunities. Compared to traditional distributed systems, mobile computing environments have distinctive characteristics which impact on the manipulation of spatial databases [2]. First, in mobile computing environments, there are usually restrictions such as a limited bandwidth, unstable wireless link and short-life battery power. All of these seriously conflict with the tremendous volume of spatial data. Second, the capabilities of mobile units vary greatly, each of which could be a powerful portable computer or a personal digital assistants (PDAs) with very small screens. Thus, users of different mobile units may require different qualities of query answers. Third, in mobile environments, spatial queries are generally not isolated to each other but lie in several typical patterns.

Semantic caching is an important method to improve the performance of mobile computing. This method maintains the semantic descriptions as well as

C. Bauzer Medeiros et al. (Eds.): SSTD 2005, LNCS 3633, pp. 382–399, 2005.

results of previous queries on the client side. Semantic caching is based on an assumption that queries are relevant to each other semantically. Interestingly, this matches the property of spatial queries in mobile environments, which will be discussed later in section 5.1. By caching pertinent and frequently queried data in mobile units, semantic caching can reduce network traffic, shorten response time and provide better scalability [10].

Generally speaking, there are two basic approaches to reduce the data size processed in spatial queries: (1) reducing the size of candidate sets; and (2) reducing the complexity of geometry objects. The filter-refinement strategy has been well-established for the former purpose, while the multiresolution method is designed for the latter purpose. Multiresolution databases break spatial objects into parts, and then fetch only those parts that may contribute to a particular resolution [13]. Hence, the multiresolution method can lead to a smaller data transmission which is crucial in mobile environments. Furthermore, it permits users to flexibly designate their required query qualities or parameters.

Therefore, combining multiresolution and semantic caching could greatly improve the performance of spatial queries in mobile environments. However, the structure and semantic description of multiresolution spatial data (MSD) is much more complex than standard data, which raises the issues in the caching mechanism, including caching granularity, query trimming strategy, cache replacement policy, etc. To the best of our knowledge, there has been no significant research in this direction to date.

In this paper, we deploy MSD in semantic cache by introducing Multiresolution Semantic Caching (MSC) to improve the performance of spatial queries in mobile environments. We analyze the characteristics of MSD and MSD queries and their impact on the semantic caching mechanism, then propose MSC inspired from the analysis. In traditional semantic caching, query processing divides the relationship between the query and the cache into three categories: 1) exact match - the query is fully answered by the cache; 2) partial match - the query is partially answered by the cache; and 3) no match - the query is not answerable from the cache. And different operations are designed toward different categories. Incorporating the characteristics of MSD, MSC enriches the relationship by two more categories: Assumable Exact Match and Approximate Match. The motivation is to provide the user acceptable query results as soon as possible. MSC improves the performance in three ways: 1) by reducing the amount and complexity of the remainder queries, and further reducing the retrieval time in the database server; 2) by avoiding the redundant transmission of spatial data that have already existed in the client cache; and 3) by providing satisfactory answers before 100% of the results have been transmitted to the client. Our extensive experiment results confirm the effectiveness of our methods for spatial query processing in mobile environments.

The remainder of this paper is organized as follows. In Section 2, we discuss previous work on semantic caching and multiresolution. In Section 3, some preliminary work is presented, followed by MSC query processing in Section 4. An extensive performance study is reported in Section 5. Finally, we conclude our paper in Section 6.

2   Related Work

One research area that is directly related to this paper is that of semantic caching, especially semantic caching in mobile environments. The other part of relevant work concerns multiresolution spatial data. In this section, we review previous works from these two fields.

2.1   Semantic Caching

The semantic caching technique was first proposed in [4] to contrast with page caching and tuple caching. The client maintains the results of previous queries as well as their semantic descriptions in its cache, which allows the tuples available locally not to be fetched from the database again. A value function based on semantic locality was also proposed in this paper, named as 'Manhattan distance'. Later, this technique was widely studied in centralized systems, OLAP, LDAP, WWW, and heterogeneous systems [10]. Semantic caching in mobile computing environments has also received extensive attentions. [3] proposed a semantic caching scheme for handling both projection and selection in a mobile environment. [9] discussed how to use semantic caching to manage location dependent data in mobile computing and extended Manhattan Distance to a new replacement policy FAR (Furthest Away Replacement). In [14], DBMS was proposed in mobile units to manage caches. Recently, processing spatial queries has begun to occur in this field, such as [15], [6]. Because mobile clients are often roaming, [15] identified the problem of the validity of previous queries and proposed the corresponding algorithms to handle neighbour and window queries. In [6], a proactive caching model, which caches the result objects as well as their R-tree index nodes, was proposed to improve the utility of local caches. However, both spatial object and location information discussed in the above papers is concerned with **Point** data. It is also the reason why [6] thought that a new query could be answered only by the cached data of the same query type. Our work differs from these previous research in that we focus on **Polygon** data, for example we focus on how to cache MSD to dynamically generalize digital maps in mobile environments.

2.2   Multiresolution

The main goal of multiresolution (also called multiscale) systems is to be able to retrieve objects with different representations to satisfy different requests from users. As different details of a particular feature are distinguishable at different scales, a geometry object should be treated as a collection of spatial elements. In [1] multiple representations of spatial objects are pre-generated and stored in the database. Queries on different scale access different representations. This technology is also known as 'multirepresentation'. Its main disadvantage is the much higher storage overhead than normal. In addition, by using pre-computed representations with a large amount of redundant data, this scheme is not capable of supporting a desirable form of data caching. Later work by [8] focused on the concept of indexing objects by not only their spatial location, but also a scale value. With the scale in the index, I/O cost can be significant reduced during retrieval. However, the approach is too simplistic and only supports graphical and content zoom. An improvement was achieved in [13], which assigns each vertex of an object to a set of map scales and vertices that share similar or equivalent scales are grouped into 'vertex layers'. The vertices in each layer form a representation of the object at a particular scale. This idea is similar to our multiresolution data structure – 'Bit-map'. However, they did not point out how to assign scales to vertices, which is a lengthy process if done manually. Later, they extended their work in [16], which proposed that the control on spatial generalization should be multi-dimensional with spatial resolution as one dimension and various types of generalisation style metrics as the other dimensions. However, they did not provide detailed algorithms or a performance study.

3 Preliminary Work

This section firstly introduces the multiresolution spatial data structure used in the remainder of the paper, then presents the constraint formula of the semantic region of MSC.

3.1 Multiresolution Spatial Databases

A spatial object O can be considered as a set of points: $\{p_1, p_2, ..., p_n\}$, where if O is a point, $n = 1$; if O is a line, $n = 2$; if O is a simple polygon (without hole), $n \geq 3$ and each point pair (p_a, p_{a+1}) forms a line on the object boundary. In traditional spatial databases, the spatial elements (points) of the object are opaque to applications and have to be accessed together. That is why only one fixed resolution can be provided by such systems. A simple method to overcome this flaw is to store and process spatial objects based on points. However, point-based data structures create undesirable overheads due to the large amount of tuples and expensive calculation required to reconstruct representations. Hence, the multiresolution spatial data structure (MSDS) is proposed to produce constituent representations effectively. A proper MSDS in mobile environments should take the following four aspects into account.

Firstly, the MSDS should make it feasible to build representation-derivation into the query process of multiresolution systems to avoid time-consuming post retrieval. Thus, when processing a query, the multiresolution database system can choose a proper resolution level and retrieve only necessary data to satisfy this resolution.

Secondly, different representations derived from the same MSD should be consistent with each other. More precisely, spatial relationships should remain the same or at least similar between different representations. Here, spatial relationships could be topological, directional or metrical.

Thirdly, different representations are convertible to each other. That is, low resolution representations can be derived from high resolution representations and high resolution representations can be produced by low resolution representations plus additional information. This characteristic is especially important for semantic caching as it can largely improve the utilization of cached data. For example, suppose that all polygons overlapping window w at resolution l_1 are cached and the most recent query is 'to find all polygons overlapping window w at resolution l_2' ($l_2 > l_1$), if representations are convertible, only the additional data between l_1 and l_2 need to be retrieved and transmitted to the mobile client. Otherwise, the data cached would not be ussful in answering this query.

Finally, as the bandwidth is so scarce in mobile environments, MSD should be well compressed before transmission in mobile severs efficiently and decompressed in mobile clients.

3.2 MSDS – *Bit_map*

In the remainder of this paper, we use *Bit_map* as our MSDS. The basic idea of *Bit_map* is to group points based on scales, then abstract the data of every group to a tuple. This method avoids the opacity of organizing data based on polygon and the high overhead of organizing data based on points.

There are three important functions in *Bit_map* scheme.

One is *Scale Value Function*, which is used to group points. This function assigns a scale value to each point, denoted as $Delta(p)$. Points holding the same value belong to the same resolution level. Thus, with scale value function, a spatial object is broken into disjointed parts: $O = \{O_n, O_{n+1}, ...O_m\}$, where $n = min(Delta(p))$ and $m = max(Delta(p))$ and $O_i = \{p|Delta(p) = i\}(n \leq i \leq m)$. As n is the lowest scale of representations users may be interested in, it is named as *Base Level*.

The second function is *Abstract Compression Function*, which is used to abstract information from grouped points and compress it to a tuple. Using this function, we get $\{Abst(O_n), Abst(O_{n+1}), ...Abst(O_m)\}$. Here, $Abst(O_n)$ is different from other $Abst()$. As it includes the data equal to or less than *Base Level*, we denote it as $blData$. Whereas $Abst(O_i)(n+1 \leq i \leq m)$ only includes the additional information to refine O from scale $i-1$ to scale i, denoted as $aData_i$. Then, spatial object O is organized as $(ObjectID, BLData, AData, delta)$ in databases, where $delta$ denotes the scale level and each O includes $m - n + 1$ tuples as $(O_{id}, blData, \emptyset, n), (O_{id}, \emptyset, aData_{n+1}, n + 1), ..., (O_{id}, \emptyset, aData_m, m)$.

The last function is *Reconstruction Function*. After retrieving data at specific resolution, $Recon()$ will reconstruct them to a specific representation of objects. In more details, when $l = n$, Reconstruction Function is $Recon(blData)$; When $l > n$, it is $Recon(blData, aData_{n+1}, ...aData_l)$.

Both Scale Value Function and Abstract Compression Function utilize the iterative decomposition property of Z-ordering. Interested readers could refer to [11], [12] for more detailed algorithms. *Bit_map* defeats the overhead problem and expensive reconstruction in point-based data structures and avoids the accessing and processing of unnecessary data.

3.3   MSC Region

A semantic cache is a collection of semantic regions, each of which groups together semantically related tuples and maintains the semantic description of these tuples. As semantic regions are created dynamically based on the queries submitted at the client [4], we need to define multiresolution spatial queries before discussing the semantic description of semantic regions. (All definitions in our model are based on *Bit_map*, however these definitions can also be easily extended to other multiresolution data structures). In this paper, our work only focuses on the multiresolution window query, the most common query in multiresolution spatial databases. Processing more complex multiresolution spatial queries with the framework of semantic caching is a potential direction for future research.

Definition 1. *(Multiresolution Window Query MWQ) Given a MSD relation R(ObjectID, BLData, AData, delta) and a window W, a MWQ finds all objects O having at least one point in common with W at resolution r from R,*

$MWQ(W, r) = \{Re(O, r) | Re(W, r) \cap Re(O, r) \neq \emptyset\}$,

where $Re(O, r)$ is the representation of O at resolution r and r should not be less than the base level n.

Multiresolution window query is a window query with resolution constraints. For the sake of describing semantic region, now we extend the concept of MWQ to be more general.

Definition 2. *(Part Multiresolution Window Query PMWQ) Given a window W and two resolution level r1 and r2, where $n \leq r1 \leq r2 \leq m$, a PMWQ finds all objects O satisfying $MWQ(W, r2)$ and only fetch their data between resolution r1 and resolution r2.*

Part multiresolution window query denotes 'select $ObjectID$, $blData$, $aData$, $delta$ from R where $Re(O, r2) \in MWQ(W, r2)$, $delta \geq r1$ and $delta \leq r2$'. Obviously, MWQ is a special kind of PMWQ when $r1 = n$. The conception of PMWQ is important in MSC as it records the result of MSC query trimming. Further, MSC region can be described as $S(W, r1, r2)$, a cached PMWQ query. And the whole cache can be described as $\sum S(W, r1, r2)$

4   MSC Query Processing

In this section, we initially analyze the characteristics of MSD and MWQ. Approaches incorporating these characteristics are then proposed to improve the performance of query processing.

4.1   Caching MSD vs. Caching Traditional Multi-dimensional Data

As mentioned in Section 2.1, there are some related works about caching standard data and point data in previous research. Then, what are the differences between caching MSD and caching those data? Specious answer is that MSD is multi-dimensional data. But point data are also 2D and queries are frequently multi-dimensional in OLAP systems as each of which covers more than one table and one attribute [5].

Then, what are the essential differences between caching MSD and caching traditional multi-dimensional data? The obvious differences of MSD from traditional data include large size, complex structures and operations etc., which have been extensively studied. Now we will analyze the characteristics of MSD which can be utilized to improve query performance.

One characteristic is that spatial objects have *extensions*. A spatial object may intersect two or more disjointed windows at the same time. Hence, the results of two disjointed window queries may overlap and caching these results may cause redundancy. The characteristic also causes another situation. Figure 1 (a) shows a window query and its result; figure 1 (b) is the results of two other small window queries. We can see that although the windows of the two small queries do not wholly cover the window of the first large query, their data can still fully answer it. That is because the window query finds all polygons having at least **one point** in common with the window and there is no polygon fully contained in the shadow area. Thus, given a window query and a cache, there are two kinds of full answer. One is **Full Window Answer**, which is that the area of cache fully covers the window of the query. Another is **Full Data Answer**, which is that the data set of cache fully contains the result of the query. A Full Window Answer must be a Full Data Answer, but the reverse is not tenable. As the constraint formula of MSC region is $(W, r1, r2)$, query trimming can only find out Full Window Answer. Finding the Full Data Answer is a key factor in improving the performance.

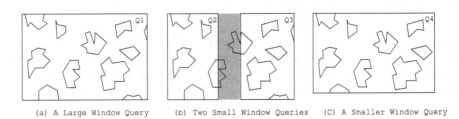

(a) A Large Window Query   (b) Two Small Window Queries   (C) A Smaller Window Query

Fig. 1. The Extension of Spatial Objects

Another characteristic is that *resolution* is not a normal axial dimension where $[a, b] \cap [c, d] = \emptyset$ $(a < b < c < d)$. To different resolutions, MSD at b includes MSD at a and MSD at a has a general feeling of MSD at b. In particular, because the data distribution on resolution is nonlinear, the difference between two consecutive high resolutions is not apparent.

Finally, as all spatial data are the digitized approximations of real world objects, *spatial queries can not be well specified*. For example, although the query window in figure 1 (c) is a bit smaller than that in figure 1(a), it does not matter greatly to most users. The other example is that users who querying resolution 18 usually can still be satisfied with a map with hybrid resolution (most area at resolution 18 and few places at resolution 17). Hence, the data do not fully answer a query, but it can satisfy the user who submitted the query. We name this case as **Approximate Satisfaction**.

As most spatial queries are not isolated to each other, **Full Data Answer** and **Approximate Satisfaction** occur quite often in MSD cache. Taking them into account when processing MWQ can reduce the amount of remainder queries posted to the database and response to the users before full answers transmitted to the client.

In the following, we investigate how to process a MWQ in MSC. We introduce the method of trimming a query with a single MSC region first, followed by extension to process a query with the whole MSC cache. We finally discuss how to retrieve the remainder query, transmit the retrieval result and coalesce the probe query and the remainder query.

4.2   Trimming a PMWQ with a MSC Region

Here, we discuss trimming PMWQ (not MWQ) with a MSC region. Given a $PMWQ(W_q, r1_q, r2_q)$ and a MSC region $S(W_s, r1_s, r2_s)$, no intersection is the simplest case, which happens when $W_q \cap W_s = \emptyset$ or $r1_q > r2_s$ or $r2_q < r1_s$. Otherwise, the PMWQ intersects the region ($W_q \cap W_s \neq \emptyset$ and $[r1_q, r2_q] \cap [r1_s, r2_s] \neq \emptyset$). In the following, we discuss how to perform trimming according to resolution ranges in detail.

Case 1: $r1_q = r1_s$ and $r2_q = r2_s$

Because $PMWQ$ and S have the same resolution range, the trimming is simplified to the clipping between two windows W_q and W_s. The remainder query may be none or single or more than one sub queries. For example, in figure 2(a), the remainder query is $(W_3, r1_q, r2_q)$; in figure 2(b), the remainder query consists of two sub queries $(W_4, r1_q, r2_q)$ and $(W_5, r1_q, r2_q)$; in figure 2(c), the window of S contains the window of $PMWQ$, no remainder query needs to be submitted to the server. However the region S will be clipped into five sub-regions.

Case 2: $r1_q \leq r1_s$ and $r2_q \geq r2_s$

In this case, the resolution range of $PMWQ$ contains the resolution range of S. Thus, the $PMSQ$ will be split into two parts $W_q \cap W_s$ and $W_q \cap \neg W_s$ first. Then, $(W_q \cap W_s, r1_q, r2_q)$ will be trimmed into three parts with disjointed resolution ranges. They are $[r1_q, r1_s)$, $[r1_s, r2_s]$ and $(r2_s, r2_q]$. Notice if the query is split by resolution first, then by window, it will cause more sub remainder queries. Figure 3(a) is a $PMWQ$ and a S fallen into case 2. After trimming, S will be simply split into 4 sub-regions. To $PMWQ$, piece 6 is the probe query, piece 5, 7, 8 together compose the remainder query.

Case 3: $r1_q \geq r1_s$ and $r2_q \leq r2_s$

In this case, the resolution range of $PMWQ$ is contained by the resolution range of S. In contrary to case 2, S will be split by window before resolution whereas $PMWQ$ is just split by window. Figure 4 is an example.

Case 4: $(r1_q \geq r1_s$ and $r2_q \geq r2_s)$ or $(r1_q \leq r1_s$ and $r2_q \leq r2_s)$

In this case, the resolution range of $PMWQ$ intersects the resolution range of S. Thus, both $PMWQ$ and S need to be split by resolution and window. Figure 5 gives an example.

After trimming a $PMWQ$ with a MSC region S, the $PMWQ$ will be divided into two parts – the probe query and the remainder query; and the S will also be divided into two parts – the intersection part and the difference part. The intersection part and the probe query are in fact the same.

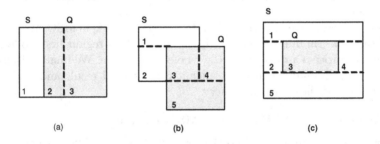

(a)   (b)   (c)

Fig. 2. Examples of Case 1

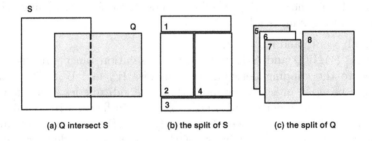

(a) Q intersect S   (b) the split of S   (c) the split of Q

Fig. 3. Example of Case 2

4.3   Processing a MWQ with a MSC Cache

As the query usually intersects more than one region in a MSC cache, processing a query with a MSC cache includes two steps.

1. The input query will be trimmed with all cache regions. This step produces three outputs. One is the probe query set Q_p; one is the remainder query set Q_r; and the last one is the difference of cache from the query, denoted as C_d. Q_p and Q_r are two sets of sub-queries, while C_d is a set of sub-regions. The cost of step 1 is usually linear in the number of regions [5].
2. According to the outputs of step 1, the query will be classified into a category and the corresponding process will be handled according to the category.

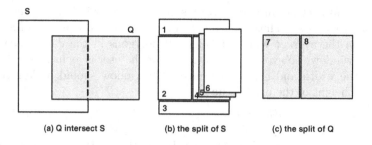

(a) Q intersect S   (b) the split of S   (c) the split of Q

Fig. 4. Example of Case 3

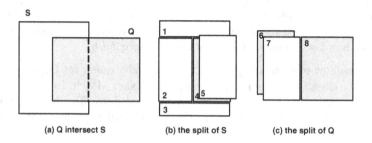

(a) Q intersect S   (b) the split of S   (c) the split of Q

Fig. 5. Example of Case 4

Traditional query processing classifies the relationship between a query and a cache into three categories: No Match, Partial Match and Exact Match. In order to improve the system performance, MSC enriches them into five categories – **No Match, Exact Match, Assumable Exact Match, Approximate Match** and **Partial Match**. Now, we will discuss each category in detail.

Category 1: No Match

It is the simplest category and happens only when $Q_p = \emptyset$. In this case, the original query will be posted to the server to get the result.

Category 2: Exact Match

In our paper, the cache exactly matches a query if and only if $Q_r = \emptyset$, which means that the cache is a Full Window Answer to the query. In this case, the result of Q_p is the query result.

Category 3: Assumable Exact Match

As discussed in section 4.1, Full Data Answer is also a kind of full answer to a query. However, due to the limitation of the constraint formula of region, it can not be figured out by query trimming directly. We propose **Assumable Exact Match** to solve this problem.

Before explaining Assumable Exact Match, **Negligible Window** needs to be introduced. According to the analysis in section 4.1, when a window is so small that it does not contain any whole polygon, the window can be ignored in point of data set because all polygons intersect this window also intersect its neighboring windows. Thus, we can assume if the width or height of a window is less than the average width or height of polygons in the data space, it is a negligible window. To avoid the situations that the user is focusing on this small window, the width and height of a negligible window should be less than the width and height of the original query.

Definition 3. *(Negligible Window) Given a MWQ (W_q, r_q) and a data space R, a window W_i is a negligible window if and only if*

$(Width(W_i) < \lambda * AvgW$ *and* $Width(W_i) < Width(W_q))$ *or* $(Height(W_i) < \lambda * AvgH$ *and* $Height(W_i) < Height(W_q))$,

where $Width()$ and $Height()$ are two functions to get the width and height of a spatial object respectively; $Avg(W)$ and $Avg(H)$ represent the average width and the average height of spatial objects in the data space R respectively; and λ is a small number to adjust the confidence of negligibility.

If the window of each sub query in the remainder query set Q_r is a Negligible Window, we classify the query into Assumable Exact Match. In other words, we assume the query is fully data answered by local data. To answer Assumable Exact Match, the local data make up the query result; Q_r does not need to be posted to the server.

Category 4: Approximate Match

Approximate Match happens when the data in local cache do not fully answer a query, but it still satisfies the user who submits it. To handle Approximate Match, the result of the probe query is represented to the user first. Then the remainder query is posted to the server. As it is not so urgent as normal remainder queries, it is assigned with a lower priority. By the chance that users are not satisfied with Approximate Match answer, they can choose to submit a finer query or to wait for the answer of the remainder query.

As a multiresolution window query is sensitive to two factors (window and resolution), both aspects should be taken into account when deciding whether a query falls into Approximate Match category. We use Algorithm 1 to make the judgement, in which λ_1 and λ_2 are used to adjust the confidence of approximation. Our experiments prove that when $\lambda_1 < 0.1$ and $\lambda_2 < 0.2$, Approximate Match can insure the quality of results.

Category 5: Partial Match

Partial Match happens when the query does not fall into any of the above four categories. In this case, the remainder query set is posted to the server and its result, together with local data, answers the user's requirement. It is worthy of note that during the transmission of the remainder query result, data in client side may become Approximate Match or Assumable Exact Match to the query. The result will be shown before the completion of the transmission.

4.4   Retrieval, Transmission and Coalescence

When the database receives the remainder query set, it retrieves these sub queries according to their window sizes and resolution ranges. The query with a larger

---

Algorithm 1. Judge_Approximate_Match(Q_r) //Q_r is the remainder query set

---

$Q_{window} = \emptyset$ {initialize two query sets}
$Q_{resolution} = \emptyset$
while $Q_r \neq \emptyset$ **do**
    $q \longleftarrow$ any sub-query in Q_r
    if $q.r2 = q.r1 + 1$ **then**
        $Q_{resolution} \longleftarrow q$
    else
        $Q_{window} \longleftarrow q$
    end if
end while
$S_W = Sum\_Area(Q_{window})$ {Sum up the area of queries in Q_{window}}
$S_R = Sum\_Area(Q_{resolution})$ {Sum up the area of queries in $Q_{resolution}$}
if $(S_W < \lambda_1)\&(S_R < \lambda_2)$ **then**
    Approximate Match
else
    No Approximate Match
end if

---

window will be executed earlier. And the data at a lower resolution will be accessed before the data at a higher resolution. Transmission also obeys the same rule. The motivation is to access and transmit minimum data to achieve Assumable Exact Match or Approximate Match in the client side. When all data arrive at the client, they will be organized together with local data to coalesce a new MSC region.

5   Performance Evaluation

In this section, we examine the performance of our MSC query processing scheme through a simulation study. Before presenting the experiment results we first introduce the simulation model.

5.1   Spatial Query Patterns in Mobile Environments

Spatial queries in mobile environments are usually not isolated to each other but lie in typical patterns as shown in the following examples.

Example 1.1. Suppose Mike is a new international student of the University of Queensland and he wants to make a travel plan of Australia. His typical search process will be querying the place near his university first, then extending his search area to the whole Australia gradually. When he finds an interesting site, he may inquire more details of this spot and inquire about areas around it.

The above example represents a typical spatial pattern, named 'Zoom in/ Zoom out Pattern' in this paper. Users focus on a small area first, then they extend their search area. At the same time, they may reduce the requirement of precision to adapt to the limit resource in mobile environments. When they find their interesting place, they will refine the precision and pan around the place. Queries belonging to 'Zoom in/Zoom out Pattern' usually jump in window sizes and resolutions. Figure 6(a) gives an illustration of such queries, where the number represents the order of query and the color represents its resolution. The darker a window is, the more details it queries.

Example 1.2. Suppose Mike chooses a scenic spot and he is travelling there now. Because he is not familiar with this place, he queries the map around his location periodically.

It is an example of another typical spatial query pattern. We name it 'Travel Pattern'. In this pattern, a user requires dynamic map generalisation according to his location. It is a special kind of location-dependent queries, which is defined as queries with constraints on the location of the mobile unit users [7]. Queries belonging to 'Travel Pattern' usually focus on small windows and high resolutions. And the resolutions of different queries are similar. Figure 6(b) is an example, in which the lines represent the route of a moving object.

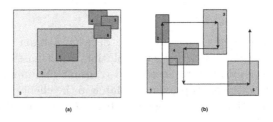

Fig. 6. Two basic query patterns.

'Zoom in/Zoom out Pattern' and 'Travel Pattern' are two typical patterns of spatial queries in mobile environments. Most spatial queries lie in one of the two patterns or the union of these two patterns, which is named as 'Hybrid Pattern' in this paper.

5.2   Workload Specification

Based on the three typical spatial query patterns, we develop three different workloads. And we assume the query issuing interval is 1 minute.

- **Workload 1 – Zoom in/Zoom out:**This workload further includes three types. a) Short Zoom in/Zoom out. This kind of query group has 3-5 queries and only zooms on one direction. b) Middle Zoom in/Zoom out. This kind of query group has 6-8 queries and pans after zoom in or zoom out. c) Long Zoom in/Zoom out. This kind of query group has 10 - 14 queries, which includes zoom in, zoom out and panning at the same time. The query window size is random from twenty times the average size of polygons to one thirtieth of the whole data space.

- **Workload 2 – Travel:** This workload also includes two types. a) Directed Movement. This kind of query group has 15-25 queries. The mobile client keeps moving in one direction with a random speed. b) Random Movement. This kind of query group has 30-50 queries. The mobile unit randomly chooses a destination and moves to it with random speed. After arrival, it randomly chooses the next destination again. According to our daily life, both two kinds of movement should be limited in a range. In our experiment, the range varies between 0.5 percent to five percent of the whole data space.
- **Workload 3 – Hybrid:** Randomly create a kind of query group in workload 1 or workload 2.

In our experiments, 100 groups of queries are randomly generated for each workload and the workload 1 has 863 queries totally, workload 2 and workload 3 have 3,040 and 2,874 queries respectively.

5.3 Simulation Model

This simulation emulates a mobile client issuing multiresolution window queries to a single server. Table 1 shows the main parameters of the model.

Table 1. Model Parameters and Default Settings

Parameter	Description	Value
ServerFrequency	server clock frequency(MHz)	2800
ClientMips	mobile client clock frequency(MHz)	996
DataSetSize	the size of data set (MB)	455.125
ClientCache	client side cache size (MB)	4
BandWidth	wireless network bandwidth (Kbps)	19.2

We use a large-scale data set, Real Estates Layer, provided by the Environmental Protection Agent, Queensland, Australia. This data set contains 398,464 polygons which are composed of 39,770,682 points. The simplest polygon has 4 points whereas the complex one has 275,531 points. For the experiment convenience, we cast away the polygons having less than 15 points or more than 10,000 points. Our experiment data set contains 309,868 polygons which are composed of 32,579,926 points. The average number of points of a polygon is 105. As the storage of the data set is 455.125 MB in Bit_map format, we choose 4MB as the cache size, around 1% of the total data set size. The client has a 19.2 Kbps wireless channel, which is standard for the mobile network.

5.4 Experiments and Results

The experiments are designed for two objectives. First, we studied the effect of adding Assumable Exact Match to the traditional query processing. Then, we compared our MSC model with the traditional cache scheme. The primary metric used was the *response time*, which was the elapsed time from MWQ being issued till an *acceptable query result* presented in the mobile client. This is a fairer metric than the response time for all data transmitted to the client, as users may be satisfied with the approximate match. Other metrics, such as the amount of queries in the remainder query set and the size of data transmitted, were also used.

Assumable Exact Match. We first studied the impact of Assumable Exact Match to traditional schemes. We choose the amount of queries in the remainder query set, the accuracy of the query result and the size of transmission data as performance metrics. Our experiment results are shown in figure 7, where x-axis represents the value of λ, to illustrate the result more visually, y-axis is the ratio of two schemes in the above three metrics respectively. In other words, $\lambda = 0$ represents the traditional scheme. As it is the baseline of our comparison, it is always 100% in y-axis. Figure 7 (b) shows the accuracy of the query result. Because Assumable Exact Match ignores Negligible Windows in the remainder query set, it might miss some query results. With the increase in λ, we can see that the query accuracy continues decreasing because the restriction of Negligible Window is looser correspondingly. However, as this is insignificant even in the worse case with a large λ value at 4.0, the lowest accuracy of the three patterns is still higher than 93%. It proves that Assumable Exact Match is reliable on the query accuracy. An interesting phenomena is that the accuracy drops slightly with lower λ whilst it drops faster with higher λ. That is because the relationship between Negligible Window and λ is nonlinear. Figure 7 (c) plots the results in the ratio of the size of data transmitted. It is apparent that with the increasing of the λ, the size of data transmitted from the server is reduced. That is because Assumable Exact Match avoids redundant transmission of data that have existed in the client cache. The ratio in the amount of remainder queries with corresponding λ is plotted in figure 7 (a). Obviously, the curve plunges with a great downward trend at lower λ, while when λ approaches 4 the amount almost does not decrease. Synthesizing these three figures together, we can determine an optimal λ value according to the users' requirements. When $\lambda = 1.2$, the accuracy of query results is higher than 99%, which is generally acceptable for uses, and the number of remainder sub-queries is reduced more than 35%. This is close to our assumption that when the width or height of a window is less than the average width or height of polygons in the data space, it can be ignored without affecting the accuracy of query results. The λ value may be affected slighting by the distribution, density and size of polygons. But it is likely close to 1. In the following experiments, we set $\lambda = 1.2$.

Comparing those curves for three different query patterns, we can see that the influences of our method on travel queries are greater than that of Zoom in/Zoom out queries. This is because of the essential differences between patterns. Compared to queries in Zoom in/Zoom out Pattern, those in Travel Pattern are generally smaller in window size and higher at resolution, which more easily produces Negligible Windows.

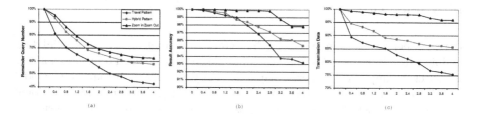

(a)   (b)   (c)

Fig. 7. Assumable Exact Match vs. λ

Five Categories vs. Three Categories. Besides Assumable Exact Match, Approximate Match also affects the performance of query processing. In this experiment, we compare the performance of three categories with five categories. Figure 8 (a) demonstrates the average response time of three different query patterns based on three categories and five categories. It can be seen that the response time of five categories is always less than that of three categories despite of the query patterns used. However the reduction in response time of five categories is greater for travel queries, which is only 73% of that of three categories. Whilst, the improvement of Zoom in/Zoom out pattern is only 9%. Moreover, the average response time of queries in Zoom in/Zoom out pattern is 43 seconds, which is much more than that of queries in Travel pattern with only 14 seconds. That is because in Zoom in/Zoom out patterns, queries are usually executed to explore interesting areas, the average data retrieved by each query is much more than that in Travel pattern. Figure 8 (b) and (c) display the average transmission time and retrieval time. We can see that the data transmission time is far more than retrieval time which implies that data transmission is the dominant factor in wireless environments (Note the different scales of y axis). As we adopt the low transmission bandwidth, 19.2 Kbps (the standard bandwidth of mobile environments), the average response time of less than 1 minute is considered to be acceptable to users.

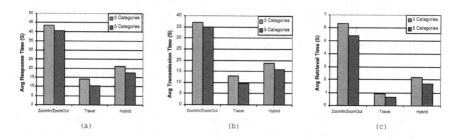

(a)   (b)   (c)

Fig. 8. Comparing the performance between 3 Categories and 5 Categories

6   Conclusion

In this paper, we have analyzed the characteristics of multiresolution spatial data and multiresolution window queries in mobile environments and proposed a new approach to improve the performance of query processing in our MSC model based on their characteristics. We refined the three categories in traditional schemes to five detailed ones and propose new processing methods against each of them. Our experiments, that were undertaken on a large real spatial data set show that the application of Assumable Exact Match can greatly reduce the amount of remainder sub-queries and the size of data transmission with very little sacrifice on the query result quality. Together with Approximate Match, it can reduce the response time by 27%. Our future work will focus on two parts: one is to extend this approach to more complex spatial queries, the other is to study the replacement policies.

Acknowledgment. The work reported in this paper has been partially supported by grant DP0345710 from the Australian Research Council.

References

1. A.U.Frank and S. Timpf. Multiple representations for cartographical objects in a multi-scale tree — an intelligent graphical zoom. *Computers and Graphics*, 18(6):823–929, 1994.
2. D. Barbara. Mobile computing and databases — a survey. *IEEE Transactions on Knowledge and Data Engineering*, 11(1):108–117, January/February 1999.
3. Ken. C.K.Lee, H.V.Leong, and Antonio Si. Semantic query caching in a mobile environment. *ACM SIGMOBILE Mobile Computing and Communications Review*, 3(2):108–117, April 1999.
4. S. Dar, M.J. Franklin, B.T. Jonsson, D. Srivatava, and M. Tan. Semantic data caching and replacement. In *Proceedings of the 22nd International Conference on Very Large Data Bases*, Bombay, India, Sept 1996.
5. P. M. Deshpande, K. Ramasamy, A. Shukla, and J. F. Naughton. Caching multi-dimensional queries using chunks. In *Proceeding of the 1998 ACM Conference on Management of Data*, pages 259–270, Seattle, USA, June 1998.
6. H. Hu, J. Xu, W. S. Wong, B. Zheng, D. L. Lee, and W.-C. Lee. Proactive caching for spatial queries in mobile environments. In *Proceedings of the 21th IEEE Int. Conf. on Data Engineering*, Tokyo, Japan, April 2005.
7. T. Imielinske and B.R. Badrinath. Querying in highly mobile and distributed environments. In *Proceedings of 18th International Conference Very Large DataBases*, Vancouver, B.C., Canada, Aug 1992.
8. M.Horhammer and M.Freeston. Spatial indexing with a scale dimension. In *Proceeding of the 6th International Symposium on Large Spatial databases*, Hong Kong, China, July 1999.
9. Q.Ren and M.H.Dunham. Using semantic caching to manage location dependent data in mobile computing. In *Proceedings of the 6th Annual International Conference on Mobile Computing and Networking*, Boston, MA, USA, August 2000.

10. Q. Ren, M. H. Dunham, and V. Kumar. Semantic caching and query process-
    ing. *IEEE Transactions on Knowledge and Data Engineering*, 15(1):192–210, Jan-
    uary/February 2003.
11. S. Sun, S. Prasher, and X. Zhou. A scaleless data model for direct and progres-
    sive spatial query processing. In *Proceedings of ER2004 Workshop on Conceptual
    Modeling for GIS*, Shanghai, China, Nov 2004.
12. S. Sun and X. Zhou. Semantic caching for web-based spatial applications. In
    Proceeding of the 7th Asia Pacific Web Conference, Shanghai, China, March 2005.
13. S.Zhou and C.B.Jones. Design and implementation of multi-scale databases. In
    7th International Symposium on Spatial and Temporal Databases, Los Angeles, CA,
    July 2001.
14. J. F. Yao and M. H. Dunham. Caching management of mobile dbms. *Journal of
    Integrated Computer-Aided Engineering*, 8(2), April 2001.
15. J. Zhang, M. Zhu, D. Papadias, Y. Tao, and D. L. Lee. Location-based spatial
    queries. In *Proceeding of the 18th ACM SIGMOD Conference*, San Diego, Califor-
    nia, USA, June 2003.
16. S. Zhou and C. B. Jones. A multi-representation spatial data model. In *Proceeding
    of the 8th International Symposium on Spatial and Temporal Databases*, Santorini
    Island, Greece, July 2003.

Probabilistic Spatial Queries on Existentially Uncertain Data*

Xiangyuan Dai[1], Man Lung Yiu[1], Nikos Mamoulis[1], Yufei Tao[2], and Michail Vaitis[3]

[1] Department of Computer Science, University of Hong Kong
{xydai, mlyiu2, nikos}@cs.hku.hk
[2] Department of Computer Science, City University of Hong Kong
taoyf@cs.cityu.edu.hk
[3] Department of Geography, University of the Aegean
vaitis@aegean.gr

Abstract. We study the problem of answering spatial queries in databases where objects exist with some uncertainty and they are associated with an *existential probability*. The goal of a *thresholding* probabilistic spatial query is to retrieve the objects that qualify the spatial predicates with probability that exceeds a threshold. Accordingly, a *ranking* probabilistic spatial query selects the objects with the highest probabilities to qualify the spatial predicates. We propose adaptations of spatial access methods and search algorithms for probabilistic versions of range queries and nearest neighbors and conduct an extensive experimental study, which evaluates the effectiveness of proposed solutions.

1 Introduction

Conventional spatial databases manage objects located on a thematic map with 100% certainty. In real-life cases, however, there may be uncertainty about the existence of spatial objects or events. As an example, consider a satellite image, where interesting objects (e.g., vessels) have been extracted (e.g., by a human expert or an image segmentation tool). Due to low image resolution and/or color definitions, the data extractor may not be 100% certain about whether a pixel formation corresponds to an actual object x; a probability E_x could be assigned to x, reflecting the confidence of x's existence. We call such objects *existentially uncertain*, since uncertainty does not refer to their locations, but to their existence. As another example of existentially uncertain data, consider emergency calls to a police calling center, which are dialed from various map locations. Depending on various factors (e.g., crime-rate of the caller's district, caller's voice, operator's experience, etc.), for each call we can generate a spatial event associated with a potential emergency and a probability that the emergency is actual. Existential probabilities are also a natural way to model *fuzzy classification* [1]. In this case, the class label of a particular object is uncertain; each class label takes an existential probability and the sum of all probabilities is 1.

We can naturally define probabilistic versions of spatial queries that apply on collections of existentially uncertain objects. We identify two types of such probabilistic

---

* Supported by grant HKU 7149/03E from Hong Kong RGC.

C. Bauzer Medeiros et al. (Eds.): SSTD 2005, LNCS 3633, pp. 400–417, 2005.

spatial queries. Given a *confidence* threshold t, a *thresholding* query returns the objects (or object pairs, in case of a join), which qualify some spatial predicates with probability at least t. E.g., given a segmented satellite image with uncertain objects, consider a port officer who wishes to find a set of vessels S such that every $x \in S$ is the nearest ship to the port with confidence at least 30%. Another example is a police station asking for the emergencies in its vicinity, which have high confidence. A *ranking* spatial query returns the objects, which qualify the spatial predicates of the query, in order of their confidence. Ranking queries can also be thresholded (in analogy to nearest neighbor queries) by a parameter m. For instance, the port officer may want to retrieve the $m = 10$ ships with the highest probability to be the nearest neighbor of the port.

Previous work on managing spatial data with uncertainty [20,15,12,21,5] focus on *locationally* uncertain objects; i.e., objects which are known to exist, but their (uncertain) location is described by a probability density function. The rationale is that the managed objects are actual moving objects with unknown exact locations due to GPS errors or transmission delays. On the other hand, there is no prior work on existentially uncertain spatial data, to our knowledge. In this paper, we fill this gap by proposing indexing and querying techniques for this important class of data. Our contributions are summarized as follows:

- We identify the class of existentially uncertain spatial data and define two intuitive probabilistic query types on them; *thresholding* and *ranking* queries.
- Assuming that the spatial attributes of the objects are indexed by 2-dimensional indexes (i.e., R–trees), we propose search algorithms for probabilistic variants of spatial range queries and nearest neighbor search.
- We show how extensions of R–trees that capture information about existential probabilities in non-leaf node entries can be used to answer probabilistic queries at lower I/O cost.

The rest of the paper is organized as follows. Section 2 provides background on querying spatial objects with rigid or uncertain locations and extents. Section 3 defines existentially uncertain data and query types on them. In Section 4 we study the evaluation of probabilistic spatial queries, when they are primarily indexed on their spatial attributes, or when considering existential probability as an additional dimension. Section 5 is a comprehensive experimental study for the performance of the proposed methods. Section 6 discusses extensions of our methods for probabilistic versions of complex query types and other (non-spatial) types of existentially uncertain data. Finally, Section 7 concludes the paper with a discussion about future work.

2 Background and Related Work

In this section, we review popular spatial query types and show how they can be processed when the spatial objects are indexed by R–trees. In addition, we provide related work on modeling and querying spatial objects of uncertain location and/or extent.

2.1 Spatial Query Processing

The most popular spatial access method is the R–tree [8], which indexes minimum bounding rectangles (MBRs) of objects. R–trees can efficiently process main spatial

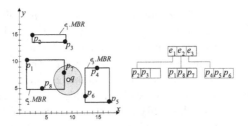

Fig. 1. Spatial queries on R–trees

query types, including spatial range queries, nearest neighbor queries, and spatial joins. Figure 1 shows a collection $R = \{p_1, \ldots, p_8\}$ of spatial objects (e.g., points) and an R–tree structure that indexes them. Given a spatial region W, a *spatial range query* retrieves from R the objects that intersect W. For instance, consider a range query that asks for all objects within distance 3 from q, corresponding to the shaded area in Figure 1. Starting from the root of the tree, the query is processed by recursively following entries, having MBRs that intersect the query region. For instance, $e_1.MBR$ does not intersect the query region, thus no object in the subtree pointed by e_1 can contain query results. On the other hand, e_2 is followed by the search algorithm and the points in the corresponding node are examined recursively to find the query result p_7.

A nearest neighbor (NN) query takes as input a query object q and returns the closest object in R to q. For instance, the nearest neighbor of q in Figure 1 is p_7. A popular generalization is the k-NN query, which returns the k closest objects to q, given a positive integer k. NN (and k-NN) queries can be efficiently processed if R is indexed by an R–tree, using the *best-first* (BF) algorithm of [10]. A *best-first* priority queue PQ, which organizes R–tree entries based on the (minimum) distance of their MBRs to q is initialized with the root entries. The top entry of the queue e is then retrieved; if e is a leaf node entry, the corresponding object is returned as the next nearest neighbor (assuming objects with no extent). Otherwise, the node pointed by e is accessed and all entries there are inserted to PQ. The process is repeated, until k objects are found. The BF algorithm is shown [10] to be no worse in terms of I/O than any NN algorithm that applies on the same R–tree. In order to find the NN of q in Figure 1, BF first inserts to PQ entries e_1, e_2, e_3, and their distances to q. Then the nearest entry e_2 is retrieved from PQ and objects p_1, p_7, p_8 are inserted to PQ. The next nearest entry in PQ is p_7, which is the nearest neighbor of q. In Section 4, we will show how BF can be extended to process probabilistic versions of nearest neighbor search on existentially uncertain data.

2.2 Locationally Uncertain Spatial Data

Recently, there is an increasing interest on the modeling, indexing, and querying of objects with uncertain location and/or extent. For instance, consider a collection of moving objects, whose positions are tracked by GPS devices. Exact locations are unknown due to GPS errors and transmission delays; e.g., if the object is in motion its location might be outdated when reaching the listening server. As a result, locations are approximated by probability density functions (PDFs), which integrate GPS error ranges and known moving object velocities. For instance, the uncertainty of a location can be modeled by a 2-dimensional Gaussian function, centered at the coordinates tracked from the GPS.

In [20], objects are assumed to move and send their positions to a centralized server. Each object o knows its last recorded location l_o; given a threshold θ, if the object finds itself θ units away from l_o, it sends an update with its new position. In this way, the server knows that objects are no more than θ away from their recorded locations. Based on this framework, a spatial region U_o (or line segment if the object's movement is constrained on a line) coupled with a PDF models the set of possible locations for each object o. The probability for o to intersect a query range W can be computed by applying the PDF to the spatial intersection of U_o and W. In this way, we can compute the result of a *probabilistic* spatial range query, which includes all pairs $\langle o, P_o \rangle$, where P_o is the probability that object o intersects W, and $P_o > 0$.

Note that the probability of an object to intersect a given query range is independent of that of other objects, a fact that makes range query processing straightforward. On the other hand, the probability of objects to be the nearest neighbor of a reference object q is not independent. Probabilistic nearest neighbor search for locationally uncertain data has been studied in [5]. The algorithm proposed there first computes fast the set of objects with $P_o > 0$, using their (indexed) uncertainty regions U_o only. Then for each object o in this set it integrates its probability to be closer to q than any other object, using the PDFs, over all possible locations of o. This process can be very expensive for arbitrary PDFs, however, [5] shows how to optimize it for basic uncertainty regions and PDFs. [19] indexes the trajectory of an object as a cylindrical volume around the tracked polyline (e.g., by a GPS), capturing uncertainty up to a certain distance from the polyline. A similar approach is followed in [15], where recorded trajectories are converted to sequences of locations connected by elliptical volumes.

[21] also models the uncertain locations of spatial objects by (circular) uncertainty regions and discuss how to process simple and aggregate spatial range queries using the fuzzy representations. In addition, they provide a methodology that sets the maximum precision error given a desired guaranteed uncertainty of the query results. [12] studies the evaluation of spatial joins between two sets of objects, for the case where the object extents are 'floating' according to uncertainty distance bounds. An extension of the R–tree that captures uncertainty in directory node entries is proposed. Both the filter and refinement steps of RJ are then adapted to process the join efficiently.

Cheng et al. [4,6] study a problem related to probabilistic spatial range queries. The uncertain data are not spatial, but ordinal (e.g., temperature values recorded from sensors). Due to measurement/sampling errors, an actual value is modeled by a range of possible values and a PDF that captures their probability. [6] indexes such uncertain data for efficient evaluation of probabilistic range queries (e.g., 'find all temperatures between 30°F and 40°F together with their probability to be in this range'). [4] classifies queries on such data to *entity-based* queries asking for the set of objects satisfying a query predicate and *value-based* queries asking for a PDF describing the distribution of a query result when it is a single aggregate value (e.g., the sum of values, the maximum value, etc.). This work also proposes generic query evaluation techniques and entropy-based measures for quantifying the quality of a probabilistic query result (e.g., how *certain* it is). Finally, [11] studies the evaluation of queries over uncertain or summarized data, where the user specifies thresholds (precision, recall, laxity) regarding the quality (i.e., accuracy) of the desired result. The query is initially applied on the uncertain data and based on how accurate the retrieved result is, some of the actual objects may be probed, in order to refine the accuracy of the result and bring its quality to the desired levels.

3   Existentially Uncertain Spatial Data

An object x is *existentially* uncertain if its existence is described by a probability E_x, $0 < E_x \leq 1$. We refer to E_x as *existential probability* or *confidence* of x. Note that since we can have $E_x = 1$, we (trivially) regard a 100% known object x as existentially uncertain. This allows us to model object collections which are mixtures of uncertain and certain data. On the other hand, $E_x = 0$ corresponds to an object x that definitely does not exist, so there is no need to store it in a database. Figure 2 shows a collection $R = \{p_1, p_2, \ldots, p_8\}$ of 8 existentially uncertain points. Next to each point label p_i, is its existential probability E_{p_i} enclosed in parentheses (e.g., $E_{p_1} = 0.2$). Given such object collections, we are interested in answering spatial queries that take uncertainty into account. We can easily define probabilistic versions of basic spatial query types:

Definition 1. Let R be a collection of existentially uncertain objects. A *probabilistic spatial range query* takes as input a spatial region W and returns all (x, P_x) pairs, such that $x \in R$ and x intersects W with probability $P_x > 0$. A *probabilistic nearest neighbor query* takes as input an object q and returns all (x, P_x) pairs, such that $x \in R$ and x is the nearest neighbor of q, with probability $P_x > 0$.

In the above definitions the output of a probabilistic query is a conventional query result coupled with a positive probability that the item satisfies the query. The case of probabilistic range queries is simple; $P_x = E_x$ for each object that qualifies the spatial predicate. Consider, for instance, the shaded window W, shown in Figure 2. Two objects p_1 and p_2 intersect W, with confidences $E_{p_1} = 0.2$ and $E_{p_2} = 0.5$, respectively. Similar to locationally uncertain data, the probability of an object x to qualify a spatial range query is independent of the locations and confidences of other objects.

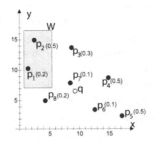

Fig. 2. NN search example

On the other hand, the probability of an object to be the nearest neighbor depends on the locations and probabilities of other objects. Consider again Figure 2 and assume that we want to find the potential nearest neighbor of q. The nearest point to q (i.e., p_7) is the actual NN iff p_7 exists. Thus, (p_7, E_{p_7}) is a query result. In order for the second nearest point p_6 to be the NN of q (i) p_7 must *not* exist and (ii) p_6 must exist. Thus, $(p_6, (1 - E_{p_7}) \cdot E_{p_6})$ is another result. By continuing this way, we can explore the whole set of points in R and assign a probability to each of them to be the NN of q.

This nearest neighbor query example not only shows the search complexity in uncertain data, but also unveils that the result of probabilistic queries may be arbitrarily large. For instance, the result of any NN query is as large as $|R|$, if $E_x < 1$ for all $x \in R$. We can define practical versions of probabilistic queries with controlled output by either *thresholding* the results of low probability to occur or *ranking* them and selecting the most probable ones:

Definition 2. Let (τ, P_τ) be an output item of a probabilistic spatial query Q. The *thresholding* version of Q takes as additional input a threshold t, $0 < t \leq 1$ and returns the results for which $P_\tau \geq t$. The *ranking* version of Q takes as additional input a positive integer m and returns the m results with the highest P_τ.

For example, a thresholding range (window) query W with $t = 0.6$ on the objects of Figure 2 returns $\varnothing$, whereas a ranking range query W with $m = 1$ returns $(p_2, 0.5)$.

4   Evaluation of Probabilistic Queries

Like spatial queries on exact data, probabilistic spatial queries can be efficiently processed with the use of appropriate access methods. In this section, we explore alternative indexing schemes and propose algorithms for probabilistic queries on them. We focus on the most important spatial query types; namely, range queries and nearest neighbor queries.

4.1   Algorithms for 2D R–Trees

The most straightforward way to index a set R of existentially uncertain spatial data is to create a 2-dimensional R–tree on their spatial attribute. The confidences of the spatial objects are stored together with their geometric representation or approximation (for complex objects) at the leaves of the tree. We now study the evaluation of probabilistic queries on top of this indexing scheme.

Range Queries. Probabilistic range queries can be easily processed in two steps; a standard depth-first search algorithm is applied on the R–tree to retrieve the objects that qualify the spatial predicate of the query. For each retrieved object x, $P_x = E_x$. If the query Q is a thresholding query, the threshold t is used to filter out objects with $P_x < t$.[1] If Q is a ranking query, a priority queue maintains the m results with the highest P_x, during search, and outputs them at the end of query processing.

---

[1] Especially for thresholding range queries of very large thresholds t, a viable alternative could be to use a B^+–tree that indexes objects based on their probability to efficiently access the objects x with $E_x \geq t$ and then filter them using the spatial query predicate.

Nearest Neighbor Search. As discussed, NN search is more complex compared to range queries, because the probability of an object to qualify the query depends on the locations and confidences of other objects. Figure 3 shows an elegant and efficient algorithm that computes the probability P_x of x to be nearest neighbor of q, for all x having $P_x > 0$.

Algorithm PNN2D(q, 2D R–tree on R)
1. $P^{first} := 1$; /*Prob. of no object before x*/
2.   **while** $P^{first} > 0$ and more objects in R **do**
3. $x :=$ next NN of q in R (use BF [10]);
4. $P_x := P^{first} \cdot E_x$;
5.   **output** (x, P_x);
6. $P^{first} := P^{first} \cdot (1 - E_x)$;

Fig. 3. Probabilistic NN on a 2D R–tree

Algorithm PNN2D applies best-first NN-search [10] on the R–tree to incrementally retrieve the nearest neighbors of q, without considering confidences. It also incrementally maintains a variable P^{first} which captures the probability that no object retrieved before the current object x is the actual NN. P^{first} is equal to $\prod(1 - E_y)$, for all objects y seen before x. Thus the probability of x to be the nearest neighbor of q is $P^{first} \cdot E_x$. In the example of Figure 2, PNN2D gradually computes $P_{p_7} = 0.1$, $P_{p_6} = (1 - 0.1) \cdot 0.1 = 0.09$, $P_{p_8} = (1 - 0.1)(1 - 0.1) \cdot 0.2 = 0.162$, $P_{p_4} = (1 - 0.1)(1 - 0.1)(1 - 0.2) \cdot 0.5 = 0.324$, etc. Note that *all* objects of R in this example are retrieved and inserted to the response set. In other words, PNN2D does not terminate, until an object x with $E_x = 1$ is found; if no such object exists, all objects have a positive probability to be the nearest neighbor.

Thresholding and ranking. As discussed in Section 3, the user may want to restrict the response set by thresholding or ranking. Figure 4 shows PTNN2D; the thresholding version of PNN2D, which returns only the objects x with $P_x \geq t$. The only differences with the non-thresholding version are the termination condition at line 2 and the filtering of results having $P_x < t$ (line 5). As soon as $P^{first} < t$, we know that the next objects, even with 100% confidence cannot be the NN of q, so we can safely terminate. For example, assume that we wish to retrieve the points in Figure 2 which are the NN of q with probability at least $t = 0.23$. First p_7 with $P_{p_7} = E_{p_7} = 0.1$ is retrieved, which is filtered out at line 5 and P^{first} is set to $0.9 \geq t$. Then we retrieve p_6 with $P_{p_6} = P^{first} \cdot E_{p_6} = 0.09$ (also disqualified) and set $P^{first} = 0.81 \geq t$. Next, p_8 is retrieved with $P_{p_8} = 0.162$ (also disqualified) and $P^{first} = 0.648 \geq t$. The next object p_4 satisfies $P_{p_4} = 0.324 \geq t$, thus $(p_4, 0.324)$ is output. Then $P^{first} = 0.324 \geq t$ and we retrieve p_3 with $P_{p_3} = 0.0972$ (disqualified). Finally, $P^{first} = 0.2268 < t$ and the algorithm terminates having produced only $(p_4, 0.324)$.

PRNN2D (Figure 5), the ranking version of PNN2D, maintains a heap H of m objects with the largest P_x found so far. Let P^m be the m-th largest P_x in H; as soon as $P^{first} < P^m$, we know that the next objects, even with 100% confidence cannot be the in the set of m most probable NN of q, so we can safely terminate. For example, assume that we wish to retrieve the point with the highest probability of being the NN of q in Figure 2. PRNN2D progressively maintains the object with the highest P_x. After each of the first 4 object accesses, P^m becomes 0.1, 0.1, 0.162, and 0.324. The algorithm terminates after the 4-th loop, when $P^{first} = 0.324$ and $P^m = P_{p_4} = 0.324$; this indicates that the next object can have P_x at most P_{p_4}, thus p_4 has the highest chances among all objects to be the NN of q.

Algorithm PTNN2D(q, 2D R–tree on R, t)
1. $P^{first} := 1$; /*Prob. of no object before x*/
2. **while** $P^{first} \geq t$ and more objects in R **do**
3. $x := $ next NN of q in R (use BF [10]);
4. $P_x := P^{first} \cdot E_x$;
5.   **if** $P_x \geq t$ **then output** (x, P_x);
6. $P^{first} := P^{first} \cdot (1 - E_x)$;

Fig. 4. Probabilistic NN on a 2D R–tree with thresholding

Algorithm PRNN2D(q, 2D R–tree on R, m)
1. $P^{first} := 1$; /*Prob. of no object before x*/
2. $H := \varnothing$; /*heap of m objects with highest P_x*/
3. $P^m := 0$; /*P_x of m-th object in H*/
4. **while** $P^{first} > P^m$ and more objects in R **do**
5. $x := $ next NN of q in R (use BF [10]);
6. $P_x := P^{first} \cdot E_x$;
7.   **if** $P_x > P^m$ update H to include x;
8. $P^{first} := P^{first} \cdot (1 - E_x)$;
9. $P^m := m$-th probability in H;

Fig. 5. Probabilistic NN on a 2D R–tree with ranking

4.2   Using Augmented R–Trees to Improve Efficiency

We can enhance the efficiency of the probabilistic search algorithms, by augmenting some statistical information to the R–tree directory node MBRs. A simple and intuitive method is to store with each directory node entry e a value e^{maxE}; the maximum E_x for all objects x indexed under e. This value can be used to prune R–tree nodes, while processing thresholding or ranking queries. Similar augmentation techniques are proposed in [12,6] for locationally uncertain data.

Table 1 summarizes the conditions for pruning R-tree entries (and the corresponding sub-trees) which do not point to any results, during range or NN thresholding and ranking queries. For range queries, we can directly prune an entry e when: (i) e.MBR does not intersect the query range, or (ii) its e^{maxE} satisfies the condition in the table. On the other hand, for NN search, a disqualified entry cannot be directly pruned, because the confidences of objects in the pointed subtree may be needed for computing the probabilities of objects with greater distances to q, but high enough probabilities to be included in the result.

Let us assume for the moment that for each non-leaf entry e we know the exact number of objects e^{num} in its subtree. Figure 6 shows the thresholding NN algorithm for the augmented 2D R-tree. BF is extended as follows. If a non-leaf entry e is de-heaped for which $P^{first} \cdot e^{maxE} < t$, the node where e points is not immediately loaded (as in PTNN2D) but e is inserted into a set L of *deleted* entries. For objects retrieved later from the Best-First heap, we use entries in L to compute P_x^{min} and P_x^{max}; lower and upper bounds for P_x. If $P_x^{min} \geq t$, we know that x is definitely a result. If $P_x^{max} < t$, we know that x is definitely not a result. On the other hand, if $P_x^{min} < t \leq P_x^{max}$ (Lines 6–16), we must refine the probability range for x. For this purpose, we pick an entry e in L and load the corresponding node n_e. If n_e is a leaf node, we access the objects e' in n_e. If q is nearer to e' than x, P^{first} is updated with the confidence of e'. Otherwise, its confidence does not affect P^{first} and we enqueue e' to the Best-First Queue. If n_e is a non-leaf node, for each entry $e' \in n_e$, we enqueue e' to the Best-First Queue if $d(q, e') > d(q, x)$, or insert e' into L otherwise. In either case, the probability range of x shrinks. The process is repeated while the range covers t.

It remains to clarify how P_x^{min} and P_x^{max} for an object x are computed. Note that L only contains entries whose minimum distance to q are smaller than $d(q, x)$. For an entry e in the list L, the confidence of each object in its subtree is in the range $(0, e^{maxE}]$. In addition, there exists at least one object in e whose confidence is exactly e^{maxE}. Thus, P_x^{min} corresponds to the case where for all objects under all entries in L are closer to q than x is and they all have the maximum possible confidences. P_x^{max} corresponds to the case, where for all $e \in L$, with maximum distance from q greater than $d(q, x)$, there is only one object with e^{maxE} confidence (for all other objects under e the confidence converges to 0):

$$P_x^{min} = P^{first} \cdot E_x \cdot \prod_{e \in L \wedge mindist(q,e) \leq d(q,x)} (1 - e^{maxE})^{e^{num}} \qquad (1)$$

$$P_x^{max} = P^{first} \cdot E_x \cdot \prod_{e \in L \wedge maxdist(q,e) \leq d(q,x)} (1 - e^{maxE}) \qquad (2)$$

In order to refine the probability range at Line 7 we must pick an entry e in L. We can use several heuristics for determining which e to select: (i) the one with the largest e^{maxE}, (ii) the one with the largest $e^{maxE} \cdot e^{num}$, (iii) the one with the smallest $d(q, e)$, or (iv) by random. By experimentation, we found that heuristic (iii) achieves the best results in most cases.

Table 1. Checking disqualified entries using augmented 2D R–trees

query type	range search	NN search
thresholding	$e^{maxE} < t$	$P^{first} \cdot e^{maxE} < t$
ranking	$e^{maxE} \leq P^m$	$P^{first} \cdot e^{maxE} \leq P^m$

Algorithm PTNN2Daug(q, augmented 2D R–tree on R, t)
1. $P^{first} := 1$; /*Prob. of no object before x*/
2. $L := \varnothing$; /*List of disqualified non-leaf entries*/
3. **while** $P^{first} \geq t$ and more objects in R **do**
4. $x :=$ next NN of q in R (use BF [10]); /* during BF-search, each non-leaf entry
              with $P^{first} \cdot e^{maxE} < t$ is removed from Best-First heap and inserted into L*/
5.   compute P_x^{min} and P_x^{max} by using P^{first}, L and E_x;
6.   **while** $P_x^{min} < t \leq P_x^{max}$ **do**
7.   pick an entry e in L;
8.   remove the entry e from L, read node n_e pointed by e;
9.   **for each** entry $e' \in n_e$
10.   **if** $mindist(q, e') > d(q, x)$ **then**
11.   enheap e' in the Best-First heap;
12.   **else if** n_e is a non-leaf node **then**
13.   insert e' into L;
14.   **else** /*e' is an object*/
15. $P^{first} := P^{first} \cdot (1 - E_{e'})$;
16.   compute P_x^{min} and P_x^{max} by using P^{first}, L and E_x;
17.   **if** $P_x^{min} \geq t$ **then output** $(x, P_x^{min}, P_x^{max})$;
18. $P^{first} := P^{first} \cdot (1 - E_x)$;

Fig. 6. Probabilistic NN on a augmented 2D R–tree with thresholding

So far, we have assumed that for each non-leaf entry e the number of objects e^{num} in its subtree is known (e.g., this information is augmented, or the tree is packed). We can still apply the algorithm for the case where this information is not known, by using an upper bound for e^{num}: $f^{level(e)}$, where $level(e)$ is the level of the entry e (leaves are at level 0) and f is the maximum R–tree node fanout. This upper bound replaces e^{num} in Equation 1.

Let us now show the functionality of the PTNN2Daug algorithm by an example. Consider the augmented R–tree of Figure 7 that indexes the pointset of Figure 2 and assume that we want to find the points that are the NN of q with probability at least $t = 0.23$. First, the entries in the root are enheaped in the Best-First heap. Next, the entry e_2 is dequeued. Since it disqualifies the query ($P^{first} \cdot e_2^{maxE} = 0.2 < t$), it is inserted into the list L. Then, the entry e_3 is dequeued. Its objects p_4, p_5, p_6 are enheaped in the Best-First Queue. The nearest object p_6 is dequeued. From Equations 1 and 2, we derive a probability range for P_{p_6} by using P^{first} and L. p_6 is disqualified as $P_{p_6}^{max} = E_{p_6} = 0.1 < t$. Then, $P^{first} = 0.9 \geq t$ and we retrieve p_4. Since $P_{p_4}^{min} = 0.9 \cdot 0.5 \cdot (1 - 0.2)^3 = 0.2304 \geq t$, p_4 is a result. Next, $P^{first} = 0.45 \geq t$ and the next entry retrieved from the priority queue of the BF algorithm is e_1. We do not access the node pointed by e_1, since we know that for each object x indexed under e_1, $P_x \leq e_1^{maxE} \cdot P^{first} = 0.225 < t$. Thus, e_1 is inserted into L. Next, p_5 is dequeued and discarded as $P_{p_5}^{max} = 0.45 \cdot 0.5 \cdot (1 - 0.2) \cdot (1 - 0.5) < t$. Now, the Best-First heap becomes empty and the algorithm terminates. Note that the PTNN2D algorithm accesses all nodes of the tree in this example, whereas PTNN2Daug saves two leaf node accesses.

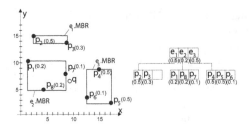

Fig. 7. Example of augmented 2D R–tree

The *ranking* NN algorithm that operates on the augmented R–tree is shown in Figure 8. It has several differences from the thresholding NN algorithm. A heap H is employed to organize objects o by their P_o^{min}. P^m denotes the m-th highest P_o^{min} in the heap. Observe that more complicated techniques are used for updating H, as the accesses to L may affect the order of objects in H. Each object o in H maintains P_o^{first}, which is the value of P^{first} when o is enheaped (line 21). At Lines 18–19, P_o^{first} (for some entries in H) is updated for each object e' found no further than o from q. The new P_o^{first} value is used to update P_o^{min} and potentially the order of objects in H at lines 23–24. Note that H may store more than m entries, since there may be objects o in it satisfying $P_o^{max} \geq P^m \geq P_o^{min}$. However, entries o are removed from H once $P_o^{max} < P^m$. The algorithm does not need to access any more objects from the Best-First heap as soon as $P^{first} < P^m$. In case H has more than m objects at that point, we need to refine the probability ranges of the objects in H (by processing entries in L) until we have the best m objects. In this case, entries e are removed from L once $mindist(q, e) > \max\{d(q, o) : o \in H\}$ because such entries cannot be used to refine the probability ranges of the objects in H.

We provide some insight for the space/time complexity of thresholding NN queries for the augmented tree approach. The worst case is that, for all disqualified entries (if any), their child nodes are accessed for refining the probability range of the objects seen. Therefore, the value of k estimated in Section 4.1 can be used as the upper bound in this case. The list L stores disqualified non-leaf entries in the tree and the cost of refining the probability range of an object is directly proportional to the size of L. Thus, the space/time complexity depends on the size of L. As the minimum distance of all entries in L from the query point is at most the distance of the last object seen (in BF-search), the maximum size of L can be estimated by using the value of k. In practice, the average size of L is quite small (10–100) and the space/time required is much less than that in the worst case.

4.3 Evaluation of Probabilistic Queries Using 3D R–Trees

An alternative method for indexing existentially uncertain data is to model the confidences E_x of objects x as an additional dimension and use a 3D R–tree to index the objects. Now, each non-leaf entry e in the tree, apart from the spatial dimensions, has a range $[e^{minE}, e^{maxE}]$ within which the existential probabilities of all objects in its subtree fall. Since every entry e still stores an e^{maxE}, the methods discussed in Section

Algorithm PRNN2Daug(q, augmented 2D R–tree on R, m)
1. $P^{first} := 1$; /*Prob. of no object before x*/
2. $L := \varnothing$; /*List of disqualified non-leaf entries*/
3. $H := \varnothing$; /*heap of objects, organized by P_o^{min}*/
4. $P^m := 0$; /*P^{min} of m-th object in H*/
5. **while** $P^{first} > P^m$ and more objects in R **do**
6. $x :=$ next NN of q in R (use BF [10]); /* during BF-search, each non-leaf entry
           with $P^{first} \cdot e^{maxE} < t$ is removed from Best-First heap and inserted into L*/
7.   compute P_x^{min} and P_x^{max} by using P^{first}, L and E_x;
8.   **while** $P_x^{min} < P^m \leq P_x^{max}$ **do**
9.   pick an entry e in L;
10.   remove the entry e from L, read node n_e pointed by e;
11.   **for each** entry $e' \in n_e$
12.   **if** $mindist(q, e') > d(q, x)$ **then**
13.   enheap e' on Best-First heap;
14.   **else if** n_e is a non-leaf node **then**
15.   insert e' into L;
16.   **else** /*e' is an object*/
17. $P^{first} := P^{first} \cdot (1 - E_{e'})$;
18.   **for each** entry $o \in H$ such that $d(q, e') \leq d(q, o)$
19. $P_o^{first} := P_o^{first} \cdot (1 - E_{e'})$;
20.   compute P_x^{min} and P_x^{max} by using P^{first}, L and E_x;
21.   **if** $P_x^{min} > P^m$ **then** enheap(H,($x,P_x^{first}:=P^{first},P_x^{min},P_x^{max}$));
22.   **if** H is changed **then**
23.   recompute, for each $o \in H$, P_o^{min} and P_o^{max} by using P_o^{first}, L and E_o;
24. $P^m := m$-th P^{min} in H;
25.   remove entries o from H with $P_o^{max} < P^m$;
26. $P^{first} := P^{first} \cdot (1 - E_x)$;
27. **while** $|H| > m$ and $|L| > 0$ **do**
28.   repeat Lines 9–19;
29.   repeat Lines 22–25;
30.   remove e from L with $mindist(q, e) > \max\{d(q, o) : o \in H\}$;

Fig. 8. Probabilistic NN on a augmented 2D R–tree with ranking

4.2 for the augmented 2D R–tree can be directly applied for the 3D R–tree. Moreover, we can utilize e^{minE} to derive tighter probability ranges:

$$P_x^{min} = P^{first} \cdot E_x \cdot \prod_{e \in L \wedge mindist(q,e) \leq d(q,x)} (1 - e^{minE})(1 - e^{maxE})^{(e^{num} - 1)} \qquad (3)$$

$$P_x^{max} = P^{first} \cdot E_x \cdot \prod_{e \in L \wedge maxdist(q,e) \leq d(q,x)} (1 - e^{minE})^{(e^{num} - 1)}(1 - e^{maxE}) \qquad (4)$$

If the exact number e^{num} of object in the subtree pointed by e is not known, we can use the fanout f and the minimum node utilization (0.4 for R*–trees) and replace e^{num} by $f^{level(e)}$ in Equation 3 and by $(0.4 \cdot f)^{level(e)}$ in Equation 4.

5   Experimental Evaluation

In this section, we evaluate the efficiency of the proposed techniques. All algorithms were implemented in C++. Experiments were run on a PC with a Pentium 4 CPU of 2.3GHz. In all experiments, the page size was set to 1Kb, unless otherwise stated. No memory buffers are used for caching disk pages between different queries; the number of node accesses directly reflects the I/O cost.

We compare the performances of five indexes and their corresponding algorithms for thresholding and ranking range queries and nearest neighbor search. The five indexes are (i) a simple 2D R–tree (denoted by 2D), (ii) a 2D R–tree, where each non-leaf entry e is augmented with e^{maxE} (denoted by 2D AUG), (iii) a 2D R–tree, where each non-leaf entry e is augmented with e^{maxE} and e^{num} (i.e., the number of objects in the subtree indexed by it), denoted by 2D AUG COUNT, (iv) a 3D R–tree (denoted by 3D), and (v) a 3D R–tree, where each non-leaf entry e is augmented with e^{num} (denoted by 3D COUNT). When comparing the indexes note that (i) captures minimum information in non-leaf entries and occupies the least space, whereas index (v) is at the other end (entries capture maximum information and the index occupies the most space). For each experiment, the measured I/O cost is the average cost of 20 queries with the same parameter values (but with different locations randomly chosen from the dataset).

5.1   Description of Data

For our experiments, we used various real datasets of different sizes and object distributions, described in Table 2. The datasets TG and SF are obtained from [2] while the other datasets are obtained from the R–tree Portal (www.rtreeportal.org).

Due to the lack of a real spatial dataset with objects having existential probabilities, we generated probabilities for the objects, using the following methodology. First we generated $K = 20$ *anchor* points randomly on the map, following the data distribution. These points model locations around which there is large certainty for the existence of data (e.g., they could be antennas of receivers close to which information is accurate). For each point x of the dataset, we (i) find the closest anchor a and (ii) assign an existential probability proportional to $\frac{1}{(c \cdot dist(x,a))^\theta}$. Thus, the distribution of probabilities around the anchors is a Zipfian one. The probabilities are normalized (using c) with respect to the maximum probability (1) corresponding to the anchor point. By changing θ (default value: 1) we can control the skew.

5.2   Experimental Results

Table 2 shows the performances of the five indexes for thresholding and ranking NN queries on different datasets. We fix $t = 0.002$ for thresholding NN queries and $m = 10$ for ranking NN queries.[2] Observe that the augmented and 3D R–trees perform better than the 2D R–tree, even though they are larger in size. The algorithms of Figures 6 and 8 manage to prune a large number of nodes that do not contain query results, which are

---

[2] A small value for t is necessary in order to observe difference between the indexes. Larger values for t will be tested in a subsequent experiment.

otherwise visited in the simple 2D R–tree index. The cost of 2D R–tree variants (i.e., methods {2D, 2D AUG, 2D AUG COUNT} does not change much with the database size. By the analysis in Section 4.1, the number of points to be examined is independent of the data size for 2D R–trees. The analysis in [18] shows that the cost increases slowly as the data size increases. On the other hand, the I/O costs of 3D R–tree variants increase slowly as the database size increases. This is due to the fact that 3D R–trees group entries using both spatial and probability dimensions, but the query algorithms mainly search for objects based on spatial dimensions.

Table 2. I/O cost of thresholding/ranking NN on different datasets, $t = 0.002$, $m = 10$

Dataset	Size	2D	2D AUG	2D AUG COUNT	3D	3D COUNT
(TG) San Joaquin roads	18623	122.7/116.7	45.2/41.0	37.3/34.2	36.5/32.9	35.4/31.6
(GR) Greece roads	23268	115.3/108.2	40.5/34.8	34.2/29.8	37.0/31.9	32.8/28.5
(LB) Long Beach roads	53145	107.5/100.1	37.3/32.7	32.4/28.2	44.7/41.1	42.0/38.0
(LA) LA streets	131461	135.4/132.3	43.1/42.3	38.1/36.9	48.4/47.4	45.6/45.2
(SF) San Francisco roads	174956	131.5/129.3	42.1/42.4	37.0/37.1	46.0/45.6	41.4/41.7
(TS) Tiger streams	194971	130.7/129.2	40.5/40.4	36.0/35.8	50.6/48.6	45.4/44.7

Figure 9 shows the I/O performance of the indexes for thresholding and ranking queries on the SF dataset. Methods {2D AUG, 2D AUG COUNT, 3D, 3D COUNT} perform much better than the simple 2D R–tree for all tested values of t and m. For $t \geq 0.02$, less than 5 accesses are required to find the query result when using the four advanced indexes and the algorithms of Figures 6 and 8. When comparing these indexes, we observe that augmenting e^{num} is not a good idea; using the fanout f gives accurate enough estimations of P^{min} and P^{max}. Thus the extra space (translated to extra accesses) required for augmenting e^{num} does not pay off. In addition, the augmented R–tree performs better than the 3D R–tree. First, the 3D R–tree occupies more space (the capacity of each non-leaf node is smaller) and results in more accesses, since the extra space is not compensated by tighter P^{min} and P^{max} (see Equations 3 and 4). Second, since the 3D R–tree groups entries to nodes using the existential probabilities as well as spatial dimensions, it does not achieve as good partitionings as the one using the spatial dimensions only; however, search is performed primarily using the spatial dimensions.

In the next experiment, we compare the performances of the indexes by varying the skewness θ of existential probability distribution of the objects (using the SF dataset). Figure 10 shows the experimental results for this case. We fix $t = 0.002$ and $m = 10$ for thresholding and ranking queries, respectively. The cost of the 2D R–tree increases much faster than the other trees when θ increases. For large θ there are a few, high probabilities around the anchors and the rest are very small. Thus, most points have low existential probabilities and the distances of the results from the query increase, causing an increase in the cost of 2D; only spatial information is used in the algorithms of Figures 4 and 5. On the other hand, the advanced NN algorithms on the augmented and 3D structures manage to prune disqualified directory nodes early.

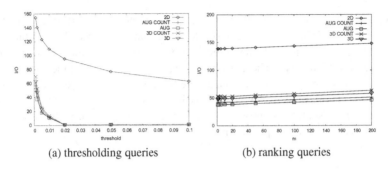

(a) thresholding queries   (b) ranking queries

Fig. 9. Queries on the SF dataset, $\theta = 1$

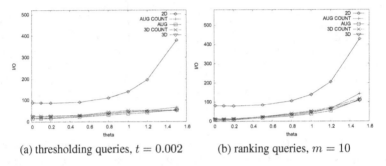

(a) thresholding queries, $t = 0.002$ (b) ranking queries, $m = 10$

Fig. 10. Queries on the SF dataset, varying θ

We also study the effect of page size on the performances of the indexes. As Figure 11 shows, the I/O costs of all indexes are inversely proportional to the page size. This is expected, due to the decrease of the number of nodes and heights of the trees.

Finally, we examine the performances of range queries on the indexes, using the SF dataset. For range queries, we use an additional parameter len, which is the extent of the query window in each dimension. The default value of len is set to 5% of values range (domain) at each dimension. Figure 12a and 12b show the cost of thresholding and ranking queries as a function of t and m respectively. Except for the simple 2D R–tree, all indexes follow similar trends as in probabilistic nearest neighbor queries. The cost of range queries on the 2D R–tree is independent of t and m as all points within the spatial range are retrieved. Observe that for very small t, the augmented and 3D indexes may perform worse than the 2D R–tree because (i) they prune no or very few directory entries that have lower e^{maxE} than t and (ii) they are larger in size than the simple 2D R–tree. Similarly, P^m decreases with m, affecting the costs of the advanced methods. The 3D R–tree performs worse than the augmented 2D R–tree also for range queries. Figure 12c shows the cost of thresholding queries as a function of len, at $t = 0.002$. As expected, the costs of all methods increase linearly with len^2. In summary, in most cases of probabilistic NN and range queries, a 2D R–tree with augmented e^{maxE} non-leaf entries achieves the best performance.

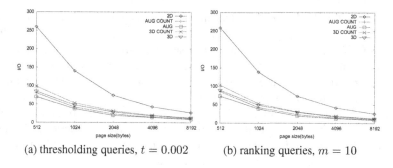

(a) thresholding queries, $t = 0.002$ (b) ranking queries, $m = 10$

Fig. 11. Queries on the SF dataset, varying page size

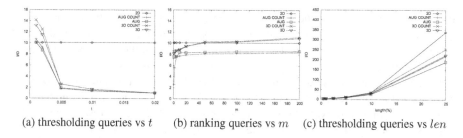

(a) thresholding queries vs t (b) ranking queries vs m (c) thresholding queries vs len

Fig. 12. Range queries on the SF dataset

6   Discussion

We have defined and studied in detail probabilistic range and nearest neighbor queries on existentially uncertain spatial data. In this section, we briefly discuss probabilistic versions of other spatial query types and queries on other (non-spatial) existentially uncertain data.

Extended query types. Given two spatial datasets R and S, a *probabilistic spatial join* returns all $(\langle r, s \rangle, P_{r \wedge s})$ pairs, such that $r \in R$, $s \in S$, and r intersects s with probability $P_{r \wedge s} > 0$. We can easily define thresholding and ranking versions of this query. Extending the well-known R–tree join algorithm [3] for probabilistic joins is straightforward, because $P_{r \wedge s}$ depends solely on E_r and E_s (i.e., $P_{r \wedge s} = E_{r \wedge s} = E_r \cdot E_s$) and is independent of the probabilities of other pairs. Given two spatial datasets R and S and a positive integer k, a closest pairs (CP) query [9,7] returns from the Cartesian product $R \times S$ the k $\langle r, s \rangle$ object pairs with the smallest distance. The probabilistic version of a CP query is challenging, due to the interdependence of the existential probabilities of qualifying pairs. The problem can be solved by extending the techniques for probabilistic NN queries and it is left for future work. Other interesting spatial query types for which we can define probabilistic versions are aggregate nearest neighbor queries [13], skyline queries [14], and reverse nearest neighbor search [17].

Spatio-temporal and ordinal data. Our methods can be easily extended for the case, where the objects also carry temporal attributes, i.e., they are spatio-temporal. In this case, the queries also include the time dimension e.g.,'find the most probable nearest neighbor at some moment in the whole past history'. R–trees that index object trajectories (e.g., [16]) can be used by our algorithms for searching. The temporal range may also be restricted to some timestamps or time interval (e.g.,'find the most probable nearest neighbor at some moment in the whole past history'). Finally, although our discussion so far has been on spatial (or spatio-temporal) data, the queries and solutions can be directly refer to ordinal data of any dimensionality (e.g., uncertain transmissions of combinations of measures, like temperature values).

7   Conclusions

In this paper, we presented the interesting problem of evaluating spatial queries for existentially uncertain data. Variants of common spatial queries, like range and nearest neighbor search, have probabilistic versions for this data model. We proposed algorithms for these probabilistic versions and several extensions of spatial access methods (i.e., R–trees) where these algorithms are applied. In addition, we discuss how more complex spatial queries can be processed in our framework. Finally, we conducted extensive experiments to evaluate the search algorithms and the corresponding spatial indexes. In most of the tested cases, the data structure that performs best is a R–tree, where non-leaf entries are augmented with maximum existential probabilities of the sub-tree they point at. In the future, we plan to study in detail more advanced query types and extend our methods to apply on data that are both existentially and locationally uncertain, as well as results of fuzzy classifiers [1].

References

1. P. M. Atkinson and N. J. Tate, editors. *Advances in Remote Sensing and GIS Analysis.* John Wiley & Sons, 1999.
2. T. Brinkhoff. A framework for generating network-based moving objects. *GeoInformatica*, 6(2):153–180, 2002.
3. T. Brinkhoff, H.-P. Kriegel, and B. Seeger. Efficient processing of spatial joins using r-trees. In *Proc. of ACM SIGMOD*, 1993.
4. R. Cheng, D. V. Kalashnikov, and S. Prabhakar. Evaluating probabilistic queries over imprecise data. In *Proc. of ACM SIGMOD*, 2003.
5. R. Cheng, D. V. Kalashnikov, and S. Prabhakar. Querying imprecise data in moving object environments. *IEEE TKDE*, 16(9):1112–1127, 2004.
6. R. Cheng, Y. Xia, S. Prabhakar, R. Shah, and J. S. Vitter. Efficient indexing methods for probabilistic threshold queries over uncertain data. In *Proc. of VLDB*, 2004.
7. A. Corral, Y. Manolopoulos, Y. Theodoridis, and M. Vassilakopoulos. Closest pair queries in spatial databases. In *Proc. of ACM SIGMOD*, 2000.
8. A. Guttman. R-trees: A dynamic index structure for spatial searching. In *Proc. of ACM SIGMOD*, pages 47–57, 1984.
9. G. R. Hjaltason and H. Samet. Incremental distance join algorithms for spatial databases. In *Proc. of ACM SIGMOD*, 1998.

10. G. R. Hjaltason and H. Samet. Distance browsing in spatial databases. *ACM TODS*, 24(2):265–318, 1999.
11. I. Lazaridis and S. Mehrotra. Approximate selection queries over imprecise data. In *Proc. of ICDE*, pages 140–152, 2004.
12. J. Ni, C. V. Ravishankar, and B. Bhanu. Probabilistic spatial database operations. In *Proc. of SSTD*, 2003.
13. D. Papadias, Q. Shen, Y. Tao, and K. Mouratidis. Group nearest neighbor queries. In *Proc. of ICDE*, 2004.
14. D. Papadias, Y. Tao, G. Fu, and B. Seeger. An optimal and progressive algorithm for skyline queries. In *Proc. of ACM SIGMOD*, 2003.
15. D. Pfoser and C. S. Jensen. Capturing the uncertainty of moving-object representations. In *Proc. of SSD*, 1999.
16. D. Pfoser, C. S. Jensen, and Y. Theodoridis. Novel approaches in query processing for moving object trajectories. In *Proc. of VLDB*, 2000.
17. Y. Tao, D. Papadias, and X. Lian. Reverse knn search in arbitrary dimensionality. In *Proc. of VLDB*, 2004.
18. Y. Tao, J. Zhang, D. Papadias, and N. Mamoulis. An efficient cost model for optimization of nearest neighbor search in low and medium dimensional spaces. *IEEE TKDE*, 16(10):1169–1184, 2004.
19. G. Trajcevski, O. Wolfson, F. Zhang, and S. Chamberlain. The geometry of uncertainty in moving objects databases. In *Proc. of EDBT Conf.*, 2002.
20. O. Wolfson, A. P. Sistla, S. Chamberlain, and Y. Yesha. Updating and querying databases that track mobile units. *Distributed and Parallel Databases*, 7(3):257–387, 1999.
21. X. Yu and S. Mehrotra. Capturing uncertainty in spatial queries over imprecise data. In *Proc. of DEXA*, 2003.

Topological Predicates Between Vague Spatial Objects

Alejandro Pauly and Markus Schneider*

University of Florida,
Department of Computer & Information Science & Engineering,
Gainesville, FL 32611, USA
{apauly, mschneid}@cise.ufl.edu

Abstract. Topological predicates are an important element of database
systems that allow manipulation of spatial data. Based on the necessity
for such systems to handle uncertainty, we introduce a general mechanism that identifies *vague topological predicates*. This definition forms
part of a formal data model referred to as *VASA* (*Vague Spatial Algebra*),
in which the data types *vague regions*, *vague lines*, and *vague points* are
defined in terms of existing definition of crisp spatial data types. Following this trend, the mechanism presented here identifies vague topological
predicates on the basis of well defined crisp topological predicates. An
example implementation of the mechanism for vague regions is given.

1 Introduction

Most, if not all, *man-made* spatial objects such as buildings, roads, pipelines and
even political divisions have a clear boundary and extension. The location of the
Eiffel tower is well known and certain, the path of the Interamerican highway
is well established and North Dakota has certainly defined boundaries and extension. These are in our words, *crisp* spatial objects. Current spatial database
models and GIS successfully implement such object types but lack modeling and
representation power when handling objects with not such crispness.

Spatial *vagueness* or *indeterminacy* is an inherent property of many objects
that are handled in the spatial database context. Point locations may not be
exactly known, paths or trails might fade and become uncertain at intervals.
The boundary of regions might not be certainly known or simply not be as
sharp as that of a building or a highway. Take as examples lakes (and rivers)
whose extension (and path) depends on pluvial activity, or take the location
of oil fields that in many cases can only be guessed. This inherent uncertainty
brings to light the necessity of more comprehensive models that are able to cope
with what we will refer to as *vague spatial objects*.

Correctly handling spatial data involves more than a good definition of the
data types. It also involves defining a complete set of operations and predicates

---

* This work was partially supported by the National Science Foundation under grant
number NSF-CAREER-IIS-0347574.

C. Bauzer Medeiros et al. (Eds.): SSTD 2005, LNCS 3633, pp. 418–432, 2005.

that make the data objects useful in their context. Topological predicates are proven to be very important in spatial data applications. In the case of crisp spatial objects, topological predicates are well studied and plenty of approaches exist for their proper definition. But this is not the case for vague spatial objects where a well defined set of topological predicates does not exist currently. The goal of this paper is to provide a comprehensive model for identifying topological predicates between vague spatial objects. The predicates to be identified have the added complexity of dealing with the vagueness present in the objects themselves, thus making the predicates vague in nature. The model we present here enhances the results of preliminary work that has been part of our own research as part of the *VASA* (Vague Spatial Algebra) project. The data types used are our own *vague spatial data types* [5] and include *vague points, vague lines* and, *vague regions*. A major benefit of *vague spatial objects* is that their is definition expressed in terms of crisp spatial operations so that they represent executable specifications. The same benefits are sought for topological predicates, and so we define them for vague spatial objects in terms of the already defined topological predicates for crisp spatial objects.

Section 2 presents related work. Section 3 introduces our vague spatial data types upon which the topological predicates are identified. Our enhanced general mechanism for identifying vague topological predicates is explained in Section 4 where we also draw a comparison to our previous approach. In Section 5 we introduce a case study based on the implementation of the identification mechanism on vague regions. Section 6 introduces the notions necessary to implement the newly identified predicates as part of common database query languages. Finally, Section 7 draws some conclusions and addresses future work.

2   Related Work

We refer to spatial vagueness as a natural feature of a spatial object. Vagueness defines object properties as uncertain or indeterminate such that it is not possible to assure whether certain components belong to the object or not. Three main alternatives have been proposed as general design methods. *Models based on fuzzy sets* (e.g., [15]) are all based on fuzzy set theory, allow a fine-grained modeling of vague spatial objects but are computationally rather expensive with respect to data structures and algorithms. *Models based on rough sets* (e.g., [16]) work with lower and upper approximations of spatial objects, which is similar to our approach. But the formal background is rather different. *Models based on exact spatial objects* (e.g., [9,10]) extend data models, type systems, and concepts for crisp spatial objects to vague spatial objects. A discussion of the differences of these approaches can be found in [15]. Vague spatial data types [5,7] leveraged in this paper belong to the latter category.

The basis of the latter category are crisp spatial data types (see [14] for a survey). We assume a very general definition of these data types and call them *complex*. Point objects are considered to be finite collection of points. Lines are assumed to be finite collections of disjoint curves which may meet in single endpoints. Regions are finite collections of disjoint faces except for single common points, and faces may have disjoint holes except for single common points.

Much research on spatial databases has been devoted to topological predicates (like *overlap*, *meet*, or *disjoint*) on crisp spatial data types. The two main solution approaches employ either spatial logic [1] or point set theory and point set topology [12]. Our definition of vague topological predicates rests on a generalization [8,13] of the latter approach (the well known 9-*intersection model* [12]) in the sense that topological predicates are defined on crisp *complex* spatial objects and not on *simple* spatial objects as in the other approaches. The difference between considering simple and complex spatial objects is important, for example, the number of topological predicates for simple regions is 8, whereas the number for complex regions is 33. Topological predicates for simplified vague regions have already been studied in [9,10]. These approaches, although already quite sophisticated, suffer from two main drawbacks. First, the crisp regions used are simple regions, i.e., they do not allow holes and only consist of a single component. Second, the vague regions defined are regions with "broad boundaries". That is, one crisp simple region, which represents the area that definitely belongs to the vague region, is located in another larger crisp simple region. The geometric difference between the larger crisp region and the smaller one is considered to be the broad, vague boundary. Operations on such simple vague spatial objects suffer greatly from a lack of closure properties and expressiveness, which are part of our goals. The simple approach they follow is insufficient for our definition and goals, hence we look to define a general method with more expressive power and that is still usable. The works in [10] and [9] have identified 44 and 46 different topological relationships, respectively. The results from these two previous approaches are not applicable to our own vague spatial data types because they represent only the topological relationships between so-called *concentric* regions (i.e.the *kernel* is always surrounded by the *conjecture*) that belong to a special case of our vague regions under which not all vague region topological relationships are covered. The previous models are only defined for vague regions and not for vague points and vague lines.

3   Vague Spatial Data Types

To motivate our definition of vague spatial data types, we illustrate an ecological scenario that is to be considered for the development of a nature preservation program. The program developers need to consider (among many other data) the extension of lakes, the paths followed by rivers, and the refuges of animals as well as their roaming routes. We can notice how the lake and river data can be uncertain due to rain activity within a time period. Roaming routes of some species might be only approximate as these can change slightly or it is not possible to record the exact path for all cases. Animal refuges might be uncertain due to their underground or cave nature. All these are examples of what we refer to as *vague spatial objects*. The animal refuge locations are specifically modeled as a *vague point* object where the precisely known locations are called the *kernel point* object and the assumed locations are denoted as the *conjecture*

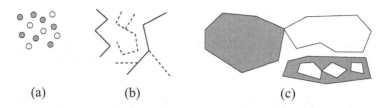

Fig. 1. Examples of a (complex) vague point object (a), a (complex) vague line object (b), and a (complex) vague region object (c). Each collection of components forms a single vague object

point object. The roaming routes and river paths can be modeled as *vague line* objects. Some routes, called *kernel line* objects, have been definitely identified and are certainly part of the river or route. Other routes can only be assumed and these are denoted as *conjecture line* objects. Knowledge about extension of lakes and other areas within the ecological system can be modeled similarly with *vague regions* formed by *kernel* and *conjecture* parts. Figure 1 gives some illustrations. Grey shaded areas, straight lines, and grey points indicate kernel parts; areas with white interiors, dashed lines, and white points refer to conjecture parts.

For the definition of vague points, vague lines, and vague regions we leverage the well known data types *point* for crisp points, *line* for crisp lines, and *region* for crisp regions [13]. These types are closed under the geometric set operations *union* ($\oplus : \alpha \times \alpha \to \alpha$), *intersection* ($\otimes : \alpha \times \alpha \to \alpha$), *difference* ($\ominus : \alpha \times \alpha \to \alpha$), and *complement* ($\rightsquigarrow \alpha \to \alpha$). The use of an exact model for constructing vague spatial data types leads to the benefit that existing definitions, techniques, data structures and algorithms need not be redeveloped but can simply be used or in the worst case slightly modified or extended as necessary.

A vague spatial object is described by a pair of two crisp complex spatial objects. Hence, the same generic definition is applicable to all vague spatial data types. That is, the extension of a crisp spatial data type to a corresponding vague type is given by a type constructor v as follows:

$$v(\alpha) = \alpha \times \alpha \qquad \forall \alpha \in \{point, line, region\}$$

This means that for $\alpha = region$ we obtain $v(region) = region \times region$, which we also name *vregion*. Accordingly, the data types *vline* and *vpoint* are defined. For a vague spatial object $w = (w_k, w_c) \in v(\alpha)$, the first crisp spatial object w_k, called the *kernel part*, describes the determinate component of w, that is, the component that definitely belongs to the vague object. The second crisp spatial object w_c, called the *conjecture part*, describes the vague component of w, that is, the component from which we cannot say with any certainty whether it or subparts of it belong to the vague object or not. *Maybe* the conjecture part or subparts of it belong to the vague object, *maybe* this is not the case. Since the kernel part and the conjecture part of the *same* vague spatial object may not share interior points, we define the following as a more general constraint from the original defined in [5]:

$$\forall \alpha \in \{point, line, region\} \ \forall w = (w_k, w_c) \in v(\alpha) : w_k^\circ \cap w_c^\circ = \varnothing$$

More details, in particular about the semantics of vague spatial data types as well as the definition of vague spatial operations, can be found in [5].

4 General Mechanism for Vague Topological Predicates

The approach we present here is based on three main goals. The first goal is to develop a formalism that works independently of the data types to which it is applied. It is desired that the formalism can be applied to two vague points equally as it can be applied to two vague lines or to the combination of a vague line and a vague region. Second, we consider important to make use of existing definitions of topological predicates for crisp spatial objects. This goal is a direct result from the definition of vague spatial objects. As noted in Section 3, vague spatial objects are constructed from crisp spatial objects. It is only consistent to let vague topological predicates be constructed from existing crisp topological predicates (see Figure 2). The final goal is to benefit from implementation advantages in such a way that VASA as a whole can make use of a preexisting crisp spatial algebra implementation as a simple executable specification.

The general method we propose, characterizes vague topological predicates on the basis of conjunctions of crisp topological predicates. The crisp topological predicates used as the underlying model are those defined in [13]. For two vague spatial object A, B, we evaluate the conjunction of the crisp topological predicates in the relationships between $(A_k, B_k), (A_k \oplus A_c, B_k), (A_k, B_k \oplus B_c)$, and $(A_k \oplus A_c, B_k \oplus B_c)$. These relationships represent the *smallest* objects which certainly exist (A_k, B_k), to the biggest possible objects represented with all uncertain features $(A_k \oplus A_c, B_k \oplus B_c)$. Given $\alpha, \beta \in \{point, line, region\}$, let $T_{\alpha,\beta}$ be the set of crisp topological predicates between the types α and β. To identify the vague topological predicates for type-combination $v(\alpha) \times v(\beta)$ we analyze all possible $|T_{\alpha,\beta}|^4$ combinations for the four relationships noted above. It is possible

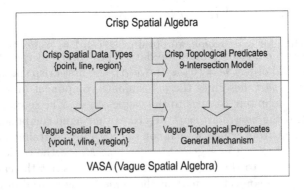

Fig. 2. Relations between crisp and vague spatial data models

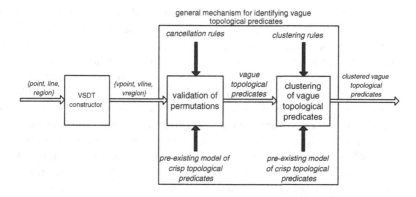

Fig. 3. General Mechanism for Identifying Vague Topological Predicates

that not all combinations are valid due to contradictions between the relationships so we proceed to apply *cancellation rules* that validate each combination (see Figure 3). The cancellation rules are defined on the basis of the point set intersections that define the crisp topological relationships between each of the components involved. This means that we do not refer to the crisp topological predicate by name, but rather specify more general cancellation rules by simply analyzing their point set intersections. Once all invalid combinations have been eliminated, we can refer to the remaining combinations as the vague topological predicates for $v(\alpha) \times v(\beta)$.

It is likely that the identified vague topological predicates will constitute a large set that could prove difficult to handle for the user. To provide for an easier management of such large sets of predicates we implement a step of what we call *clustering*. At this step, we define *clustering rules* which are in charge of grouping single vague topological predicates (as identified in the previous step) into meaningful clusters. Due to the involvement of the *conjecture* in the definition of the topological relationship, it becomes insufficient for the predicates representing these relationships to simply result in either *true* or *false*. As previously proposed in [6], each cluster results in a new data type $vbool = \{true, false, maybe\}$ that extends the regular boolean type and allows handling the inherent uncertainty. This *three-valued* logic can be adapted to boolean logic through a simple conversion that extends the set of available predicates and is detailed in Section 6.

The method we just detailed, is similar to the preliminary approach we expose in [6]. Based on the experiences learned from that preliminary work, we identified two key features that make our current approach better. First, we only take into account four combinations instead of the nine used in the preliminary approach. We eliminated all relationships from the preliminary approach in which the *conjecture* of an object might be considered alone without the *kernel*. The reasoning behind this change rests on the fact that in no situation the conjecture of an object will be considered without considering the kernel. From the definition of vague spatial objects, the kernel is always part of the object and the conjecture might not be part of it. This reasoning relates to the rough set idea of lower approximation $(A_k, A \in v(\alpha))$ and an upper approximation $(A_k \oplus A_c)$, noticing that no consideration of $(A_k \oplus A_c - A_k = A_c)$ is made. The second key feature improvement lies in the specification of cancellation and clustering rules. In the preliminary approach, we specified cancellation and clustering rules on the basis of the named crisp topological predicates. This was possible with a small set of easily identified predicates like those for complex points. When dealing with a large set of crisp topological predicates, such as the 33 predicates between complex regions and the 82 predicates between complex lines, it becomes unfeasible to understand the semantics of each unnamed crisp topological predicate and define the rules this way. This is why the rules are now specified on the basis of the point set intersections that represent the individual topological predicates. This removes the necessity of dealing with the large unnamed set, which reduces the probability for errors, and simplifies the set of rules.

We move on in the next section to show how the approach can be implemented. This is done by using a case study involving the identification of topological predicates between vague regions.

5   Topological Predicates Between Vague Regions

In this section we use vague regions to illustrate the mechanism described in Section 4. The example implementation shown here identifies the topological predicates between two vague region objects. The underlying crisp topological predicates used are those defined in [13] for two complex crisp regions. A summary of such definition follows.

5.1   Topological Predicates Between Complex Spatial Objects

Originally described in [4,12], the 9-intersection model (4-intersection model previously), defines topological predicates between simple regions on the basis of the intersection between the parts (interior, boundary, exterior) of the regions involved. Later the model is extended in [3] to account for simple regions with holes and in [11] to work with regions made up of multiple components. Finally, in [13], a comprehensive definition of topological predicates for complex regions is proposed. Complex regions can contain both holes and multiple components. The proposed method works by simply applying the 9-intersection model to the point sets belonging to the complex regions.

The topological predicate definition from [13] initially analyzes all possible 3×3 matrices (for a total of 512 matrices). Each matrix entry contains either a 1 or a 0 that represent whether that intersection is *non-empty* or *empty* respectively. A type-combination dependent set of *constraint rules* is applied to the original set of 512 matrices. The constraint rules are in charge of eliminating all non-constraint satisfying matrices that represent invalid scenes. Once all invalid matrices are eliminated, the remaining ones are considered the topological predicates between objects of the type-combination in question. For the purpose of this case study it is only necessary to consider complex regions, but the reference identifies topological predicates between all type combinations of complex points, complex lines and complex regions.

In the case of complex regions, a total of nine constraint rules result in 33 possible topological predicates between two complex regions. Such a large number of predicates presents problems of manageability for the user. This is the reason for the *clustering* of topological predicates presented in [13]. So-called *topological cluster predicates* are defined by means of clustering rules that are some kind of relaxed constraint rules. The clustering rules define a cluster by taking into account not all the nine intersections in the matrix causing a slight generalization that results in possibly more than one of the original predicates to form part of a single cluster. The authors define eight clustered predicates with semantics similar to the original topological predicates for simple regions identified by the 9-intersection model.

5.2   Cancellation Rules

Now we proceed to identify the topological predicates between vague regions. The first step is the definition of the cancellation rules that eliminate all contradictory information within the combinations explored as part of the general mechanism detailed in Section 4. The underlying set of topological predicates used is, as mentioned before, as shown in [13] and identifies 33 topological predicates between complex regions. We originally deal with a total of $33^4 = 1185921$ combinations. To make the formal rules more compact, we make use of variables $w \in \{B_k, B_k \oplus B_c\}$, $v \in \{A_k, A_k \oplus A_c\}$.

Lemma 1. *Any part (interior, boundary, exterior) of a single component (i.e. kernel or conjecture) or the union of components of the first object that intersects the interior of at least one component from the second object, must also intersect the interior of the union of components from the second object, i.e.,*

Lemma 1.1 $\forall p, q \in T_{\alpha, \beta}$:
$\neg(p(A_k, w) \land q(A_k \oplus A_c, w))$ s.t.
$\exists\, r \in \{\circ, \partial, -\}$:
$(p(A_k, w) \Rightarrow A_k^\circ \cap w^r \neq \varnothing \land$
$q(A_k \oplus A_c, w) \Rightarrow$
$(A_k \oplus A_c)^\circ \cap w^r = \varnothing)$

Lemma 1.2 $\forall p, q \in T_{\alpha, \beta}$:
$\neg(p(v, B_k) \land q(v, B_k \oplus B_c))$ s.t.
$\exists\, r \in \{\circ, \partial, -\}$:
$(p(w, B_k) \Rightarrow B_k^\circ \cap v^r \neq \varnothing \land$
$q(v, B_k \oplus B_c) \Rightarrow$
$(B_k \oplus B_c)^\circ \cap v^r = \varnothing)$

Proof. We know that for any vague region C, $C_k^\circ \subseteq (C_k \oplus C_c)^\circ$. From this fact, we can derive that for any point set x, it is always true that $C_k^\circ \cap x \neq \varnothing \Rightarrow (C_k \oplus C_c)^\circ \cap x \neq \varnothing$. Thus, Lemma 1 takes care of those combinations in which this implication is contradicted.   □

Lemma 2. *Any part of a single component or the union of components of the first object that does not intersect the exterior of at least one component from the second object must also not intersect the exterior of the union of components from the second object, i.e.,*

Lemma 2.1 $\forall p, q \in T_{\alpha,\beta}$:
$\neg(p(A_k, w) \;\wedge\; q(A_k \oplus A_c, w))$ s.t.
$\exists\, r \in \{\circ, \partial, -\}$:
$(p(A_k, w) \Rightarrow A_k^- \cap w^r = \varnothing\; \wedge$
$q(A_k \oplus A_c, w) \Rightarrow$
$(A_k \oplus A_c)^- \cap w^r \neq \varnothing)$

Lemma 2.2 $\forall p, q \in T_{\alpha,\beta}$:
$\neg(p(v, B_k) \;\wedge\; q(v, B_k \oplus B_c))$ s.t.
$\exists\, r \in \{\circ, \partial, -\}$:
$(p(w, B_k) \Rightarrow B_k^- \cap v^r = \varnothing\; \wedge$
$q(v, B_k \oplus B_c) \Rightarrow$
$(B_k \oplus B_c)^- \cap v^r \neq \varnothing)$

Proof. We know that for any vague region C, $C_k^- \supseteq (C_k \oplus C_c)^-$. From this fact, we can derive that for any point set x, it is always true that $C_k^- \cap x = \varnothing \Rightarrow (C_k \oplus C_c)^- \cap x = \varnothing$. Thus, Lemma 2 takes care of those combinations in which this implication is contradicted. □

Lemma 3. *A single component or the union of components of the first object that is not* disjoint *from at least one component from the second object, must also not be* disjoint *from the union of both components from the second object, i.e.,*

Lemma 3.1 $\forall p, q \in T_{\alpha,\beta}$:
$\neg(p(A_k, w) \;\wedge\; q(A_k \oplus A_c, w))$ s.t.
$\exists\, r, s \in \{\circ, \partial\} : (p(A_k, w) \Rightarrow$
$A_k^r \cap w^s \neq \varnothing)\; \wedge$
$\forall\, t, u \in \{\circ, \partial\}$:
$(q(A_k \oplus A_c, w) \Rightarrow$
$(A_k \oplus A_c)^t \cap w^u = \varnothing))$

Lemma 3.2 $\forall p, q \in T_{\alpha,\beta}$:
$\neg(p(v, B_k) \;\wedge\; q(v, B_k \oplus B_c))$ s.t.
$\exists\, r, s \in \{\circ, \partial\} : (p(v, B_k) \Rightarrow$
$B_k^r \cap v^s \neq \varnothing)\; \wedge$
$\forall\, t, u \in \{\circ, \partial\}$:
$(q(v, B_k \oplus B_c) \Rightarrow$
$(B_k \oplus B_c)^t \cap v^u = \varnothing))$

Proof. Due to the nature of the $\oplus$ (geometric union) operation, any intersection of the interiors of the complex regions a and b will remain untouched, and any intersection between boundaries or between an interior and a boundary will either remain untouched or be replaced by an interior-interior intersection when the union operation is applied and instead the intersections between $a \oplus d, b$ are analyzed. This means that there is no possibility for any such intersection to disappear or be replaced by an intersection with an exterior that could result in disjointment of the objects involved. Thus, we can imply that $\neg(disjoint(a, b)) \Rightarrow \neg(disjoint((a \oplus d), b))$ where *disjoint* refers to the clustered predicate as defined in [13]. □

Lemma 4. *If we assume that some τ representing a component or the union of components of the first object is not* contained *by a single component of the second object but is* contained *by the union of components of the second object, then τ must not* contain *the interior of the union of the components from the second object, i.e.,*

Lemma 4.1 $\forall p, q \in T_{\alpha,\beta}$:
$\neg(p(A_k, w) \;\wedge\; q(A_k \oplus A_c, w))$ s.t.
$(p(A_k, w) \Rightarrow \partial A_k \cap \partial w \neq \varnothing)\; \wedge$
$(q(A_k \oplus A_c, w) \Rightarrow ((A_k \oplus A_c)^\circ \subseteq w^\circ)$
$\wedge\; (\partial(A_k \oplus A_c) \cap \partial w = \varnothing))$

Lemma 4.2 $\forall p, q \in T_{\alpha,\beta}$:
$\neg(p(v, B_k) \;\wedge\; q(v, B_k \oplus B_c))$ s.t.
$(p(v, B_k) \Rightarrow \partial B_k \cap \partial v \neq \varnothing)\; \wedge$
$(q(v, B_k \oplus B_c) \Rightarrow ((B_k \oplus B_c)^\circ \subseteq v^\circ)$
$\wedge\; (\partial(B_k \oplus B_c) \cap \partial v = \varnothing))$

Proof. We show the first sublemma; the proof for the second sublemma is similar. Given $(A_k \oplus A_c)^\circ \subseteq w^\circ \Rightarrow A_k^\circ \subseteq w^\circ$, then the rule represents a contradiction because, for it to be true that $\partial A_k \cap \partial w \neq \varnothing \;\wedge\; \partial(A_k \oplus A_c) \cap \partial w = \varnothing$, A_c must share the same boundary with A_k as A_k shares with w, i.e.($\partial A_k \cap \partial A_c = \partial A_k \cap \partial w$). This means that A_c makes A_k *grow* towards the exterior of w which in turn makes it impossible for $(A_k \oplus A_c)^\circ \subseteq w^\circ$ to hold. $\square$

Lemma 5. *If some τ which represents a single component or the union of components from the first object, is completely contained within the interior of a single component from the second object, then the boundary of τ must not intersect the boundary of the union of components from the second object, i.e.,*

Lemma 5.1 $\forall p, q \in T_{\alpha,\beta}$:
$\neg(p(A_k, w) \;\wedge\; q(A_k \oplus A_c, w))$ *s.t.*
$(p(A_k, w) \Rightarrow \partial A_k \cap \partial w = \varnothing \;\wedge$
$(A_k \oplus A_c)^\circ \supseteq w^\circ) \;\wedge$
$(q(A_k \oplus A_c, w) \Rightarrow \partial(A_k \oplus A_c) \cap \partial w \neq \varnothing))$

Lemma 5.2 $\forall p, q \in T_{\alpha,\beta}$:
$\neg(p(v, B_k) \;\wedge\; q(v, B_k \oplus B_c))$ *s.t.*
$(p(v, B_k) \Rightarrow \partial B_k \cap \partial v = \varnothing) \;\wedge$
$(B_k \oplus B_c)^\circ \supseteq v^\circ \;\wedge$
$(q(v, B_k \oplus B_c) \Rightarrow (\partial(B_k \oplus B_c) \cap \partial v \neq \varnothing)$

Proof. This case represents a similar situation to that in Lemma 4. The difference here is that the boundaries of the subsets do not intersect but, when the conjecture is added, the boundaries intersect. Such a situation is impossible when the first object is completely contained in the second that is being expanded by the conjecture. The reason is that the conjecture would expand the object towards the inside of its own region's kernel which is in direct contradiction with the definition of vague regions. $\square$

Lemma 6. *If it can be inferred that the conjecture of the first object is empty ($\bot$), then it must be true that the predicates defined by the relationships between the kernel of the first object and any component or union of the components of the second object, and between the union of the components of the first object and the same component or whole of the second object, are the same, i.e.,*

Lemma 6.1 $\forall p, q, r, s \in T_{\alpha,\beta}$:
$\neg(p(A_k, B_k) \;\wedge\; q(A_k \oplus A_c, B_k) \;\wedge$
$r(A_k, B_k \oplus B_c) \;\wedge$
$s(A_k \oplus A_c, B_k \oplus B_c))$ *s.t.*
$(p(A_k, B_k) \;\wedge\; q(A_k \oplus A_c, B_k) \Rightarrow$
$A_c = \bot \;\wedge$
$(r(A_k, B_k \oplus B_c) \neq s(A_k \oplus A_c, B_k \oplus B_c))$

Lemma 6.2 $\forall p, q, r, s \in T_{\alpha,\beta}$:
$\neg(p(A_k, B_k) \;\wedge\; q(A_k \oplus A_c, B_k) \;\wedge$
$r(A_k, B_k \oplus B_c) \;\wedge$
$s(A_k \oplus A_c, B_k \oplus B_c))$ *s.t.*
$(p(A_k, B_k) \;\wedge\; r(A_k, B_k \oplus B_c) \Rightarrow$
$B_c = \bot \;\wedge$
$(q(A_k \oplus A_c, B_k) \neq s(A_k \oplus A_c, B_k \oplus B_c))$

Proof. If a conjecture is known to be empty, then it does not add any features to the kernel of the object, in other words $(A_k \oplus A_c = A_k)$ for some vague region A. Thus, any crisp topological predicates p and q of the whole object and the kernel by itself, respectively, with some other region w must be the same $(p(A_k \oplus A_c, w) = q(A_k, w))$. $\square$

After applying all rules to the over 1.1 million original combinations, only 69682 remain valid. Such a large number is difficult to manage at any level of usability, thus we present clustering rules to reduce the predicates to a workable set. We consider important to note that we have not provided a formal proof that all the remaining combinations are valid. It is our opinion that this is a weakness of the case study presented here but does not take any merit away from the general mechanism for identifying vague topological predicates.

5.3 Clustering Rules

Being able to identify topological predicates by name is necessary for the user. The original names of the eight topological predicates between simple regions (*disjoint*, *meet*, *inside*, *contains*, *coveredBy*, *covers*, *equal* and *overlap*) as detailed in [12] seem to be appropriate names thus we name the clustered vague topological predicates alike. We simply capitalize their names to differentiate them from the original. The clusters presented here represent only one of many ways in which the clustering of the predicates can be performed. We attempt to define each cluster in a way so that it has similar semantics as those that identified the topological predicates between simple regions in the 9-intersection model. It is important that the clusters are mutually exclusive in terms of the *true* results. This means that, after the cancellation rules are applied, each of the resulting combinations should result in *true* for one and only one of the following clusters.

Disjoint. Two vague regions as truly disjoint if none of their components have intersections of interior or boundaries between each other. The vague regions are truly not disjoint if the interiors or boundaries of their kernel parts intersect. Any other situation leaves the topological relationship uncertain; thus the predicate result is *maybe*. Formally:

- $Disjoint(A, B) = true \Leftrightarrow ((A_k \oplus A_c)^\circ \cap (B_k \oplus B_c)^\circ = \varnothing) \wedge ((A_k \oplus A_c)^\circ \cap \partial(B_k \oplus B_c) = \varnothing) \wedge$
  $(\partial(A_k \oplus A_c) \cap (B_k \oplus B_c)^\circ = \varnothing) \wedge (\partial(A_k \oplus A_c) \cap \partial(B_k \oplus B_c) = \varnothing)$
- $Disjoint(A, B) = false \Leftrightarrow (A_k^\circ \cap B_k^\circ \neq \varnothing) \vee (A_k^\circ \cap \partial B_k \neq \varnothing) \vee$
  $(\partial A_k \cap B_k^\circ \neq \varnothing) \vee (\partial A_k \cap \partial B_k \neq \varnothing)$
- $Disjoint(A, B) = maybe \Leftrightarrow \neg(Disjoint(A, B) = true \vee Disjoint(A, B) = false)$

Meet. Two vague regions certainly meet when the boundaries of their kernels intersect but the interiors of all components do not intersect. They certainly do not meet when the interiors of their kernels intersect or when they are certainly *Disjoint*. Formally:

- $Meet(A, B) = true \Leftrightarrow (\partial A_k \cap \partial B_k \neq \varnothing) \wedge ((A_k \oplus A_c)^\circ \cap (B_k \oplus B_c)^\circ = \varnothing)$
- $Meet(A, B) = false \Leftrightarrow (A_k^\circ \cap B_k^\circ \neq \varnothing) \vee$
  $(((A_k \oplus A_c)^\circ \cap (B_k \oplus B_c)^\circ = \varnothing) \wedge ((A_k \oplus A_c)^\circ \cap \partial(B_k \oplus B_c) = \varnothing) \wedge$
  $(\partial(A_k \oplus A_c) \cap (B_k \oplus B_c)^\circ = \varnothing) \wedge (\partial(A_k \oplus A_c) \cap \partial(B_k \oplus B_c) = \varnothing))$
- $Meet(A, B) = maybe \Leftrightarrow \neg(Meet(A, B) = true \vee Meet(A, B) = false)$

Inside. A vague region is certainly inside another one if all parts from all components of the first vague region are inside the kernel component of the second vague region. On the other hand, a vague region is certainly not inside another one when the interior of its kernel intersects the exterior of the second region or their boundaries intersect. Formally:

- $Inside(A, B) = true \Leftrightarrow ((A_k \oplus A_c)^\circ \cap B_k^\circ \neq \varnothing) \ \wedge \ (\partial(A_k \oplus A_c) \cap B_k^\circ \neq \varnothing) \ \wedge \ ((A_k \oplus A_c)^\circ \cap B_k^- = \varnothing) \ \wedge$
  $(\partial(A_k \oplus A_c) \cap B_k^- = \varnothing) \ \wedge \ ((A_k \oplus A_c)^\circ \cap \partial B_k = \varnothing) \ \wedge \ (\partial(A_k \oplus A_c) \cap \partial B_k = \varnothing)$
- $Inside(A, B) = false \Leftrightarrow (A_k^\circ \cap (B_k \oplus B_c)^- \neq \varnothing) \ \vee \ (\partial A_k \cap \partial(B_k \oplus B_c) \neq \varnothing)$
- $Inside(A, B) = maybe \Leftrightarrow \neg(Inside(A, B) = true \ \vee \ Inside(A, B) = false)$

Contains. The result for the *Contains* cluster is symmetric to *Inside*. Formally, $Contains(A, B) \Leftrightarrow Inside(B, A)$.

CoveredBy. A vague region is certainly covered by another one if all parts from all components of the first vague region are inside the kernel component of the second vague region and the boundary of the kernel of the first region intersects the boundary of the kernel of the second region. On the other hand, a vague region is certainly not covered by another one if the interior of its kernel intersects the exterior of the second region or their boundaries do not intersect. Formally:

- $CoveredBy(A, B) = true \Leftrightarrow ((A_k \oplus A_c)^\circ \cap B_k^\circ \neq \varnothing) \ \wedge$
  $(\partial(A_k \oplus A_c) \cap B_k^\circ \neq \varnothing) \ \wedge \ ((A_k \oplus A_c)^\circ \cap B_k^- = \varnothing) \ \wedge$
  $(\partial(A_k \oplus A_c) \cap B_k^- = \varnothing) \ \wedge \ ((A_k \oplus A_c)^\circ \cap \partial B_k = \varnothing) \ \wedge \ (\partial(A_k \oplus A_c) \cap \partial B_k \neq \varnothing)$
  $\wedge \ (\partial(A_k \oplus A_c) \cap \partial(B_k \oplus B_c) \neq \varnothing) \ \wedge \ (\partial A_k \cap \partial B_k \neq \varnothing) \ \wedge \ (\partial A_k \cap \partial(B_k \oplus B_c) \neq \varnothing)$
- $CoveredBy(A, B) = false \Leftrightarrow (A_k^\circ \cap (B_k \oplus B_c)^- \neq \varnothing) \ \vee \ (Inside(A, B) = true)$
- $CoveredBy(A, B) = maybe \Leftrightarrow \neg(CoveredBy(A, B) = true$
  $\vee \ CoveredBy(A, B) = false)$

Covers. The result of *Covers* cluster is symmetric to *CoveredBy*. Formally, $Covers(A, B) \Leftrightarrow CoveredBy(B, A)$.

Equal. The only way two vague regions are certainly equal is if their kernels are equal and their conjectures are empty. They are definitely not equal when the interior of one kernel touches the exterior of the other region or if one region is contained inside the kernel of another. Formally:

- $Equal(A, B) = true \Leftrightarrow (A_k^\circ \cap \partial B_k = \varnothing) \ \wedge \ (A_k^\circ \cap B_k^- = \varnothing) \ \wedge$
  $(\partial A_k \cap B_k^\circ = \varnothing) \ \wedge \ (\partial A_k \cap B_k^- = \varnothing) \ \wedge$
  $(A_k^- \cap \partial B_k = \varnothing) \ \wedge \ (A_k^- \cap B_k^\circ = \varnothing) \ \wedge$
  $(A_k^\circ \cap \partial(B_k \oplus B_c) = \varnothing) \ \wedge \ (A_k^\circ \cap (B_k \oplus B_c)^- = \varnothing) \ \wedge$
  $(\partial A_k \cap (B_k \oplus B_c)^\circ = \varnothing) \ \wedge \ (\partial A_k \cap (B_k \oplus B_c)^- = \varnothing) \ \wedge$
  $(A_k^- \cap \partial(B_k \oplus B_c) = \varnothing) \ \wedge \ (A_k^- \cap (B_k \oplus B_c)^\circ = \varnothing) \ \wedge$
  $((A_k \oplus A_c)^\circ \cap \partial B_k = \varnothing) \ \wedge \ ((A_k \oplus A_c)^\circ \cap B_k^- = \varnothing) \ \wedge$
  $(\partial(A_k \oplus A_c) \cap B_k^\circ = \varnothing) \ \wedge \ (\partial(A_k \oplus A_c) \cap B_k^- = \varnothing) \ \wedge$

$((A_k \oplus A_c)^- \cap \partial B_k = \varnothing) \wedge ((A_k \oplus A_c)^- \cap B_k^\circ = \varnothing) \wedge$
$((A_k \oplus A_c)^\circ \cap \partial(B_k \oplus B_c) = \varnothing) \wedge ((A_k \oplus A_c)^\circ \cap (B_k \oplus B_c)^- = \varnothing) \wedge$
$(\partial(A_k \oplus A_c) \cap (B_k \oplus B_c)^\circ = \varnothing) \wedge (\partial(A_k \oplus A_c) \cap (B_k \oplus B_c)^- = \varnothing) \wedge$
$((A_k \oplus A_c)^- \cap \partial(B_k \oplus B_c) = \varnothing) \wedge ((A_k \oplus A_c)^- \cap \int (B_k \oplus B_c) = \varnothing)$

- $Equal(A, B) = false \Leftrightarrow (A_k^\circ \cap (B_k \oplus B_c)^- \neq \varnothing) \vee ((A_k \oplus A_c)^- \cap B_k^\circ \neq \varnothing) \vee$
  $(((A_k \oplus A_c)^\circ \cap B_k^\circ \neq \varnothing) \wedge (\partial(A_k \oplus A_c) \cap B_k^\circ \neq \varnothing) \wedge ((A_k \oplus A_c)^\circ \cap B_k^- = \varnothing) \wedge$
  $(\partial(A_k \oplus A_c) \cap B_k^- = \varnothing) \wedge ((A_k \oplus A_c)^\circ \cap \partial B_k = \varnothing)) \vee$
  $((A_k^\circ \cap (B_k \oplus B_c)^\circ \neq \varnothing) \wedge (A_k^\circ \cap \partial(B_k \oplus B_c) \neq \varnothing) \wedge (A_k^- \cap (B_k \oplus B_c)^\circ = \varnothing) \wedge$
  $(A_k^- \cap \partial(B_k \oplus B_c) = \varnothing) \wedge (\partial A_k \cap (B_k \oplus B_c)^\circ = \varnothing))$
- $Equal(A, B) = maybe \Leftrightarrow \neg(Equal(A, B) = true \vee Equal(A, B) = false)$

Overlap. Two vague regions surely overlap if their kernel interiors intersect each other and also intersect their whole exteriors. We can certainly say the vague regions do not overlap if any of the other 7 clusters holds *true*, which leaves a large number of possibilities for the regions to maybe overlap. Formally:

- $Overlap(A, B) = true \Leftrightarrow (A_k^\circ \cap B_k^\circ \neq \varnothing) \wedge (A_k^\circ \cap B_k \oplus B_c^- \neq \varnothing) \wedge (A_k \oplus A_c^- \cap B_k^\circ \neq \varnothing)$
- $Overlap(A, B) = false \Leftrightarrow (Disjoint(A, B) = true) \vee (Meet(A, B) = true) \vee (Inside(A, B) = true) \vee$
  $(Contains(A, B) = true) \vee (CoveredBy(A, B) = true) \vee (Covers(A, B) = true) \vee$
  $(Equal(A, B) = true)$
- $Overlap(A, B) = maybe \Leftrightarrow \neg(Overlap(A, B) = true \vee Overlap(A, B) = false)$

In the next section, we show by using examples, how these clusters can be used in common database queries.

6   Querying

Popular database query language such as SQL understand predicates as boolean expressions. This means that the three values resulting from the clusters must be translated to boolean logic in order to use the vague topological predicates in SQL-like query languages. The translation is done by the following general definition for any clustered predicate P:

$$
\begin{aligned}
True\_P(A, B) = true &\Rightarrow P(A, B) = true \\
True\_P(A, B) = false &\Rightarrow P(A, B) = maybe \vee P(A, B) = false \\
Maybe\_P(A, B) = true &\Rightarrow P(A, B) = maybe \\
Maybe\_P(A, B) = false &\Rightarrow P(A, B) = true \quad \vee P(A, B) = false \\
False\_P(A, B) = true &\Rightarrow P(A, B) = false \\
False\_P(A, B) = false &\Rightarrow P(A, B) = true \quad \vee P(A, B) = maybe
\end{aligned}
$$

For the sample queries, we assume a scenario (similar to that in Section 3) of an ecological application. We define roaming areas for species as vague regions with sections that definitely delimit the area in which some species lives. The representation also includes some areas for which we are not sure whether or not the species roam around.

In the first query, we are interested in the roaming areas of all groups of an imaginary endangered species *pastuzo* whose main predator is the tiger. We need to know of all roaming areas for pastuzos, that *might* be contained inside roaming areas of tigers. This would help establish whether or not the majority of pastuzos left are in danger of being eaten by tigers.

```
SELECT  p.id
FROM    groups p, groups t
WHERE   p.species = "pastuzo"
and     t.species = "tiger"
and     Maybe_Inside(p.area,t.area);
```

In the next query, we want to establish all pastuzo groups that *certainly* live in areas outside those of the tigers. The purpose of this is to allocate resources and provide reproductive help to these pastuzo groups.

```
SELECT  p.id
FROM    groups p, groups t
WHERE   p.species = "pastuzo"
and     t.species = "tiger"
and     True_Disjoint(p.area,t.area);
```

7   Conclusions and Future Work

We provide a mechanism that is able to identify topological predicates between vague spatial objects. We believe the strength of the mechanism stems from having accomplished all three goals imposed on its design. First, the mechanism described is type-independent. That is, given as input a set of cancellation rules and a set of clustering rules the mechanism identifies all vague topological predicates for the respective vague spatial data type combination. Second, the mechanism makes use of existing definitions for crisp topological predicates by defining vague topological predicates on their basis. Third, the definition of vague spatial data types, operations and predicates can be regarded as an executable specification based on crisp concepts. The accomplishment of these goals is improved by the lessons learned from a preliminary approach resulting in a mechanism that is both powerful and simple. We also showed how using the vague topological predicates in common query languages requires only a simple conversion that allows for common sense handling of uncertainty in spatial data through querying.

The obvious step to follow now is the identification of topological predicates between all type combinations of vague spatial data types. Also we consider important to finalize this approach by a full implementation of the concepts that are part of *VASA*.

References

1. Z. Cui and A. G. Cohn and D. A. Randell. Qualitative and Topological Relationships. In *3rd Int. Symp. on Advances in Spatial Databases*, LNCS 692, pages 296–315, 1993.
2. P. A. Burrough and A. U. Frank, editor. *Geographic Objects with Indeterminate Boundaries*. GISDATA Series, vol. 2. Taylor & Francis, 1996.
3. M.J. Egenhofer and E. Clementini and P. Di Felice. Topological Relations between Regions with Holes. *Int. Journal of Geographical Information Systems*, 8(2):128–142, 1994.
4. M. J. Egenhofer and J. Herring. Categorizing Binary Topological Relations Between Regions, Lines, and Points in Geographic Databases. Technical Report 90-12, National Center for Geographic Information and Analysis, University of California, Santa Barbara, 1990.
5. A. Pauly and M. Schneider. Vague Spatial Data Types, Set Operations, and Predicates. In *8th East-European Conf. on Advances in Databases and Information Systems*, pages 379–392, 2004.
6. A. Pauly and M. Schneider. Identifying Topological Predicates for Vague Spatial Objects. In *20th ACM Symp. for Applied Computing*, pages 587–591, 2005.
7. M. Erwig and M. Schneider. Vague Regions. In *5th Int. Symp. on Advances in Spatial Databases*, LNCS 1262, pages 298–320. Springer-Verlag, 1997.
8. T. Behr and M. Schneider. Topological Relationships of Complex Points and Complex Regions. In *Int. Conf. on Conceptual Modeling*, pages 56–69, 2001.
9. A. G. Cohn and N. M. Gotts. *The 'Egg-Yolk' Representation of Regions with Indeterminate Boundaries*, pages 171–187. In and A. U. Frank [2], 1996.
10. E. Clementini and P. Di Felice. *An Algebraic Model for Spatial Objects with Indeterminate Boundaries*, pages 153–169. In and A. U. Frank [2], 1996.
11. E. Clementini and P. Di Felice and G. Califano. Composite Regions in Topological Queries. *Information Systems*, 20(7):579–594, 1995.
12. M. J. Egenhofer and R. D. Franzosa. Point-Set Topological Spatial Relations. *Int. Journal of Geographical Information Systems*, 5(2):161–174, 1991.
13. M. Schneider and T. Behr. Topological Relationships between Complex Spatial Objects. Technical Report, CISE Dept., University of Florida, 2004.
14. M. Schneider. *Spatial Data Types for Database Systems - Finite Resolution Geometry for Geographic Information Systems*, volume LNCS 1288. Springer-Verlag, Berlin Heidelberg, 1997.
15. M. Schneider. Uncertainty Management for Spatial Data in Databases: Fuzzy Spatial Data Types. In *6th Int. Symp. on Advances in Spatial Databases*, LNCS 1651, pages 330–351. Springer-Verlag, 1999.
16. M. Worboys. Computation with Imprecise Geospatial Data. *Computational, Environmental and Urban Systems*, 22(2):85–106, 1998.

Author Index

Lecture Notes in Computer Science

For information about Vols. 1–3511

please contact your bookseller or Springer